A DICTIONARY OF SCIENTIFIC BIRD NAMES

This Dictionary gives the derivation and meaning of all currently accepted scientific bird names. Some birds have been named for their appearance, some for a person or place, or for some aspect of their habitat, behaviour, food, or voice, or with reference to their native-language name. Each part of the name is defined separately so that *Passer domesticus*, for example, will not be found as such, but both *Passer* and *domesticus* are explained in their respective alphabetical places. The usefulness of the Dictionary will thus not be affected by future taxonomic revisions of generic or species names. About 8500 names are defined, including a selection of historical synonyms.

This book will find a permanent place on every ornithologist's shelf and will be a valuable reference source for everyone whose work or interests brings them into contact with birds.

The author, James Jobling, is a civil servant. He was born in Buckinghamshire, brought up in London, and now lives in Welwyn Garden City. Interested in languages and ornithology, he first began working on this Dictionary to fill a gap he perceived in his own library. He is a member of the British Ornithologists' Union, the American Ornithologists' Union, and the British Ornithologists' Club, and a fellow of the Royal Society for the Protection of Birds. Correspondence may be directed to him c/o The Zoology Editor, Oxford University Press, Walton Street, Oxford OX2 6DP, UK.

A DICTIONARY OF SCIENTIFIC BIRD NAMES

✦

JAMES A. JOBLING

Illustrations by

RICHARD FOWLING

and

THE AUTHOR

Oxford New York Tokyo
OXFORD UNIVERSITY PRESS
1991

Oxford University Press, Walton Street, Oxford OX2 6DP

Oxford New York Toronto
Delhi Bombay Calcutta Madras Karachi
Petaling Jaya Singapore Hong Kong Tokyo
Nairobi Dar es Salaam Cape Town
Melbourne Auckland

and associated companies in
Berlin Ibadan

Oxford is a trade mark of Oxford University Press

Published in the United States
by Oxford University Press, New York

A catalogue record for this book is available from the British Library

Library of Congress Cataloging in Publication Data

Jobling, James A.
A dictionary of scientific bird names/James A. Jobling.
1. Birds—Nomenclature.
2. Birds—Dictionaries—Latin, Medieval and modern. I. Title.
QL677.J55 1991 598'.014—dc20 91–7675

ISBN 0–19–854634–3

Set by
Pentacor PLC, High Wycombe, Bucks.
Printed in Great Britain
by Bookcraft (Bath) Ltd
Midsomer Norton, Avon

Acknowledgements

THIS dictionary is the culmination of an enduring interest in birds and latent curiosity in language kindled thirty years ago, when I received R. D. Macleod's *Key to the Names of British Birds* as a school prize.

To assemble the facts offered here I have consulted, to varying degrees, all of the volumes in the Bibliography as well as many original diagnoses and descriptions too numerous to acknowledge individually. In the search for the latter I cannot praise too highly J. L. Peters' *Check-list*, which has been a boon in the location of source material.

I owe a great debt to Anne Vale and Frances 'Effie' Warr (who was a mine of useful information and brought Herbert Langton's and Lewis Conisbee's manuscripts to my attention), ornithology librarians at the Natural History Museum in Tring; to Carol Gokce, deputy librarian (Zoology) and Carolyn Smith, library assistant (General), at the Natural History Museum in London; and to Ian Dawson, senior librarian at the Royal Society for the Protection of Birds, who gave me every facility and full access to the collections in their care. On the occasions when I have had to examine skins at Tring I have received generous guidance and courtesy from Graham S. Cowles and Michael Walters.

My thanks, also, to Emmet R. Blake of the Field Museum of Natural History, Chicago; Walter E. Bolle of the Australian Museum; Peter Boyce of the Royal Botanic Gardens at Kew; Vladimir Loskot of the Zoologichyeskiy Institut, Leningrad; Michel Louette of the Koninklijk Museum voor Midden-Afrika in Belgium; Brian Lym of the California Academy of Sciences; Annegret Ogden of the Bancroft Library, University of California; Frank Rozendaal of the Rijksmuseum van Natuurlijke Historie, The Netherlands; Richard Schodde of the CSIRO, Australia; Helmut Sick of the Academia Brasileira de Ciências; and H. E. Wolters of the Zoologisches Forschungsinstitut und Museum Alexander Koenig, Germany, for giving their time and replying to enquiries so fully.

For continued support, constructive comments on drafts of the Introduction and main text, and the loan of reference material I am indebted to John Byrne, Brian Fletcher, Alan Ford, the late Richard Litster and his widow, Doris, Gary Player, Arthur T. Smith, and Alan

Starr. My work was made easier by the late Robert Simpson who translated German texts so ably, and Stephen Woolmington, who similarly dealt with Spanish texts.

The jacket has been enlivened by the delicate brush of Anne Hopkinson, and the text by the striking scenic pen-and-ink drawings of Richard Fowling. To both of them I am grateful for their cheerful enthusiam and talent.

The experienced pen of John Farrand, Jr. brought an extensive knowledge to bear on my manuscript, and his thorough critique not only ensured that inconsistencies of treatment, content, and style were pin-pointed for eradication, but also led to renewed enquiries into the etymologies of some names.

Thanks are due also to Walter J. Bock, who painstakingly reviewed the nomenclatural and historical sections of the Introduction and suggested important improvements.

The staff of the Oxford University Press have shown great dedication in transforming my manuscript into a Dictionary worthy of the Oxford tradition.

My brother Barry C. Jobling must be acknowledged for his interest in the project, and for his contacts in providing American reference material.

Finally, to my wife, Patricia, and children, Lorna and Paul, all my love and thanks for their patience, assistance, puzzled understanding, and unswerving support during the prolonged gestation and birth pangs of the Dictionary. Without them my 'Great Work' would still be a manuscript gathering fingermarks on a shelf.

Welwyn Garden City J. A. J.
November 1990

Falco subbuteo European Hobby

Contents

Semnornis ramphastinus Toucan Barbet

Introduction

*'When **I** use a word,' Humpty Dumpty said in rather a scornful tone, ' it means just what I choose it to mean —neither more nor less.'*

LEWIS CARROLL (1872). Through the looking-glass.

IN 1758 the tenth edition of Linnaeus's *Systema Naturae* was published in Stockholm. It described and diagnosed the natural world as then known to him, including 63 genera and 556 species of birds. This edition of Linnaeus is now accepted as the beginnings of scientific nomenclature in zoology, including ornithology. Since then, more than 30 000 scientific bird names, of genera, species, and subspecies, have been proposed and arranged in a hierarchy of taxa far more complex than that envisaged and used by Linnaeus.

Linnaeus's aims were to describe relationships, systematize the natural world, and promote world-wide understanding and discussion by providing simple two-part names for each species, using words taken directly from classical Latin or transliterated from Greek or other, mainly European, languages. Today the concept is still relevant and useful in a polyglot world made smaller by technology and instant communication. The object of this book is to explain the meaning of the international language created by Linnaeus and his successors in so far as it applies to the genera and species of birds of the world.

The importance of a system which identifies a species in any tongue is apparent when one considers, as examples, the various species world-wide sharing the names 'robin', 'wren', 'blackbird', 'warbler', 'sparrow', or 'catbird', the variety and limitations of vernacular names (Common Gull, Mew Gull, Racek bourní, Stormmåge, Láros ó tephrókhrous, Stormmöwe, Gaviota cana, Goéland cendré, SæDingur, Gavina, Viharsirály, Fiskemåke, Mewa pospolita, Sizaya chayka, Kalalokki, Küçük marti, Kamome, all for *Larus canus*), or the debates of English-speakers over the preferred vernacular names of even common birds. When the British Ornithologists' Union Records Committee (1988) suggested standardizing English names and incurring the demise, amongst others, of Dunnock and Bearded Tit, strong emotions were aroused, the

Committee offered their resignation, and the project was placed on hold whilst responses were evaluated! As my dictionary is about scientific names, governed by an agreed set of rules, and not English names, rightly unfettered by codes, I have not followed any single authority on the occasions when English names are introduced.

THE PARTS OF SCIENTIFIC NAMES

The full scientific name of a bird species consists of four parts; *Parus major* Linnaeus 1758, is the scientific name of the Great Tit. The first two parts, the binomen *Parus major*, are written in a Latin or neo-Latin form and traditionally printed in italics.

The first part of the name denotes the genus, distinguishing a group of related species or an isolated, distinctive species. It must be in the form of a noun (*Parus* is the Latin word for a tit), must be unique in the zoological world, and is always capitalized.

The second, uncapitalized, part of the name is the specific name (previously also called the trivial name), distinguishing the several species within a genus. The specific name is commonly in the form of an adjective (*major* is from the Latin for larger). Only in combination with a generic name does it have any validity or make any sense, and it can be used in more than one genus: *Parus major* Linnaeus 1758 (Great Tit); *Dendrocopos major* (Linnaeus) 1758 (Great Spotted Woodpecker); *Podiceps major* (Boddaert) 1783 (Great Grebe); *Tinamus major* (Gmelin) 1789 (Great Tinamou); and *Crotophaga major* Gmelin 1788 (Greater Ani). Within the genus, however, no two species, subspecies, or forms (taxa) may bear the same specific name. For example, in 1843 the American explorer William Gambel described a chickadee he collected in the mountains of New Mexico as *Parus montanus*. That name, however, was already occupied by *Parus montanus*, the familiar Willow Tit of Europe, described by Conrad von Baldenstein in 1827. As a result Gambel's bird had to be renamed and the Mountain Chickadee is now distinguished as *Parus gambeli* Ridgway 1886, in honour of the original collector.

The third and fourth parts of a species scientific name give the author of the specific name and the year in which the name was first properly published. The author's name placed in parentheses after a specific name indicates that the current generic classification differs from the genus assigned by the original author. For example, the House Sparrow *Passer domesticus* (Linnaeus)1758 was originally described in the genus *Fringilla*.

Systematic publications, such as check-lists, may also give full publication details and the type locality (i.e. the location where the type specimen that defines the species was collected or is believed to have been collected). Non-systematic scientific works and more popular works, such as field-guides and magazines, usually give only the first two names, the 'binomen'; only this part of the scientific name will be considered further in this Dictionary.

Many species are divided by systematists into subspecies or geographic races which are populations of the species occupying a distinct geographic range and distinguished by recognizable morphological criteria from other such populations of the species. Species subdivided into subspecies are 'polytypic'; those for which no subspecies are recognized are 'monotypic'. Because of the continuity of the evolutionary process, some subspecies are so divergent that they are considered by some workers to have crossed the specific threshold and become species in their own right. In scientific nomenclature, subspecies are designated by adding a third name to the binomen, creating a trinomen.

Further, in zoological nomenclature, species and subspecies names belong to the class of species-group names and are subjected as a group to the same rules of nomenclature. When proposed, species-group names should be attached to a definite specimen, the type, which serves as the name-bearer for the species-group name. Any question as to the taxon to which a species-group name applies is resolved by reference back to the type specimen. These specimens are therefore of special value in zoological nomenclature and are specially designated and usually stored separately in museum collections. The subspecies of the Great Tit breeding in continental Europe and western Siberia is known as *Parus major major* (the nominotypical subspecies, whose name is created by repeating the specific name) to distinguish it from populations breeding in the British Isles *Parus major newtoni,* north-west Africa *Parus major excelsus,* the Holy Land *Parus major terraesanctae,* and others found elsewhere in this species' large Palaearctic and Indomalayan range. At least one of these forms, *Parus major minor* of Japan and China, has been elevated by some workers to specific rank, *Parus minor*, whilst the Turkistan Tit is considered by some workers as a distinct species, *Parus bokharensis,* and by others as a subspecies of the Great Tit, *Parus major bokharensis.* An extreme case is the Golden Whistler *Pachycephala pectoralis* of the south Pacific, for which more than 70 distinct and usually easily recognizable races have been described.

Neither generic nor specific names need to be descriptive, accurate, or relevant, and they cannot be rejected if found to be erroneous in these respects. For example, *papua* need not refer to a bird found in Papua New Guinea, and *rufus* need not refer to the colour red. Many names coined in the early years of systematic zoology are now known to be inappropriate. These names may be the result of badly coloured plates or figures (*Frederickena viridis*—described from an inaccurate and garish plate, *Le Batara vert*, of the mainly black and slate (male) or brown and grey (female) Black-throated Antshrike), specimens affected by poor storage conditions (*Cotinga maculata*—named from a cabinet specimen whose plumage had suffered prolonged exposure to strong light), lay descriptions (*Aerodramus fuciphagus*—named in the belief that the swiftlets built their nests from seaweed), or mistaken provenance (*Pygoscelis papua*—based on Sonnerat's brash claim that he had seen three different species of penguin in New Guinea!).

CODES OF NOMENCLATURE

Although Linnaeus sought to lay down guidelines in his *Philosophica botanica*, 1751, there were no generally accepted rules governing the formation, use, and priorities of names in zoology, including ornithology, for nearly one hundred years afterwards. Many naturalists rushed to embrace Linnaeus's simple binominal system, but chose to do so in an uncoordinated and chaotic fashion. Men of independent spirit such as François Levaillant and the Comte de Buffon refused to recognize the order heralded by the Swedish botanist, but their works were eagerly scanned and used as bases for catalogues, classificatory systems, or nomenclators by subsequent cabinet authors. The names of Levaillant, Buffon, Azara, Marcgrave, Catesby, and others will not be found amongst the ranks of Linnaean descriptive authors, but their works are vital sources for the etymologist.

As European imperialism opened new portions of the globe to scientific exploration in the late eighteenth and early nineteenth centuries, and as methods of preservation of specimens and storage of collections improved, naturalists became overwhelmed by the sheer numbers of new species brought back by soldiers, mariners, missionaries, explorers, and colonial administrators. The Linnaean binominal system, now almost universally adopted by naturalists, was threatened with collapse as authors independently described the same species under different names, unaware of, or perhaps without regard for, the work of others. Often males, females, immature individuals, and colour morphs of the same species were described as different species. Moreover, authors differed in their approaches to the Linnaean binominal nomenclatural system, disagreeing as to whether inappropriate names, original misspellings, names based on the local vernacular, and so on, should be corrected or changed. As zoological taxonomy improved and species became better known, the earlier errors were gradually sorted out and corrected. The result, however, was a plethora of names, disagreement on availability and usage of names for individual species, and differences on the starting date for binominal nomenclature (whether it should be pre-Linnaean, Linnaeus's tenth edition of 1758, or Linnaeus's twelfth edition of 1766). Great instability in use of names and a collapse of binominal nomenclature loomed less than 100 years after Linnaeus introduced his concept of an efficient international system of biological names essential for communication between all biologists.

Clearly this situation could not continue, and individual zoologists developed sets of rules of nomenclature to bring order to the impending chaos. The most successful of these early attempts to bring uniformity to zoological nomenclature was the Strickland Code, originally conceived in 1835 by the British ornithologist, zoologist, and palaeontologist Hugh E. Strickland, presented to the British Association for the Advancement of Science in 1842, and adopted by the Scientific Congress in Padua in 1843, by the American Society of Geologists and Naturalists in 1845, and by the British Association for the Advancement of Science in 1846. The Strickland Code was the basis of subsequent codes, including the American Ornithologists' Union Code and eventually the *Règles internationales*. The Strickland Code adopted the twelfth (1766) edition of the *Systema Naturae* as the starting date for zoological nomenclature, as this was the first time Linnaeus had consistently used binominal nomenclature. Although this decision to regard 1766 as the beginning was broadly accepted at the time in Britain and parts of continental Europe, many workers in North America and Europe argued that the tenth (1758) edition of Linnaeus should be used. The nomenclature code adopted by the American Ornithologists' Union in 1886 accepted the tenth edition of *Systema Naturae* (1758) as the start of zoological nomenclature. This new code was enthusiastically adopted by most workers except the British, who remained isolated until 1901, when the Fifth International Congress of Zoology at Berlin accepted the tenth edition and promulgated the first set of rules of zoological nomenclature to be recognized internationally, the *Règles internationales de la Nomenclature zoologique*, first officially published in 1905. These rules have been modified and clarified at various times and underwent a major revision during the 1950s, resulting in publication of the new *International Code of Zoological Nomenclature* in 1961 (3rd edition, 1985).

The fundamental goal of the Code is to ensure stability and universality in scientific names of animals, a basic necessity for communication between zoologists throughout the world and over time. Three basic principles need to be presented for an understanding of the names treated in this Dictionary. These are: priority, homonymy, and preservation of well-established names. These principles will be discussed in turn.

Priority

This principle was developed gradually in the early decades of the nineteenth century as a method of dealing with the different names available for the same species or genus, these names being the result of zoologists intentionally or unintentionally describing the same taxon independently, the merger of taxa, correction of inappropriate names for taxa, and so on. *The principle of priority states simply that the earliest name applied properly to a taxon of animals is the correct scientific name, with the date of publication determined by the stated date on the publication or by other means if that information is not reliable.* Priority now dates from 1 January 1758, the date fixed for the publication of the tenth edition of Linnaeus's *Systema Naturae.* If two species or two genera are merged for whatever reason, the correct name is the earliest one proposed. If taxonomic investigation indicates that a species or a genus should be divided into two, then the former name remains with the type and a new name must be proposed for the other taxon if a name does not already exist for it. Many changes in zoological nomenclature resulted from the application of priority, especially when sorting out the work of the early taxonomists. Major changes stemmed from the decision to change the beginning date for zoological nomenclature from the twelfth edition (1766) of Linnaeus to the earlier tenth edition (1758). Most of these changes in ornithology were resolved many decades ago, but they still cause difficulties when searching the literature of the last century and in the nomenclature of avian family-group names. The latter will not be treated in this Dictionary.

Homonymy

This principle states that a particular name can be used only once in zoological nomenclature. Hence a generic name or a family-group name can be used only once in the animal kingdom — it must be unique. Within a genus, a species-group name can be used only once. Thus when the generic name *Atrichia,* which had been applied to the Australian scrub-birds, was shown to be a junior objective homonym, it had to be replaced with *Atrichornis.* The rules for generic homonyms have changed over the years. Formerly genera differing only in their gender terminations, such as the feminine duck genus *Polysticta* Eyton 1836 and the masculine tyrant flycatcher genus *Polystictus* Reichenbach 1850, were regarded as homonyms or identical, and the junior name (i.e. *Polystictus*, the one last published) was replaced. *Polystictus* was renamed *Habrura* by Cabanis and Heine in 1859. This rule for homonymy is no longer accepted, and such names are now considered distinct and available. Names like *Polystictus*, that were replaced earlier, have been reinstated. The generic name *Habrura* is not a junior synonym of *Polystictus.* A specific (including subspecific) name can be used only once within a genus under the principle of homonymy, but the same name can be used in more than one genus, as explained earlier. If, through error or omission, two species or subspecies (species-level taxa) within a genus bear the same name, or if taxonomic research results in the submergence of one genus into another resulting in two species-level taxa bearing the same name, the name proposed later becomes the junior subjective homonym and that taxon must take the next available name by precedence of the date of publication or be given a new name.

Preservation of well-established names

The third principle, dealing with preservation of well-established names, is relatively recent, dating only from 1953. It is still not clearly articulated within the Code and is not fully accepted by all zoologists. *This concept is concerned with preservation of stability and universality in zoological nomenclature. It operates by protecting well-established names from being replaced by long-forgotten and hence unused senior synonyms.* At present, such protection must be achieved by action of the International Commission on Zoological Nomenclature, often under its plenary powers. For example, strong disagreement has existed amongst ornithologists as to the proper application of the generic name *Colymbus* Linnaeus 1758, i.e. whether it applies to the divers (loons) as claimed by British and some other European ornithologists or to the grebes as claimed by North American workers. Because a clear resolution of this dispute was not possible, the ICZN declared the name *Colymbus* suppressed and hence objectively invalid, and ruled that the generic name *Gavia* J. A. Forster 1788 (type *Colymbus immer* Linnaeus 1758) would apply to the divers and the name *Podiceps* Latham 1787 (type *Colymbus cristatus* Linnaeus 1758) would apply to the grebes.

Despite more than two hundred years of intensive scientific study, the family limits and relationships of birds, and to a lesser extent, generic and specific limits and relationships, are still the subject of considerable investigation. Most attempts to achieve some degree of consensus have foundered on conservatism, individual interpretation of the scientific evidence, still insufficient study, or, in the eighteenth and nineteenth centuries, just plain perversity. Subspecies are incipient species, and hence may be evolving intrinsic isolating mechanisms and other attributes which separate fully evolved species. These geographic races range from poorly differentiated forms to well-marked geographic entities, often considered to be allospecies—members of a superspecies. No fixed morphological or other phenotypic criteria exist by which one may reach undisputed conclusions on full species or subspecies status. While one ornithologist (a 'lumper') considers a geographic form to be only a subspecies, another ornithologist ('a splitter') may consider the same taxon to be a full species. So whereas the lumper recognizes only one pantropical, polytypic species of darter (or anhinga) *Anhinga anhinga*, or a broad-based genus *Erithacus*, the splitter would discern four species of anhinga (*Anhinga anhinga, A. rufa, A.melanogaster,* and *A. novaehollandiae*); limit *Erithacus* to the familiar robins of Europe and Japan; and resurrect *Luscinia, Pseudaedon, Tarsiger, Cyanosylvia, Sheppardia,* and so on, for the nightingales, bluetails, rubythroats, bluethroat, and akalats. Differences of opinion such as those just illustrated—and there are many more in ornithology—are based upon individual interpretation of the same evidence, although most workers agree that little scientific knowledge is to be gained by disputing the subjective minutiae of generic and specific limits.

This is a book about bird names, not bird classification, and in order to be as comprehensive as possible I have allied myself with the splitters by including a larger than usual number of names of genera and species. Using Peters' *Check-list of Birds of the World* as a foundation, I have included those subgenera and subspecies that have been given generic or specific status by at least one of the authors mentioned in the Bibliography, although I have not adopted all of the conclusions of, for example, Mathews (1927 and 1930), or Sclater (1924 and 1930). For historical interest I have also included names such as *Colymbus, Cuncuma, Tanagra, helvola, asha,* and others now synonymized or suppressed, although the sheer volume of avian names prohibits the inclusion of all synonyms and subspecific names in this initial survey.

Throughout the text the word 'synonymized' (usually in the abbreviated form 'syn.') is used in its broadest sense, to include names preoccupied elsewhere in zoological nomenclature, coined separately for the same taxon, submerged as a result of new research into generic and specific relationships, or rejected and suppressed by the International Commission on Zoological Nomenclature. The resultant list of names is, of necessity, arbitrary and uncritical and will not meet with the approval of all workers, but I have had neither the time nor the facilities to undertake an assessment of every form. By highlighting the diversity of birds in a comprehensive list, and explaining the origins of their names, I offer my hope that enquiry and further investigation will lead to a better understanding of their life histories, relationships, and ecology before man's indiscriminate development of 'his' world destroys the unique system we know as Earth.

GRAMMAR AND GENDER

The details of classical grammars are beyond the scope of this Dictionary, the interested reader being referred to Stearn (1983) or a standard primer for further clarification, but the following basic remarks will be useful.

All scientific names, regardless of their origin, are treated grammatically as Latin. Most are derived from classical Latin and its successors or from ancient Greek. Greek words are usually transliterated in accordance with generally accepted rules (*k* becomes c, *u* becomes y, final *-os* becomes -us, *kh* becomes ch; see Coues (1882), pp. 12–14; Stearn (1983), pp. 261–262; or International Code of Zoological Nomenclature, 3rd edition, (1985), appendix B, for complete listings). General Recommendation 16 of the ICZN states, 'A zoologist should give the etymology and gender of a new genus-group name.' When the *Systema Naturae* was published, classical Latin was still the medium of learning and international communication between savants and it was considered unnecessary to elaborate further on the scientific names published by Linnaeus and his adherents. In the nineteenth century authors such as Jean Cabanis and Harry Oberholser took pains to provide etymologies for their newly created genera, but those of the stamp of Prince Bonaparte and Gregory Mathews seldom threw light on the origins of the names they coined. Not all naturalists and ornithologists were classical scholars, however, and whilst some were poorly served by their printers, others were early students of Humpty Dumpty, so the purist will not have to search far to discover errors of grammar, structure, or transliteration. Indeed, earlier authors tried, unsuccessfully, to purge nomenclature of improperly formed names and those barbarisms considered unsuitable, erroneous, or without a classical pedigree.

Latin nouns are declined and verbs are conjugated; that is, their terminations change according to their case, tense, person, and number, or, more simply, the manner in which they are used. In this Dictionary nouns are indicated in the nominative singular (*ager,* the field), and, where the derivation is from the stem of the noun, they are also indicated in the genitive or possessive case (*agri,* of the field), although the genitive is shown only once in the first of a series of epithets sharing the initial combining form (e.g. see *albicapilla* to *albiventer*). Latin verbs are shown in the present infinitive (*clamare,* to shout) rather than the present indicative (*clamo,* I shout).

Adjectival epithets or trivial names have to agree in gender with the genus to which they are assigned. If a species is transferred from a masculine genus to a feminine one, or vice versa, then the specific termination must be changed accordingly. For example, Bonelli's Eagle was originally described in the feminine genus *Aquila* by Vieillot (in 1822) as *Aquila fasciata.* Later regarded as sufficiently distinct to warrant inclusion in the masculine genus *Hieraaetus*, it became known as *Hieraaetus fasciatus,* the binomen it still holds. Some species names that may look like adjectives (such as *arada* in *Cyphorhinus arada* or *cirlus* in *Emberiza cirlus*, which are based on native vernacular names) are, in fact, nouns in apposition given an adjectival function, and their terminations do not change to agree with the gender of the generic name.

The most common terminations are:

(1) *-us* (masculine), *-a* (feminine), *-um* (neuter) (e.g. *auritus, aurita, auritum,* long-eared);

(2) *-is* (masculine), *-is* (feminine), *-e* (neuter) (e.g. *agilis, agilis, agile,* nimble);

(3) *-er* (masculine), *-era* (feminine), *-erum* (neuter) (e.g. *pulcher, pulchra, pulchrum,* beautiful).

Although the Dictionary lists all accepted described forms in alphabetical order, in the etymologies only the masculine forms (*auritus, agilis, pulcher*) are shown, but the gender of the epithet and, thus, the generic name may also be identified from the examples of endings given here.

ANALYSIS OF NAMES

The scientific names of birds can be conveniently divided into categories according to their meaning, as follows: 1. appearance, 2. eponym, 3. native name, 4. toponym, 5. classification, 6. habitat, 7. behaviour, 8. food, 9. voice. These are analysed briefly below.

1. *Appearance.* The plumage, colours, and physical characteristics of birds form the largest category, accounting for over 55 per cent of all specific names and nearly 45 per cent of all generic names. Genera are more obvious candidates for names highlighting physical features (*Oxyura,* stifftail, *Recurvirostra,* recurved-bill, *Heteroscelus,* uneven-legs, *Lagopus,* hare-foot), whilst specific epithets incline more towards colour or pattern (*rufus* occurs in twenty-eight genera, *striatus* in twenty-four), although there are many exceptions to these generalizations.

Specific names of a general nature, such as those last mentioned, carried by more than a few species, may apply to only parts of the plumage or solely to the female or male bird. For example, of those species bearing the epithet *badius* (bay-coloured), *Accipiter badius* has rufous barred underparts, *Caprimulgus badius* has a tawny half-collar, *Ducula badia* has reddish-purple and dark brown upperparts which fade to chestnut in worn plumage, *Halcyon badia* has dark chocolate head and back, *Molothrus badius* has only rufous chestnut wings, *Phodilus badius* has rich chestnut upperparts, and only the breeding male *Ploceus badius* has chestnut mantle and underparts. Considerations of space prevent such detailed analysis under each entry in the main text.

2. *Eponym.* A popular form in nomenclature, representing nearly 20 per cent of all specific names, an eponym commemorates a real person or a mythical or a fictional character. Most perpetuate the name of the collector or discoverer of the species (*Microgoura meeki,* after the intrepid and far from eponymous Albert Meek who braved disease, cannibals, and hurricanes in the quest for specimens; *Cettia cetti,* after the Jesuit priest Francesco Cetti who wrote authoritatively on the natural history of Sardinia) or, the modern trend, a fellow ornithologist specializing in the appropriate group or area (*Atlapetes paynteri,* after Raymond A. Paynter, Jr., for his work on the biology of the genus *Atlapetes*; *Pyrrhura orcesi,* after Gustavo Orces V., in recognition of his many contributions to Ecuadorean ornithology).

Generally Latin genitives are formed from personal names by adding *-i* to a man's name if it ends in a vowel or -er (the vowel itself often being changed to *i*), *-ii* if it ends in a consonant, has been Latinized, or has a Latin form (e.g. von Martius), and by adding *-ae* to a woman's name. Plural forms (e.g. of the cousins Sarasin) are given the terminations *-orum* (masculine) or *-arum* (feminine). Personal names may also be converted into adjectives agreeing in gender with the generic name (e.g. *webbianus*).

Rules for the formation of eponyms have changed and been flouted over the years, hence the hotch-potch of the genitive terminations *-i* and *-ii* commemorating the same persons which can be found throughout the Dictionary. Nineteenth-century authors frequently named birds after members of their own family (*Ducula zoeae* after the wife of French naturalist René Lesson), or for royal patrons (*Paradisaea rudolphi* after the ill-fated Archduke Rudolf, Crown Prince of Austria–Hungary), but just as often gave no further details of the person commemorated, believing explanation unnecessary within their own closeted world. Over one hundred years ago Prince Bonaparte could afford to deride the fashion of naming birds after kings and princes in the hope of patronage. As a most prolific author of new genera and species he cast his net wide in search of new names, sometimes being inspired by bouts of republicanism (*Masius, Diphyllodes respublica*) or even sheer caprice (*Starnoenas, Graydidascalus*), but even he succumbed to the unimaginative eponym in the genitive (*Ptilinopus greyii*) or adjectival (*Macropygia emiliana*) forms.

Linnaeus supported the use of classical and mythical eponyms, regardless of their relevance, and authors such as Reichenbach expressed relationships between groups or types of birds by this means (e.g. the hummingbird genera *Archilochus, Atthis, Damo-*

phila, Doricha, Klais, Myrtis, Rhodopis, and *Sappho*). With regretably few, but notable, exceptions including the paradise kingfishers *Tanysiptera nympha, T. galatea, T. nais, T. sabrina, T. doris, T. acis,* and *T. danae,* classical eponyms have been used in an arbitrary fashion since the middle of the last century. In the text I have given the barest details of the person or character commemorated; further sources may be found in the essential Wynne (1969) or in the bibliographic references indicated. The attributed works mentioned are the most important and accessible or indicate the relevant activities and interests of the author. For reasons of space co-authors are not cited. Where eponyms can refer to more than one individual, the appropriate genera in which the individuals are commemorated are placed in parentheses after each brief biographical entry. For example, under *halli* the name of Robert Hall is commemorated in *Macronectes halli,* but it is Major Harold Wesley Hall whose name is perpetuated by *Pomatostomus halli.*

3. *Native name.* As well as those names adopted from over twenty-five modern languages (Nepalese *Ducula*; Malay *Ketupa*; Tupí Indian *Ajaia*; Russian *pleschanka*; Zulu *impetuani*; Japanese *komadori*), I also include here those taken or modified directly from classical Latin (*Asio, Columba*) and ancient Greek (*Circus, Drepanis*). The principal classical and late classical sources (Aristotle, Pliny, Aristophanes, Dionysius, and Hesychius) are referred to by name only. Their works may be found in the Bibliography. The identification of many birds mentioned in the classics has never been fully resolved because the details given in the original passages are often obscured by fantasy or myth, or are insufficient for accurate identification. Even such obviously common birds as *spiza* and *turannos* cannot be satisfactorily identified, and mediaeval and later writers allotted names in a haphazard way to the birds they knew. For example, the Greek substantive *trokhilos* was given by the ancients to the wren *Troglodytes* and also to the 'crocodile–bird' (perhaps the Egyptian Plover *Pluvianus* or the Spur-winged Plover *Hoplopterus*), but in modern ornithology the Jamaican hummingbird *Trochilus polytmus* bears the name. The origin of some names, although unrecorded and lost, can be assumed to be aboriginal because their authors, the likes of Thomas Horsfield, Andrew Smith, and Brian Hodgson, frequently made use of native names. Early Linnaean authors reaped rich harvests of aboriginyms from the works of Marcgrave, Buffon, Levaillant, and Azara, and although

names such as *Ara, arada, aguimp,* and *Suiriri* were deplored as barbarisms by classicists they have well served their purpose as the labels of nomenclature.

4. *Toponym.* These have proved more popular as specific names than generic names, there being over 1100 specific toponyms or geographical epithets. In the eighteenth and early nineteenth centuries much use was made of classical names for those parts of the Old World known to the Greeks and Romans (*Caprimulgus europaeus, C. indicus, C. aegyptius*). As exploration and collecting proceeded a spate of new toponyms was created to reflect the origins of the new avifaunas and discoveries. Some were created imaginatively (*Vitia, Sipia, Patagona*), but most often the addition of adjectival suffixes such as *-ensis, -anus,* or *-icus* to the type locality sufficed (*Tetraogallus himalayensis, T. tibetanus, T. caucasicus*). Certain epithets were used in the broadest sense, designating a continent or part of a continent for species of otherwise unknown provenance. Specifically *cayennensis* (Cayenne, French Guiana) was sometimes used for tropical America or the Neotropics, *americanus* for both North and South America, *ludovicianus* (Louisiana) for the interior of North America, *capensis* (Cape of Good Hope) for southern tropical Africa (the Afrotropics), *senegalensis* (Senegal) generally for Guinea or the western Afrotropics, *abyssinicus* (Abyssinia, Ethiopia) for the eastern and north-eastern Afrotropics, and *novaehollandiae* (New Holland) for Australia. Erroneous type localities were not infrequent in the early days, authors being misled as to the origins of the species they were cataloguing (*Touit batavica,* a Neotropical parakeet mistakenly thought to have come from Batavia in Java), or printers being confused by misreading an author's original manuscript notes or shorthand (*Bucco capensis,* from Linnaeus's shorthand *cayensis* for *cayennensis*). Such names, however, are not invalidated merely because they are incorrect.

5. *Classification.* A variety of names is included here, all suggestive of relationship: anagrams (*Dacelo* and *Lacedo* from *Alcedo*); diminutives (*Glaucidium, Calandrella, Tyranniscus*); generic combinations (*Ninox* from *Nisus* and *Noctua*); combinations of generic and substantive names (*Fregilupus* from *Fregilus* and *huppe*); and names indicating resemblance (*Merops bullockoides, Tryngites*), comparison (*Circus assimilis, Hypochera*), or affiliation (*Garrulax affinis, Propyrrhula, Pseudoscops*). The use of the Greek combining form *-oides,* resembling, appearing, featuring (from *eidos,* form, appearance, species; *eido,* to see) is especially common. Epithets such as *aberrans, affinis,* and *similis* imply a comparative or relative degree (i.e. the species so designated are considered different from or similar to another species or group of species). Where such names are held by numerous species (e.g. *affinis*, which occurs in twenty-two different genera), it has not been possible to list all the birds referred to, and the reader should try to consult original descriptions or contact a specialist library for further details.

6. *Habitat.* Birds, being the most mobile of animals, occur in every type of habitat and environmental condition. At first epithets relating to habitat were confined to those current in classical times (*Monticola, Acrocephalus palustris*). Exploration brought about the use of other, mainly native, names for specific local habitats (*Anas puna, Cisticola dambo, Picumnus varzeae*). Various suffixes indicate breeding, inhabiting, or occurring: *-manes* (passionately fond of); *-bates* (walker, dweller); *-philus* (loving, fond of); and *-cola* (inhabitant), to mention a few.

7. *Behaviour.* Most names in this category refer to display (*Philomachus pugnax*), behaviour in defence of the breeding territory (*Tyrannus*), or temperament (*Polemaetus bellicosus*). Others refer to mode of flight (*Aerodramus elaphrus*), nest-building (*Mniotilta*), or more tranquil pursuits (*Nystalus, Agapornis*).

8. *Food.* The names in this category reflect the variety of food items or prey of birds, including spiders, flies, bananas, seeds, carrion, fish, and monkeys. Most names here can be identified by the use of the suffixes *-phagus,* a glutton; *-thera,* a hunter; and *-vorus,* eating.

9. *Voice.* Despite the fact that the songs and calls of birds have elicited admiration, imitation, and comment throughout the ages, they have yielded few names in nomenclature, although aboriginal names tend to be based on onomatopoeia. Most epithets describe the sound of the voice or call itself (*Upupa, Cuculus, Aburria*), whilst others are more general in nature (*Baryphthengus, Larus ridibundus, Haliaeetus vocifer*).

To enable nearly 8500 names to be comprehensively treated, free use has been made of standardized phraseology and abbreviations (see page xxv). To avoid constant repetition of introductory phrases such as ' Coined from. . .', 'Adopted from. . .', and so on, all described names, unless otherwise indicated, may be taken as being derived, adopted, abbreviated, corrupted, or combined from the origins indicated. Many Greek and Latin words have a variety of shades of meaning. For example, the Greek word *leptos* can mean narrow, thin, slender, fine, delicate, light, subtle, slight, refined, insignificant, small, or meagre, according to context. Obviously such a catalogue could not be repeated after each entry as appropriate. The etymology, therefore, identifies only the shade of meaning relevant to each name, and readers should consult Liddell and Scott (1961) or Lewis and Short (1962) for the fullest range.

All generic and specific names in this Dictionary are arranged alphabetically, with forms differing in gender but separated by one or more other entries being cross-referenced. Specific epithets identical to generic names (*phoenicurus* and *Phoenicurus*) are not shown separately, but are referred to under the generic heading if the relevance of the name is not immediately apparent. For example, *Aquila* is the genus of great eagles, but the specific epithets *Fregata aquila* and *Eutoxeres aquila* refer to the predatory habits of the Ascension Frigatebird and the hooked bill of the White-tipped Sicklebill (a hummingbird) respectively, and these are mentioned under *Aquila*. Names which are apposite (*melanogaster* for a black-bellied species, *brachyrhynchus* for a short-billed species, *stagnatilis* for a marsh-dwelling species) or subjective (*splendidus,* brilliant, *egregius*, admirable) are not elaborated upon further.

Allowing for these constraints my aims have been to provide a complete and accurate etymology for all currently valid generic and specific bird names described between January 1758 and June 1990 together with an eclectic selection of historical synonyms; and a record of reliable information for the busy professional ornithologist, the ecologist and conservationist, the science historian, the curious birdwatcher, and those who, like myself, enjoy browsing through a good dictionary.

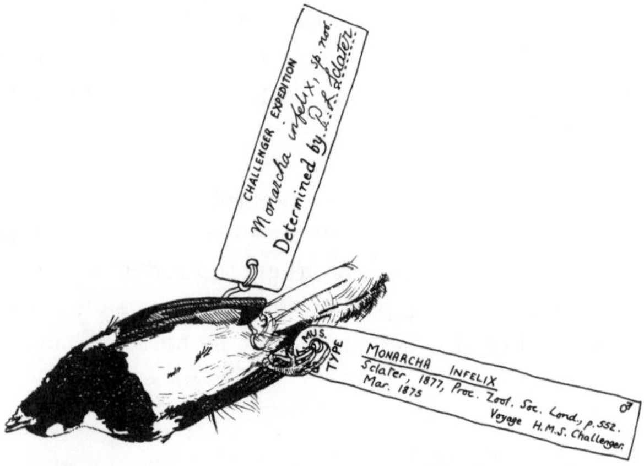

Monarcha infelix Admiralty Pied Monarch

Abbreviations

GENERAL

b.	born
cf.	compare (Latin *confer*)
comp.	comparative degree or form
d.	died
dim.	diminutive degree or form
e.g.	for example (Latin *exempli gratia*)
fide	according to (Latin *fides,* reliance, confidence)
fl.	flourished, lived (Latin *floruit*)
gen.	genitive or possessive case
Gr.	Greek (ancient or classical)
i.e.	that is (Latin *id est*)
L.	Latin (classical: approx. 200 BC to AD 160)
Late L.	Late Latin (approx. AD 160 to AD 600)
Med. L.	Mediaeval Latin (approx. AD 600 to AD 1500)
Mod.	Modern
Mod. L.	Modern Latin (approx. AD 1500 to the present)
myth.	mythology
No expl.	No explanation for this name is given in the original diagnosis/ description (original reference).
orn.	ornithology/ornithological
p.	page number (abbreviation not shown in serial works)
pl.	plate number
pp.	page numbers (abbreviation not shown in serial works)
q.v.	which see (Latin *quod vide*)
ref.	reference / refers to
sp.	species (singular)
spp.	species (plural)
super.	superlative degree or form
subsp.	subspecies (singular)
syn.	synonymized see p. xvii
v.	see (Latin *vide*)
Vol.	volume/book/livre/ fascicle number (shown in **bold** type)

CITED WORKS AND SERIALS

Abh. natur. Ver. Bremen	Abhandlungen herausgegeben vom naturwissenschaftlichen Vereine zu Bremen
Amer. Mus. Nov.	American Museum Novitates
Analyse	Analyse d'une nouvelle Ornithologie élémentaire (Vieillot, 1816)
Ann. Mag. Nat. Hist.	Annals and Magazine of Natural History
Av. syst. nat.	Avium Systema Naturale, das Natürliche System der Vögel (Reichenbach, 1849–1853)
Bull. Brit. Orn. Cl.	Bulletin of the British Ornithologists' Club
Bull. Mus. Comp. Zool.	Bulletin of the Museum of Comparative Zoology, Harvard
Bull. Sci. Nat.	Bulletin des Sciences Naturelles et de Géologie

Classif. birds	On the Natural History and Classification of Birds (Swainson, 1836–1837)
Consp. gen. avium	Conspectus Generum Avium (Bonaparte, 1850–1857)
Fauna Bor.-Amer.	Fauna Boreali-Americana (Swainson and Richardson, 1831–1832)
Gen. synop.	A General Synopsis of Birds (Latham, 1781–1785) (supplements 1787, 1801)
Gen. zool.	General Zoology—Aves (Shaw and Stephens, 1809–1826)
Icon. faun. Ital. uccelli	Iconografia della Fauna Italica - Uccelli (Bonaparte, 1841)
Illus. orn.	Illustrations of Ornithology (Jardine and Selby, 1827–1843)
Illustr. zool. south Africa	Illustrations of the Zoology of South Africa (Smith, 1838–1849)
Index orn.	Index Ornithologicus (Latham, 1790) (supplement 1801)
Izvest. Imper., Moskva	Izvestiya Imperatorskogo Obshchestva Lyubitelei Estestvoznaniya Antropologii i Etnografii, Moskva [Moscow]
Journ. Asiat. Soc. Bengal	Journal of the Asiatic Society of Bengal
Journ. für Orn.	Journal für Ornithologie
Madras Journ. Lit. Sci.	The Madras Journal of Literature and Science
Mag. Zool.	Revue et magazin de Zoologie
Mém. Soc. Imp. Sci. Nat. Cher.	Mémoires de la Société Impériale des Sciences Naturelles de Cherbourg
Ned. Tjdschr. Dierk.	Nederlandsch Tijdschrifj voor de Dierkunde
Nomen. av. mus. zool. Berol.	Nomenclator Avium Musei Zoologici Berolinensis (Lichtenstein, 1854)
Notes Leyden Mus.	Notes from the Leyden Museum
Nouv. dict. hist. nat.	Nouveau Dictionnaire d'Histoire Naturelle (Vieillot, 1816–1819)
Nov. Zool.	Novitates Zoologicae
Proc. Acad. Nat. Sci.	Proceedings of the Academy of Natural Sciences of Philadelphia.
Proc. Biol. Soc. Washington	Proceedings of the Biological Society, Washington.
Proc. U S Nat. Mus.	Proceedings of the United States National Museum
Proc. Zool. Soc. London	Proceedings of the Zoological Society of London
Remarques Obs. Blanchard Psittac.	Remarques à propos des Observations de M. Émile Blanchard sur les Caractères ostéologiques chez les oiseaux de la famille des Psittacides, et Tableau des

	genres de Perroquets disposés en séries parallèles (Bonaparte, 1857)
Rep. exped. centr. Africa	Report of the Expedition for exploring Central Africa from the Cape of Good Hope. . . (Smith, 1836)
Rev. Amer. birds	Review of American Birds in the Museum of the Smithsonian Institution (Baird, 1866)
Rev. Zool.	Revue Zoologique
Riv. Contemp. Torino	Rivista Contemporanea, Torino [Turin]
Smithson. Misc. Coll.	Smithsonian Miscellaneous Collections
Synop. bds. Austr.	A Synopsis of the Birds of Australia and the adjacent islands (Gould, 1837–1838)
Trans. Linn. Soc. London	Transactions of the Linnean Society of London
Trans. N. Z. Inst.	Transactions and proceedings of the New Zealand Institute
Wilson Bull.	The Wilson Bulletin
Zool. Journ.	Zoological Journal

Xenopirostris xenopirostris Razor-billed Vanga

How to use this Dictionary

THE arrangement of entries is alphabetical with generic and specific names in one sequence. Each part of a bird's scientific name is entered separately, so that the scientific name for Lonely Kingfisher, *Halcyon monacha*, for example, will not be found as such, but both *Halcyon* and *monacha* are explained in their respective alphabetical places. English or other-language names are mentioned when they have a bearing on a scientific name.

Here are some typical entries with notes to clarify their meanings.

> **coryphaeus** **1** French *le Coriphée,* the name given to the Karoo Scrub Robin by Levaillant (1802), ref. its beauty and voluptuous song (Gr. myth. Coryphaea, an epithet of the goddess Diana) (*Erythropygia).* **2** Gr. *koruphaia,* of the head, a tuft on the head (*koruphe,* the crown of the head) (*Pogoniulus).*

This is a specific name, as shown by its lower-case first letter, and it is in Latin masculine form, as indicated by the ending *-us*. (Taxonomic terminology and Latin endings are explained in the Introduction, pp. xviii–xix.) It has two derivations. The first one comes from a usage by Levaillant in 1802. (The full citation for works thus cited by author and date will be found in the Bibliography under the author's surname.) Levaillant coined this name in French from Greek mythology, and this gave rise to the scientific name *coryphaeus*. It refers to a bird classified in the genus *Erythropygia*, shown in parentheses. The second derivation is directly from the Greek word *koruphaia*, which in turn derives from the Greek term *koruphe*. This second derivation applies to a bird in the genus *Pogoniulus*.

> **Erythropygia/erythropygius** Gr. *eruthros*, red; Mod. L. *pygius*, rumped (Gr. *-pugos*, rumped; *puge*, the rump).

This name occurs both as a generic name and as a specific name (see Introduction, p. x). Both forms are shown in this entry as they share the same derivation. The capitalized generic form, shown first, happens to be in the Latin feminine gender, while the lower-case specific name is masculine in this instance (ending *-us*). The first three syllables of the name come from the Greek word for red. The last three syllables derive from a Latin word formulated after the year 1500 (see Abbreviations, p. xxv) which was based on a Greek word-ending derived from a Greek noun. The composite meaning of this scientific name is red-rumped.

> **Pogoniulus** Dim. of syn. genus *Pogonias*, barbet (Gr. *pogonias*, bearded).

This is a generic name, a diminutive form of another generic name, *Pogonias*, which has been synonymized (no longer in use for any genus of bird). The abbreviation 'syn' is listed in the Abbreviations (p. xxv) and directs you to an explanation of synonymy in the Introduction (p. xvii). The origin of the name is a Greek term for bearded, and the generic name means 'little barbet' or 'little bearded-one'.

> **nelicourvi** No expl. (Scopoli (1786), *Deliciae florae et faunae insubricae*, fasc. 2, p. 96); doubtless from the Tamil (Sri Lanka) name *nellukuruvi* for a finch or waxbill, the Madagascan Nelicourvi Weaver *Ploceus* mistakenly thought to have come from Indomalaya.

This specific name does not come from Latin or Greek. No expl. (see Abbreviations, p. xxv) indicates that no explanation of its source appears in the original description of the bird given this specific name. The author and full reference for the original description then follow. There is however, a likely derivation based on a Tamil name. Reference is made to the particular bird because it helps explain the circumstances of naming.

A

aalge Danish *aalge*, an auk (cf. Norwegian *alke* ; Old Norse *alka).*

abbas L. *abbas*, an abbot; after J. Abbott (1751–1840) US philologist.

abbotti After **1** Lieutenant-Colonel J. R. Abbott (1811–1888) Assistant Commissioner of the Arakan, Burma 1837–1845 (*Trichastoma*). **2** William L. Abbott (1860–1936) US explorer, naturalist, and collector (*Coracornis, Papasula, Spilornis).*

abdimii After El Arnaut Abdim Bey (1780–1827) Egyptian governor of Dongola, Sudan, 1821–1825.

abeillei /Abeillia After M. Abeillé (fl. 1839) and his wife, Félice, French naturalists and collectors.

aberdare After the Aberdare mountain range in central Kenya.

aberrans L. *aberrans*, wandering, deviating (i.e. different) (*aberrare*, to wander).

aberti After James W. Abert (1820–1897) US army collector and explorer with the Fremont expedition.

abingoni After Montagu Bertie, 5th Earl of Abingdon (1784 –1854) .

abnormis L. *abnormis*, irregular (i.e. different, unorthodox).

Abrornis Gr. *habros*, delicate, soft; *ornis*, a bird.

Abroscopus Gr. *habros*, delicate, soft; *skopos*, a watcher, a seeker (*skeptomai*, to look about).

aburri /Aburria Colombian Indian names *aburri* and *burria* for the Wattled Guan *A. aburri*, after its rolling siren call ' ba...reeeeer...ah! '

abyssinica /abyssinicus After Abyssinia, the term widely used before 1939 for Ethiopia (Arabic *habish*, a mixture, with reference to the mixed religious and ethnic populations of the country).

acadicus After Acadia, a former French colony in eastern Canada centred on Nova Scotia and ceded to Britain in 1763 at the end of the French and Indian Wars.

Acanthagenys Gr. *akantha*, a thorn, spine; *genus*, the cheek; the Spiny-cheeked Honeyeater *A. rufogularis* has stiff, brush-like cheek feathers.

Acanthidops Gr. *akanthis, akanthidos*, thorny, spiky (*akantha*, a thorn, spine, thistle); *ops*, the face; ref. the narrow, pointed bill of the Peg-billed Finch *A. bairdii.*

Acanthis Gr. *akanthis*, a small bird mentioned by Aristotle and other authors, not further identified, but since associated with the Linnet *A. cannabina* or Goldfinch *Carduelis carduelis* (*akantha*, a thistle, a common food plant).

Acanthisitta Gr. *akantha*, a thorn; genus *Sitta*, nuthatch; ref. the narrow, pointed bill of the nuthatch-like New Zealand Rifleman *A. chloris.*

Acanthiza Gr. *akantheon*, a thorn brake; *zao*, to live; ref. habitat of the small Australasian thornbills.

acanthizoides Genus *Acanthiza*, thornbill; Gr. *-oides*, resembling (*eidos,* shape, species, appearance; *eido,* to see).

Acanthopneuste Gr. *akantha*, a thorn; *pneo*, to live (*pneustiao*, to breathe hard).

Acanthorhynchus Gr. *akantha*, a thorn, a spine; *rhunkhos*, the bill .

Acanthornis Genus *Acanthiza*, thornbill; *ornis*, a bird; ref. the thin, spiky bill of the Tasmanian Scrubtit *A. magnus*.

accentor Syn. genus *Accentor*, dunnock, accentor (L. *ad*, towards; *cantor*, a singer); ref. the sober plumage and ground-loving habits of the Bornean Friendly Warbler *Bradypterus*.

Accipiter L. *accipiter*, a hawk (*accipere*, to grasp, take; the original meaning was 'to understand' rather than 'to seize').

accipitrinus L. *accipitrinus*, hawk-like (*accipiter, accipitris*, a hawk; *-inus*, resembling, possessing).

acer L. *acer*, sharp, shrill.

Aceros Gr. *akeros*, hornless, without horns (privative *a*-, lacking; *keras*, a horn); the bill of the Rufous-necked Hornbill *A. nipalensis* is not surmounted by a casque.

Acestrura Gr. *akestra*, a needle; *oura*, the tail.

Achaetops L. *ad*- (*a*- before c), resembling; genus *Chaetops*, rockjumper.

acis Gr. myth. Acis, a Sicilian shepherd beloved by Galatea and changed into a stream on his death.

Acridotheres Gr. *akris, akridos*, a locust; *theres*, a hunter.

Acrocephalus Gr. *akros*, pointed; *-kephalos*, headed (*kephale*, the head); from the shape of the male warbler's head when singing.

Acrochordopus Gr. *akrokhordon*, a type of wart; *pous*, the foot; ref. the coarse, scaly appearance of the legs of the Rough-legged Tyrannulet *A. burmeisteri*.

Acropternis Gr. *akros*, pointed; *pternis*, the ankle; ref. the very straight, long hindtoe and straight, spiky hind-claw of the Ocellated Tapaculo *A. orthonyx*.

Acryllium Dim. from Gr. *akris*, a point, a peak; from the numerous lanceolate feathers on the underparts of the Vulturine Guineafowl *A. vulturinum*.

Actenoides Gr. *aktis, aktinos*, brightness, splendour; *-oides*, resembling.

Actinodura Gr. *aktis, aktinos*, brightness, splendour; *odouros*, a watcher, a guardian.

Actitis Gr. *aktites*, a coast dweller.

Actophilornis Gr. *akte*, the seashore, the beach; *philos*, loving, fond of; *ornis*, a bird; ref. the aquatic habitats of the Afrotropical jacanas.

acuflavidus L. *acus*, a needle (i.e. the bill); *flavidus*, yellowish.

acuminata L. *acuminatus*, pointed, tapered (*acuere*, to sharpen).

acunhae After Tristan da Cunha Island, discovered by the Portuguese navigator Tristão da Cunha in 1506.

acuta L. *acutus*, acute, sharp-pointed (*acuere*, to sharpen to a point).

acuticauda L. *acutus*, sharp-pointed; *cauda*, the tail.

acuticaudatus L. *acutus*, sharp-pointed; *caudatus*, tailed (*cauda*, the tail).

acuticaudus see **acuticauda.**

acutipennis L. *acutus*, sharp-pointed; *pennis*, winged (*penna*, feather).

acutirostris L. *acutus*, sharp-pointed; *-rostris*, billed (*rostrum*, the bill).

adalberti After Admiral Prince Heinrich Wilhelm Adalbert of Prussia (1811–1873).

Adamastor Gr. *adamastos*, untameable; ref. the wild seas inhabited by the Grey Petrel *A. cinereus*.

adamsi /adamsii After **1** Edward Adams (1824–1856), British surgeon explorer in the Arctic, 1849–1856 (*Gavia*). **2** A. Leith Adams (1826–1882) Army surgeon in India, 1848 (*Montifringilla*).

adansonii After Michel Adanson (1727–1806), French botanist in West Africa 1748–1753, collector, and author (*Histoire naturelle du Sénégal*, 1757; *Familles naturelles des plantes*, 1763).

addendus L. *addendus*, added (*addere*, to add, join).

addita L. *additus*, added (i.e. a new species).

adela Gr. *adelos*, ignoble; ref. plain coloration of the Wedge-tailed Hillstar *Oreotrochilus*.

adelaidae After **1** the city of Adelaide in South Australia, named for Queen Adelaide, consort to King William IV

(*Aprosmictus*) **2** Adelaide Swift (fl. 1865) daughter of US financier R. Swift (*Dendroica*).

adelberti After Vice-Admiral M. C. Adelbert le Barbier de Tinan (1803–1876) French explorer and conchologist.

adeliae After Adélie Dumont d'Urville (1790–1842) wife of French explorer Admiral Vicomte Dumont d'Urville who discovered the Adélie coast of Antarctica in 1840.

Adelomyia Gr. *adelos*, ignoble; *muia*, a fly (i.e. a very small bird); ref. the dull plumage of the Speckled Hummingbird *A. melanogenys*.

admiralitatis After the Admiralty Islands, Bismarck Archipelago, Papua New Guinea; named for the British Admiralty, the government department that administered the Royal Navy, 1832–1964 (Old French *amiral*, admiral; Arabic *amir al bahr*, lord of the sea).

adolphinae After Adolfina Bruijn (fl. 1875) wife of J. Bruijn, Dutch botanist, zoologist and explorer.

adorabilis L. *adorabilis*, adorable, admirable (*adorare*, to worship, to honour).

adscitus L. *adscitus* (a form of *ascitus*), approved (*asciscere*, to approve, to adopt).

adsimilis L. *adsimilis* (a form of *assimilis*), similar, like.

adspersus L. *adspersus* (a form of *aspersus*), a sprinkling (*aspergere*, to sprinkle); ref. the black spotted wings and mantle of the Red-billed Francolin *Francolinus*.

adusta L. *adustus*, burnt, sunburnt (*adurere*, to set fire to).

Aechmolophus Gr. *aikhme*, a spear; *lophos*, a crest; ref. the prominent crest of the Pileated Flycatcher *A. mexicanus*.

Aechmophorus Gr. *aikhmophoros*, a spearman, a spear carrier; ref. the long, spear-like bills of these American grebes.

Aechmorhynchus Gr. *aikhme*, a spear; *rhunkhos*, the bill.

aedon L. *aedon*, the nightingale (Gr. *aedon*). The ICZN has banned the use of diacritic marks in scientific names, but, in transliteration, the Gr. *aedon* should be spelt *aëdon*, the diaeresis indicating that the second of the two vowels be pronounced separately. Similarly, the Gr. combining form *-oides*, amongst further examples, should be spelt *-oïdes*. Unhappily, whereas the removal of the German umlaut (ü) is compensated for by the addition of an *e* after the modified vowel, the removal of the diaeresis, a useful guide to pronunciation, leaves no trace of its passing.

Aegintha Gr. *aiginthos*, a variant form of *aigithos*, a mythical bird mentioned by Aristotle, Pliny, and other authors, not further identified. In modern times associated with the Linnet *Acanthis cannabina* or the Long-tailed Tit *Aegithalos caudatus*.

aegithaloides Genus *Aegithalos*, long-tailed tit; Gr. *-oides*, resembling.

Aegithalos Gr. *aigithalos*, a tit.

Aegithina Dim. of Gr. *aigithos*, a mythical bird (v. *Aegintha*).

Aegolius L. *aegolius*, a bird of prey, a screech-owl (Gr. *aigolios*, a bird of ill omen, an owl).

Aegotheles Gr. *aigothelas*, the nightjar (*aix, aigos*, a goat; *thelas*, a sucker; *thelazo*, to suckle).

Aegypius Gr. *aigupios*, a vulture.

aegyptiacus L. *Aegyptiacus*, Egyptian (*Aegyptus*, Egypt; Gr. *Aiguptos*).

aegyptius L. *Aegyptius*, an Egyptian (Gr. *Aiguptios*).

aenea L. *aeneus*, of a bronze colour (*aes, aeris*, copper, bronze).

aeneocauda L. *aeneus*, bronze-coloured; *cauda*, the tail.

aeneum /aeneus see **aenea**.

aenigma L. *aenigma*, a mystery, a riddle (Gr. *ainigma*).

Aenigmatolimnas Gr. *ainigma, ainigmatos*, a mystery; Mod. L. *limnas*, a crake or rail (Gr. *limnas*, a form of *limnaios*, of the marsh; *limne*, a mere or marsh).

aenobarbus L. *aenobarbus*, red-bearded (*aeneus*, bronze-coloured; *barba*, the beard).

Aepypodius Gr. *aipus*, steep, lofty; *pous, podos*, the foot; ref. the long legs and big feet of the New Guinea brush-turkeys.

aequatoriale /aequatorialis Late L. *aequatorialis*, equatorial (*aequator*, the equator); **1** after Ecuador (Spanish *ecuador*, the equator) (*Androdon, Cistothorus, Xiphorhynchus*). **2** more generally of equatorial latitudes of Africa, in Cameroun (*Apaloderma*), Abyssinia (Ethiopia) (*Apus*), and Kenya (*Sheppardia*).

aequinoctialis L. *aequinoctialis*, equinoctial, relating to the equinox (*aequus*, equal; *nox, noctis*, night); ref. equatorial latitudes of Christmas Island, Line Islands (Kiritimati, Republic of Kiribati) (*Conopoderas*), and Cayenne (French Guiana) (*Buteogallus, Geothlypis*), or to the stormy waters of the Cape of Good Hope commonly associated with the equinox (*Procellaria*).

aereus L. *aereus*, made or fitted with copper or bronze (*aes, aeris*, copper, bronze).

Aerocharis Gr. *aer, aeros*. the air, mist; *kharis*, grace, beauty.

Aerodramus Gr. *aer, aeros*, the air; *dromos*, running (*trekho*, to run).

Aeronautes Gr. *aer, aeros*, the air; *nautes*, a sailor or pilot (*naus*, a ship).

Aerops Gr. *aerops* (a form of *merops*), the bee-eater *Merops*.

Aerornis Gr. *aer, aeros*, the air; *ornis*, a bird.

aeruginosum /aeruginosus L. *aeruginosus*, covered with copper rust, rusty (*aerugo*, copper rust; *-osus*, abundance, full of).

aesalon Gr. *aisalon*, a kind of hawk.

aestiva L. *aestivus*, summery, relating to summer (*aestus*, summer).

aestivalis L. *aestivalis*, summer-like, pertaining to summer (*aestus*, summer).

aethereus L. *aetherius*, heavenly, on high (Gr. *aitherios*, the ether or upper air); given to the Red-billed Tropicbird *Phaethon* for its ethereal plumage and graceful flight, and to the Large-tailed Potoo *Nyctibius* because it is supposed to rise high in the air and hover like a falcon.

Aethia Gr. *aithuia*, a seabird mentioned by Aristotle, Hesychius, and other authors, not further identified. In modern times associated with a variety of seabirds, including a gull *Larus*, a skua *Stercorarius*, a grebe *Podiceps*, a diver *Gavia*, and, in nomenclature, an auklet.

aethiopica /aethiopicus L. *Aethiopicus*, an Ethiopian. Classical Ethiopia was an ill-defined part of Africa south of Egypt, in contrast to Africa proper which referred to the area around Carthage (Tunis).

aethiops L. *Aethiops*, an Ethiopian, a Negro.

Aethiopsar L. *Aethiops*, a Negro; Gr. *psar*, the starling (i.e. a black starling).

Aethocorys Gr. *aethes*, unusual; Mod. L. *corys*, a lark (Gr. *korudalos* or *korudos*, a lark. The syn. genus *Corys* was diagnosed by Reichenbach in 1850).

Aethopyga Gr. *aithos*, fiery, fire coloured; *puge*, the rump.

Aethorhynchus Gr. *aethes*, strange, unusual; *rhunkhos*, the bill; ref. long, heavy bill of the Great Iora *A. lafresnayei*.

afer L. *Afer*, African, from Africa.

affine /affinis L. *affinis*, related, allied.

afra see **afer.**

Afribyx L. *Afer*, African; Mod. L. *ibyx*, a plover or lapwing (Gr. *ibuo*, to shout; ref the persistent loud cries of lapwings when protecting nest and young. Gr. *ibux* is a variant form of *ibis*, an ibis).

africana L. *Africanus*, African, from Africa.

africanoides Specific name *Mirafra africana* ; Gr. *-oides*, resembling.

africanus see **africana.**

Afropavo L. *Afer*, African; genus *Pavo*, peacock.

Afrotis L. *Afer*, African; genus *Otis*, bustard.

agami /Agamia Cayenne Indian name *agami* for a forest bird, perhaps a trumpeter *Psophia*, but associated with the Chestnut-bellied Heron *A. agami* by d'Aubenton in 1765.

Agapeta Gr. *agapetos*, dearly beloved.

Agapornis Gr. *agape*, love; *ornis*, a bird; the substantive name 'lovebird' reflects the strong pair bond and sociable nature of these small Afrotropical parrots.

Agelaius Gr. *agelaios*, flocking, gregarious; the blackbirds of this genus breed in large colonies and gather into immense flocks in winter.

Agelastes Gr. *agelastos*, sullen, grave; ref. the solitary disposition of the White-breasted Guineafowl *A. meleagrides*.

agile /agilis L. *agilis*, nimble, agile.

Aglaeactis Gr. *aglaia*, splendour; *aktis*, a sunbeam.

aglaiae Gr. myth. Aglaia, one of the Graces (*aglaia*, splendour, beauty).

Aglaiocercus Gr. *aglaios*, splendid; *kerkos*, the tail.

agraphia Gr. *agraphos*, unwritten (privative *a-*, lacking; *graphe*, writing; *grapho*, to write or draw); ref. the plain plumage of the Unstreaked Tit-Tyrant *Uromyias*.

agricola L. *agricola*, a farmer, a countryman; ref. winter habitat of the Paddyfield Warbler *Acrocephalus*.

Agriocharis Gr. *agrios*, living in fields (*agros*, a field); *kharis*, grace, loveliness.

Agriornis Gr. *agrios*, living in fields; *ornis*, a bird.

Agrobates Gr. *agrobates*, a country dweller (*agros*, a field; *-bates*, a dweller; *baino*, to go, to tread).

aguimp Namaqua (Namibian) native name *a-guimp* meaning beach- or strand-runner for the African Pied Wagtail *Motacilla*.

Agyrtrina Syn. genus *Agyrtria*, hummingbird (Gr. *agurtria*, a gatherer or collector; *ageiro*, to collect); L. *-inus*, resembling.

ahantensis After the region of Ashanti in the Gold Coast (Ghana).

Aidemosyne Gr. *aidemosune*, modest, bashful.

aikeni After Charles Edward Howard Aiken (1850–1936) US taxidermist, pioneer, and collector.

Ailuroedus Gr. *ailouros*, a cat; *odos*, a singer; ref. the yowling, cat-like calls of the Australasian catbirds.

Aimophila Gr. *aimos*, thorn; *philos*, loving, fond of; ref. habitat of these American sparrows.

Aix Gr. *aix*, a diving bird, perhaps some sort of duck, mentioned by Aristotle, not further identified, but since conjectured to be a goose *Branta* or a grebe *Podiceps*.

Ajaia /ajaja Tupí (Brazilian) Indian name *ajaja* for the Roseate Spoonbill *A. ajaja*. This is one of many names originally found in Marcgrave (1648). Georg Marcgrave (or Marcgraf), a German doctor and mathematician, wrote comprehensively, from first hand knowledge and in Latin, on the birds of the Pernambuco district of Brazil, giving for each sp. its names in the local Tupí Indian vernacular and, where known, in Portuguese also. For many years his classic book provided the sole source of information on Brazilian birds, and over forty specific names by Linnaeus and Gmelin were based on Marcgrave's detailed descriptions.

ajax Gr. myth. Ajax or Aias, hero of the Trojan War; the Painted Quail-thrush *Cinclosoma* was so described because of its large size.

akahige Japanese name *akahige,* meaning 'red beard', for the Ryukyu Robin *Erithacus komadori*, erroneously applied to the Japanese Robin *Erithacus* by Temminck who believed it came from the Ryukyu Islands (v. *komadori).*

akool Sinhala (Sri Lankan) name *kukkula* for the moorhen *Gallinula* and watercock *Gallicrex* (cf. Malayalam *kula*, waterhen).

Alaemon Gr. *alemon*, a wanderer, a rover (i.e. cursorial).

alai Hawaiian name *alae,* a generic term for coots *Fulica* and moorhens *Gallinula* (e.g. the Hawaiian Coot is called *alae ke'oke'o* and the Moorhen *alae 'ula).*

alario L. *alarius*, on the wing (originally the wing of an army) (i.e. a flier).

alaschanicus After the Ala Shan mountain range in Ningxia Zizhiqu, northern China.

Alauda L. *alauda*, a lark; according to Pliny the Gaulish name for the Crested Lark *Galerida*.

alaudina / alaudinus L. *alauda*, a lark; *-inus*, resembling.

alaudipes L. *alauda*, a lark; *pes*, the foot; because of its curved bill and black and white wing pattern, the Hoopoe Lark *Alaemon* was originally described as a species of *Upupa*, hoopoe.

alba L. *albus*, white.

albatrus German *albatros*, the albatross; 'albatross' has equivalents throughout Europe, all modifications from Portuguese *alcatras*, a form of *alcatruz*, the pelican *Pelecanus*, from Arabic *al qadus*, water-bucket, ref. the capacious bill. The name was mistakenly identified and transferred to other large waterbirds, firstly by English voyagers to the frigatebirds *Fregata* (then known as *algatross)* and then to the present species of *Diomedea* (*albitrosse*; altered by L. *albi* -, white, to distinguish it from the frigatebirds, which are black).

albellus Dim. from L. *albus*, white.

albeola Dim. from L. *albus*, white (cf. *albere*, to be white).

alberti After **1** Prince Albert of Saxe-Coburg-Gotha (1819–1861) Prince Consort to Queen Victoria (*Crax, Menura)*. **2** King Albert I of the Belgians (1875–1934; reigned 1909–1934) (*Prionops)*. **3** King Albert or Albrecht of Saxony (1828–1902; reigned 1873–1902) (*Pteridophora)*.

albertinae After Albertina Schlegel (fl. 1869) wife of German ornithologist Hermann Schlegel.

albertinum After the Albertine Rift Valley, in the region of Lake Albert, north-eastern Zaïre.

albertisi / albertisii After Cavaglieri Luigi M. d'Albertis (1841–1901) Italian botanist, zoologist, and ethnologist in New Guinea, 1871–1877.

albescens L. *albescens*, whitish, turning white (*albescere*, to become white).

albicapilla / albicapillus L. *albus, albi*, white; *-capillus*, capped, headed (*capillus*, the hair of the head).

albicauda L. *albus*, white; *cauda*, tail.

albicaudatus L. *albus*, white; *caudatus*, tailed (*cauda*, tail).

albiceps L. *albus*, white; *-ceps*, headed (*caput*, the head).

albicilla L. *albus*, white; Mod. L. *cilla*, the tail (v. *Motacilla)*. In the case of the Whitehead *Mohoua* the epithet is an error for *albicapilla* (q.v.).

albicollis L. *albus*, white; Mod. L. *-collis*, necked (L. *collus* or *collum, colli*, the neck).

albidinuchus L. *albidus*, whitish; Med. L. *nuchus*, the nape, or (medical use) *nucha*, the medulla oblongata (Arabic *nukha*, spinal marrow).

albidiventris L. *albidus*, whitish; *venter, ventris*, the belly.

albifacies L. *albus*, white; *facies*, the face.

albifrons L. *albus*, white; *frons*, the forehead, brow.

albigula L. *albus*, white; *gula*, the throat.

albigularis L. *albus*, white; Mod. L. *gularis*, throated (L. *gula*, the throat).

albilatera L. *albus*, white; *lateralis*, lateral, of the side (*latus*, the side, flank).

albilinea L. *albus*, white; *linea*, a line.

albilineata L. *albus*, white; *lineatus*, lined, marked with lines (*linea*, a line).

albilora / albiloris L. *albus*, white; Mod. L. *loris*, lored, of the lores (the area on a bird's face between the upper mandible and the eye) (L. *lorus* or *lorum*, a thong, strap, or rein).

albina L. *albinus*, a plasterer; ref. whitish speckled plumage of the Kolombangara Fantail *Rhipidura*.

albinucha L. *albus*, white; Med. L. *nucha*, the nape (Arabic *nukha*, spinal marrow).

albipectus L. *albus*, white; *pectus*, the breast.

albipennis L. *albus*, white; *pennis*, winged (*penna*, a feather).

albirostris L. *albus*, white; *-rostris*, billed.

albispecularis L. *albus,* white; *specularis*, like a mirror (*speculum*, a mirror); ref. white, window-like wing-patches of the Ashy Robin *Heteromyias*.

albistriata L. *albus,* white; Mod. L. *striatus*, streaked, striped (L. *stria*, a furrow).

albitarsus L. *albus,* white; Gr. *tarsos*, the flat of the foot (i.e. the visible 'leg' of a bird); ref. the white legs of the Rufous-banded Owl *Ciccaba*.

albitorques L. *albus,* white; *torques*, a collar, a necklace.

albiventer /albiventre /albiventris L. *albus,* white; *venter, ventris*, the belly.

alboauricularis L. *albus*, white; *auricularis*, the ear.

albobrunneus L. *albus*, white; Mod. L. *brunneus*, brown (Med. L. *brunus*, brown).

albocinctus L. *albus*, white; *cinctus*, girdled, encircled (*cingere*, to encircle).

albocoronata L. *albus*, white; *coronatus*, crowned (*coronare*, to crown).

albocristata /albocristatus L. *albus*, white; *cristatus*, crested (*crista*, a crest).

albofasciata /albofasciatus L. *albus*, white; Late L. *fasciatus*, banded, striped (L. *fascia,* a band, a streak).

albofrenatus L. *albus*, white; *frenatus*, bridled (*frenare*, to bridle); ref. face pattern.

albofrontata L. *albus*, white; Mod. L. *frontatus*, fronted, browed (L. *frons*, the forehead, brow).

albogriseus L. *albus*, white; Med. L. *griseus*, grey.

albogulare /albogularis L. *albus*, white; Mod. L. *gularis*, throated (L. *gula*, the throat).

albolarvatus L. *albus*, white; *larvatus*, masked (*larva*, a mask).

albolimbata /albolimbatus L. *albus*, white; *limbatus*, edged, bordered (*limbus*, a border, a fringe); ref. white-edged or fringed wing feathers.

albolineatus L. *albus*, white; *lineatus*, lined, marked with lines (*linea*, a line).

alboniger L. *albus*, white; *niger*, black, dark coloured.

albonotata /albonotatus L. *albus*, white; *notatus*, marked (*notare*, to mark).

alboscapulatus L. *albus*, white; Mod. L. *scapulatus*, of the scapulars, shouldered (L. *scapulae*, the shoulder blades).

albosignata L. *albus*, white; *signatus*, marked (*signare*, to mark, inscribe).

albospecularis L. *albus*, white; *specularis*, like a mirror (*speculum*, a mirror); ref. broad, white wing-panels.

albosquamatus L. *albus*, white; *squamatus*, scaled (*squama*, a scale); ref. shape of the white spots on the underparts of the White-wedged Piculet *Picumnus*.

albostriatus L. *albus*, white; Mod. L. *striatus*, streaked, striped (L. *stria*, a furrow).

alboterminatus L. *albus*, white; *terminatus*, a boundary, limit; ref. the white tail tips of the Crowned Hornbill *Tockus*.

albotibialis L. *albus*, white; *tibialis*, pertaining to the shin, a legging (*tibia*, the shin).

albovittatus L .*albus*, white; *vittatus*, banded, ribboned (*vitta*, a head band).

albus see **alba.**

Alca Norwegian name *alke* for the Razorbill *A. torda* (Old Norse *alka*, the Razorbill).

Alcedo L. *alcedo*, the kingfisher (Gr. *alkuon* ; v. *Halcyon).*

alchata Arabic name *al kattar* for a sandgrouse *Pterocles*.

alcinus Corrupted from Gr. *alkimos*, strong, stout, brave.

Alcippe Gr. myth. Alcippe or Alkippe, daughter of Aries the god of war.

alcyon L. *alcyon*, the kingfisher (Gr. *alkuon* ; v. *Halcyon).*

aldabranus After the Aldabra Atoll in the Indian Ocean.

Aleadryas Gr. *alea*, the heat of the sun; *druas*, a wood-nymph, a dryad; an allusion to the 'sunburnt' nape of the Rufous-naped Whistler *A. rufinucha*.

alecto Gr. myth. Alecto or Allecto, one of the three Furies, the deities who avenged crimes and tormented criminals.

alector Gr. *alektor*, the domestic fowl.

Alectoris Gr. *alektoris*, the domestic hen.

Alectroenas Gr. *alektruon*, the domestic cock; *oinas*, a dove; ref. the bare red face and coxcomb of the Mauritius Blue Pigeon *A. nitidissima*.

Alectrurus Gr. *alektruon*, the domestic cock; *oura*, the tail; ref. the long, broad, erect tail feathers of the Cock-tailed Tyrant *A. tricolor*.

Alectura Gr. *alektor*, the domestic cock; *oura*, the tail; the Australian Brush Turkey *A. lathami* has a prominent tail flattened sideways.

Alethe No expl. (Cassin (1859), *Proc. Acad. Nat. Sci.*,**11**, 43); perhaps based on Gr. *alethes*, actual, sincere (cf. Gr. myth. Alethes, an Heraclid king of Corinth).

aleutica /aleuticus After the Aleutian Islands, Alaska, in the North Pacific Ocean.

alexandrae After Princess Alexandra (later Queen Alexandra) (1844–1925) wife of Edward Prince of Wales (later Edward VII).

alexandri After **1** Captain Boyd Alexander (1873–1910) British ornithologist and explorer (*Apus*). **2** Dr Alexandre (fl. 1846) who practised in Mexico and sent specimens to France (*Archilochus*). **3** Alexander the Great (356–323 BC) King of Macedonia, whose conquering armies introduced eastern parakeets to Greece (*Psittacula*).

alexandrinus L. *Alexandrinus*, of Alexandria, Egypt.

alfredi After **1** Alfred Newton (1829–1907) British ornithologist and author (*A dictionary of birds*, 1893) (*Bradypterus*). **2** Alfred H. Everett (1848–1898) English ornithologist, collector and explorer in the East Indies (*Otus*). **3** Sir Alfred Sharpe (1853–1935) British colonial administrator and Governor of Nyasaland (Malawi), 1907–1910 (*Phyllastrephus*). **4** Alfred (fl. 1855), an otherwise unknown child, probably a relation of Des Murs or the Comte de Castelnau (*Psarocolius*).

alice No expl. (Bourcier and Mulsant (1848), *Rev. Zool.*, 274); a female eponym.

aliciae After Alice Robinson (fl. 1895) wife of US collector and explorer Colonel W. Robinson.

alienus L. *alienus*, strange, foreign .

alinae After **1** Aline Bourcier (fl. 1842) wife of French naturalist and Consul-General to Ecuador Jules Bourcier (*Eriocnemis*). **2** Lady Aline Jackson (d. 1966) wife of English ornithologist and Governor of Uganda Sir Frederick Jackson (*Nectarinia*).

alisteri /Alisterus After Alister William Mathews (b. 1907) son of Australian ornithologist, collector, and author Gregory MacAlister Mathews.

alius L. *alius*, different, another, other.

alixii After Édouard Alix (1823–1893) French zoologist.

Alle Local Swedish name *alle* for a seabird on the island of Öland.

alleni After **1** Arthur A. Allen (1885–1964) US ornithologist, field worker, and collector in Panama and Colombia, 1911 (*Grallaria*). **2** Rear Admiral William Allen (1793–1864) who led the Niger expedition, 1841–1842 (*Porphyrula*).

Allenia After Joel Asaph Allen (1838–1921) US ornithologist and curator of American Museum of Natural History, New York, 1885–1921.

alligator After the South Alligator River, Arnhem Land, Australia.

Allocotopterus Gr. *allokotos*, unusual, strange; *-pteros*, winged (*pteron*, a wing, a feather; *ptera*, wings); the inner remiges of the Club-winged Manakin *A. deliciosus* are very peculiar, the shafts being thick, curved and club-like at their extremities.

alnorum L. *alnorum*, of the alders (*alnus*, the alder tree).

Alopecoenas Gr. *alopex, alopekos*, the fox; *oinas*, a pigeon; ref. the chestnut upperparts of the Chestnut-backed Ground Dove *Gallicolumba hoedtii*.

alopex Gr. *alopex*, the fox; ref. rich tawny plumage.

Alophoixus Gr. *alophos*, crestless; genus *Ixos*, bulbul; coined to contrast the uncrested Yellow-bellied Bulbul *A. phaeocephalus* with its crested relatives.

Alopochelidon Gr. *alopos*, fox-like (originally meant sly, but here ref. colour) (*alopex*, the fox); *khelidon*, the swallow; ref. the coloration of the Tawny-headed Swallow *A. fucata*.

Alopochen Gr. *alopos*, fox-like; *khen*, the goose; the Egyptian Goose *A. aegyptiacus* was known as *khen alopex* in ancient Greece because of its fox-brown plumage.

alpestris L. *alpestris*, belonging to the Alps or high mountains (*Alpes*, the Alps or any high mountains).

alphonsianus After Alphonse Milne-Edwards (1835–1900) French zoologist.

alpina/alpinus L. *alpinus*, alpine (*Alpes*, the Alps or any high mountains).

Alseonax Gr. *alsos, alseos*, a grove; *onax*, a king, a lord.

altaicus After the Altai mountain range on the borders of Kazakhstan, USSR, western Mongolia, and Xinjiang Zizhiqu, China.

altera L. *alter*, second, another (i.e. related to); the White-ruffed Manakin *Corapipo* was originally described as a second subspecies of the White-throated Manakin *C. gutturalis*.

althaea Gr. myth. Althaea, wife of King Oenus of Calydon and mother of Meleager.

alticola L. *altus*, high; *-cola*, a dweller (*colere*, to inhabit); ref. montane habitat.

altiloquus L. *altus*, shrill, high (*alere*, to strengthen); *loqui*, to speak.

altirostris L. *altus*, high, great; *-rostris*, billed.

aluco Italian names *allocco* or *alocho* for the Tawny Owl *Strix* (L. *ulucus*, a screech-owl.)

amabilis L. *amabilis*, lovable, amiable, lovely.

Amadina Corrupt dim. of genus *Ammodramus*, sparrow; the Cut-throat Finch *A. fasciata* was formerly considered to form a link between *Ammodramus* (Emberizidae) and *Estrilda* (Estrildidae).

Amalocichla Gr. *amalos*, soft, weak; *kikhle*, a thrush (in orn. *cichla* is used for an assortment of stout-billed, thrush-like birds); ref. soft, loose plumage of the New Guinea ground robins.

amami After Amami-o-shima, Ryukyu Islands, Japan.

Amandava Corrupted from Ahmadabad, a town of Gujerat, India, whence the first examples of the Red Munia *A. amandava* were brought.

Amauresthes Gr. *amauros*, dark; *esthes*, clothes, dress.

amaurocephala/amaurocephalus Gr. *amauros*, dark, gloomy; *-kephalos*, headed.

amaurochalinus Gr. *amauros*, dark, dusky, gloomy; *khalinos*, a bridle; ref. dark lores of the Creamy-bellied Thrush *Turdus*.

Amaurocichla Gr. *amauros*, dark, dusky, gloomy; *kikhle*, a thrush .

Amaurolimnas Gr. *amauros*, dark, dusky; Mod. L. *limnas*, a crake or rail.

amauroptera Gr. *amauros*, dark, dusky; *-pteros*, winged.

amauropteryx Gr. *amauros*, dark, dusky; *pterux*, a wing.

Amaurornis Gr. *amauros*, dark, dusky; *ornis*, a bird.

Amaurospiza Gr. *amauros*, dark, dusky; *spiza*, a common sort of finch, identified by most authors as the Chaffinch *Fringilla*. In orn. sometimes misused instead of *spizias*, a hawk.

amaurotis Gr. *amauros*, dark, dusky; *-otis*, eared (*ous, otos*, the ear).

Amazilia After Amazili, an Inca heroine in Jean François Marmontel's novel *Les Incas, ou la destruction de l'Empire du Pérou*, 1777.

Amazona After the Amazonian rain-forests whence many of these Neotropical parrots were described. The Amazon was probably first seen in 1500 by the Spanish commander Vicente Yáñez Pinzón, but real exploration of the river came with the voyage of Francisco de Orellana in 1540–1541. His florid tales of female warriors, perhaps based on a misunderstood Tupí Indian word *amassona* meaning 'boat destroyer', gave the river its name. The Amazon Kingfisher *Chloroceryle* was described from Cayenne, and ranges widely through lowland tropical America.

Amazonetta After the Amazon River; Gr. *netta*, a duck.

amazonica /amazonicus After the Amazon River; L. *-icus*, belonging to.

amazonina After the Amazon River; L. *-inus*, belonging to.

ambigua /ambiguus L. *ambiguus*, doubtful, uncertain (*ambigere*, to doubt); usually a reference to status as a distinct species or uncertain systematic position.

Amblycercus Gr. *amblus*, dull, blunt; *kerkos*, the tail.

Amblyornis Gr. *amblus*, dull, blunt; *ornis*, a bird; ref. plain coloration of the gardener bowerbirds.

Amblyospiza Gr. *amblus*, dull, blunt; *spiza*, a finch; ref. the heavy bill of the Thick-billed or Grosbeak Weaver *A. albifrons*.

Amblyramphus Gr. *amblus*, dull, blunt; *ramphos*, the bill.

amboimensis After the area of Porto Amboim, Cuanza Sul, Angola.

amboinensis After Amboina Island (Ambon), in the Moluccas, Indonesia.

ameliae After Marquise Amélie de Tarragon (fl. 1845) wife of French explorer Marquis Léone de Tarragon who visited South Africa, 1840–1841.

amelis Gr. *ameles*, unheeded, overlooked (*ameleo*, to be neglected).

americana /americanus After America. America is generally agreed to have received its name after Amerigo Vespucci (1451–1512), a Florentine merchant who promoted and sailed on several voyages to the New World. The name was first used by the cartographer Martin Waldseemüller in 1507. Coues (1882), p. 34, gives counter-arguments against this etymology and refers to a Nicaraguan Indian name *Americ* for a local range of mountains.

amethysticollis L. *amethystus*, the amethyst; Mod. L. *-collis*, necked.

amethystina /amethystinus L. *amethystinus*, amethyst coloured, amethystine.

Ametrornis Gr. *ametros*, immense, boundless; *ornis*, a bird; ref. the very long bills of these plain coloured hummingbirds.

amherstiae After Sarah, Countess Amherst (1762–1838) first wife of William Pitt Amherst, 1st Earl Amherst, Governor-General of Bengal, 1822–1828.

amicta Gr. *amiktos*, unsociable.

ammodendri Botanical genus *Ammodendron*, the sand acacia (Gr. *ammos*, sand, desert; *dendron*, a tree); ref. desert habitat of the Saxaul Sparrow *Passer*.

Ammodramus Gr. *ammos*, sand, desert; *dromos*, running (*trekho*, to run).

Ammomanes Gr. *ammos*, sand, desert; *-manes*, passionately fond of (*mainomai*, to rage, to be frenzied). It has been mistakenly suggested that the termination *manes* is from Gr. *manes*, a type of cup, and refers to the form and shape of a nest.

Ammoperdix Gr. *ammos*, sand, desert; *perdix*, a partridge.

Ammospiza Gr. *ammos*, sand, desert; *spiza*, a finch.

amnicola L. *amnicola*, that dwells upon a river (*amnis*, a stream; *-cola*, a dweller).

amoena /amoenus L. *amoenus*, delightful, pleasant.

Amoropsittaca Gr. *amoros*, unlucky, wretched; *psittakos*, a parrot; Bonaparte first proposed *Bolborhynchus* in *Comptes Rendus de l'Académie des Sciences, Paris* (1857), 596, but there the name is a nomen nudum (i.e. not made properly available). A subsequent separate (Bonaparte (1857), *Remarques Obs. Blanchard Psittac.*, 6) validated the name, but included further detail that restricted the genus to two spp. now synonymized with *B. lineola* and, in effect, left the sp. *aymara* without a generic name. Richmond (1915), *Proc. Biol. Soc. Washington*, **28**, 183, explained the ensuing confusion and rescued the unfortunate *aymara* from its nomenclatural limbo.

Ampeliceps Syn. genus *Ampelis*, the waxwing (Gr. *ampelis*, a bird mentioned by Aristophanes, not further identified); L. *-ceps*, headed, capped; ref. crested appearance of the Golden-crested Mynah *A. coronatus*.

ampelinus Syn. genus *Ampelis*, the waxwing (v. *Ampeliceps*); L.*-inus*, belonging to.

Ampelioides Syn. genus *Ampelion*, cotinga; Gr. *-oides*, resembling.

Ampelion Gr. *ampelion*, a small bird mentioned by Dionysius, not further identified.

amphichroa Gr. *amphi-*, on all sides, around; *khroa*, complexion, colour of the skin; ref. uniform brown coloration of the Thick-billed Newtonia *Newtonia*.

Amphilais Gr. *amphi-*, near; Mod. L. *lais*, a warbler (**1** Gr. *elaios*, a small bird that frequented the olivegroves, mentioned by Alexandrios, perhaps some kind of tit *Parus*, but since conjectured to be a warbler *Sylvia* or *Hippolais* . **2** Gr. *elea* or *elaia*, a small bird mentioned by Aristotle, Aristophanes, and Callimachus, identified as a warbler *Acrocephalus*).

Amphispiza Gr. *amphi-*, near, around; *spiza*, a finch; coined to reflect relationship.

amplus L. *amplus*, handsome, splendid.

amsterdamensis After Amsterdam Island in the southern Indian Ocean.

amurensis After Amuria or Amurland, the drainage area of the River Amur forming the border between Heilongjiang, China, and Amur Oblast, USSR.

Amydrus Gr. *amudros*, dark, dim.

Amytornis Syn. genus *Amytis*, grasswren (Gr. myth. Amytis, daughter of Astauges); Gr. *ornis*, a bird.

Anabacerthia Syn. genus *Anabates*, foliage-gleaner (Gr. *anabates*, a climber); genus *Certhia*, treecreeper.

Anabathmis Gr. *anaba*, a climber; *bathmis*, a step; ref. the sunbirds feeding habits.

anabatina/anabatinus Syn. genus *Anabates*, foliage-gleaner (Gr. *anabates*, a climber); L.*-inus*, belonging to.

anachoreta L. *anachoreta*, a hermit, a recluse (Gr. *anakhoretes*, a hermit).

Anabazenops Syn. genus *Anabates*, foliage-gleaner (Gr. *anabates*, a climber); genus *Xenops*, xenops; ref. the arboreal habits and xenops-like, wedge-shaped bill of the White-collared Foliage-gleaner *A. fuscus*.

anaethetus Gr. *anaisthetos*, senseless, stupid; because nesting birds allowed seafarers close enough to kill them (v. *Anous*).

Anaimos Gr. *anaimos*, bloodless; the flowerpeckers in this genus lack the red spots or blotches of *Dicaeum*.

Anairetes Gr. *anairetes*, a destroyer, a murderer; a typical epithet for a tyrant flycatcher.

anais After Anaïs Lesson (1827–1838) daughter of French naturalist and author René Primavère Lesson.

analis Mod. L. *analis*, relating to the undertail-coverts (L. *anus*, the anus, the fundament).

analoga L. *analogus*, analogous, proportionate (i.e. related) (Gr. *analogos*, proportionate).

Anaplectes Gr. *anapleko*, to wreath, to entwine; ref. the elaborate retort-shaped nest of the Red-winged Weaver *A. melanotis*.

Anarhynchus Gr. *ana-*, backwards; *rhunkhos*, the bill; ref. the unique bent bill of the New Zealand Wrybill *A. frontalis*.

Anas L. *anas*, a duck.

Anastomus Gr. *anastomoo*, to furnish with a mouth; the openbill storks have long, specialized bills with a conspicuous gap or 'mouth' between the central portions of the upper and lower mandibles, being an adaptation for feeding on freshwater molluscs.

anchietae After J. de Anchieta (d. 1897) Portuguese naturalist and collector in Moçambique.

Ancistrops Gr. *ankistron*, a fish hook; *ops*, the face (as an adjective *ops* can also mean appearance); the Chestnut-winged Hookbill *A. strigilatus* has a heavy bill with a small hook.

andaecola After the Andes Mountains (French *Andes*) of South America; L.*-cola*, a dweller, an inhabitant.

andamanensis After the Andaman Islands in the Bay of Bengal, Indian Ocean.

andecola /andecolus see **andaecola.**

anderssoni After Carl Johan Andersson (1827–1867) Swedish explorer, collector in South West Africa (Namibia) 1851–1867 and author (*Lake Ngami; or explorations and discoveries during four years wandering in the wilds of south western Africa*,1856; *The Okavango River; a narrative of travel, exploration and adventure*,1861).

andicola After the Andes Mountains (Mod. L. *Andium)*; L. *-cola*, a dweller (*colere*, to inhabit).

Andigena After the Andes Mountains (Mod. L. *Andium)*; Gr. *genus*, genus, descendant.

andina /andinus After the Andes Mountains (Mod. L. *Andium)*; L.*-inus*, belonging to.

andrei After Eugène André (1861–1922) French naturalist, author, and collector in Venezuela, 1897–1900.

andrewsi After C. W. Andrews (1866–1924) British geologist and collector on Christmas Island, 1897 and 1908.

Androdon Gr. *aner, andros*, a man; *odon*, a tooth; ref. bill serrations of the Tooth-billed Hummingbird *A. aequatorialis*.

andromedae Gr. myth. Andromeda or Andromede, daughter of Cepheus and Cassiopeia, was exposed to a sea-monster.

Andropadus Gr. *aner, andros*, a man; *opados*, an attendant; Levaillant (1802), gave the name *L'Importune* (French *importune*, tiresome, obtrusive) to the Sombre Greenbul *A. importunus* because it followed hunters and warned other birds by its loud calls.

Androphobus Gr. *aner, andros*, a man; *phobos*, terror (*phobeo*, to fear); ref. kinship to syn. genus *Androphilus*, bush warbler (Gr. *aner, andros*, a man; *philos*, loving, fond of).

anerythra Gr. *aneruthriao*, to blush, to begin to blush.

angelae After Angela Kay Kepler (fl. 1972) US ornithologist, field-worker, and collector.

angelinae After Angeline Bartels (fl. 1912) wife of Dutch zoologist and author Max Bartels.

angolensis After Angola (Mbundu native name *ngola*, title of the rulers of the kingdom of Ndongo). In the case of *Oryzoborus* the toponym is erroneous.

anguitimens L. *Anguitinens*, the constellation Ophiuchus, the serpent-holder; ref. the supposed prey of the Southern White-crowned Shrike *Eurocephalus*.

angulata L. *angulatus*, angular, cornered (*angulus*, a corner); the frontal shield of the Lesser Moorhen *Gallinula* has a pointed peak, not rounded as in the Common Moorhen *G. chloropus*.

angusticauda L. *angustus*, narrow, small; *cauda*, the tail.

angustifrons L. *angustus*, narrow, small; *frons*, the forehead, brow; ref. the smaller, unswollen frontal shield of the Russet-backed Oropendola *Psarocolius*.

angustipluma L. *angustus*, narrow, small; *pluma*, a featherlet, a plume; the extinct Kioea *Chaetoptila* had '... all the plumage of the head and body remarkably narrow, and open in texture...' (Peale (1848), *US Exploring Exped.*, (Ed.1), p.147).

angustirostris L. *angustus*, narrow, small; *-rostris*, billed .

Anhima Tupí (Brazilian) Indian name *anhuma* or *anhima* for the screamer or some other large bird.

Anhinga Tupí (Brazilian) Indian name *anhingá* or *anhangá* for the devil bird, an evil spirit of the woods. The name was first used as an English substantive name in 1818 for the Afrotropical darter.

ani Tupí (Brazilian) Indian name *anim* or *anum* for these cuckoos.

Anisognathus Gr. *anisos*, uneven, unequal; *gnathos*, lower jaw; ref. the strongly arched, almost bullfinch-like, upper mandible of the mountain tanagers.

ankoberensis After the town of Ankober, Shewa Province, Ethiopia.

anna After **1** Princess Anna d'Essling (1806–1896) wife of French collector François Victor Prince d'Essling and Duc de Rivoli *(Calypte)*. **2** Anna Dole (d. 1918) wife of Judge Sanford Dole, President of the Hawaiian Republic,1893–1898 *(Ciridops)*.

annae After **1** Anna A. Weber van Bosse (1852–1942) Dutch botanist and collector in the East Indies 1888–1890, 1899–1900 with her husband M. W. C. Weber *(Dicaeum)*. **2** No expl. (Hartlaub and Finsch (1868), *Proc. Zool. Soc. London*, 5); probably after a member of the Saxon royal family (e.g. Anna-Maria (1836–1859) Grand Duchess of Tuscany and daughter of Johann I King of Saxony (d. 1873), or Maria Anna (1843–1884) wife of Crown Prince Georg of Saxony (d. 1904) and daughter of Maria II da Glória, Queen of Portugal (d. 1853)), or an unknown relation of the authors (*Psamathia*).

annaea After Anna Elliot (fl. 1865) wife of US ornithologist Daniel Giraud Elliot.

annamarulae After Mrs Forbes-Watson (fl. 1970) wife of British ornithologist and collector A. D. Forbes-Watson.

annectans/annectens L. *annectans*, connecting (*annectere*, to bind, to connect) (i.e. similar or related to). Of the Rufous-backed Sibia *Heterophasia annectens* Blyth (1847), *Journ. Asiat. Soc. Bengal*, **16**, 450, wrote, ' This form serves to connect *Sibia* ... with *Yuhina* but cannot be placed satisfactorily with either '.

annumbi see **Anumbius.**

Anodorhynchus Gr. *anodon*, toothless; *rhunkhos*, the bill; ref. the unnotched bill of the macaws in this group.

anomala L. *anomalus*, irregular, anomalous (Gr. *anomalos*, irregular, strange).

Anomalophrys Gr. *anomalos*, strange, irregular; *ophrus*, eyebrow; the Brown-chested Plover *A. superciliosus* has a yellow wattle in front of the eye.

Anomalospiza Gr. *anomalos*, strange, irregular; *spiza*, a finch; ref. the parasitic breeding behaviour of the Parasitic Weaver *A. imberbis*.

anomalus see **anomala.**

anonyma Gr. *anonumos*, anonymous, nameless.

Anorrhinus Gr. *ano-*, lacking; *rhis, rhinos*, the nose; ref. the very small bill-casque of the Bushy-crested Hornbill *A. galeritus*.

Anous Gr. *anous*, foolish; the noddies were considered stupid by early mariners because they had no fear of man and allowed themselves to be easily caught.

anoxanthus Gr. *ano*, above; *xanthos*, yellow, golden; ref. the yellow upperparts of the Yellow-shouldered Grassquit *Loxipasser*.

anselli After H. F. Ansell (fl. 1874) a merchant in the Gaboon (Gabon) who supplied natural history specimens.

Anser L. *anser*, a goose.

Anseranas Genus *Anser*, a goose; genus *Anas*, a duck.

ansorgei After W. J. Ansorge (1850–1913) traveller, collector in Angola 1905, 1909, and author (*Under the African sun*,1899).

antarctica/antarcticus L. *antarcticus*, southern, opposite to the north; alludes to species with a southern distribution in South America *(Anthus, Catharacta, Cinclodes, Geositta, Rallus)* and Australia *(Lopholaimus)* as well as those (*Pygoscelis, Thalassoica)* that actually breed on the Antarctic continent.

antelius Gr. *antelios*, eastern.

Anthipes Genus *Anthus*, pipit; L. *pes*, the foot.

Anthobaphes Gr. *anthobaphes*, bright-coloured, flower-coloured.

Anthocephala Gr. *anthos*, a flower, a bloom; *-kephalos*, headed .

Anthochaera Gr. *anthos*, a flower, a bloom; *khairo*, to enjoy; the wattlebirds feed on nectar.

Anthocincla Genus *Anthus*, pipit; Mod. L. *cinclus*, a thrush (Gr. *kinklos*, a waterside bird that bobs its tail, mentioned by Aristotle, Aelian, Hesychius, and other authors, but never properly identified. Variously treated by later authors as a wagtail *Motacilla*, a sandpiper *Tringa*, a dipper *Cinclus*, or a thrush or ouzel *Turdus*. In orn. *cinclus* is associated with a great variety of unrelated thrush-like or dipper-like birds, but rarely with the true thrushes); ref. the straight, pipit-like claws and plain plumage of the Eared Pitta *Pitta phayrei*.

anthoides Genus *Anthus*, pipit; Gr. *-oides*, resembling.

anthonyi After Alfred W. Anthony (1865–1939) US ornithologist and collector.

anthopeplus Gr. *anthos*, a flower, a bloom; *peplos*, a cloak, a robe; ref. deep yellow and blue-black upperparts of the Regent Parrot *Polytelis*.

anthophilus Gr. *anthos*, a bloom, a flower; *philos*, loving, fond of.

Anthornis Gr. *anthos*, a flower, a bloom; *ornis*, a bird; ref. nectar diet of the New Zealand Bellbird *A. melanura*.

Anthoscopus Gr. *anthos*, a flower, a bloom; *skopos*, a searcher, an examiner.

anthracinus L. *anthracinus*, coal-black (Gr. *anthrakinos*, carbuncled; *anthrax, anthrakos*, charcoal, coal).

Anthracoceros Gr. *anthrax, anthrakos*, coal; *keros*, a horn; ref. the black casques of these Indomalayan hornbills.

Anthracothorax Gr. *anthrax, anthrakos*, coal (i.e. black); *thorax*, the chest.

Anthreptes Gr. *anthos*, a flower, a bloom; *threpter*, a feeder; these sunbirds feed on nectar and minute insects gathered at blooming trees and flowers.

Anthropoides Gr. *anthropos*, a woman; *-oides*, resembling; in classical times the crane, by reason of its graceful dancing, intelligence, vigilance, and piety was commonly associated with women.

Anthus L. *anthus*, a small bird that inhabited grasslands mentioned by Pliny, not further identified, but probably the Yellow Wagtail *Motacilla flava* (Gr. *anthos*, a small, brightly coloured bird mentioned by Aristotle).

Antichromus Gr. *antios*, different; *khroma*, colour, complexion.

antigone Gr. myth. Antigone, daughter of King Laomedon of Troy, transformed into a stork or crane by the goddess Hera.

antillarum After the Lesser Antilles chain of islands in the Caribbean Sea.

Antilophia Gr. *antios*, different; *lophos*, a crest; the Helmeted Manakin *A. galeata* has a crimson crest sweeping forward over the bill.

antinorii After Marchese Orazio Antinori (1811–1882) Italian zoologist, traveller, and collector in Ethiopia, 1876–1882.

antipodes L. *antipodes*, the antipodes; ref. the position of the Auckland Islands, New Zealand.

antiquus L. *antiquus*, ancient; ref. the white-streaked collar and neck of the Ancient Murrelet *Synthliboramphus*, imparting an 'old' look.

antisianus/antisiensis After the Andes Mountains, deriving their name from an Inca word *anti* meaning copper, which was mined there.

Antiurus Gr. *antios*, different; *oura*, the tail; the tail feathers of the Pit-sweet *Caprimulgus maculicaudus* are much broader and wider terminally than basally (the reverse is the case in related genera) and their pattern was considered peculiar.

antoniae After Antonia Zeledón (fl. 1884) sister of collector José Zeledón.

Antrostomus Gr. *antron*, a cavern; *stoma*, the mouth.

Anumbius Güaraní (Paraguayan) Indian name *añumbí* or *güirá anumbí* for the Añumbi or Firewood-gatherer *A. annumbi* (v. *mocino*).

Anurolimnas Gr. privative *an-*, lacking; *oura*, the tail; Mod. L. *limnas*, a crake or rail; the Chestnut-headed Crake *A. castaneiceps* has a very short tail.

Anurophasis Gr. privative *an-*, lacking; *oura*, the tail; Mod. L. *phasis*, a pheasant.

Apalis No expl. (Swainson (1833), *Zool. illustr.*, ser. **2**[3], p. 119). Swainson was notorious for his careless misspellings, often compounded by his publishers and printers, and the name is undoubtedly from Gr. *hapalos*, delicate, gentle.

Apaloderma Gr. *hapalos*, delicate, tender; *derma*, the skin; the skin of trogons is delicate and easily torn, collected specimens being notoriously difficult to prepare as the feathers readily drop out.

Apalopteron Gr. *hapalos*, delicate, tender; *pteron*, a feather; ref. the soft, fluffy plumage of the White-eyed Honeyeater *A. familiare*.

apertus L. *apertus*, clear, uncovered; ref. the plain, unmarked chest of the Plain-breasted Brush-Finch *Atlapetes*.

Aphanapteryx Gr. *aphanes*, vanished, invisible; *pterux*, a wing.

Aphanolimnas Gr. *aphanes*, hidden, secret; Mod. L. *limnas*, a crake or rail; the extinct Kusaie Crake *A. monasa* is known only from two specimens collected over one hundred years ago from shadowy swamp forest in the Caroline Islands.

Aphantotriccus Gr. *aphantos*, obscure, secret (i.e. new, distinctive); Mod. L. *triccus*, a tyrant flycatcher (Gr. *trikkos*, a small wren-like bird mentioned by Hesychius, not further identified. The syn. genus *Triccus* was diagnosed by Cabanis in 1845).

Aphantochroa Gr. *aphantos*, obscure; *khroa*, skin colour, complexion; ref. unusual dull plumage of the Sombre Hummingbird *A. cirrhochloris*.

Aphelocephala Gr. *apheles*, smooth, plain; *kephale*, the head; ref. plain, silky appearance of the heads of the Australian whitefaces.

Aphelocoma Gr. *apheles*, smooth, plain; *kome*, the hair; these jays are uncrested.

Aphrastura Gr. *aphrastos*, strange, extraordinary; *oura*, the tail; the tail feathers of the rayaditos terminate in rather stiff, almost bare, shafts, about 12 mm long, the tips being thorn-like.

Aphriza Gr. *aphros*, sea foam; *zao*, to live; ref. littoral habitat of the Surfbird *A. virgata*.

apiaster L. *apiastra*, the bee-eater *Merops* (*apis*, a bee).

apiata No expl. (Vieillot (1816), *Nouv. Dict. Hist. Nat.*, Vol.**1**, p. 342), where the Clapper Lark *Mirafra* is called *Alouette bateleuse*. Perhaps a L. rendition of the bird's cry 'pi-ouit', but probably an error for L. *apicatus*, capped (cf. *apiatus*, parsley-like).

apicalis Mod. L. *apicalis*, apical, of the crown (L. *apex, apicis*, the crown, the cap).

apicauda L. *apex, apicis*, a point; *cauda*, the tail.

apivorus L. *apis*, a bee; *-vorus*, eating (*vorare*, to devour, to eat greedily).

Aplonis Coined arbitrarily by Gould (1836), *Proc. Zool. Soc. London*, 73, who believed these starlings, 'to approximate, in his opinion, in nearly an equal degree to the genera *Lanius, Turdus* and *Lamprotornis*: but believed that they might with propriety be arranged among the Thrushes'; perhaps derived from Gr. *haploos*, sheer, simple, and *ornis*, a bird, with ref. to metallic plumage.

Aplopelia Gr. *haploos*, simple, sheer, plain; *peleia*, a dove.

apoda Gr. privative *a-*, lacking; *pous, podos*, the foot; 'The first birds of Paradise transported from the Australasian region had no feet, because the natives of New Guinea and the neighbouring islands, where these birds appeared exclusively to exist, used to make certain ornaments of them, and deprive them of those limbs which could not answer that special purpose. The quantity of superabundant feathers with which the sides of these birds are covered, must have necessarily concealed over the dried skin, the places whose parts had been mutilated. A little attention, however, might have discovered the traces of them, in raising the sub-alar feathers; but the love of the marvellous, and the fondness for conjecture prevailed, and absurd theories were formed, to explain how the birds could live and propagate in the air. As they were rarely seen at the epochs of incubation, they were dismissed to nestle in the terrestrial Paradise, from which, without doubt, their name was derived; as in the same way, from certain virtues attributed to them by the soothsayers and priests of the country, was derived their native name of *Manucode*, which signifies *Bird of God*. ' (Gray (1829), in Griffiths' *Animal Kingdom*, p. 339).

Apoia After Mount Apo, an active volcano on the island of Mindanao, Philippines.

apolinari After Brother H. Apolinar (fl. 1914) Colombian missionary and zoologist.

apolites Gr. *apolites* (a form of *apopolites)*, an outlaw, an exile.

apperti After the Revd. Fr. O. Appert (fl. 1972) German missionary, collector in Madagascar, and author.

approximans L. *approximans*, nearing, drawing near to (i.e. related to) (*approximare*, to draw near to).

apricaria L. *apricus*, exposed to the sun (i.e. sun-kissed); ref. the Golden Plover's *Pluvialis* gold-spangled upperparts.

Aprosmictus Gr. *aprosmiktos*, solitary, unsociable; from supposed sullen disposition of the Red-winged Parrot *A. erythropterus* in captivity.

Aptenodytes Gr. privative *a-*, lacking; *ptenos*, winged (cf. *apten*, unfledged); *dutes*, a diver; the penguins are flightless.

Apteryx Gr. *apterugos*, without wings; the flightless kiwis have vestigial wings.

Apus L. *apus*, the swift, a kind of swallow said to have no feet (Gr. *apous*, a bird of the swallow kind; privative *a-*, lacking; *pous*, a foot).

aquatica /aquaticus L. *aquaticus*, aquatic, water-living (*aqua*, water).

Aquila L. *aquila*, an eagle. Used also in a specific sense for species with rapacious habits (*Fregata)* or hooked bills (*Eutoxeres)*.

Ara Tupí (Brazilian) Indian word *ara*, a component of various bird names such as *ararauna*, a macaw, *aratica*, a hummingbird, and *araracanga*, a parrot.

arabs L. *Arabs*, an Arabian (Gr./L. *Arabia)*.

aracari Tupí (Brazilian) Indian name *arasari* for a small toucan (Portuguese spelling *araçari)*.

Arachnothera Gr. *arakhnes*, a spider; *-thera*, a hunter (*therao*, to hunt, to seek); from a supposed favourite food of the Indomalayan spiderhunters.

arada Cayenne Indian name *arada* for the Musician Wren *Cyphorhinus*. This epithet is often wrongly made to agree in gender with *Cyphorhinus*, as if it were an adjective rather than a noun in apposition.

araea L. *arae*, rocks at sea; the Seychelles Kestrel *Falco* was formerly more widespread, breeding on rocky outcrops throughout the islands.

araguayae After the Rio Araguaía, Goiás, Brazil.

Aramides Genus *Aramus*, limpkin; Gr. *-ides*, resembling (*eidos*, form, shape, appearance).

Aramidopsis Genus *Aramides*, woodrail; Gr. *opsis*, appearance (as an adjective *opsis* means faced).

Aramus Gr. *aramos*, a kind of heron mentioned by Hesychius.

ararauna Tupí (Brazilian) Indian name *ararauna* for a macaw.

Aratinga Tupí (Brazilian) Indian name *arucatinga* for a parrot.

araucana After the Araucana Indians of Chile, who offered stout resistance to the Spanish conquerors.

araucuan Tupí (Brazilian) Indian names *aracua* or *arecua* for the chachalacas.

arausiaca Prestwich (1963), Appendix C, conjectures that this name is based on L. *Arausio*, the Roman name for the town of Orange in the south of France. Müller

(1776), *Des Ritters Carl von Linné vollständige naturgeschichte, Suppl.*, p. 79, calls the Red-necked Amazon *Amazona*, 'Pomeranzen parfit' (German *pomeranze*, orange), and the epithet is probably a misprint for *aurantiaca* (Mod. L. *aurantiacus*, orange-coloured).

arborea L. *arboreus*, arboreal, relating to trees (*arbor*, a tree).

Arborophila L. *arbor, arboris*, a tree; Gr. *philos*, loving, fond of; the hill partridges inhabit hill forests and roost in trees at night.

arcaei After Enrique Arcé (fl. 1869) collector in Costa Rica.

Arcanator L. *arcanus*, secret, hidden; *-ator* (in composition), one who or that which; ref. the confused and obscure taxonomic history of the Dappled Mountain Robin *A. orostruthus* .

arcanus L. *arcanus*, secret, hidden; the Bare-eyed Fruit-Dove *Ptilinopus* is still only known from a single specimen shot in 1955.

archboldi /Archboldia After Richard Archbold (b. 1907) US zoologist at the American Museum of Natural History and leading light behind the Archbold Expeditions to various parts of Australasia.

archeri After **1** Sir Geoffrey Francis Archer (1882–1964) British explorer, Governor-General of the Sudan, 1924–1926, and author (*The birds of British Somaliland and the Gulf of Aden*, 1937) (*Dryocichloides*). **2** A. L. Archer (b. 1933) hunter in Kenya (*Mirafra*).

Archilochus After Archilochus or Arkhilokhos, a Thracian poet of the sixth century BC famous for his savage wit and flaunting of conventions (he seems to have been the first soldier to run away to fight again another day!) One of eight subsisting hummingbird genera diagnosed by Reichenbach in 1854 commemorating classical artisans, following the precedents set by Linnaeus, Brisson, and other eighteenth century authors who culled many names from the classics and mythology to express close relationships between groups of species in large, unwieldy genera (e.g. in the Lepidoptera).

archipelagus After the Malay Archipelago (Italian *arcipelago*, an archipelago; in classical times *Arkhipelagos* was the name given to the Aegean Sea (Gr. *arkhi*, chief; *pelagos*, a sea). The term is now extended to any sea like it, thickly interspersed with islands, or rather to the islands themselves).

Archiplanus Gr. *arkhiplanos*, a nomad chieftain.

arctica /arcticus L. *arcticus*, northern, arctic (Gr. *arktikos*, northern).

arctitorquis L. *arctus*, narrow; *torquis*, a collar, a necklace.

arctoa L. *arctous*, northern (*arctos*, the north; Gr. *arktos*, the north).

Arctonetta L. *arctos*, northern; Gr. *netta*, a duck.

arcuata L. *arcuatus*, bow-shaped, curved (*arcuare*, to shape like a bow; *arcus*, a bow).

Ardea L. *ardea*, a heron.

Ardeirallus Genus *Ardea*, heron; genus *Rallus*, rail.

ardens L. *ardens*, glowing, burning (*ardere*, to burn).

Ardeola L. *ardeola* or *ardiola*, a little heron (dim. of *ardea*).

Ardeotis Genus *Ardea*, heron; genus *Otis*, bustard.

ardesiaca /ardesiacus Mod. L. *ardesiacus*, slate-coloured (French *ardoise*, slate-coloured, blue–grey).

ardosiaceus Mod. L. *ardosiacus*, lead-coloured (cf. French *ardoise*, slate-coloured).

Arenaria L. *arenarius* or *harenarius*, relating to sand (*arena* or *harena*, sand); ref. seashore habitat of the turnstones.

arenarum L. *arenarum* or *harenarum*, of the sands (*arena* or *harena*, sand); ref. arid habitat of the Scrub Flycatcher *Sublegatus*.

arequipae After the city of Arequipa, Peru.

arfaki /arfakiana /arfakianus After the Arfak Mountains, Irian Jaya (Indonesian New Guinea).

argentata /argentatus L. *argentatus*, ornamented with silver (*argentum*, silver).

argentauris L. *argentum*, silver; *auris*, the ear.

argentea L. *argenteus*, of silver, silvery.

argenticeps L. *argentum, argenti*, silver; *-ceps*, capped, headed.

argentifrons L. *argentum,* silver; *frons*, the forehead, the brow.

argentigula L. *argentum,* silver; *gula*, the throat.

argentina L. *argentum,* silver; *-inus,* belonging to, possessing (cf. Roman myth. Argentinus, the god of silver money).

argoondah Assamese (Indian) native name *goondri* for the Manipur Bush Quail *Perdicula manipurensis* (cf. Bengali *gundri).*

argus Gr. myth. Argus or Argos, the 100-eyed guardian of Io slain by Mercury; ref. the elaborately patterned secondaries of the Great Argus *Argusianus*.

Argusianus Specific name *Phasianus argus* ; L. - *anus,* relating to.

arguta L. *argutus*, noisy, rattling (*arguere*, to declare).

Argya L. *argutus*, noisy, penetrating (*arguere*, to declare).

argyrofenges Gr. *argurophenges*, silver-shining (*arguros*, silver; *phengo*, to shine).

argyrotis Gr. *arguros*, silver; *-otis*, eared (cf. *argurodes*, rich in silver).

aricomae After the Aricoma Pass, Cordillera de Carabaya, Puno, Peru.

aridula L. *aridula*, somewhat dry (dim. of *aridus*, dry, arid); ref. arid habitat of the Desert Cisticola *Cisticola*.

ariel Med. folklore Ariel, a spirit or sylph of the air. Of the Ariel Toucan *Ramphastos ariel* Vigors (1826), *Zool. Journal*, **2**, 479, wrote, '...his appearance is not only graceful, but his movements, as he glides from perch to perch, are light and sylphlike; so much so as to have suggested to an intelligent friend who witnessed them, the specifick name which I have ventured to assign him '.

aristotelis After Aristotle (384–322 BC) Greek philosopher, historian, and naturalist whose work *Peri ta zoa historiai* (*Enquiry concerning animals* or *Historia animalium*) in ten books can be considered the beginning of descriptive zoology .

Arizelocichla Gr. *arizelos*, very conspicuous; *kikhle*, a thrush.

arizonae After the Arizona Territory, USA (admitted to statehood in 1912).

armandii After Abbé Armand David (1826–1900) French missionary to China, ornithologist, collector, and author (*Les oiseaux de Chine*,1877).

armata /armatus L. *armatus*, armed, armoured (*arma*, weapons, armour); ref. carpal spurs.

armenicus Med. L. *Armenicus,* Armenian, of Armenia (L. *Armenius*, Armenian).

armenti After T. Arment (fl. 1851) collector in Colombia.

armillaris L. *armilla*, a bracelet; ref. orange-yellow collar of the Blue-crowned Barbet *Megalaima*.

armillata L. *armillatus*, adorned with a bracelet (*armilla*, a bracelet); ref. the red-gartered legs of the Red-gartered Coot *Fulica*.

arminjoniana After Vice-Admiral Vittorio Arminjon (1830–1897) Italian explorer, first Italian circumnavigation of the world in the warship *Magenta,* 1865 (v. *magentae).*

arnaudi After A. d'Arnaud (fl. 1843) French big-game hunter and explorer in the Sudan and Ethiopia, 1839–1843.

arnotti After David Arnott (1822–1894) South African attorney.

aromatica L. *aromaticus*, fragrant, aromatic (*aroma*, a spice); the Grey-winged Green Pigeon *Treron* was described from the Moluccas or Spice Islands.

aroyae After La Aroya, Inambari Valley, Marcapata, Peru.

arquata **1** Med. L. *arquata*, the curlew (L. *arcuatus* or *arquatus*, bow-shaped, curved); ref. shape of bill (*Numenius*). **2** L. *arquatus*, a jaundice sufferer; a yellowing of the skin and eyes is symptomatic of jaundice, and here the name refers to the yellowish coloration of the Morning Warbler (*Cichladusa*).

arquatrix L. *arquatus*, a jaundice sufferer; ref. the yellow orbital skin of the Olive Pigeon *Columba* (v. *arquata* **2**).

Arremon Gr. *arrhemon*, silent, speechless (*a-*, lacking; *rema*, speech); genus based on Buffon's (1770) name *Oiseau silencieuse* for the Pectoral Sparrow *A. taciturnus*.

Arremonops Genus *Arremon*, sparrow; Gr. *ops*, appearance.

Arrenga Javanese native name *arrengan* for a whistling thrush *Myiophoneus*.

Arses After Arses, king of the Persians (fl. 337 BC), great great great grandson of Xerxes I, raised to the throne by eunuchs and poisoned after a reign of only three years.

arsinoe After ancient Arsinoë or Crocodilopolis (El Faiyûm), Egypt.

Artamella Dim. of syn. genus *Artamia*, vanga (French name *Artamie* for the White-headed Vanga *A. viridis*, from a supposed relationship with the woodswallows *Artamus*).

Artamus Gr. *artamos*, a butcher, a murderer; the woodswallows or swallow-shrikes were formerly thought to be allied to the shrikes or butcherbirds (v. *Lanius*).

arthus After Arthus Bertrand (fl. 1832) French publisher of the work (Lesson (1832), *Illustrations de zoologie*) in which the Golden Tanager *Tangara* was described.

Artisornis L. *ars, artis,* skill, cunning (cf. Gr. *artisis,* an adorning); Gr. *ornis,* a bird; ref. nest-building skills of the African Tailorbird or Red-capped Warbler *A. metopias*.

Artomyias No expl. (Verreaux and Verreaux 1855, *Journ. für Orn.*, **3**, 103); perhaps from genus *Artamus*, wood-swallow, and Mod. L. *myias*, flycatcher, ref. the long broad wings and short, swallow-like bill of the African Sooty Flycatcher *Muscicapa infuscata*, or misspelt from L. *artire*, to be cunning or artful, or Gr. *artios*, active, quick, fitting.

aruensis After the Aru Islands, Indonesia.

arundinaceus L. *arundinaceus*, like a reed (i.e. of the reeds) (*arundo*, a reed; *-aceus*, resembling).

Arundinicola L. *arundo, arundinis*, a reed; *-cola*, a dweller (*colere*, to inhabit).

arvensis L. *arvensis*, of the field (*arvum, arvis*, a field).

asbenaicus After the Asben or Aïr Mountains, Agadèz, Niger.

ascalaphus Gr. *askalaphos*, a bird mentioned by Aristotle, usually identified as an owl. In Gr. myth. Ascalaphus, son of Acherone and Orphne, was changed by Proserpine into an owl because he betrayed her to Pluto.

asha No expl. (Sykes (1832), *Proc. Zool. Soc. London* , 157). Probably a name from Hindu myth. (cf. Assamese *ay-jan,* a bittern).

Ashbyia After Edwin Ashby (1861–1941) English conchologist and ornithologist, settled in Australia and founder member of the Royal Australasian Ornithologists Union.

ashi After John S. Ash (fl. 1981) English ornithologist and collector, specializing in the birds of Somalia.

asiatica /asiaticus L. *Asiaticus*, Asiatic; in orn. almost synonymous with Indian, but erroneous in *Zenaida*.

Asio L. *asio* or *axio*, a type of eared or horned owl mentioned by Pliny.

aspasia Gr. *aspasios*, pleasing, welcome.

Aspatha Gr. privative *a-*, lacking; *spathe*, a spatula; the Blue-throated Motmot *A. gularis* lacks the racquet tips to the tail found in most members of its family.

aspersiventer L. *aspersus*, sprinkled (*aspergere*, to sprinkle upon); *venter*, the belly.

assamica After the state of Assam, north-eastern India.

assimilis see **adsimilis.**

astec After the Aztecs, the dominant tribe and rulers of much of modern Mexico when the Spanish conquistadores arrived in 1519.

Asthenes Gr. *asthenes*, insignificant; used as an epithet of the bird *boudutes* (v. *Budytes)* or the Yellow Wagtail *Motacilla flava*, to which the Neotropical canasteros were formerly, but erroneously, considered related.

Astrapia Gr. *astrapios*, flashing, of lightning (*astrape*, a flash of lightning) (cf. L. *astrapias*, a precious black stone with gleams of light crossing the middle of it).

Astrapimachus Genus *Astrapia*, long-tailed bird of paradise; genus *Epimachus*, sickle-billed bird of paradise.

astreans L. *astreans*, gleaming like a star (Gr./ L. *aster*, a star).

astrild No expl. (Linnaeus (1758), *Systema Naturae*, (Ed.10), book **1**, p. 173). Perhaps from Gr. myth. Astraea, daughter of Jupiter, and goddess of justice, or Gr. *aster*, a star, ref. the bright crimson bill of the African Waxbill *Estrilda*.

Astur Late L. *astur*, a kind of hawk mentioned by Firmicus Maternus (fl. AD 340), not further identified (Gr. *asterias*, a kind of hawk mentioned by Aristotle and Claudius Aelianus, usually identified with the Northern Goshawk *Accipiter gentilis)*.

Asturina /asturinus Syn. genus *Astur*, goshawk; L. *-inus*, resembling.

Asyndesmus Gr. privative *a* -, lacking; *sun*, together; *desmos*, bond; ' feathers of the underparts and of a nuchal collar with the fibres on their terminal portion disconnected, loosened, enlarged in calibre, stiffened, almost bristle-like, with a peculiar glistening silicious hardness, destitute of fibrillae whereby to interlock...My name has reference to the disconnection of the fibres of the feather ' (Coues (1866), *Proc. Acad. Nat. Sci.*, 55); Lewis' Woodpecker *A. lewis* is an aberrant North American sp. of bizarre plumage and unusual habits.

atacamensis After the desert town of San Pedro de Atacama, Antofagasta, Chile.

Atalotriccus Gr. *atalos*, delicate; Mod. L.*triccus*, tyrant flycatcher; the Pale-eyed Pygmy Tyrant *A. pilaris* has the four 'outer primaries greatly reduced in size...the feathers very narrow and acuminate ' (Ridgway (1905), *Proc. Biol. Soc. Washington*, **18**, 208).

Ateleodacnis Gr. *ateleos*, imperfect; genus *Dacnis*, dacnis; ref. the dark colour and small size of this group of conebills *Conirostrum*, formerly considered to be dacnises.

Atelornis Gr. *ateleos*, imperfect; *ornis*, a bird ; ref. the very short toes of the Madagascan ground-rollers.

ater L. *ater*, black, dead black (*niger* is shining black).

ateralbus L. *ater*, black; *albus*, white; ref. to pied plumage.

aterrima /aterrimus Mod. L. *aterrimus*, very black (super. of L. *ater*, black).

Athene Gr. myth. Athene, goddess of wisdom, war and the liberal arts, whose favourite bird was the owl (*glaux)*, an ancient association from her primitive role as goddess of the night.

athertoni After Lieutenant J. Atherton (1797–1827) British Army in India.

atlanticus L. *Mare Atlanticus*, the Atlantic Ocean ; the Plata Gull *Larus* was described from the Atlantic coast of Argentina as a subsp. of the Pacific Band-tailed Gull *L. belcheri*.

atlantis After the fabled Atlantis (v. *Atlantisia)*; the Atlantis Gull *Larus* was described from the Azores.

Atlantisia After Atlantis, the celebrated vanished island, traditionally located in the Atlantic Ocean, which disappeared as the result of a natural catastrophe. Modern research suggests that the tradition may have had its origin in the destruction of ancient Thera (Santorini), in the Greek Cyclades, by volcanic eruption in c.1450 BC.

Atlapetes Gr. myth. Atlas, Titan king of Mauretania, who was changed into a mountain; Gr. *petes*, a flier (i.e. a bird); ref. montane habitats of the Neotropical brush-finches.

atra see **ater.**

atrata /atratus L. *atratus*, clothed in mourning (i.e. black).

atricapilla /atricapillus L. *atricapillus*, black-haired (i.e. black-capped, black-headed).

atricaudus L. *ater, atri*, black; *cauda*, the tail.

atriceps L. *ater*, black; *-ceps*, capped, headed.

Atrichornis Syn. genus *Atrichia* (Gr. *atrikhos*, without bristles or hair); Gr. *ornis*, a bird; the scrub-birds are distinguished by the absence of bristles around the gape.

atricilla L. *ater*, black; Mod. L. *cilla*, the tail (v. *Motacilla)*.

atricollis L. *ater*, black; Mod. L. *-collis*, necked.

atrifrons L. *ater*, black; *frons*, the forehead, the brow.

atrifusca L. *ater*, black; *fuscus*, dark.

atrigularis L. *ater*, black; Mod. L. *gularis*, throated, of the throat (L. *gula*, the throat)

atrimaxillaris L. *ater*, black; *maxilla*, the jaw or jawbone (i.e. the bill).

atripennis L. *ater*, black; *pennis*, winged (*penna*, a feather).

atripunctatus L. *ater*, black; Mod. L. *punctatus*, spotted (L. *punctum, puncti*, a spot).

atrirostris L. *ater*, black; *-rostris*, billed.

atrocaerulea Mod. L. *atrocaeruleus*, dark blue (L. *ater*, black; *caeruleus*, blue). The use of *o* as a connecting vowel instead of *i* is not an error but standard practice in ornithology, as in botany (v. Stearn (1983), p. 287).

atrocapillus see **atricapilla.**

atrocaudata L. *ater*, black; *caudatus*, tailed (*cauda*, tail).

atrochalybeia Mod. L. *atrochalybeus*, dark blue (L. *ater*, black; *chalybeius*, steely; *chalybs*, steel).

atrococcineus Mod. L. *atrococcineus*, deep crimson (L. *ater*, black; *coccineus*, scarlet-coloured).

atrocyaneum Mod. L. *atrocyaneus*, dark greenish-blue (L. *ater*, black; *cyaneus*, dark blue, sea blue).

atroflavus L. *ater*, black; *flavus*, yellow, gold-coloured.

atrogularis L. *ater*, black; Mod. L. *gularis*, throated, of the throat (L. *gula*, throat).

atronitens L. *ater*, black; *nitens*, glittering, shining (*nitere*, to glitter).

atropileus L. *ater*, black; *pileus*, a skull-cap.

atropurpurea Mod. L. *atropurpureus*, deep purple, dark purple.

atrosuperciliaris L. *ater*, black; Mod. L. *superciliaris*, the eyebrow (L. *supercilium*, the eyebrow).

atrothorax L. *ater*, black; *thorax*, a breastplate (Gr. *thorax)*.

atroviolacea Mod. L. *atroviolaceus*, deep violet, dark violet.

atrovirens Mod. L. *atrovirens*, deep green.

Attagis Gr. *attagis* or *attagas*, the Black Francolin *Francolinus francolinus*, although some authors identified the bird with the Grey Partridge *Perdix perdix*.

Atthis After Atthis, a beautiful young woman of Lesbos and favourite of the poetess Sappho.

Atticora No expl. (Boie (1844), *Isis von Oken*, col.172), but probably from Gr. myth. Atthis, another name for Procne, who was metamorphosed into a swallow (cf. *Attikos*, Athenian), and Gr. *kora*, a maiden.

Attila After Attila (406–453) the 'Scourge of God', king of the Hunnish hordes that ravaged Europe in the fifth century.

atypha L. *atypus*, stammering (Gr. *atupos)*; ref. cadence of song.

aubryana After C. E. Aubry-Lecomte (1821–1879) French colonial administrator in New Caledonia.

aucklandica After the Auckland Islands, south of New Zealand.

audax L. *audax*, bold, courageous.

audeberti After M. Audebert (fl. 1878) collector in Madagascar.

audouinii After Jean Victoire Audouin (1797–1841) French entomologist, ornithologist, and author (*Dictionnaire classique d'histoire naturelle*, 1822; *Explication sommaire des planches d'oiseaux de l'Égypte et de la Syrie*, 1826).

auduboni After John James Audubon (1785–1851) French/US naturalist, painter, and author (*The birds of America*, 1827; *The viviparous quadrupeds of North America*, 1849).

auga Gr. *auge*, sun rays, beams; ref. yellow auriculars of immature Yellow-eared Meliphaga *Meliphaga*.

Augastes Gr. *augasteira*, light-giving (*auga*, a sunbeam).

augur L. *augur*, an augur, a soothsayer (*avis*, a bird); an honoured office in ancient times, when the future was foretold by observing the heavens, birds, mammals, sacred objects, and minor calamities. The observation of eagles, hawks, and owls was particularly significant.

auguralis Specific name *Buteo augur* ; L. *-alis*, belonging to (i.e. related to).

augusti After Auguste Sallé (1820–1896) French entomologist and traveller in tropical America, 1846–1856.

Aulacorhynchus Gr. *aulax, aulakos*, a furrow; *rhunkhos*, the bill; the toucanets have grooved bills.

aura Mexican Indian name *aurouá* for the Turkey Vulture *Cathartes* (v. *urubitinga)*.

aurantia Mod. L. *aurantius*, orange-coloured, tawny (*aurantia*, an orange).

aurantiaca /aurantiacus Mod. L. *aurantiacus*, orange colour (*aurantia*, an orange).

aurantiifrons Mod. L. *aurantius, aurantii*, orange-coloured; L. *frons*, the forehead or brow.

aurantiigula Mod. L. *aurantius*, orange-coloured; L. *gula*, the throat.

aurantiirostris Mod. L. *aurantius*, orange-coloured; L. *-rostris*, billed (*rostrum*, the bill).

aurantiivertex Mod. L. *aurantius*, orange-coloured; L. *vertex*, the crown of the head.

aurantioatrocristatus Mod. L. *aurantius*, orange-coloured; L. *ater, atri*, black; *cristatus*, crested (*crista*, a crest).

aurantium /aurantius see **aurantia.**

aurata /auratus L. *auratus*, gilded, ornamented with gold (*aurum, auri*, gold).

aurea /aureus L. *aureus*, golden (*aurum*, gold).

aureata L. *aureatus*, decorated with gold (*aurum*, gold).

aureliae After Aurélie Hénon (fl. 1846) wife of French agronomist J. L. Hénon.

aureocincta L. *aureus*, golden; *cinctus*, girdled, wreathed (*cingere*, to encircle).

aureodorsalis L. *aureus*, golden; Med. L. *dorsalis*, dorsal, -backed (L. *dorsum*, the back).

aureolus L. *aureolus*, golden, gold-coloured, brilliant (dim. of *aureus*, golden).

aureolimbatum L. *aureus*, golden; *limbatus*, edged, bordered (*limbus*, a border or fringe); ref. yellow flanks or sides of the Yellow-sided Flowerpecker *Dicaeum*.

aureonucha L. *aureus*, golden; Med. L. *nucha*, the nape (Arabic *nukha*, spinal marrow).

aureopectus L. *aureus*, golden; *pectus*, the breast.

aureoventris L. *aureus*, golden; *venter, ventris*, the belly.

aurescens L. *aurescens*, golden-yellowish (*aurescere*, to become gold).

aureus see **aurea.**

auricapilla /auricapillus L. *aurum, auri*, gold; *-capillus*, capped, headed .

auriceps L. *aurum,* gold; *-ceps,* crowned, headed.

auricollis L *aurum,* gold; Mod. L. *-collis,* necked.

auricularis Med. L. *auricularis*, pertaining to the ear (L. *auricula*, the ear).

auriculata Med. L. *auriculatus*, eared, having long ears.

aurifrons L. *aurum,* gold; *frons*, the forehead, the brow.

aurigaster L. *aurum,* gold; *gaster*, the belly.

Auriparus L. *aurum,* gold; genus *Parus*, tit.

Auripasser L. *aurum,* gold; genus *Passer*, sparrow.

aurita /auritum /auritus L. *auritus*, eared, having long ears (*auris*, the ear).

auriventris L. *aurum,* gold; *venter, ventris*, the belly.

aurocapillus see **auricapilla.**

auropalliata L. *aurum,* gold; *palliatus*, mantled, hooded (*pallium*, a coverlet or mantle).

aurorae After Aurora Island (Makatea Island), Tuamotu Islands, in the Pacific Ocean.

auroreus Mod. L. *auroreus*, eastern (L. *aurora*, the east, dawn).

aurovirens L. *aurum,* gold; *virens*, green.

aurulenta /aurulentus L. *aurulentus*, of the colour of gold (*aurum*, gold).

austeni After Lieutenant-Colonel Henry Haversham Godwin-Austen (1834–1923) British army surveyor who gave his name to the Himalayan peak K2 or Mount Godwin-Austen.

australasia L. *australis,* southern; *Asia*, Asia; the Cinnamon-banded Kingfisher *Halcyon* was described from Timor, Indonesia.

australe /australis L. *australis*, southern (*auster, austri*, the south).

Automolus Gr. *automolos*, solitary, a deserter.

autumnalis L. *autumnalis*, autumnal (*autumnus*, autumn); ref. orange, yellow, and brown coloration.

averano French name *L'Averano* given to the Bearded Bellbird *Procnias* by Buffon (1778) (Brazilian Portuguese name *ave de verão,* bird of summer, because it was believed to call only during the six weeks of high summer). Snow (1982) considers that the local name should be properly attributed to the Bare-throated Bellbird *P. nudicollis.*

Aviceda L. *avis*, a bird; *caedere,* to kill.

Avocettula Dim. of syn. genus *Avocetta*, avocet (v. *avosetta)*; ref. the upturned, avocet-like bill of the fiery-tailed Awlbill *A. recurvirostris*.

avosetta Local Venetian (Italian) name *avosetta* for the Pied Avocet *Recurvirostra*.

awokera Japanese name *ao-gera,* meaning green woodpecker (*ao*, green; *gera*, woodpecker) for the Wavy-bellied Woodpecker *Picus*.

axillaris L. *axillaris*, of the armpit, shoulders or wing (*axilla*, the armpit, shoulder or wing; *ala*, the wing).

ayacuchensis After the Department of Ayacucho in south-central Peru, scene of the battle in 1824 that secured Peruvian independence from Spain.

aylmeri After G. P. V. Aylmer (1856–1936) British traveller, collector in Somaliland.

aymara After the Aymará, the native Indian peoples of southern Peru, Bolivia, and northern Chile, whose lands were conquered by the Inca in the fifteenth century.

ayresi /ayresii After Thomas Ayres (1828–1918) naturalist, collector in South Africa.

Aythya Gr. *aithuia*, a seabird mentioned by Aristotle, Hesychius, and other authors, not further identified, but since associated with a variety of different seabirds (v. *Aethia)*.

azarae After Felix Manuel de Azara (1746–1811) Spanish officer commanding the Paraguayan frontier, 1781–1801, naturalist, and author (*Apuntiamentos para la historia natural de los páxaros de Paragüay y Río de la Plata*, 1805).

azurea Med. L. *azureus*, azure-coloured (Persian *lazhward*, lapis lazuli).

azureocapilla Med. L. *azureus*, azure-coloured; L. *-capillus*, capped, headed.

azureus see **azurea.**

Acrocephalus arundinaceus Great Reed Warbler

B

Babax Gr. *babax*, a chatterer, a babbler (*bazo*, to speak).

baboecalus Dim. from Gr. *babax*, a chatterer.

bacchus L. *bacchus*, wine (myth. Bacchus, god of wine); ref. colour of head, neck, and breast of the Chinese Pond Heron *Ardeola*.

bachmani /bachmanii After John Bachman (1790–1874) US naturalist and author.

bactrianus L. *Bactrianus*, Bactrian, of Bactria or Bactriana, a country of central Asia, whose warlike, barbarian inhabitants lived on plunder. Bactria corresponded to modern Afghanistan and northern Iran.

badeigularis L. *badius*, chestnut-coloured; Mod. L. *gularis*, throated (L. *gula*, the throat).

badia L. *badius*, chestnut-coloured, reddish-brown.

badiceps L. *badius*, chestnut-coloured; *-ceps*, capped, crowned.

badius see **badia.**

Baeopogon Gr. *baios*, short, small; *pogon*, a beard.

baeticatus Mod. L. *baeticus* or *boeticus*, brown; L. *-atus*, possessing.

baeri After **1** Karl Ernst von Baer (1792–1876) Prussian embryologist, geographer, and explorer in Siberia under Russian auspices (*Aythya*). **2** G. A. Baer (1839–1918) French naturalist and collector in Peru and Brazil (*Asthenes, Compsospiza, Leucippus, Paroaria*).

baglafecht French name *Le Baglafecht* given to the Baglafecht Weaver *Ploceus* by Buffon (1770); probably from a native Abyssinian name.

bahamensis After the Bahama Islands in the West Indies.

baileyi After V. O. Bailey (1864–1942) US field naturalist and collector.

bailleui After Pierre Étienne Ballieu (1829–1900) French Consul to Hawaii and collector.

bailloni After Jean François Emmanuel Baillon (1744–1802) French lawyer, naturalist, and collector.

Baillonius After Jean Baillon (1744–1802) (v. *bailloni)* and his son Louis Antoine François Baillon (1778–1855) French naturalist and collector.

bailunduensis After Bailunduland, Benguella, southern Angola.

bairdi /bairdii After Spencer Fullerton Baird (1823–1887) US ornithologist, collector, and author (*Catalogue of North American birds*, 1858).

bakeri After **1** J. R. Baker (b. 1900) British zoologist in the New Hebrides, 1922–1934 (*Ducula).* **2** George Fisher Baker, Jr (1878–1937) US banker and trustee of American Museum of Natural History (*Sericulus).* **3** Edward Charles Stuart Baker (1864–1944) British ornithologist, oologist, Indian Police 1883–1912, collector, and author (*The nidification of the birds of the Indian Empire*, 1932; *Fauna of British India: birds*, 1922; *Game birds of India, Burmah and Ceylon*, 1921; *Cuckoo problems*, 1942) (*Yuhina).*

bakkamoena Sinhala (Sri Lankan) name *bakamuna* for the the White Barn Owl *Tyto alba* and the Brown Fish Owl *Ketupa zeylonensis*.

balaenarum After Walvis Bay, Damaraland, South West Africa (Namibia), from L. *balaena*, a whale (Dutch name *Walvisch Baai*, Whale Bay, anglicized to Walvis Bay).

Balaeniceps L. *balaena*, a whale; *-ceps*, headed; ref. the grotesque head-shape and huge bill of the Shoebill or Whale-headed Stork *B. rex*.

balasiensis Named the Balasian Swift, *Cyp. Balasiensis*, by Gray (1829), in Griffiths' *Animal Kingdom*, **7**, p. 60. The original toponym is probably after the town and district of Balasore, Orissa, where the British made their first permanent settlement in the Bengal region. However, the name has since been corrected to *batasiensis*, as if from the Bengali (Indian) name *bathasi* for the Asian Palm Swift *Cypsiurus*.

Balearica **1** L. *grui Balearicae*, the Balearic crane with a tufted head (Pliny); it is not known to which sp. Pliny refers, or even if it was a type of crane, although the Demoiselle Crane *Anthropoides virgo* formerly occurred in Spain and it is possible that the Black Crowned Crane *B. pavonina* bred in Egypt or the marshes of Tunisia in classical times. **2** L. *Balearicus*, Balearic, of the Balearic Islands (Mallorca, Menorca); the Mediterranean Crossbill *Loxia* inhabits the Balearics, Corsica, Sardinia, northern Algeria, and Tunisia.

balfouri After Sir Isaac Bayley Balfour (1853–1922) Scottish botanist, explorer, and collector on Rodrigues, 1874 and Sokotra, 1880.

balicassius Filipino (Philippines) native name *bali-cassio* for the Philippine Drongo *Dicrurus*.

balli After V. Ball (1843–1895) Irish naturalist and collector.

balliviani After José Ballivían (1804–1852) President of Bolivia 1841–1847.

balstoni After R. J. Balston (1839–1916) English businessman and patron of natural history.

bambla French *Le Bambla*, the name given to the Wing-banded Wren *Microcerculus* by Buffon (1770), with reference to the white wing band (*bande*, band or stripe; *blanc*, white). Buffon wrote that his aim was to conserve native names wherever possible, but Stresemann (1975, p. 58), regarded the Count as, '...an enemy of formalism in nomenclature he chose his specific names as unsystematically as possible – they were intended to be not instructive, but amusing '.

bamendae After the town of Bamenda, British Cameroons (Cameroun).

Bambusicola Botanical genus *Bambusa*, bamboo (Malay *bambu*, bamboo); L. *-cola*, dweller, inhabitant.

Bananivorus French *banane*, banana (Portuguese *banano*, banana tree); L. *-vorus*, devouring.

bangsi /Bangsia After Outram Bangs (1863–1932) US ornithologist and author.

banksiana After the Banks Islands, New Hebrides. The islands received their name after Sir Joseph Banks (v. *banksii*).

banksii After Sir Joseph Banks (1743–1820) British botanist and explorer with Captain Cook 1768–1771, President of the Royal Society 1778–1820.

bannermani After David Armitage Bannerman (1886–1979) Scottish ornithologist, collector, and author (*Birds of tropical west Africa*, 1930; *The birds of the British Isles*, 1953; *Birds of the Atlantic Islands*, 1963).

banyumas After the province of Banyumas, Java.

barakae After Baraka (d. 1911) native African hunter and skinner for Sir Frederick Jackson in Kenya and Uganda.

baraui After Armand Barau (fl. 1964) French naturalist and collector in Réunion.

barbadensis After the island of Barbados (Portuguese *Ilha de los Barbados* (*barbado*, bearded) from the number of bearded fig-trees found there by early explorers), in the Lesser Antilles, West Indies. An erroneous toponym in the case of *Amazona*.

barbara After the Barbary Coast of North Africa. Barbary (from the original *Berber* inhabitants) comprised the coastal North African states of Morocco, Algeria, Tunis, Tripolitania, and Cyrenaica, and spawned the Moslem pirates who terrorized Christian shipping up until the 1830s.

barbarus **1** L. *barbarus*, barbarous, savage; ref. the savage mien and 'blood-stained' underparts of the Gonolek (*Laniarius*). **2** after Santa Barbara, Vera Paz, Guatemala (*Otus*).

barbata/barbatus L. *barbatus*, bearded (*barba*, the beard); ref. to feather formation or a distinctive throat patch. In *Pycnonotus* the epithet is a misprint for *barbarus* (i.e. of the Barbary Coast [v. *barbara*]).

barbirostris L. *barba*, the beard; *-rostris*, billed; the Sad Flycatcher *Myiarchus* has prominent bristles springing from the base of the bill.

baritula Dim. from Gr. *barites*, a small bird mentioned by Dionysius, not further identified.

barnardi/Barnardius After Edward Barnard (1786–1861) English zoologist, horticulturalist, and Agent-General for the Crown Colonies.

barnesi After Wilfred Barnes (fl. 1948) Australian naturalist.

baroni After O. T. Baron (1847–1926) German engineer in Peru and Ecuador.

barrabandi After Jacques Barraband or Barraban (1767–1809) French bird and flower illustrator.

barratti After F. A. Barratt (fl. 1875) collector in the Transvaal, South Africa.

barringeri After Brandon Barringer (b. 1899) US banker and trustee of Academy of Natural Sciences in Philadelphia.

barroti After Adolphe Barrot (1801–1870) French diplomat in Colombia, 1831–1835 and the Philippines, 1835–1838.

bartelsi After Max E. G. Bartels (1871–1936) Dutch plantation owner and naturalist in Java, 1895–1936.

bartletti After Abraham Dee Bartlett (1812–1897) English zoologist.

Bartramia After William Bartram (1739–1823) US naturalist, collector, and explorer.

bartschi After P. Bartsch (1871–1960) US naturalist and collector.

Baryphthengus Gr. *baruphthongos*, deep-voiced, lowing (*barus*, deep, bass; *phthengomai*, to speak); the Rufous-capped Motmot *B. ruficapillus* utters deep, hooting, owl-like calls.

basalis Mod. L. *basalis*, basal (L. *basis*, a base, a pedestal); from the distinctive rufous-based tail feathers of Horsfield's Bronze Cuckoo *Chrysococcyx*.

basilanica After the island of Basilan, Sulu Archipelago, Philippines.

Basileuterus Gr. *basileuteros*, more kingly (comp. of *basileus*, a king; applied in classical times to a small bird mentioned by Aristotle, usually identified as the wren *Troglodytes*, but also conjectured to be a type of warbler *Phylloscopus* or a goldcrest *Regulus*).

basilica/basilicus L. *basilicus*, royal, splendid (Gr. *basilikos*, royal; *basileus*, a king).

Basilinna Gr. *basilinna*, a queen.

Basilornis Gr. *basileus*, a king; *ornis*, a bird.

bassana After the Bass Rock, Firth of Forth, Scotland, site of a gannetry known to have been established by the early 1500s.

Batara Güaraní (Paraguayan) Indian name *batará* for various small birds.

batasiensis see **balasiensis.**

batavica Erroneous toponym after Batavia (Djakarta), Java. The Seven-coloured Parakeet *Touit* is confined to Venezuela and the Guianas.

batesi After George Latimer Bates (1863–1940) British naturalist in tropical West Africa, 1895–1931, and author (*Handbook of the birds of west Africa*, 1930).

Bathilda After Bathilde Aloïse Léonie Comtesse de Cambacérès (1840–1861) daughter of Charles Lucien Prince Bonaparte.

Bathmocercus Gr. *bathmos*, a step, a stair; *kerkos*, the tail; from the graduated tail feathers of these Afrotropical warblers.

Batis Gr. *batis*, a worm-eating bird mentioned by Aristotle, not further identified but later conjectured, without much evidence, to be the Western Stonechat *Saxicola torquata*.

Batrachostomus Gr. *batrakhos*, a frog; *stoma*, the mouth; ref. the wide, boat-like bills of the Indomalayan frog-mouths.

battyi After J. H. Batty (1846–1906) collector in tropical America, 1896–1906.

baudi After J. C. Baud (1789–1859) official of the Dutch East India Company.

baudinianus/baudinii After Captain Thomas N. Baudin (1754–1803) French mariner, explorer, and collector.

baumanni After E. Baumann (1865–1895) German collector and botanist in Togoland (Togo), 1893–1895.

baweanus After Bawean Island, in the Java Sea between Java and Borneo, Indonesia.

bayleyi After Sir Edward Clive Bayley (1821–1894) British statesman and archaeologist in India, 1842–1878.

Baza Hindi *baaz*, a hawk (applied, without qualification, to the female Northern Goshawk *Accipiter gentilis)*.

beani After T. H. Bean (1846–1917) US ichthyologist, collector in Mexico, 1884, and author.

beaudouini After M. Beaudouin (fl. 1862) who collected in Portuguese Guinea (Guinea Bissau) for the Verreaux dealership.

beauharnaesii After Auguste Beauharnais (1810–1835) 2nd Prince of Eichstadt, collector, and patron of the sciences.

Beauharnaisius After Maximilian Eugène Joseph Napoléon Beauharnais (1817–1852) 3rd Prince of Eichstadt and Duke of Leuchtenberg, patron of the sciences.

Bebrornis No expl. (Sharpe (1883), *Catalogue of the birds in the British Museum*, Vol. **7**, p. 102); probably from Gr. *bebros*, eating (*bibrosko*, to eat); *ornis*, a bird (cf. *bebaios*, firm; *bebros*, stupid).

beccarii After Odoardo Beccari (1843–1920) Italian botanist, explorer, collector, and author.

becki After Rollo H. Beck (1870–1950) US ornithologist and collector in the Galápagos Islands, 1897–1898, South Africa, 1914 and New Guinea, 1928.

bedfordi After Herbrand Arthur, 11th Duke of Bedford (1858–1940), President of the Zoological Society of London, 1899–1936 and trustee of the British Museum, 1906–1927.

beecheii After Rear Admiral Frederick William Beechey (1796–1856) in the Pacific, 1825–1828 and President of the Royal Geographical Society,1855–1856.

behni After Wilhelm Friedrich Georg Behn (1808–1878) German explorer who crossed South America in 1847.

bejaranoi After Gastón Bejarano B. (fl. 1981) Bolivian zoologist.

belcheri After Admiral Sir Edward Belcher (1799–1877) British naval explorer on the Pacific coast of America, 1825–1828 and the Arctic, 1852–1854.

beldingi After L. Belding (1829–1917) US naturalist and collector.

belfordi After George Belford (d. 1906) son of a Samoan chieftain and collector for Sir William McGregor (v. *Macgregoria)*.

bella L. *bellus*, pretty, handsome, charming.

belli After John G. Bell (1812–1889) US taxidermist and collector in Missouri with Audubon, 1843 and California, 1849.

bellicosa/bellicosus L. *bellicosus*, aggressive, martial (*bellum, belli*, war); ref. the bright red underparts of the Peruvian Meadowlark *Sturnella* (v. *militaris)*, and the aggressive habits of the Martial Eagle *Polemaetus*.

bellii see **belli.**

bellulus L. *bellulus*, pretty, elegant (dim. of *bellus*, pretty).

bellus see **bella.**

Belonopterus Gr. *belone*, a needle; *-pteros*, winged; the Southern Lapwing *B. chilensis* has sharp carpal spurs.

bendirei After Major Charles Emil Bendire (1836–1897) German/US soldier, field naturalist, oologist, collector, and author.

bengalensis After the states of East (Bangla Desh) and West Bengal, India. Historically Bengal (native name *Bangala)* comprised the greater part of northern India.

bengalus French name *le Bengali* given to the Red-cheeked Cordon-bleu *Uraeginthus* by Brisson (1760) in the mistaken belief that it came from Bengal (v. *bengalensis)*.

benghalense / benghalensis After Benghal, a variant spelling of Bengal (v. *bengalensis*).

benjamini After Benjamin Leadbeater (d. 1837) English natural history dealer.

bennetti / bennettii After **1** George Bennett (1804–1893) British surgeon, zoologist, and botanist who settled in Australia in 1836, and author (*Wanderings in New South Wales, Batavia*,1834; *Gatherings of a naturalist in Australia*,1860) (*Aegotheles, Casuarius)*. **2** Edward Turner Bennett (1797–1836) British zoologist and secretary of Zoological Society of London, 1831–1836 (*Campethera)*. **3** K. H. Bennett (d. 1891) Australian ornithologist and collector (*Corvus)*.

bensbachi After J. Bensbach (fl. 1893) resident on the island of Ternate, Moluccas.

benschi After J. H. E. Bensch (b. 1868) French colonial administrator in Madagascar, 1903.

bensoni After Constantine Walter Benson (1909–1982) English colonial administrator, ornithologist, collector, and author (*The birds of Zambia*, 1971).

Berenicornis After Berenice, wife of Ptolomey Euergetes king of Egypt c.248 BC, whose hair, vowed by her to Venus, was stolen from the temple and, afterwards, taken to heaven and placed in a constellation; Gr. *ornis*, a bird; ref. the bushy crest of the White-crested Hornbill *B. comatus*.

bergii After C. H. Bergius (1790–1818) German naturalist and collector in South Africa, 1816–1818.

berigora Aboriginal (Australian) name *berigora* for the Brown Falcon *Falco*.

berlepschi / Berlepschia After Hans Karl Hermann Ludwig Graf von Berlepsch (1850–1915) German ornithoogist, collector specializing in the Neotropics, and author.

berliozi After J. Berlioz (b.1891) French ornithologist and author.

bernardi After Captain Bernard (fl. 1844) French sea captain from Bordeaux and collector.

bernicla Med. L. *bernicla*, the barnacle; until the middle of the eighteenth century the Brent Goose *Branta* and Barnacle Goose *B. leucopsis* were often regarded as one species.

bernieri / Bernieria After Chevalier J. A. Bernier (fl. 1834) French naval surgeon, botanist, and collector in Madagascar, 1831–1834.

bernsteini After Heinrich Agathon Bernstein (1828–1865) German physician, zoologist, and collector.

berthelotii After Sabin Berthelot (1794–1880) French Consul at Tenerife, naturalist, and author (*Histoire naturelle des Iles Canaries*, 1835).

bertrandi After Captain Bertrand L. Sclater (1866–1897) British Army in Uganda, 1897.

beryllina /beryllinus Mod. L. *beryllinus*, emerald green (Med. L. *beryllinus*, berylline, crystalline).

Bessonornis Gr. *bessa*, a wooded glen; *ornis*, a bird.

bewickii After Thomas Bewick (1753–1828) British naturalist, artist, and author (*History of British birds*, 1797).

bewsheri After C. E. Bewsher (fl. 1876) resident on Anjouan in the Comoro Islands.

Bhringa Bengali (Indian) native name *bhringraj* for the Greater Racquet-tailed Drongo *Dicrurus paradiseus*.

biarcuatum L. *bi* -, double, two; *arcuatus*, bow-shaped, curved (*arcuare*, to shape like a bow).

biarmicus Perhaps a corruption of Mod. L. *beardmanica*, a bearded manikin; ref. black moustachial streak (Macleod (1954), p. 39 under *Panurus)*.

Bias Gr. *buas*, a bird of evil omen, usually identified with the Great Eagle Owl *Bubo bubo*.

Biatas Gr. *biatas*, a tyrant, a violent man; ref. shrike-like appearance of the White-bearded Antshrike *B. nigropectus*.

bicalcarata / bicalcaratum / bicalcaratus Mod. L. *bicalcaratus*, two-spurred (L. *bi-*, twice, double; *calcar, calcaris*, a spur; *calx, calcis*, the heel).

bichenovii After J. E. Bicheno (1785–1851) English barrister and Colonial Secretary Tasmania, 1842.

bicincta /bicinctus L. *bi-*, two, double; *cinctus*, girdled, encircled (*cingere*, to encircle).

bicolor L. *bicolor*, bicoloured, two coloured (*bi-*, twice, double; *color*, colour).

bicornis L. *bicornis*, two-horned (*bi-*, twice, double; *cornum, corni*, a horn); the huge yellow bill of the Great Hornbill *Buceros* is surmounted by a casque.

biddulphi After Colonel John Biddulph (1840–1921) British ornithologist with Yarkand Mission, 1873–1874.

bidentata /bidentatus Med. L. *bidentatus*, having two teeth, double-toothed (L. *bi-*, twice, double; *dentatus*, toothed; *dens, dentis*, a tooth); ref. notches in upper mandible.

bido Javanese name *bido* for the Black Serpent Eagle *Spilornis*.

bieti After Monsignor F. Biet (1838–1901) French missionary to China.

bifasciata /bifasciatus Mod. L. *bifasciatus*, double-barred, two-barred (L. *bi-*, twice, double; Late L. *fasciatus*, banded, striped; L. *fascia*, a band, a streak).

bilineata /bilineatus Mod. L. *bilineatus*, double-striped, two-lined (L. *bi-*, twice, double; *linea*, a line).

bilopha L. *bi-*, twice, double; Gr. *-lophos*, crested (*lophos*, a crest).

bimaculata /bimaculatus Mod. L. *bimaculatus*, two-spotted, double-spotted (L. *bi-*, twice, double; *maculatus*, spotted, blotched; *maculare*, to blotch, to spot).

binotata /binotatus L. *bi-*, twice, double; *notatus*, marked (*nota*, a mark; *noscere*, to acknowledge).

bipunctatus L. *bi-*, twice, double; Mod. L. *punctatus*, spotted (L. *punctum*, a spot).

birostris L. *bi-*, twice, double; *-rostris*, billed; ref. appearance of casqued bill of the Indian Grey Hornbill *Tockus*.

biscutatus L. *bi-*, twice, double; *scutatus*, armed with a shield (*scutum*, a shield); ref. shape and appearance of white breast patch and nuchal collar of the White-breasted Swift *Streptoprocne*.

bishopi After **1** Charles Reed Bishop (1822–1915) US banker and husband of an Hawaiian princess (*Moho)*. **2** L. B. Bishop (1865–1950) US ornithologist and collector (*Catharopeza)*.

bispecularis L. *bi-*, twice; *specularis*, like a mirror (*speculum*, a mirror); ref. white wing patches of the Red-crowned Jay *Garrulus*.

bistriatus L. *bi-*, twice, double; Mod. L. *striatus*, streaked, striped (L. *stria*, a furrow, a streak).

bistrigiceps L. *bi-*, twice, double; *striga*, a furrow; *-ceps*, capped, headed; ref. distinctive eyebrow pattern of the Black-browed Reed Warbler *Acrocephalus*.

bitorquata /bitorquatus Mod. L. *bitorquatus*, double-collared, -banded (L. *bi-*, twice, double; *torquatus*, collared; *torques*, a collar or necklace).

bivittata /bivittatus L. *bi-*, twice, double; *vittatus*, banded, ribboned (*vitta*, a ribbon).

Biziura Gr. *buzen*, thick, matted; *oura*, the tail; ref. the pincushion-like undertail coverts exposed in display by the drake Musk Duck *B. lobata*.

blainvillii After Henri Marie Ducrotay de Blainville (1777–1850) French zoologist, comparative anatomist, and author (*Cours de physiologie générale et comparée*, 1829).

blakei After Emmet Reid Blake (b. 1908) US ornithologist, collector, and author (*Birds of Mexico*, 1953; *Manual of Neotropical birds*, 1977).

blakistoni After Captain A. W. Blakiston (1832–1892) British ornithologist and explorer in Japan.

blanchoti After P. Blanchot (fl. 1790) French Governor of Senegal.

blanfordi After W. T. Blandford (1832–1905) English geologist, zoologist, and collector in India, 1845 and Abyssinia (Ethiopia), 1867.

blasii After W. A. H. Blasius (1845–1912) German ornithologist and director of Brunswick Museum, 1870.

Bleda After Bleda (d. 445), brother to Attila and joint chief of the Huns (v. *Attila)*; these bulbuls were originally thought to be allied to the tyrant flycatchers.

blewitti After W. T. Blewitt (b. 1816) Indian civil service, Collector of Customs for the Punjab, 1867.

blighi After Samuel Bligh (fl. 1887) British coffee planter in Ceylon (Sri Lanka), 1872–1887.

blissetti After H. F. Blissett (1847–1916) British colonial administrator in tropical West Africa, 1869–1885 and collector.

bloodi After Major Neptune B. N. Blood (b. 1907) New Guinea police 1943, and District Commissioner Rabaul, New Britain, 1958.

blumenbachii After J. F. Blumenbach (1752–1841) German anatomist and naturalist.

blythii After Edward Blyth (1810–1873) English zoologist, curator of Museum of the Asiatic Society of Bengal, Calcutta, 1841–1862, collector, and author (*Catalogue of the birds of the Asiatic Society*, 1849; *The natural history of the cranes*, 1881).

Blythipicus After Edward Blyth (v. *blythii)*; genus *Picus*, woodpecker.

boanensis After the island of Boano, near Seram, in the Moluccas, Indonesia.

bocagei /Bocagia /bocagii After José Vicente Barboza du Bocage (1823–1907) Portuguese ornithologist, father of Angolan ornithology, and author (*Ornithologie d'Angola*, 1881).

bodessa After the village of Bodessa, Sidamo Province, Ethiopia.

boehmi After Richard Boehm (1854–1884) German ornithologist, collector in tropical Africa, 1880–1884, and author (*Von Sansibar zum Tanganjika*,1888).

bogotensis After the city of Santa Fé de Bogotá, capital of Colombia (v. *granadense)*.

bohmi see **boehmi.**

boiei After Heinrich Boie (1794–1827) German ornithologist and explorer.

Boissonneaua After A. Boissonneau (fl. 1839) French ornithologist and author.

boissonneautii Misspelt after A. Boissonneau (v. *Boissonneaua)*.

bojeri After Wenzel Bojer (1800–1856) Czech naturalist and collector in tropical Africa and Madagascar.

bokharensis After the city of Bokhara or Bukhara, Uzbekistan (Uzbekskaya SSR), USSR.

Bolbopsittacus Gr. *bolbos*, a bulb; genus *Psittacus*, parrot; ref. the very large, broad bill of the Philippine Guaiabero *B. lunulatus*.

Bolborhynchus Gr. *bolbos*, a bulb; *rhunkhos*, the bill; ref the broad, blunt bills of the Neotropical mountain parakeets.

boliviana /bolivianus After Bolivia. On independence from Spain in 1825 the former Upper Peru took the name Bolivia, in honour of Simón Bolívar (1783–1830) South American statesman, soldier, and liberator.

bollei /bollii After Carl August Bolle (1821–1909) German naturalist, collector, and author (*Meiner zweiter beiträge zur vögelkunde der Canarischen Inseln*, 1857).

boltoni After Lieutenant E. C. Bolton (fl. 1905) US army administrator in the Philippines.

bombus L. *bombus*, a buzzing or humming; ref. the buzzing, droning flight and very small, almost bee-like, size of the Little Woodstar *Acestrura*.

Bombycilla Gr. *bombux, bombukos*, silk; Mod. L. *cilla*, the tail (v. *Motacilla)*; after the German name *seidenschwanz*, silk-tail, for the Bohemian Waxwing *B. garrulus* ; ref. the bright yellow tail tip.

Bombylonax Gr. *bombulios*, a bumble-bee; *onax*, a king, a lord; the Black-headed Bee-eater *B. breweri* feeds mainly on wasps, honeybees, and beetles.

bonana English name *banana bird* for the various spp. of orioles *Icterus* in the West Indies (banana formerly being spelt bonana or bonano; Portuguese *banano*, the banana tree), from the birds' occurrence in banana plantations.

bonapartei After Jules Laurent Lucien, later called Charles Lucien Jules Laurent, 2nd Prince of Canino and Musignano and Prince Bonaparte (1803–1857), French ornithologist, collector, and author (*American ornithology*, 1825; *Iconografia della fauna Italica - uccelli*, 1832; *Conspectus generum avium*, 1850).

bonariensis After Buenos Aires (Mod. L. *Bonaria)*, Argentina (L. *bonum*, good; *aer*, air).

Bonasa L. *bonasus*, a type of maned bull from Paeonia, Macedonia; alluding to the neck ruffs and drumming displays of the Ruffed Grouse *B. umbellus*.

bonasia Italian name *bonasia* for the Hazel Hen *Tetrastes*.

bonelli After Franco Andrea Bonelli (1784–1830) Italian naturalist, collector, and author (*Catalogue des oiseaux du Piémont*, 1811).

bonensis After the Bone Mountains of northern Sulawesi, Indonesia.

bonthaina After Bonthain Peak (Gunung Lompobattang), of southern Sulawesi, Indonesia.

boobook Aboriginal (Australian) native name *boobook* for the Australian Boobook Owl *Ninox*, in imitation of its call.

boraquira No expl. (Spix (1825), *Avium species novae itinere Brasiliam*, Vol. **2**, p. 63).

borbae After the town of Borba on the Rio Madeira, Amazonas, Brazil.

borbonica/borbonica Former French name *Ile de Bourbon* for Réunion in the Mascarene Islands.

borealis L. *borealis*, northern (*boreas*, north). Used of species described from Canada (*Numenius, Nuttallornis)* or Siberia (*Phylloscopus)*, but erroneous in the case of the Red-cockaded Woodpecker *Dendrocopos*, of the southeastern USA, mistakenly described from the northern USA in 1808.

borin Genovese (Italian) name *borin* for a type of warbler related to the *beccafigo* or fig-eater.

bornea Erroneous toponym after Borneo; the Red Lory *Eos* is confined to the Moluccas.

boschas Gr. *boskas,* a kind of wild duck.

Bostrychia Gr. *bostrikhion*, a small curl, lock of hair (dim. of *bostrukhos*, a curl of hair); ref. the hairy crest of the Wattled Ibis *B. carunculata*.

Botaurus Med. L. *butaurus* or *bootaurus*, the bittern (Med. English *botor*, the bittern; L. *bos*, oxen; *taurus*, the bull; ref. the characteristic booming calls).

Botha After General Louis Botha (1862–1919) South African soldier and statesman; Prime Minister of the Transvaal, 1907; Prime Minister of the Union of South Africa, 1910–1919.

bottae After P. E. Botta (1802–1870) French archaeologist and naturalist in Arabia and Iran, 1832–1846.

botterii After Mateo Botteri (1808–1877) Dalmatian naturalist, explorer, and collector in Turkey, Greece, and Mexico, and founder of a museum at Orizaba, Mexico.

boucardi After A. Boucard (1839–1904) French collector and author (*The Hummingbird*, 1891).

bougainvillei/bougainvillii **1** After Bougainville Island, Solomon Islands (named for Vice Admiral Baron Louis Antoine de Bougainville [1729–1811] French world navigator) (*Actenoides, Stresemannia*). **2** after Admiral Baron Hyacinthe Yves Philippe Potentien de Bougainville (1781–1846) French navigator and scientist (*Phalacrocorax*).

bougueri After Pierre Bouguer (1698–1758) French mathematician and hydrographer in Peru, 1736–1742.

boulboul Persian name *bulbul* for the bulbul *Pycnonotus* or nightingale of the East.

bourbonnensis Former French name *Ile de Bourbon* for Réunion in the Mascarenes. The island took its name from the Bourbon dynasty of France (named after the Bourbonnais region of central France).

bourcieri/bourcierii After Jules Bourcier (1797–1873) French Consul-General to Ecuador 1849–1850, naturalist, and collector specializing in humming-birds.

bourkii After Major-General Sir Richard Bourke (1777–1855) Governor of New South Wales, Australia, 1831–1837; Bourke's Parrot *Neophema* was discovered on the banks of the River Bogan, near a stockade named Fort Bourke after the governor.

bouroensis After the island of Bourou (Buru), in the Moluccas.

bouvieri After Aimé Bouvier (fl. 1870) French zoologist and collector.

bouvreuil French name *bouvreuil* for the bullfinch *Pyrrhula*.

bouvronides French name *bouvreuil bouvron* for the Lined Seedeater *Sporophila lineola* ; Gr. *-ides*, resembling. Lesson (1831), *Traité d'Ornithologie*, called this sp. *Bouvreuil faux bouvron*, the 'false bouvron'. Buffon (1770) coined the name *bouveron* from *bouvreuil* (bullfinch) and *bec-rond* (seed-eater).

Bowdleria After Richard Bowdler Sharpe (1847–1909) English ornithologist at British Museum (Natural History) 1872–1909, and author (*Monograph of the Alcedinidae*, 1869; *A history of the birds of Europe*, 1871; *A monograph of the Hirundinidae*, 1885; *Monograph of the Paradiseidae and Ptilonorhynchidae*, 1891; *A monograph of the Turdidae*, 1898; *A hand-list of the genera and species of birds*, 1899).

boweri After Captain T. H. Bowyer-Bower (1862–1886) English explorer and collector in Australia, 1884–1885.

boyciana After Robert Henry Boyce (1834–1909) civil servant at Shanghai, China.

boyeri After Captain J. E. P. Boyer (b.1815) French navigator in the Pacific with Dumont d'Urville.

boydi After Captain Boyd Alexander (1873–1901) British ornithologist, explorer, and collector.

braccatus L. *braccatus*, wearing breeches or trousers (*braccae*, breeches).

Brachycope No expl. (Reichenow (1900), *Journ. für Orn.*, **48**, 249); probably from Gr. *brakhus*, short, little; *ops, opos*, appearance (cf. *kope*, a shaft or handle); ref. the small size and very short tail of the Bob-tailed Weaver *B. anomala*.

brachydactyla Gr. *brakhudaktulos*, short-fingered (i.e. short-toed) (*brakhus*, short; *daktulos*, a finger).

Brachygalba Gr. *brakhus*, short; contraction of genus *Galbula*, jacamar; ref. the shorter tails and smaller size of these jacamars.

brachyptera Gr. *brakhupteros*, short-winged (*brakhus*, short; *-pteros*, winged).

Brachypteracias Gr. *brakhupteros*, short-winged; genus *Coracias*, roller; the Madagascan ground-rollers are distinguished from the true rollers by their short wings, large eyes and head, and stout bills.

brachypterus see **brachyptera.**

Brachypteryx Gr. *brakhus*, short; *pterux*, a wing.

Brachyramphus Gr. *brakhus*, short; *rhamphos*, the bill.

brachyrhyncha / brachyrhynchos / brachyrhynchus Gr. *brakhus*, short; *rhunkhos*, the bill.

Brachyspiza Gr. *brakhus*, short; *spiza*, a finch; ref. the short tail of the Rufous-collared Sparrow *Zonotrichia capensis*.

brachyura /brachyurus Gr. *brakhus*, short; *-ouros*, tailed (*oura*, the tail).

bracteatus L. *bracteatus*, golden, covered with gilt or gold plate (*bractea*, gold leaf).

bradfieldi After **1** R. D. Bradfield (1882–1949) South African farmer, collector, and naturalist (*Tockus*) and his wife **2** Mrs M. Bradfield (fl. 1935) (*Apus*).

Bradornis Gr. *bradus*, slow, sluggish; *ornis*, a bird.

Bradypterus Gr. *bradus*, slow, sluggish; *-pteros*, winged; the bush warblers prefer to seek escape by creeping or scuttling away.

brama French name *Chouette brame* given to the Spotted Owlet *Athene* by Temminck, ref. its Indian habitat (Brahma, the Hindu supreme spirit).

brandti After Johann Friedrich von Brandt (1802–1879) German zoologist in the Russian service 1831–1879 and explorer in Siberia.

branickii After **1** Konstanty Graf von Branicki (1824–1884) Polish zoologist, cofounder of Branicki Zoological Museum, Warsaw (*Heliodoxa, Odontorchilus*). **2** Wladyslaw Graf von Branicki (1848–1914) Polish ornithologist, collector, and cofounder of Branicki Zoological Museum, Warsaw (*Leptosittaca*).

Branta Old Norse *brandgas*, burnt goose, alluding to the black plumage of the Barnacle *B. leucopsis* and Brent Geese *B. bernicla*.

brasiliana / brasilianum / brasiliensis After Brazil (Portuguese *Brasil*). Brasil was the name given to a dyewood long imported from the East. Pedro Cabral's expedition in 1500 discovered a strange tree of similar hue and gave it the same name, brasil or brazil, the name gradually attaching itself to the country (1530, *Terra do Brasil*).

brassi After L. J. Brass (b. 1900) Australian botanist and collector in New Guinea, 1933–1939.

brauni After R. H. Braun (b. 1908) German collector in Angola.

brazzae After Pierre Paul François Camille Savorgnan de Brazza (1859–1888) French naturalist, collector, and explorer in tropical Africa.

brehmeri After Herr Brehmer (fl. 1865) German trader in Gabon.

brehmii After Alfred Edmund Brehm (1829–1884) German zoologist, traveller, collector, and author (*Das leben der vögel*, 1861; *Illustrirtes thierleben*, 1864).

brenchleyi After Julius Lucius Brenchley (1816–1873) English traveller in the Solomon Islands and author (*The cruise of the Curaçoa among the South Sea islands*, 1865).

bres Javanese native name *bres* for the Grey-cheeked Bulbul *Alophoixus* (cf. the name *bressi* for the Fairy Bluebird *Irena puella*).

bresilius French name *Brésil* for Brazil (v. *brasiliana*).

brevicauda L. *brevis*, short; *cauda*, the tail.

brevicaudata L. *brevis*, short; *caudatus*, tailed (*cauda*, the tail).

brevipennis L. *brevis*, short; *pennis*, winged (*penna*, a feather).

brevipes L. *brevis*, short; *pes*, the foot.

brevirostris L. *brevis*, short; *-rostris*, billed.

brevis L. *brevis*, short; ref. shorter bill (*Ramphastos*) or casque (*Bycanistes*).

breviunguis L. *brevis*, short; *unguis*, a claw or nail.

breweri After Thomas Mayo Brewer (1814–1880) US politician, journalist, naturalist, and author (*North American oology*, 1857; *History of North American birds*, 1874).

brewsteri After William Brewster (1851–1919) US ornithologist, field naturalist, collector, and author (*Birds of* *the Cape regions of Lower California*, 1902; *Birds of the Cambridge region of Massachusetts*, 1906).

bridgesi/bridgesii After Thomas C. Bridges (1807–1865) traveller and collector in tropical America, 1822–1865.

brissonii After Mathurin Jacques Brisson (1723–1806) French ornithologist and author (*Règne animal*, 1756; *Ornithologie*, 1760).

broadbenti After Kendall Broadbent (1837–1911) Australian taxidermist and collector.

brodiei After Sir Benjamin Collins Brodie (1783–1862) sergeant-surgeon to Queen Victoria, and President of the Royal College of Surgeons, 1844.

brookii After Sir Charles Johnson Brooke (1829–1917) second Rajah of Sarawak, 1868–1917.

Brotogeris Gr. *brotogerus*, with human voice (an epithet of the parrot) (*brotos*, a man; *gerus*, a voice, speech).

browni After **1** Revd. G. Brown (b. 1835) missionary to Melanesia (*Monarcha*). **2** W. W. Brown (fl. 1904) US collector in Colombia, 1888, Panama, 1900–1904 and Mexico, 1922 (*Thryorchilus*).

brucei After Revd. Henry James Bruce (1835–1909) US missionary to India, 1862–1909.

bruijni/bruijnii After Anton August Bruijn (d. 1885) Dutch merchant of Ternate in the Moluccas, engaged in the New Guinea plume trade, whose collectors supplied many specimens.

bruneli After J. Brunel (fl. 1980) French ornithologist and collector.

bruniceps Med. L. *brunus*, brown; L. *-ceps*, capped, headed.

brunnea Mod. L. *brunneus*, brown (Med. L. *brunus*, brown).

brunneata Mod. L. *brunneus*, brown; L. *-atus*, possessing.

brunneicapilla/brunneicapillum/brunneicapillus Mod. L. *brunneus*, brown; L. *-capillus*, capped.

brunneicauda Mod. L. *brunneus*, brown; L. *cauda*, the tail.

brunneicaudalis Mod. L. *brunneus*, brown; L. *cauda*, the tail; *-alis*, pertaining to.

brunneiceps Mod. L. *brunneus*, brown; L. *-ceps*, headed, crowned.

brunneicollis Mod. L. *brunneus*, brown; *-collis*, necked.

brunneinucha Mod. L. *brunneus*, brown; Med. L. *nucha*, the nape (Arabic *nukha*, spinal marrow).

brunneiventris Mod. L. *brunneus*, brown; L. *venter, ventris*, the belly.

brunneopectus Mod. L. *brunneus*, brown; L. *pectus*, the breast.

brunneopygia Mod. L. *brunneus*, brown; *-pygius*, rumped (Gr. *puge*, the rump).

brunnescens Mod. L. *brunnescens*, brownish (Mod. L. *brunneus*, brown; Med. L. *brunus*, brown).

brunneus see **brunnea.**

brunnicephalus Mod. L. *brunneus*, brown; Gr. *-kephalos*, headed.

brunniceps see **bruniceps.**

brunnifrons Mod. L. *brunneus*, brown; L. *frons*, brow, forehead.

brutus L. *brutus*, heavy, weighty.

bryantae After Mrs E. B. Bryant (fl. 1876) wife of US ornithologist H. Bryant.

Buarremon Gr. *bou-*, huge (*bous*, an ox); genus *Arremon*, sparrow.

Bubalornis Gr. *boubalos*, the buffalo; *ornis*, a bird; early travellers in Africa observed the buffalo-weavers following the herds of buffalo *Synceros*.

Bubo L. *bubo*, the eagle-owl.

Bubulcus Med. L. *bubulcus* or *bobulcus*, a cow-herd; the Cattle Egret *B. ibis* usually feeds among grazing cattle or other livestock or game.

Bucanetes Gr. *bukanetes*, a trumpeter.

Buccanodon Mod. L. *bucco*, barbet; Gr. *an-*, lacking; *odous, odontos*, a tooth; the Yellow-spotted Barbet *B. duchaillui* has a 'toothless', unnotched bill.

buccinator L. *bucinator*, a trumpeter (*bucina*, a military trumpet).

Bucco L. *bucca*, the cheek, especially when puffed out; originally applied to the toucans and barbets because of the fullness of their cheeks, but latterly to the puffbirds because, 'the disproportionate size of the head is rendered more conspicuous by the bird raising its feathers so as to appear not unlike a puffball' (Newton and Gadow (1896), p. 749). The name was first given by Brisson (1760) to the barbets (Capitonidae) and the puffbirds (Bucconidae), then considered united in the same family, but subsequently separated. For example, Reichenbach (1850) used *Bucco* for the barbets and *Tamatia* for the puffbirds. In orn. *bucco* is generally used to indicate barbet (Capitonidae) and is applied to the puffbirds only in its present generic sense.

buccoides Mod. L. *bucco*, barbet; Gr. *-oides*, resembling; ref. heavy bill and green plumage of the White-eared Catbird *Ailuroedus*.

Bucephala/bucephalus Gr. *boukephalos*, bull-headed, large-headed (*bous*, an ox; *kephale*, the head).

buceroides Genus *Buceros*, hornbill; Gr. *- oides*, resembling.

Buceros L. *bucerus*, horned like an ox (*bos*, an ox); ref. the large upturned casque on the bill of the Rhinoceros Hornbill *B. rhinoceros*.

buchanani After **1** Francis Hamilton Buchanan (1762–1829) Scottish horticulturalist and collector in India and Nepal (*Emberiza*). **2** Walter Buchanan (d. 1856) English naturalist (*Prinia*). **3** Captain Angus Buchanan (1886–1954) Scottish explorer and collector (*Serinus*).

bucinator see **buccinator.**

buckleyi After Charles Buckley (fl. 1878) entomologist and collector in Bolivia, 1874 and Ecuador, 1878.

Bucorvus Gr. *bou-*, huge (*bous*, an ox); L. *corvus*, a raven; ref. the large size, booming calls, and black plumage of the ground hornbills.

budongoensis After the Budongo Forest of western Uganda.

Budytes Gr. *boudutes*, a small bird mentioned by Dionysius, not further identified.

budytoides Syn. genus *Budytes*, yellow wagtail; Gr. *-oides*, resembling.

buergersi After T. J. Bürgers (1881–1954) German zoologist in New Guinea 1912–1913. Under the current Code the German umlaut is deleted and replaced by the letter e after the modified vowel (v. *aedon*).

Buettikoferella Dim. after Johann Büttikofer (1850–1927) Swiss botanist and zoologist.

buettikoferi see **Buettikoferella.**

buffoni/buffonii After George Louis Leclerc Comte de Buffon (1707–1788) French naturalist, director of Jardin du Roi in Paris, and author (*Histoire naturelle des oiseaux*, 1770).

Bugeranus Gr. *bous*, an ox; *geranos*, a crane; the Wattled Crane *B. carunculatus* has a loud, sonorous, bell-like call.

bulleri After Sir Henry Lawry Buller (1838–1906) New Zealand ornithologist, collector, and author (*A history of the birds of New Zealand*, 1873).

bulliens L. *bulliens*, bubbling (*bullire*, to bubble); ref. the lively, bubbling song of the Bubbling Cisticola *Cisticola*.

bullockii After William Bullock (b. 1775) English naturalist, archaeologist, and entrepreneur.

bullockoides Specific name *Merops bulocki*; Gr. *-oides*, resembling.

bulocki After William Bullock (v. *bullockii*).

bulweri After Sir Henry Ernest Gascoyne Bulwer (1836–1914) British diplomat, Governor of Labuan, Borneo, 1871–1875.

Bulweria/bulwerii After the Revd. James Bulwer (1794–1879) Scottish naturalist and collector, chaplain at Madeira, 1828.

Buphagus Gr *bous*, an ox; *-phagos*, eating (*phagein*, to eat); ref. the oxpecker habit of clambering over game animals (buffalo *Synceros*, giraffe *Giraffa*, etc.) to rid them of ticks and other parasites.

burchelli/burchellii After William John Burchell (1782–1863) British naturalist, explorer in South Africa, 1811 and author (*Travels in the interior of southern Africa*, 1822).

Burhinus Gr. *bous*, an ox; *rhis, rhinos*, the nose; ref. the short, thick bill and bull-headed appearance of the Stone Curlew *B. oedicnemus*

burkii After William Augustus Burke (1769–1837) Inspector-General of Hospitals, Bengal, 1825.

burmannicus After Burma (Myanmar), named for the indigenous Burman inhabitants ; L. - *icus*, belonging to.

burmeisteri After Karl Hermann Konrad Burmeister (1807–1892) German ornithologist in Argentina, 1856–1892 and author (*Reise nach Brasilien*,1853).

burnesi /Burnesia After Sir Alexander Burnes (1804–1841) British political officer in Afghanistan, 1839–1841.

burnieri After Eric Burnier (fl. 1986) British doctor, field naturalist, and collector in Africa.

burra L. *Burrus*, an old form of the personal name *Pyrrhus* (Gr. *purrhus*, red, flame-coloured); ref. the bright chestnut upperparts of the Red Lark *Mirafra*.

burrovianus After M. Burrough (fl. 1839), a doctor about whom nothing else is known.

burtoni After **1** H. Burton (fl. 1838) English taxidermist and collector (*Callacanthis*). **2** Sir Richard Francis Burton (1821–1890) British explorer, linguist, and author (*Kama sutra*,1883; *Arabian nights*, 1885; *The perfumed garden*,1886) (*Serinus*).

buruensis see **bouroensis.**

buryi After George Wyman Bury (1874–1920) English explorer, political officer, Arabist, naturalist, and author (*Arabia infelix*, 1915).

Busarellus Dim. from French *buse*, buzzard, and *rayée*, streaked.

Butastur Genus *Buteo*, buzzard; syn. genus *Astur*, goshawk.

Buteo L. *buteo*, the buzzard.

Buteogallus Genus *Buteo*, buzzard; L. *gallus*, a farmyard cock; the Rufous Crab Hawk *B. aequinoctialis*, by reason of its bare face, small head, heavy body, and long, reticulated legs was considered to resemble a chicken.

Buthraupis Gr. *bous*, an ox; genus *Thraupis*, tanager; ref. the large size and 'bull-headed' appearance of the mountain tanagers.

butleri After **1** Lieutenant Colonel Edward Arthur Butler (1842–1916) British collector, ornithologist, and author (*Birds of Sind, Cutch and neighbourhood*, 1879) (*Strix)* and his son **2** Arthur Lennox Butler (1873–1939) zoologist, ornithologist, curator Selangor State Museum, Malaya, and Superintendent of Game Preservation, Sudan, 1901–1915 (*Accipiter)*.

Butorides Syn. genus *Butor*, bittern (Med. L. *butor*, the bittern); Gr. *-ides*, resembling.

Butreron Gr. *bou-*, huge (*bous*, an ox); genus *Treron*, green pigeon; ref. large size and thick-set appearance of the Large Green Pigeon *Treron capellei*.

Bycanistes Gr. *bukanistes*, a trumpeter.

Bubo capensis Mountain Eagle Owl

C

Cabalus No expl.(Hutton (1874), *Trans. N. Z. Inst.*, **6**, 108); perhaps from English *cabal*, a secret society (French *cabale*); ref. the secretive habits of the extinct Chatham Rail *C. modestus*.

cabanisi After Jean Louis Cabanis (1816–1906) German ornithologist, founding editor of *Journal für Ornithologie*, 1853, and author (*Museum Heineanum*, 1850).

caboti After S. Cabot (1815–1885) US zoologist.

Cacatua Malay names *kakatua* or *kakak-tua* for the cockatoo.

cachinnans L. *cachinnans*, laughing (*cachinnare*, to laugh aloud, to guffaw).

Cacicus Spanish name *cacique* for the caciques or orioles (Carib (Haitian) Indian *cazique*, a chieftain).

Cacomantis Gr. *kakos*, evil, ill-boding; *mantis*, a prophet, a soothsayer; the monotonous cries of these cuckoos are supposed to predict ill-fortune and bad weather.

cactorum Mod. L. *cactorum*, of the cacti (L. *cactus*, a prickly plant, either the Spanish artichoke *Cynara* or the oyster plant *Scolymus* ; Gr. *kaktos*, a prickly plant).

Cactospiza Mod. L. *cactus*, the cactus; Mod. L. *spiza*, a finch.

Caenotriccus Gr. *kainos*, new, strange; Mod. L. *triccus*, tyrant flycatcher.

caerulata /caerulatus L. *caerulatus* (poetic form of *caeruleatus*), dark blue, sky blue.

caerulea L. *caeruleus*, cerulean, azure blue, dark blue (can also mean dark, black, or dark green).

caeruleiceps L. *caeruleus,* azure blue; *-ceps,* headed.

caeruleirostris L. *caeruleus*, azure blue; *-rostris*, billed.

caeruleocapilla L. *caeruleus*, azure blue; *-capillus*, capped, crowned.

caeruleocephalus L. *caeruleus*, azure blue; Gr. *-kephalos*, headed.

caeruleogaster L. *caeruleus*, azure blue; *gaster*, the belly.

caeruleogrisea Mod. L. *caeruleogriseus*, sky-grey (L. *caeruleus*, azure blue; Med. L. *griseus*, grey).

caerulescens Inferred L. *caerulescens*, bluish; *caerulescere* and *caerulescens* have not been found in Classical Latin, but their existence may be inferred from the base *caeruleus*, azure blue, by the example, amongst others, of *rufus*, red, ruddy, giving *rufescere*, to become reddish, and *rufescens,* reddish.

caeruleus see **caerulea** (cf. L. *caeruleum*, steel blue).

caesar L. *Caesar*, the crown-prince (i.e. imperial, princely); ref. large size of the Chestnut-bellied Mountain Finch *Poospizopsis*.

caesia /caesius L. *caesius*, bluish-grey; also had the additional meaning of grey-eyed (cf. Mod. L. *caesius*, eye-blue colour).

cafer /caffer /caffra /cafra Mod. L. *Caffer,* South African, of South Africa (*Caffraria*); from the terms Kaffir, Caffre, or Caffer applied to certain indigenous peoples of that country (Arabic *kafir*, an infidel). The toponym is applied erroneously in the cases of *Colaptes* (where the Bay of Good Hope, Nootka Sound,

Canada, was misread as the Cape of Good Hope, South Africa), *Conopoderas* (where *caffra* was used in the general sense of southern), and *Pycnonotus*.

cahow Onomatopoeic names *cahow*, *cahowe* or *cowhaw* given to the Cahow Petrel *Pterodroma* by early settlers in Bermuda.

caica Carib (Haitian) Indian name *caica* for a parrot.

cailliautii After Frédéric Cailliaud (1787–1869) French naturalist and collector in Egypt and the Sudan, 1815–1822.

Cairina Italian *Cairino*, a Cairene, a native of Cairo, Egypt, in the mistaken belief that the Muscovy Duck *C. moschata* came from Egypt (it was called the Cairo Duck by Ray (1678)); in fact the Muscovy Duck comes neither from Muscovy, in Russia, nor Egypt but from tropical America.

cajanea After Cayenne or French Guiana.

cajeli Of the Silver-blue Kingfisher *Ceyx* Wallace (1863), *Proc. Zool. Soc. London*, 25, wrote,'I have named this species after the town or fort of Cajeli in Bouru, to which island this pretty bird is most probably strictly confined'.

Calamanthus Gr. *kalame*, a reed stalk; genus *Anthus*, pipit; ref. damp grass habitat and streaked, pipit-like appearance of the Australian fieldwrens.

Calamocichla Gr. *kalamos*, a reed; *kikhle*, a thrush.

Calamodus Gr. *kalamos*, a reed; *odos*, a singer.

Calamonastes Gr. *kalamos*, a reed (also used for any plant neither tree nor bush); *astes*, a singer.

Calamospiza Gr. *kalamos*, a reed; *spiza*, a finch.

calandra Gr. *kalandros*, the Calandra Lark *Melanocorypha* (cf. Med. L. *kalandra* and *calendris*, a lark).

Calandrella Dim. from Gr. *kalandros*, the Calandra Lark.

Calcarius L. *calcaria*, spurs (*calcar*, a spur); these buntings have very straight, long hindclaws, and are known in America as longspurs.

calcostetha Gr. *khalkostethos*, bronze-breasted.

caledonica/caledonicus After New Caledonia in the Pacific Ocean, sighted by Captain Cook in 1774 and so named because its forested and mountainous aspect reminded him of Scotland (L. *Caledonia*, the highlands of Scotland).

Calendula Dim. from Gr. *kalandros*, the Calandra Lark *Melanocorypha*. Specifically the name refers to the Ruby-crowned Kinglet's *Regulus* remarkably loud song, which has been compared to that of the Skylark *Alauda* (cf. L. *caliendrum*, a head-dress; *calere* to glow; Med. L. *calendula*, a 'plover').

Calicalicus Malagasy name *cali-calic* for the Red-tailed Vanga *C. madagascariensis* .

Calidris Gr. *kalidris* (variant form of *skalidris*), a grey-coloured, waterside bird mentioned by Aristotle, not further identified, but later conjectured to be a sandpiper *Tringa* or a wagtail *Motacilla*.

Caliechthrus Gr. *kaleo*, to call; *ekhthros*, odious; ref. the loud, monotonous calls of the White-crowned Koel *C. leucolophus*, which according to tradition can foretell bad weather and calamities.

californianus After **1** the Spanish Governorate of Alta California (Upper California), New Spain (Mexico) (*Gymnogyps*). **2** the province of Alta California, Mexico (*Geococcyx*).

californica/californicus After **1** the Spanish Governorate of Alta California, New Spain (*Callipepla*). **2** the province of Alta California, Mexico (*Aphelocoma*). **3** the state of California, USA (admitted to statehood 1850) (*Larus*, *Polioptila*).

caligata/caligatus L. *caligatus*, booted (*caliga*, a stout shoe or boot); ref. smooth, unscutellated tarsi.

Callacanthis Gr. *kalli-*, beautiful; genus *Acanthis*, linnet, finch.

Callaeas Gr. *kallaia*, a cock's wattles; the New Zealand Kokako *C. cinerea* has orange or blue wattles at the base of the bill.

callainus L. *callainus*, turquoise-coloured (Gr. *kallainos*, like the turquoise; *kallais*, the turquoise).

Callene L. *callainus*, turquoise-coloured.

Callichelidon Gr. *kalli-*, beautiful; *khelidon*, the swallow.

calligyna Gr. *kalli-*, beautiful; *gune*, a maiden or woman.

callinota Gr. *kalli-*, beautiful; *-notos*, backed (*noton*, the back).

calliope Gr. myth. Calliope or Kalliope, the chief of the Muses, fine-voiced and goddess of poetry.

calliparaea Gr. *kallipareos*, beautiful cheeked (*kalli-*, beautiful; *pareia*, the cheek).

Callipepla Gr. *kallipeplos*, beautifully clad (*kalli-*, beautiful; *peplos*, a robe).

Calliphlox Gr. *kalliphlox*, auspiciously burning, beautiful blazing.

calliptera Gr. *kalli-*, beautiful; *-pteros*, winged.

Calliste Gr. *kallistos*, very beautiful (super. of *kalos*, beautiful).

callizonus Gr. *kallizonos*, with beautiful girdles, beautifully banded.

Callocephalon Gr. *kallos*, beauty; *kephale*, the head.

Callonetta Gr. *kallos*, beauty; *netta*, a duck.

callonotus Gr. *kalli-*, beautiful; *-notos*, backed (*noton*, the back).

callophrys Gr. *kalli-*, beautiful; *-ophrus*, browed (*ophrus*, the eyebrow).

callopterus Gr. *kalli-*, beautiful; *-pteros*, winged.

Calochaetes Gr. *kalos*, beautiful; *khaite*, a mane, long hair.

Calocitta Gr. *kalos*, beautiful; Mod. L. *citta*, a jay (v. *Cissa*).

Caloenas Gr. *kalos*, beautiful; *oinas*, a dove.

calolaema Gr. *kalos*, beautiful; *laimos*, the throat.

Calonectris Gr. *kalos*, good, beautiful; syn. genus *Nectris*, shearwater (Gr. *nektris*, a swimmer).

Caloperdix Gr. *kalos*, beautiful; *perdix*, a partridge.

Calophasis Gr. *kalos*, beautiful; Mod. L. *phasis*, pheasant.

calophrys Gr. *kalos*, beautiful; *-ophrus*, browed (*ophrus*, the eyebrow).

calopterum /calopterus Gr. *kalos*, beautiful; *-pteros*, winged.

Calorhamphus Gr. *kalos*, beautiful; *rhamphos*, the bill.

calorhynchus Gr. *kalos*, beautiful; *rhunkhos*, the bill.

Calospiza Gr. *kalos*, beautiful; *spiza*, a finch.

Calothorax Gr. *kalos*, beautiful; *thorax*, the breast.

calthorpae After Miss Calthorp (fl. 1849) first wife of Edgar Layard (v. *layardi*).

calurus Gr. *kalos*, beautiful; *-ouros*, tailed.

calva /calvus L. *calvus*, bald.

calyorhynchus Gr. *kalos*, beautiful; *rhunkhos*, the bill.

Calypte Gr. *kaluptre*, a woman's veil (*kalupto*, to cover); ref. the glittering gorgets of the male hummingbirds in this genus.

Calyptocichla Gr. *kaluptos*, covered, buried; *kikhle*, a thrush; the Serine Greenbul *C. serina* is the sole member of this '...genus notable for its closely and completely feathered nasal operculum' (Oberholser (1905), *Smithsonian Misc. Coll.*, **48**, 165).

Calyptomena Gr. *kaluptos*, covered, buried; *mene*, the moon; from the crescent-shaped fan of feathers covering the lores and bills of the Indomalayan green broadbills.

Calyptophilus Gr. *kaluptos*, concealed, covered; *philos*, loving, fond of; the Chat Tanager *C. frugivorus* occurs in dense scrub and is difficult to observe.

Calyptorhynchus Gr. *kaluptos*, concealed, covered; *rhunkhos*, the bill.

Calyptura Gr. *kalupto*, to conceal; *oura*, the tail; the Kinglet Calyptura *C. cristata* has the '…tail very short, divaricated, almost concealed' (Swainson, in Swainson and Richardson (1832), *Fauna Bor.-Amer*., Vol. **2**, p. 491).

Camarhynchus Gr. *kamara*, arched, vaulted; *rhunkhos*, the bill; ref. the strongly arched culmen of the Large Tree Finch *C. psittacula*.

Camaroptera Gr. *kamara*, arched, vaulted; *ptera*, wings.

cambodiana After Cambodia, a former province of French Indochina.

camelus L. *camelus*, a camel (Gr. *kamelos*, a camel); the Ostrich *Struthio* was called the camel-bird (*struthio-camelus)* because of its large size.

cameronensis After British Cameroon, now part of Cameroun.

camerunensis After Mount Cameroon, British Cameroon, now part of Cameroun.

Camiguinia After Camiguin Island, north of Luzon, Philippines.

campanisona Late L. *campana*, a bell; L. *sonus*, sounding; ref. hollow, sonorous callnotes.

campbelli After Robert Watt Campbell (fl. 1981) Australian ornithologist and bird-ringer or bander.

Campephaga Gr. *kampe*, a caterpillar; *phagos*, a glutton.

Campephilus Gr. *kampe*, a caterpillar; *philos*,loving, fond of.

campestris L. *campester, campestris*, of the plains, of flat country (*campus, campi*, a field).

campestroides Specific name *Colaptes campestris* ; Gr. *-oides*, resembling.

Campethera Gr. *kampe*, a caterpillar; *-thera*, a hunter.

Campochaera Gr. *kampe*, a caterpillar; *khairo*, to delight in.

Camptorhynchus Gr. *kamptos*, bent, curved; *rhunkhos*, the bill.

Camptostoma Gr. *kamptos*, curved, bent; *stoma*, the mouth (i.e. the bill).

Campylopterus Gr. *kampulos*, bent, curved; *-pteros*, winged; the sabrewing hummingbirds have thickened, sabre-like outer primary shafts.

Campylorhamphus Gr. *kampulos*, bent, curved; *rhamphos*, the bill.

Campylorhynchus G.r *kampulos*, bent, curved; *rhunkhos*, the bill (cf. *kampulorhin*, hook-nosed).

camurus L. *camur*, hooked, curved.

cana L. *canus*, grey, hoary.

Canachites Syn. genus *Canace,* grouse (Gr. myth. Canace, daughter of King Aeolus of Etruria, who committed incest with her brother Macaraeus and, after the birth of their child, was ordered to kill herself; ref the polygamous habits of grouse); Gr. *-ites,* resembling (cf. *kanakhedes,* noisy, resounding; *kanakhe*, a noise).

canadensis After Canada (Iroquois Indian *kanata,* a village or community); erroneously applied in the cases of *Caryothraustes* and *Sakesphorus*, where *cayennensis* was intended.

canagicus After Kanaga Island in the Aleutian Islands, Alaska.

canaria L. *Canaria insula*, one of the Canary Islands named for its large dogs (*canis*, a dog).

cancellatus L. *cancellatus*, trellis-like, lattice-like (*cancellare*, to lattice; *cancelli*, a lattice); from appearance of buff-fringed feathers of upper- and underparts of the Tuamotu Sandpiper *Aechmorhynchus*.

cancrominus Specific name *Platyrinchus cancromus* (syn. genus *Cancroma,* Boat-billed Heron (L. *cancer, cancri,* a crab); ref. bill-shape) (now treated as a subsp. of *P. mystaceus)*; L. *-inus,* resembling.

candei After Admiral A. M. F. de Candé (1801–1867) French naval explorer in South America, 1841.

candicans L. *candicans*, whitish (*candicare*, to be white).

candida /candidus L. *candidus*, shining white, pure white (cf. *albus*, matt white).

canente L. *canens, canentis*, grey, hoary, white.

canescens L. *canescens*, rather hoary, greyish (*canescere*, to become hoary; *canus*, grey).

canicapilla/canicapillus L. *canus*, grey; *-capillus*, capped, crowned.

caniceps L. *canus*, grey; *-ceps*, crowned, capped.

canicollis L. *canus*, grey; Mod. L. *-collis*, necked.

canicularis L. *Canicularis*, of the bright star Sirius or Canicula, in the constellation Canis Major.

canifrons L. *canus*, grey; *frons*, the forehead, the brow.

canigularis L. *canus*, grey; Mod. L. *gularis*, throated, of the throat (L. *gula*, the throat).

caninde Güaraní (Paraguayan) Indian name *canindé* for the Blue-throated Macaw *Ara*. This is one of many names culled from Azara (1805). Azara, a military surveyor in Paraguay, did not use Linnaean names, but gave all spp. a Spanish name, which was either based on a familiar Spanish sp. or taken from the Güaraní Indian vernacular. Subsequent authors gave these names permanence in Linnaean nomenclature.

Canirallus L. *canus*, grey; genus *Rallus*, rail.

canivetii After E. Canivet de Carentan (fl. 1876) French ornithologist and collector.

cannabina L. *cannabinus*, hempen (*cannabis*, hemp); the Brown Linnet *Acanthis* feeds mainly on the seeds of weeds.

canningi After Charles John Earl Canning (1812–1862) Governor-General of India, 1856–1862.

canora/canorus L. *canorus*, melodious (*canere*, to sing).

cantans L. *cantans*, singing (*cantare*, to sing).

cantator L. *cantator*, a singer, a musician.

cantillans L. *cantillans*, warbling, humming (*cantillare*, to warble).

cantonensis After the city of Canton (Guangzhou), Guangdong Province, China.

cantoroides Syn. specific name *Aplonis cantor* (L. *cantor*, a singer); Gr. *-oides*, resembling.

canturians L. *canturians*, chirping (*canturire*, to chirp).

canus see **cana.**

canutus After Canute or Cnut (995–1035; reigned 1016–1035) king of England, Denmark, and Norway, who, according to legend, regarded the Red Knot *Calidris* as a delicacy.

Capella L. *capella*, a she-goat; ref. the bleating noise produced by the extended tail-feathers of the Marsh Snipe *Gallinago* during its diving display flight.

capellei After Baron Godert Alexander van der Capellen (1778–1848) Governor-General of the Dutch East Indies, 1816–1826.

capense/capensis After the Cape of Good Hope (Mod. L. *Caput Bonae Spei*; Portuguese *Cabo de Boa Esperança*), South Africa. Erroneous epithets in the cases of *Bucco* and *Zonotrichia* (where *cayennensis* or *cayennsis* were meant), and *Turnagra* (q.v.).

capicola After Cape Colony, South Africa; L. *-cola*, inhabitant.

capillatus L. *capillatus*, hairy, with a fine head of hair.

capistrata/capistratum/capistratus L. *capistratus*, haltered; ref. face pattern.

capitale/capitalis L. *capitalis*, capital, of the head (*caput*, the head).

capitata L. *capitatus*, headed (*caput, capitis*, the head).

Capito L. *capito*, big-headed, a man with a large head.

caprata Local Spanish (Philippines) name *Maria-capra* for the Pied Bush Chat *Saxicola* on Luzon.

Caprimulgus L. *caprimulgus*, the nightjar or goatsucker (*capra*, a nanny goat; *mulgere*, to milk); '... thereby testifying to the widespread belief in the malpractice it attributes to its unfortunate bearer' (Newton and Gadow (1896), p. 365).

caprius No expl. (Boddaert (1783), *Table des planches enluminées*, p. 40). Based on d'Aubenton's name *Coucou verd du Cap de bonne-Espérance* for the Diederik Cuckoo *Chrysococcyx,* and either a misprint for *cupreus* (L. *cupreus,* coppery) or for *capensis* (q.v.).

Capsiempis Gr. *kapto,* to gulp down; *empis,* a gnat, a mosquito.

capueira Brazilian Portuguese name *capueira* for a toothed forest partridge or quail (Tupí Indian *caapora* or *corupira,* a spirit of the woods, a forest dweller).

carabayae After the Cordillera de Carabaya, Puno, Peru.

caracae After Caracas, the capital city of Venezuela.

Caracara Tupí (Brazilian) Indian name *caracará* for the noisy, scavenging caracaras.

carbo L. *carbo,* charcoal.

carbonaria /carbonarius L. *carbonarius,* a charcoal burner; alluding to dark plumage.

Cardellina Dim. of genus *Carduelis,* goldfinch.

Cardinalis Med. L. *cardinalis,* a cardinal, a senior bishop of the Roman church who wears scarlet robes; ref. the red plumage of the Northern Cardinal *C. cardinalis.*

cardis French name *Merle carde* given to the Japanese Thrush *Turdus* by Temminck (1831), *Planches coloriées,* Vol. **87**, pl. 518 (probably from Gr. *kardia,* the stomach; ref. white belly and black-spotted white flanks and lower breast).

cardonai After P. F. Cardonai (b. 1903) Spanish explorer and collector in Venezuela, 1930.

Carduelis L. *carduelis,* the goldfinch.

Cariama Portuguese name *çariama* for the Crested Seriema *C. cristata,* based on Güaraní (Paraguayan) Indian name *saria* and Tupí (Brazilian) Indian name *sariama* (Güaraní and Tupí are members of a language group, Tupian, widespread in the Amazonian forests.)

caribaea /caribaeus /caribbaea After the Caribbean islands of Cuba (*Contopus*), Jamaica (*Columba, Pterodroma*), St John (*Fulica*), and San Andrés (*Vireo*). The Caribbean Sea, discovered by Columbus in 1493, was named after the indigenous Carib Indians, since enslaved and exterminated.

Caridonax Gr. *karis, karidos,* a shrimp or prawn; *anax,* a lord or king; from supposed prey of the White-rumped Kingfisher *C. fulgidus.*

carinatum L. *carinatus,* keeled, ridged; ref. the distinctive keel-like ridges on the bill of the Keel-billed Motmot *Electron.*

caripensis After the Caripé Caverns, Cumaná, Venezuela.

carmioli After **1** Julian Carmiol (fl. 1865) ornithologist and collector in Costa Rica (*Vireo*) and his son **2** F. Carmiol (fl. 1865) collector in Costa Rica (*Chlorothraupis*).

carneipes Mod. L. *carneus,* flesh-coloured (L. *caro, carnis,* flesh); L. *pes,* the foot.

carneus Mod. L. *carneus,* flesh-coloured (L. *caro, carnis,* flesh).

carnifex L. *carnifex,* an executioner, a murderer (i.e. blood-stained).

carnipes see **carneipes.**

carola Late L. *Carola,* Carol, Charlotte; after Charlotte Honorine Joséphine Pauline Contessa Primoli di Foglia (1832–1891) daughter of French ornithologist Prince Charles Bonaparte.

carolae Late L. *Carola,* Carol, Charlotte; after Queen Carola of Saxony (1833–1907) wife of King Albert of Saxony (v. *alberti* **3**).

carolina After the Carolina colonies, North America. Opinions differ as to whether the colonies, which originally covered a larger area than the present states, were named for Charles IX King of France (1550–1574; reigned 1560–1574), for Charles I King of England (1600–1649; reigned 1625–1649), or for Charles II King of England (1630–1685; reigned 1660–1685) (Late L. *Carolus,* Charles).

carolinae After **1** Caroline Ash (fl. 1981) daughter of English ornithologist J.

S. Ash (*Turdoides*). **2** Caroline Rozendaal (fl. 1987) wife of Dutch ornithologist Frank G. Rozendaal (*Cettia*). **3** No expl. (Schlegel (1871), *Ned. Tjdschr. Dierk.*, 4, 13) (*Tanysiptera*).

carolinensis After **1** the Carolina colonies, North America (*Anas, Caprimulgus, Conuropsis, Dumetella, Sitta*). **2** the state of South Carolina, USA (*Parus*) (v. *carolina*).

carolinus After **1** the Carolina colonies, North America (*Melanerpes*). **2** the states of Carolina, USA (*Euphagus*) (v. *carolina*).

carpalis Mod. L. *carpalis*, of the wrist (i.e. bend of the wing) (Gr. *karpos*, the wrist).

Carphibis Gr. *karphe*, a blade of straw; *ibis*, an ibis; ref. the straw-like breast plumes of the Straw-necked Ibis *C. spinicollis*.

carpi After Bernard Carp (1901–1966) South African naturalist.

Carpococcyx Gr. *karpos*, fruit; *kokkux*, the cuckoo; from assumed food of these Indomalayan forest cuckoos.

Carpodacus Gr. *karpos*, fruit; *dakos*, a biter (strictly an animal whose bite is dangerous) (*dakno*, to bite); the rosefinches feed mainly on berries, fruits, buds and seeds.

Carpodectes Gr. *karpos*, fruit; *dektes*, a biter (*dakno*, to bite); the white cotingas are primarily fruit-eaters.

Carpornis Gr. *karpos*, fruit; *ornis*, a bird; the berry-eaters feed exclusively on fruit.

Carpospiza Gr. *karpos*, the wrist; *spiza*, a finch; ref. carpal bar of the Pale Rock Sparrow *Petronia brachydactyla*.

carrikeri After M. R. Carriker (b. 1915) US biologist.

carruthersi After Alexander Douglas Mitchell Carruthers (1882–1962) English explorer, big-game hunter, and author (*Unknown Mongolia*, 1913; *Arabian adventure*, 1935).

carteri After Thomas Carter (1863–1931) English ornithologist and pastoralist in Australia, 1886–1921.

carunculata/carunculatus Mod. L. *carunculatus*, carunculated, wattled (L. *caruncula*, a small piece of flesh).

caryocatactes Gr. *karuokataktes*, the nutcracker (*karuon*, a nut; *kataseio*, to shatter).

caryophyllacea Mod. L. *caryophyllaceus*, pink; ref. deep pink head and hind neck of the Pink-headed Duck *Rhodonessa* (cf. pre-Linnaean Botanical genus *Caryophylla*, cultivated pink; L. *carus*, precious, esteemed; *phyllon*, a kind of plant).

Caryothraustes Gr. *karuon*, a nut; *thraustes*, a breaker (*thrauo*, to break); a group of large-billed American cardinals that feed on seeds, fruit and insects.

Casarca Italian name *casarca* for the Ruddy Shelduck *Tadorna ferruginea* (Russian *kazarka*, a small goose *Branta*).

Casiornis Gr. *kasia*, the cinnamon tree; *ornis*, a bird; ref. cinnamon underparts of the Rufous Casiornis *C. rufa*.

casiquiare After the Río Casiquiare, Amazonas, Venezuela.

Casmerodius Gr. *kasis*, a brother or sister (i.e. related); *erodios*, a heron.

caspia/caspius L. *Caspius*, Caspian, of the Caspian Sea (*Caspium mare*).

cassicus French name *Le Cassican de la Nouvelle Guinée* given to the Hooded Butcherbird *Cracticus* by Buffon (1770) (*cassique*, cacique, oriole; *toucan*, toucan).

Cassidix L. *cassis, cassidis*, a metal helmet; ref. iridescent sheen on head of the Great-tailed Grackle *C. mexicanus*.

Cassinaetus After John Cassin (v. *cassini*); Gr *aetos*, an eagle.

cassini/cassinii After John Cassin (1813–1869) US ornithologist, collector, and author (*Birds of Chili*, 1855; *Illustrations of the birds of California, Texas, Oregon, British and Russian America*, 1865).

castanea L. *castaneus*, chestnut-coloured (*castanea*, the chestnut; Gr. *kastanon*).

castaneceps/castaneiceps L *castaneus*, chestnut-coloured; *-ceps*, capped, crowned.

castaneicollis L. *castaneus*, chestnut-coloured; Mod. L. *-collis*, necked.

castaneiventris L. *castaneus*, chestnut-coloured; *venter*, *ventris*, the belly.

castaneocapillus L. *castaneus*, chestnut-coloured; *-capillus*, capped, headed.

castaneocoronata L. *castaneus*, chestnut-coloured; *coronatus*, crowned (*coronare*, to crown).

castaneofuscus L. *castaneus*, chestnut-coloured; *fuscus*, dusky, dark, black, tawny.

castaneothorax L. *castaneus*, chestnut-coloured; *thorax*, a breastplate (Gr. *thorax*, the chest).

castaneoventris see **castaneiventris.**

castaneum /castaneus see **castanea.**

castaniceps see **castaneceps.**

castanilius L. *castaneus*, chestnut-coloured; *ilia*, the groin, the flanks.

castanonota L. *castaneus*, chestnut-coloured; Gr. *-notos*, backed (*noton*, the back).

castanops L. *castaneus*, chestnut-coloured; Gr. *ops*, the face.

castanopterus L. *castaneus*, chestnut-coloured; Gr. *-pteros*, winged.

castanota L. *castaneus*, chestnut-coloured; Gr. *-notos*, backed .

castanotis L. *castaneus*, chestnut-coloured; Gr. *-otis*, eared.

castanotum /castanotus see **castanota.**

castelnau After François Louis Nompar de Caumont Comte de Laporte de Castelnau (1812–1880) French explorer in tropical America, 1843–1847.

castelnaudii After the Comte de Castelnau (v. *castelnau*).

castro Harcourt (1851), *A sketch of Madeira*, p. 123, wrote of the Madeiran Storm Petrel *Oceanodroma*, 'There is another petrel called by the natives "Roque de Castro", and pronounced "Roque de Crasto", which is likewise an inhabitant of the Dezerta Islands'. I cannot find the meaning of this name, although Macleod (1954), p. 37, opines "'rook' de Castro" (perhaps meaning rock crow or shrimp crow).

castus L. *castus*, clean, spotless.

Casuarius Malay native name *kesuari* (or *suwari* according to Newton and Gadow (1896)) for the cassowary.

Catamblyrhynchus Gr. *katambluno*, to blunt; *rhunkhos*, the bill; the Plush-capped Finch or Plushcap *C. diadema* has a short, thick, swollen bill.

catamene /Catamenia Gr. *katamenia*, menstrual, menstruous (*katemenos*, monthly); ref. chestnut or red undertail coverts or vent of the Band-tailed Seedeater *C. analis* and the orange-red undertail coverts of the Sangihe Hanging Parrot *Loriculus*.

Cataponera Gr. *kataponeo*, to wear out, to exhaust; in describing the Sulawesi Mountain Thrush *C. turdoides* Hartert (1896), *Nov. Zool.*, **3**, 71, wrote, 'The generic name is given in reference to the difficulty and hardships connected with the shooting of these birds in the great altitudes they inhabit'.

Catharacta Gr. *katarrhaktes*, a rapacious seabird mentioned by Aristotle and Dionysius, not further identified. In modern times variously identified as a grebe *Podiceps*, a cormorant *Phalacrocorax*, or the Great Skua *C. skua*.

Catharopeza Gr. *katharos*, bright, clean; *peza*, the foot; the Whistling Warbler *C. bishopi* has bright yellow feet and legs.

Cathartes Gr. *kathartes*, a cleanser, a purifier; the American vultures act as very useful scavengers in the Neotropics.

Catharus Gr. *katharos*, pure, clear; ref. the immaculate brown and white plumage of the Orange-billed Nightingale Thrush *C. aurantiirostris*.

Catherpes Gr. *katherpes*, a creeper (*katherpo*, to creep down).

cathpharius Hindi name *katphora* for a woodpecker.

Catoptrophorus Gr. *katoptron*, a mirror; *-phoros*, carrying, wearing; the Willet *C. semipalmatus* has striking white wing-patches.

Catreus Gr. *katreus*, a bird mentioned

by Aelianus, not further specified and probably mythical, but identified as a type of pheasant by later commentators.

caucae After the department of Cauca, south-western Colombia

caucasicus After the Caucasus Mountains in southwest USSR, lying between the Black and Caspian Seas.

caudacuta /caudacutus L. *cauda*, the tail; *acutus*, sharpened, pointed.

caudata /caudatus L. *caudatus*, tailed (i.e. long-tailed). *Spelaeornis caudatus* has a very short tail, but was originally described in the virtually tail-less genus *Tesia* .

caudifasciatus L. *cauda*, the tail; Late L. *fasciatus*, banded, striped (L. *fascia*, a band, a girdle).

caurensis After the Río Caura, Bolívar, Venezuela.

caurinus L. *caurus*, the northwest wind (Gr. *khoros,* the north-west wind); - *inus,* belonging to; ref. the Washington Territory of northwest USA.

cauta /cautus L. *cautus*, shy, wary (*cavere*, to beware).

cayana /cayanensis /cayanus / cayennensis After Cayenne or French Guiana. In early ornithology usually indicated a Neotropical species of otherwise unknown provenance.

cearae After the state of Ceará, north-eastern Brazil.

Ceblepyris Gr. *keblepuris,* the redcap (*keble,* the head; *pur,* fire), a bird mentioned by Aristophanes, not otherwise identified, but latterly associated with the Common Redpoll *Acanthis flammea.*

cebuensis After the island of Cebu in the Philippines.

ceciliae After **1** Cécile Gautrau (fl. 1845) daughter of French naturalist René Primavère Lesson (*Metriopelia).* **2** Cecilia Torres (1952–1985) wife of Brazilian ornithologist Dante Martins Teixeira (*Phylloscartes).*

Cecropis Gr. *Kekropis*, an Athenian woman (*Kekrops,* an ancient king and founder of Cecropia (Athens)) (v. *Atticora).*

cedrorum L. *cedrorum*, of the cedars (*cedrus*, the cedar tree; Gr. *kedros,* the cedar tree).

cela No expl. (Linnaeus (1758), *Systema Naturae*, (Ed.10), Book **1**, p. 191); perhaps a shorthand form from Gr. *kelainos*, black.

celaeno Gr. myth. Celaeno or Kelaino, a daughter of Atlas placed in the stars amongst the Pleiades after her death (cf. *kelainos*, dark-coloured, black with blood).

celaenops Gr. *kelainops*, swarthy, dark.

celata L. *celatus*, secret, hidden (*celare*, to hide, keep secret); ref. concealed crown patch of the Orange-crowned Warbler *Vermivora*.

celebense /celebensis /celebicum After the island of Celebes (Malay *Sulawesi*) in Indonesia.

Celeus Gr. *keleos*, the green woodpecker *Picus*.

cenchroides Syn. specific name *Falco cenchris* (former name for the Lesser Kestrel *F. naumanni)* (Gr. *kenkhris*, a kestrel); Gr.-*oides*, resembling.

centralasicus L. *centralis*, central; *Asius*, Asiatic; Vaurie's Nightjar *Caprimulgus* was described from Sinkiang (Xingjiang Zizhiqu), western China.

Centrocercus Gr. *kentron*, a spike or point; *kerkos*, the tail.

Centropelma Gr. *kentron*, a spike or point; *pelma*, the sole of the foot (i.e. the visible 'leg' of a bird); the Short-winged Grebe *C. micropter*a has two rows of projecting spines (formed by compressed scales) running up the back of the tarsus.

Centropus Gr. *kentron*, a spike, a spur; *pous*, the foot; ref. the long, straight hind-claw of the Palaeotropical coucals.

Centurus Gr. *kentron,* a spike, a point; *oura,* the tail.

Cephalophoneus Gr. *kephale,* the head; syn. genus *Phoneus,* shrike.

Cephalopterus Gr. *kephale*, the head; *pteros,* feathered (*pteron*, a feather); ref the exotic head plumes of the Neotropical umbrellabirds.

Cephalopyrus Gr. *kephale*, the head; *pur, puros*, fire.

cephalotes Gr. *kephale*, the head; *-oides*, resembling; ref. the bushy-crested appearance of the Pale-edged Flycatcher *Myiarchus*.

Cepphus Gr. *kepphos*, a waterbird mentioned by Aristotle, Dionysius, Hesychius, and other authors, not further identified, but later conjectured to be a storm petrel *Hydrobates*, a gull *Larus*, a gannet *Sula*, or, in modern nomenclature, a guillemot.

ceramensis After the island of Ceram or Seran (Seram) in the Moluccas, Indonesia.

Ceratogymna Gr. *keras, keratos*, a horn (i.e. a hornbill); *gumnos*, naked ; ref. the bare faces and throats of the African wattled hornbills.

Ceratopipra Gr. *keras*, a horn; genus *Pipra*, manakin; ref. the double crest of the male Scarlet-horned Manakin *C. cornuta*.

Ceratotriccus Gr. *keras*, a horn; Mod. L. *triccus*, tyrant flycatcher; ref. distinctive curved and forked tail '...almost like the arms (horns) of anchors' of the Fork- tailed Pygmy Tyrant *C. furcatus* (Cabanis (1874), *Journ. für Orn.*, **22**, 87).

Cerchneis Gr. *kerkhneis*, a kestrel *Falco*.

Cercibis Gr. *kerkos*, the tail (i.e. long-tailed); *ibis*, an ibis .

Cercococcyx Gr. *kerkos*, the tail (i.e. long-tailed) ; *kokkux*, the cuckoo.

Cercomacra Gr. *kerkos*, the tail; *makros*, long.

Cercomela Gr. *kerkos*, the tail; *melas*, black.

Cercotrichas Gr. *kerkos*, the tail (i.e. long-tailed); *trikhas*, a thrush, probably the Song Thrush *Turdus philomelos*.

Cereopsis Gr. *kerinos*, waxen (*keros*, wax); *opsis*, the face; the Cape Barren Goose *C. novaehollandiae* has an extensive waxy, greenish-yellow cere almost covering the short bill.

Cerorhinca Gr. *keros*, a horn; *rhunkhos*, the bill; ref. yellow horn at the base of the bill of the Rhinoceros Auklet *C. monocerata*.

cerritus L. *cerritus*, frantic, mad.

Certhia Gr. *kerthios*, a small, tree-dwelling bird mentioned by Aristotle, not otherwise identified, but associated with the treecreeper by subsequent authors. In orn. *certhia* is used for an assortment of brown, curve-billed, but otherwise unrelated species.

Certhiaxis Genus *Certhia*, treecreeper; genus *Synallaxis*, spinetail.

Certhidea Dim. of genus *Certhia*, treecreeper (Gr. dim.*-idion)*.

Certhilauda Genus *Certhia*, treecreeper (ref. curved bill); genus *Alauda*, lark.

certhioides Genus *Certhia*, treecreeper; Gr.*-oides*, resembling; ref. curved bill of the Chaco Earthcreeper *Upucerthia*.

certhiola Dim. of genus *Certhia*, treecreeper; ref. streaked brown plumage of Pallas's Grasshopper Warbler *Locustella*.

Certhionyx Genus *Certhia*, treecreeper; Gr. *onux*, a nail or claw; ref. treecreeper-like claws of the Pied Honeyeater *C. variegatus*.

cerulea L. *caeruleus*, cerulean, azure blue.

cerverai After Fermin Z. Cervera (fl. 1926) Cuban landowner.

cervicalis L. *cervicalis*, cervical, of the neck (*cervix, cervicis*, the neck).

cervinicauda Mod. L. *cervinus*, yellowish-brown (L. *cervinus*, cervine, relating to a stag, stag-coloured; *cervus*, a stag); L. *cauda*, the tail.

cerviniventris Mod. L. *cervinus*, yellowish-brown (L. *cervinus*, cervine, relating to a stag, stag-coloured); L. *venter, ventris*, the belly.

cervinus Mod. L. *cervinus*, yellowish-brown (L. *cervinus*, cervine, relating to a stag, stag-coloured)

Ceryle Gr. *kerulos*, a bird mentioned by Aristotle, Moschus, Hesychius, and other authors, not further identified but probably mythical and associated with the halcyon (v. *Halcyon)*.

cetti/Cettia After Fr. Francesco Cetti (1726–1778) Italian mathematician, naturalist, and author (*Storia naturale della Sardegna*, 1774).

Ceuthmochares Gr. *keuthmos,* a hiding place, a hole; *khairo,* to delight in; the Yellowbill *C. aereus* is a secretive species favouring dense undergrowth.

ceylonensis After Ceylon (Sri Lanka).

Ceyx Gr. *keux*, a seabird considered identical to the halcyon mentioned by Dionysius and Lucian. In Gr. myth. Ceyx, husband to Alcyone, was drowned and metamorphosed into a kingfisher along with his desolated wife.

chabert Malagasy native name *tchachert-bé* for the Black-and-white Vanga *Leptopterus* .

chacoensis After Chaco Province, Argentina.

chacuru Onomatopoeic Güaraní (Paraguayan) Indian name *chacurú* for the White-eared Puffbird *Nystalus* .

Chaetocercus Gr. *khaite*, long flowing hair; *kerkos*, the tail; ref. forked tail feathers of the Rufous-shafted Woodstar *C. jourdanii*.

Chaetops Gr. *khaite*, long flowing hair; *ops*, the face; ref. soft, fluffy plumage and long rictal bristles of the Rufous Rock-jumper *C. frenatus*.

Chaetoptila Gr. *khaite*, long flowing hair; *ptilon*, feather, plumage; ref. decomposed feathers of breast and belly of the extinct Hawaiian Kioea *C. angustipluma*.

Chaetorhynchus Gr. *khaite*, long flowing hair; *rhunkhos,* the bill; ref. long rictal bristles (modified hair-like feathers around the base of the bill) of the Mountain Drongo *C. papuensis*.

Chaetornis Gr. *khaite*, long flowing hair; *ornis*, a bird; ref. strong, protective rictal bristles of the Bristled Grass Warbler *C. striatus*.

Chaetura Gr. *khaite*, long flowing hair; *oura*, the tail; these American swifts have stiff hairs or spines projecting from the ends of the tail feathers.

Chaimarrornis Gr. *kheimarrhos*, a torrent, a mountain stream; *ornis*, a bird.

chalcauchenia Gr. *khalkos,* copper, bronze; *aukhen, aukhenos,* the neck.

Chalcites Gr. *khalkitis*, containing copper (*khalkos*, copper, bronze).

chalcolophus Gr. *khalkolophos*, bronze-crested (*khalkos*, copper, bronze; *lophos*, a crest).

chalcomelas Gr. *khalkos*, copper, bronze; *melas*, black.

Chalcomitra Gr. *khalkos*, copper, bronze; *mitra*, a head-band, a cap.

chalconota/chalconotus Gr. *khalkonotos*, brass-backed.

Chalcophaps Gr. *khalkos*, copper, bronze; *phaps*, a wild pigeon, probably the Wood Pigeon *Columba palumbus*.

Chalcopsitta Gr. *khalkos*, copper, bronze; Mod. L. *psitta*, a parrot (Gr. *psittake*, a parrot).

chalcoptera/chalcopterus Gr. *khalkopteros*, metallic-winged (*khalkos*, copper, bronze; *-pteros*, winged).

chalcospilos Gr. *khalkos*, copper, bronze; *spilos*, a spot.

Chalcostigma Gr. *khalkos*, copper, bronze; *stigme*, a spot, a mark; ref. glittering gorget or beard of the Bronze-tailed Thornbill *C. heteropogon*.

chalcothorax Gr. *khalkos*, copper, bronze; *thorax*, the breast.

chalcurum/chalcurus Gr. *khalkos*, copper, bronze; *-ouros,* tailed.

chalybaeus L. *chalybeius*, steely (Gr. *khalups, khalubos*, steel).

chalybatus Mod. L. *chalybeatus*, steely (L. *chalybeius*, steely; *-atus,* resembling); based on the French name *Le Calybé de la Nouvelle Guinée* given to the Crinkle-collared Manucode *Manucodia* by d'Aubenton (1765), *Les planches enluminéez d'histoire naturelle*.

chalybea L. *chalybeius*, steely (*chalybs*, steel; Gr. *khalups, khalubos*, steel).

chalybeata/chalybeatus see **chalybatus.**

chalybeus see **chalybea.**

Chalybura Gr. *khalups, khalubos*, steel; *oura*, the tail.

Chamaea Gr. *khamai*, on the ground; the distinctive Wrentit *C. fasciata* stays on or near the ground and is more often heard than seen.

Chamaepetes Gr. *khamaipetes*, falling to the ground, prostrate; the Sickle-winged Guan *C. goudotii* is mainly arboreal, but sometimes seeks escape on the ground.

Chamaethlypis Gr. *khamai*, on the ground; Mod. L. *thlypis*, a warbler (Gr. *thlupis*, a form of *thraupis*, a type of finch mentioned by Aristotle, not further identified. In orn. *thlypis* is used for the American warblers (Parulidae) and warbler-billed tanagers (Thraupidae)).

Chamaeza Gr. *khamaze*, on the ground; the ant-thrushes are terrestrial.

chapalensis After Lake Chapala, Jalisco state, Mexico.

chapini After James P. Chapin (1889–1964) US ornithologist, collector in the Belgian Congo (Zaïre), curator American Museum of Natural History, and author (*Birds of the Belgian Congo*, 1932).

chaplini After Sir Francis Drummond Percy Chaplin (1866–1933) British colonial administrator in Rhodesia (Zimbabwe) and Nyasaland (Malawi).

chapmani After Frank Michler Chapman (1864–1945) US ornithologist and author (*The distribution of bird life in Colombia*, 1917; *The distribution of bird life in Ecuador*, 1926).

Chaptia Nepalese name *chaptia* for the Bronzed Drongo *Dicrurus aeneus*.

Charadrius Late L. *charadrius*, a yellowish bird mentioned in the Vulgate Bible (AD 383) (Gr. *kharadrios*, a nocturnal waterbird, the sight of which was supposed to cure jaundice, mentioned by Aristotle, Aristophanes, and Suidas, not further identified, but later conjectured to be the Stone Curlew *Burhinus oedicnemus).*

chariessa Gr. *khariessa*, graceful, beautiful.

Charitornis Gr. *kharis, kharitos*, grace, loveliness; *ornis*, a bird.

Charitospiza Gr. *kharis,* grace, loveliness; *spiza*, a finch.

charlottae No expl. (Finsch (1867), *Journ. für Orn.*, **15**, 19); a female eponym.

charltonii After Lieutenant-Colonel Andrew Charlton (1803–1888) of the Bengal Light Infantry.

Charmosyna Gr. *kharmosunos*, glad, joyful.

Chasiempis Gr. *khasma*, the open mouth, yawning chasm (*khaino*, to open wide, to gape); *empis*, a mosquito or gnat.

chathamensis After the Chatham Islands in the Pacific Ocean east of South Island, New Zealand, named for HMS *Chatham*, commanded by Lieutenant Broughton, who discovered the islands in 1791.

Chaulelasmus Gr. *khaulios,* prominent, projecting; *elasmos,* a plate, lamella; ref. the denticulations of the bill of the Gadwall *Anas strepera.*

Chauna Gr. *khaunos*, silly, foolish; ref. the noisy behaviour and clumsy appearance of the Neotropical screamers.

Chaunonotus Gr. *khaunos,* loose; *-notos,* backed (*noton,* the back).

Chaunoproctus Gr. *khaunos*, loose, flaccid; *proktos*, the tail or back (cf. *khaunoproktos*, wide trousered); no expl. (Bonaparte (1851), *Consp. gen. avium*, Vol.**1**, p. 526), but doubtless ref. the soft, full feathered rump of the Bonin Island Grosbeak *C. ferreirostris.*

chavaria Tupí (Brazilian) Indian name *javariá* for some sort of bird, perhaps a woodpecker.

cheela Hindi name *cheel* for a kite or hawk.

cheleensis After Petcheli Bay, near Port Arthur, Manchuria (Ta-lien Bay, near Lü-ta, Liaoning Province, China).

Chelictinia Gr. *khelidon*, a swallow; *iktinos*, a kite.

chelicuti After Chelicut (?village), Abyssinia (Ethiopia).

Chelidoptera Gr. *khelidon*, a swallow; *ptera*, wings (*pteron*, a feather or wing).

Chelidorhynx Gr. *khelidon*, a swallow; *rhunkhos*, the bill; ref. the very short, broad, swallow-like bill of the Yellow-bellied Fantail *Rhipidura hypoxantha*.

Chen Gr. *khen*, a goose.

cheniana Probably adapted from some native African name like Singuni *tsiyana* for a small brown warbler, but no expl. (Smith (1843), *Illustr. zool. south Africa, aves*, pl. 89, fig. 2).

Cheniscus Dim. of syn. genus *Chen*, goose.

Chenonetta Syn. genus *Chen*, goose; genus *Netta*, duck; ref. the goose-like appearance of the Australian Maned Duck *C. jubata*.

Chenorhamphus Gr. *khen*, a goose; *rhamphos*, the bill; ref. broad bills of the New Guinea wrens.

Cheramoeca Gr. *kheramos*, a hole or hollow; *-oika*, living (*oikeo*, to inhabit); the White-backed Swallow *C. leucosternum* burrows in the ground and builds its nest at the end of a tunnel.

cherina No expl. (Smith (1843), *Illustr. zool. south Africa, aves*, pl. 77, fig. 2 and text); perhaps based on Gr. *khera*, a widow, ref. melancholy callnotes, or some unrecorded native name.

cheriway Carib (Venezuelan) Indian name *cheriway* for the noisy and familiar Crested Caracara *Polyborus*.

chermesina Mod. L. *chermesinus*, carmine red (Persian *qirmiz*, a red dyestuff).

cherriei After George K. Cherrie (1865–1948) US field ornithologist, collector in Costa Rica 1894–1897, and author.

cherrug Hindi name *charg* for the female Saker Falcon *Falco* (the tiercel, sakret or male being known as *chargela)*.

Chersomanes Gr. *khersos*, a desert, wasteland; *-manes*, passionately fond of.

Chersophilus Gr. *khersos*, a desert, a wasteland; *philos*, loving, fond of.

Chettusia No expl. (Bonaparte (1841), *Icon. faun. Ital. uccelli*, Intr., p. 12), but it is doubtless his Italian rendering of the Russian name *keptuschka* for the Sociable Plover *C. gregaria*. The modern Italian name *chettusia* for a plover appears to date only from Bonaparte's usage. The name has been amended to *Chaetusia* by some authors, as if from Gr. *khaite*, long flowing hair.

chiapensis After the state of Chiapas, Mexico.

Chibia Nepalese name *chibey* for the Hair-crested Drongo *Dicrurus hottentottus*.

chicquera French name *Le Chicquera* given to the Red-headed Falcon *Falco* by Levaillant (1799) (Hindi names *shikra* or *shikara* for a hawk or hunter).

chiguanco Aymará (Peruvian) Indian name *chiguanco* for the Chiguanco Thrush *Turdus*.

chihi Onomatopoeic French name *chihi* given to the White-faced Ibis *Plegadis* by Vieillot (1817), *Nouv. Dict. Hist. Nat.*, Vol. **8**, p. 303.

chilensis After Chile; coined erroneously in the case of *Tangara* .

Chilia After Chili, a former spelling of Chile, to which country the Crag Chilia *C. melanura* is confined.

chimachima Onomatopoeic local Argentinian name *chimachima* for the Yellow-headed Caracara *Milvago* .

chimaera Gr. myth. Chimaera or Khimaira, the fabulous Lycian monster of varied and hideous aspect; ref. variegated, hybrid appearance of the Long-tailed Ground-roller *Uratelornis*.

chimango Local Argentinian name *chimango* for the Chimango Caracara *Milvago* .

chimborazo After Mount Chimborazo, an inactive volcano in Chimborazo province, Ecuador.

chinchorrensis After the Chinchorro Bank, Quintana Roo, Mexico.

chinensis After China (named after the first imperial dynasty, the Ch' in).

chiniana No expl. (Smith (1843), *Illustr. zool. south Africa, aves*, pl. 79, and text), but probably based on a native name such as Singuni *tsiyana* for a small brown warbler.

Chionis Gr. *khion, khionos*, snow; ref. the snow-white plumage of the sheathbills.

chionogaster Gr. *khion*, snow; *gaster*, the belly.

chionopectus Gr. *khion*, snow; L. *pectus*, the breast.

chionura Gr. *khion*, snow; *-ouros*, tailed.

chirindensis After the Chirinda Forest, Gazaland, Southern Rhodesia (Zimbabwe).

chiriquensis After **1** Volcán de Chiriquí, Chiriquí province, Panama (*Geothlypis*). **2** Chiriquí province, New Granada (Panama) (*Elaenia*).

chiriri Güaraní (Paraguayan) Indian name *chiriri* for the Yellow-chevroned Parakeet *Brotogeris*.

Chirocylla Gr. *kheir, kheiros*, the hand or arm; *kullos*, curved, bent; ref. the short, recurved primaries of the Scimitar-winged Piha *C. uropygialis*. Genus amended by purists to *Chiroxilla*, as if a dim, from Gr. *kheir* and *oxus*, sharp, pointed.

Chiroxiphia Gr. *kheir*, the hand or arm (i.e. the primary feathers); *xiphion*, a small sword, a dagger (dim. of *xiphos*, a sword); ref. the modified primary feathers of the Swallow-tailed Manakin *C. caudata*.

chirurgus Med. L. *chirurgus*, a surgeon; from the French name *le chirurgien* given to the jacanas by Brisson (1760) with ref. to their scalpel-like carpal spurs.

chivi French onomatopoeic name *la fauvette chivi* given to the Chivi Vireo *Vireo* by Vieillot (1817), *Nouv. Dict. Hist. Nat.*, Vol. **11**, p. 174.

Chlamydera Gr. *khlamus*, a short cloak; *dere*, the neck; ref. fan-shaped nuchal crest of the Spotted Bowerbird *C. maculata*.

Chlamydochaera Gr. *khlamus, khlamudos*, a short cloak or mantle; *khairo*, to rejoice in, to welcome; the Black-breasted Triller *C. jefferyi* has a pearl-grey mantle.

Chlamydotis Gr. *khlamus, khlamudos*, a horseman's short cloak with weights sewn into the corners; *otis*, a bustard; ref. the fringed neck crests or frills of the Houbara Bustard *C. undulata*.

Chlidonias Gr. *khelidonios*, swallow-like (*khelidon*, a swallow); terns were formerly known as sea-swallows.

Chloebia Gr. *khloe*, grass (strictly the tender shoots of young grass); *bios*, life, livelihood; the Gouldian Finch *C. gouldiae* occurs in flat grasslands of northern Australia.

Chloephaga Gr. *khloe*, grass; *phagos*, a glutton (*phagein*, to eat); the sheldgeese feed on grass and seaweed.

Chlorestes Gr. *khloros*, green; *esthes*, dress, clothes.

chloricterus Gr. *khloros*, green; Mod. L. *icterus*, yellow.

Chloridops Syn. genus *Chloris*, greenfinch (Gr. *khloris*, the greenfinch); Gr. *ops*, the face; ref. the heavy finch-like bill and olive green plumage of the extinct Kona Grosbeak *C. kona*.

chlorigula Gr. *khloreis*, pale green; L. *gula*, the throat.

chloris Gr. *khloris*, the greenfinch (i.e. green or yellowish-green).

chlorocephalus Gr. *khloros*, green or yellow; *-kephalos*, headed.

chlorocercus Gr. *khloros*, green or yellow; *kerkos*, the tail.

Chloroceryle Gr. *khloros*, green; genus *Ceryle*, kingfisher.

Chlorocharis Gr. *khloros*, green; *kharis*, grace, loveliness.

Chlorochrysa Gr. *khloros*, green; *khrusos*, gold.

Chlorocichla Gr. *khloros*, green or yellow; *kikhle*, a thrush.

chlorolepidota/chlorolepidotus Gr. *khloros*, green or yellow; *lepidotos*, scaly (*lepis, lepidos*, a scale).

chlorolophus Gr. *khloros*, green or yellow; *-lophos*, crested (*lophos*, a crest).

chloromeros Gr. *khloros*, green or yellow; *meros*, the thigh.

chloronota Gr. *khloros*, green or yellow; *-notos*, backed .

chloronothos Gr. *khloros*, green or yellow; *nothos*, adulterated, spurious; ref. plain olive plumage of the Mauritius Olive White-eye *Zosterops*.

chloronotus see **chloronota.**

Chloropeta Gr. *khloros*, green or yellow; Mod. L. *petes*, a bird (Gr. *petes*, flying).

chloropetoides Genus *Chloropeta*, yellow warbler; Gr. *-oides*, resembling.

chlorophaea Gr. *khloros*, green; *phaea*, the eyes; ref. pale green, bare orbital patch of the Little Malkoha *Rhinortha*.

Chlorophanes Gr. *khloros*, green; *-phanes*, shining, appearing (*phaino*, to shine forth).

Chlorophoneus Gr. *khloros*, green; syn. genus *Phoneus*, shrike (Gr. *phoneus*, a killer; *phonos*, murder).

Chlorophonia Gr. *khloros*, green; genus *Euphonia*, euphonia.

Chloropipo Gr. *khloros*, green; Mod. L. *pipo*, manakin (Gr. *pipo*, a form of *pipra*, a kind of bird mentioned by Aristotle and Nicander, identified by later authors as a pied woodpecker *Dendrocopos*. In orn. used only once in the latter sense (v. *Sapheopipo)*, but otherwise associated with the manakins as a synonym of *pipra)*.

Chloropsis Gr. *khloros*, green; *opsis*, appearance.

chloroptera /chloropterus Gr. *khloros*, green or yellow; *-pteros*, winged.

chloropus Gr. *khloros*, green or yellow; *pous*, the foot.

chloropygia Gr. *khloros*, green or yellow; Mod. L. *pygius*, rumped (Gr. *-pugos*, rumped; *puge*, the rump).

chlororhynchos /chlororhynchus Gr. *khloros*, green or yellow; *rhunkhos*, the bill.

Chlorornis Gr. *khloros*, green; *ornis*, a bird.

Chlorospingus Gr. *khloros*, green; *spingos*, a form of *spinos*, probably identical to *spiza*, a common sort of finch identified by most authors as the Chaffinch *Fringilla coelebs*. In orn. *spingos* and *spiza* are not confined to the finches (Fringillidae) but used indiscriminately for an assortment of finch-billed or finch-like birds (e.g. Thraupidae, Emberizidae, Cardinalidae, Ploceidae, Drepanididae).

Chlorostilbon Gr. *khloros*, green; *stilbon*, glistening (*stilbo*, to glitter).

Chlorothraupis Gr. *khloros*, green; genus *Thraupis*, tanager.

chlorotica Mod. L. *chloroticus*, greenish (Gr. *khlorotes*, greenness).

Chlorura /chlorurus Gr. *khloros*, green; *oura*, the tail.

chocolatina /chocolatinus Mod. L. *chocolatinus*, chocolate brown-coloured (Spanish *chocolate*, chocolate; Nahuatl Indian *chocolatl*, a food made from cacao and the pochotl tree).

choiseuli After Choiseul Island in the Solomon Islands, named for French statesman Étienne François Duc de Choiseul (1719–1785) .

choliba Local Aragonese (Spanish) name *chóliba* for the European Scops Owl *Otus scops*, given to the Tropical Screech Owl *Otus* by Azara (1805), because he considered the two species almost identical.

choloensis After the Cholo Mountains, Nyasaland (Malawi).

Chondestes Gr. *khondros*, grain, seed; *edestes*, an eater (*edo*, to devour).

Chondrohierax Gr. *khondros*, grain, seed; *hierax*, a hawk; the Hook-billed Kite *C. uncinatus* does not occur in cornfields but in open wet woods, swamps, reedbeds and plantations.

chopi Onomatopoeic Güaraní (Paraguayan) Indian name *chopi* for the Chopi Blackbird *Gnorimopsar*.

Chordeiles Gr. *khoreia*, dancing; *deiles*, in the evening; ref. the bounding flight and nocturnal habits of the Common Nighthawk *C. minor*.

Choriotis Specific name *Ardeotis kori* ; Gr. *otis*, bustard.

christinae After Christina Swinhoe (fl. 1869) wife of English naturalist Robert Swinhoe.

christophori After St Christopher Island (Spanish *San Cristóbal)* (Makira) in the Solomon Islands.

chrysaea Late L. *chryseus*, golden (Gr. *khruseos*, golden).

chrysaetos Gr. *khrusaetos*, the Golden Eagle *Aquila* (*khrusos*, gold; *aetos*, an eagle).

chrysaeus see **chrysaea.**

chrysater Late L. *chryseus*, golden; L. *ater*, black.

chrysauchen Gr. *khrusos*, gold; *aukhen*, the neck or throat.

chrysia see **chrysaea.**

chrysocaulosus Gr. *khrusos*, gold; *kaulos*, a spear shaft (i.e. the wing-shafts).

chrysocephalum /chrysocephalus Gr. *khrusokephalos*, golden-headed (*khrusos*, gold; *-kephalos*, headed).

chrysochloros Gr. *khrusos*, gold; *khloros*, green or yellow.

Chrysococcyx Gr. *khrusos*, gold; *kokkux*, the cuckoo.

Chrysocolaptes Gr. *khrusos*, gold; genus *Colaptes*, flicker, woodpecker.

chrysocome Gr. *khrusokomos*, golden-haired (*khrusos*, gold; *kome*, the hair) (cf. *khrusokomes,* the golden-haired, an epithet of Apollo).

chrysoconus Gr. *khrusos*, gold; L. *conus,* the apex of a helmet (i.e. the crown).

chrysocrotaphum Gr. *khrusos*, gold; *krotaphos*, the sides of the head, the temples.

Chrysoena Gr. *khrusos*, gold; *oinas*, a kind of pigeon, perhaps the Rock Dove *Columba livia*.

chrysogaster Gr. *khrusos*, gold; *gaster*, the belly.

chrysogenys Gr. *khrusos*, gold; *genus*, the cheek.

chrysoides Gr. *khrusoeides*, like gold, golden.

Chrysolampis Gr. *khrusolampis*, the glow-worm.

chrysolaus Gr. *khrusos*, gold; *lais*, a kind of thrush mentioned by Aristotle, not further identified, but conjectured to be a rock thrush *Monticola*.

Chrysolophus Gr. *khrusolophos*, golden-crested (*khrusos*, gold; *lophos*, a crest).

chrysomela /chrysomelas Gr. *khrusos*, gold; *melas*, black.

Chrysomma Gr. *khrusos*, gold; *omma*, the eye (*ommai*, to see) (cf. *khrusoma*, gold wrought).

chrysonotus Gr. *khrusonotos*, golden-backed (*khrusos*, gold; *-notos*, backed).

chrysopareia Gr. *khrusos*, gold; *-pareos*, cheeked .

chrysopasta Gr. *khrusopastos*, shot with gold, golden.

chrysopeplus Gr. *khrusopeplos,* golden-robed (*khrusos*, gold; *peplos*, a robe).

chrysophrys Gr. *khrusophrus*, golden-browed (originally a type of fish with a golden spot over each eye) (*khrusos*, gold; *ophrus*, eyebrow).

chrysopogon Gr. *khrusos*, gold; *pogon*, the beard.

chrysops Gr. *khrusops*, gold-coloured, shining like gold.

chrysoptera /chrysopterus Gr. *khrusopteros*, golden-winged (*khrusos*, gold; *-pteros*, winged).

chrysopterygius Gr. *khrusos*, gold; *pterux*, a wing.

Chrysoptilus Gr. *khrusos*, gold; *ptilon*, a feather.

chrysorrheum Gr. *khrusos*, gold; *orrhos*, the rump or vent, the area where the body meets the tail.

chrysorrhoa Gr. *khrusos*, gold; *orrhos*, the rump (cf. *khrusorrhoes*, streaming with gold).

chrysostoma Gr. *khrusostomos*, golden-mouthed (i.e. golden-billed) (*khrusos*, gold; *stoma*, the mouth or tongue).

Chrysothlypis Gr. *khrusos*, gold; Mod. L. *thlypis*, a warbler .

chrysotis Gr. *khrusos*, gold; *-otis*, eared.

Chrysotrogon Gr. *khrusos,* gold; genus *Trogon*, trogon.

chrysura Gr. *khrusos*, gold; *-ouros*, tailed.

Chrysuronia Apparently an error for *Chrysuriscus* (dim. from Gr. *khrusos*, gold; *oura*, the tail) (cf. *ouron*, a boundary).

chthonia Gr. *khthonios*, in the earth (i.e. terrestrial).

Chthonicola Gr. *khthon, khthonos*, the earth, the ground; L. *-cola*, dweller.

chuana After Bechuanaland (Botswana), named after the native Chuana peoples.

chubbi After Ernest Charles Chubb (b. 1884) South African zoologist and collector.

Chubbia After Charles Chubb (1851–1924) British ornithologist at the British Museum (Natural History), 1877–1920.

chukar Onomatopoeic Hindi name *chukor* for the Chukar Partridge *Alectoris*.

chunchotambo After the Chuncho Indians of Peru in whose territory the Chunchos Woodcreeper *Xiphorhynchus* was first found.

Chunga Local Argentinian name *chunga* for the Black-legged Seriema *C. burmeisteri* .

cia Local Genovese (Italian) name *cia* for the Rock Bunting *Emberiza*, being imitative of its callnotes.

Ciccaba Gr. *kikkabe*, a sort of owl mentioned by Aristophanes and Hesychius, perhaps the Little Owl *Athene noctua*.

Cichladusa Gr. *kikhle*, a thrush; *adousa*, a singer.

Cichlherminia Gr. *kikhle*, a thrush; after Félix Louis l'Herminier (1779–1833) French naturalist exiled to Guadeloupe in the West Indies.

Cichlocolaptes Gr. *kikhle*, a thrush; genus *Colaptes*, woodpecker; ref. brown plumage and forest habitat of the Pale-browed Treehunter *C. leucophrus*.

Cichlopasser Gr. *kikhle*, a thrush; L. *passer*, the sparrow or any small bird.

Cichlopsis Gr. *kikhle*, a thrush; *opsis*, appearance.

Cichlornis Gr. *kikhle*, a thrush; *ornis*, a bird.

Cicinnurus Gr. *kikinnos*, a curled lock of hair; *-ouros*, tailed; ref. the racquet-tipped tail-wires of the King Bird of Paradise *C. regia*.

Ciconia L. *ciconia*, the stork.

cinchoneti No expl. (Tschudi (1844), *Archiv für Naturgeschichte*, **10** [1], 272).

Cinclidium Dim. of Mod. L. *cinclus*, a thrush (Gr. dim. *-idion)*.

Cinclocerthia Mod. L. *cinclus*, a thrush; genus *Certhia*, treecreeper .

Cinclodes Genus *Cinclus*, dipper; Gr. *-odes*, excess (i.e. close resemblance).

Cincloramphus Mod. L. *cinclus*, thrush; Gr. *rhamphos*, the bill.

cinclorhynchus Mod. L. *cinclus*, thrush; Gr. *rhunkhos*, the bill; the Blue-capped Rock Thrush *Monticola* was '...observed to bear an affinity to the family of Merulidae by the strength and thrush like form of the bill' (Vigors (1832), *Proc. Zool. Soc. London*, 172).

Cinclosoma Mod. L. *cinclus*, thrush; Gr. *soma*, the body (i.e. appearance).

Cinclus Gr. *kinklos*, a small waterside bird mentioned by Aristotle, Aristophanes, Aelianus, and other authors, never satisfactorily identified, but conjectured to be a wagtail *Motacilla* or a sandpiper *Actitis*. In orn. *cinclus* is used for a great variety of otherwise unrelated thrush-like or dipper-like birds (e.g. Furnariidae, Dendrocolaptidae, Mimidae, Pachycephalidae).

cincta L. *cinctus*, girdled, encircled, wreathed (*cingere*, to encircle).

cincturus L. *cinctus*, girdled (i.e. banded); Gr. *-ouros*, tailed (cf. L. *cinctura*, a girdle).

cinctus see **cincta.**

cineracea/cineraceus L. *cineraceus*, ash-grey, ashy (*cinis, cineris*, ashes).

cinerascens Late L. *cinerescens*, ashen (*cinerescere*, to turn to ashes; L. *cinis*, ashes)

cinerea L. *cinereus*, ash-coloured, ash-grey (*cinis*, ashes).

cinereicapillus L. *cinereus*, ash-coloured; *-capillus*, capped .

cinereicauda L. *cinereus*, ash-coloured; *cauda*, the tail.

cinereiceps L. *cinereus*, ash-coloured; *-ceps*, crowned, headed.

cinereicollis L. *cinereus*, ash-coloured; Mod. L.*-collis*, necked.

cinereifrons L. *cinereus*, ash-coloured; *frons*, the forehead, the brow.

cinereigulare L. *cinereus*, ash-coloured; Mod. L. *gularis*, throated, of the throat (L. *gula*, the throat).

cinereiventris L. *cinereus*, ash-coloured; *venter, ventris*, the belly.

cinereocapilla see **cinereicapillus.**

cinereocephalus L. *cinereus*, ash-coloured; Gr. *-kephalos*, headed .

cinereola Dim. from L. *cinereus*, ash-coloured.

cinereovinacea L. *cinereus*, ash-coloured; *vinaceus*, vinaceous, wine-coloured.

cinereum /cinereus see **cinerea.**

cinnamomea Mod. L. *cinnamomeus*, cinnamon-coloured (L. *cinnamum*, cinnamon).

cinnamomeipectus Mod. L. *cinnamomeus*, cinnamomeous; L. *pectus*, the breast.

cinnamomeum /cinnamomeus see **cinnamomea.**

cinnamomeiventris Mod. L. *cinnamomeus*, cinnamomeous; L. *venter, ventris*, the belly.

cinnamomina L. *cinnamum*, cinnamon; *-inus*, possessing, resembling.

Cinnamopteryx Gr. *kinnamomon*, cinnamon; *pterux*, a wing.

Cinnycerthia Syn. genus *Cinnyris*, sunbird; genus *Certhia*, treecreeper; ref. brown, creeper-like plumage and curved bill of the Rufous Wren *C. unirufa*.

Cinnyricinclus Syn. genus *Cinnyris*, sunbird; Mod. L. *cinclus*, thrush; ref. the brilliant, sunbird-like plumage of the Amethyst Starling *C. leucogaster* (the glossy starlings are called *merles bronzés*, bronzed thrushes, by the French) .

Cinnyris Gr. *kinnuris*, a small bird mentioned by Hesychius, not further identified.

cioides Specific name *Emberiza cia* ; Gr. *-oides*, resembling.

Circaetus Genus *Circus*, harrier; Gr. *aetos*, an eagle.

circumcinctus L. *circumcinctus*, enclosed, surrounded (*circumcingere*, to enclose, to surround); ref. spotted nuchal collar and wings of the Spot-winged Falconet *Spiziapteryx*.

Circus Gr. *kirkos*, a partly mythical hawk mentioned by many classical authors and later identified with the Hen Harrier *C. cyaneus*.

Ciridops Specific name *Passerina ciris* ; Gr. *-ops*, appearance, colouring; the Ula-ai-hawane *C. anna* was '...so named because its bright coloration recalls the well known *Emberiza ciris* of Linnaeus, the Painted Bunting of authors, or "Nonpareil" of bird dealers' (Newton (1892), *Nature*, **45**, 469).

ciris L. *ciris*, a mythical bird (Gr. *kiris* or *kirris*, a mythical bird mentioned by Dionysius and Hesychius; Ciris or Kiris was the mysterious bird into which Scylla was metamorphosed).

cirlus Local Bolognese (Italian) name *cirlo* for some sort of bunting, perhaps the Rock Bunting *Emberiza cia* or the Yellowhammer *E. citrinella*.

cirratus /cirrhata /cirrhatus L. *cirratus*, curly headed, having ringletted hair (*cirrus*, a curl of hair) (cf. Med. L. *cirratus*, crested).

cirrhocephalus Gr. *kirrhos*, tawny ; *-kephalos*, headed, crowned .

cirrhochloris Mod. L. *cirrhus*, grey (from a supposed Gr. *kirrhos*, grey); Gr. *khloris*, green.

cirrocephalus Mod. L. *cirrhus*, grey (from a supposed Gr. *kirrhos*, grey); Gr. *-kephalos*, headed .

Cissa Gr. *kissa*, the Acorn Jay *Garrulus glandarius*. Thompson (1895), p. 85, considered that the ancient Greeks also applied this name, as well as the form *kitta*, to the Northern Magpie *Pica pica*, and in orn. it has been used indiscriminately for jay and magpie.

Cissilopha Gr. *kissa*, the jay; *lophos*, a crest.

Cissomela Genus *Cissa*, magpie; genus *Myzomela*, honeyeater; ref. pied, magpie-like plumage of the Banded Honeyeater *C. pectoralis*.

Cissopis Gr. *kissa*, the magpie; *-ops*, appearance; from the pied plumage and long tail of the Magpie Tanager *C. leveriana*.

Cisticola Botanical genus *Cistus*, rock rose (Gr. *kistos*, a flowering shrub, cistus); L. *-cola*, a dweller.

Cistothorus Gr. *kistos*, a shrub; *thouros*, leaping (*throsko*, to leap).

citrea L. *citreus*, citrine, belonging to the citrus tree (*citrus, citri*, the citrus tree).

citreogularis L. *citreus*, citrine; Mod. L. *gularis*, throated L. *gula*, the throat).

citreola/citreolus Dim. from L. *citreus*, citrine, yellow.

citrina L. *citrinus*, citrine, of the citrus tree (i.e. yellow).

citrinella Italian name *citrinella* for a small yellow bird (dim. of L. *citrinus*, citrine).

citrinelloides Specific name *Serinus citrinella* ; Gr. *-oides*, resembling.

citrinellus see **citrinella.**

citrinipectus L. *citrinus*, citrine; *pectus*, the breast.

citriniventris L. *citrinus*, citrine; *venter, ventris*, the belly.

citrinocristatus L. *citrinus*, citrine (*citrus*, the citrus tree); *cristatus*, crested.

Cittura Gr. *kitta*, the jay or magpie (v. *Cissa)*; *oura*, the tail.

Cladorhynchus Gr. *klados*, a twig or thin branch (i.e. thin); *rhunkhos*, the bill (cf. *kladarorhunkhos*, the clapper-bill; *kladaros*, flexible, pliant; *rhunkhos*, the bill).

clamans L. *clamans*, shouting, clamorous (*clamare*, to shout).

Clamator L. *clamator*, a shouter (*clamare*, to shout); also used for a type of screech owl mentioned by Pliny as a bird of ill omen.

clamosa/clamosus L. *clamosus*, noisy, clamorous (*clamare*, to shout).

clanga Gr. *klangos* (a variant form of *plangos)*, a kind of eagle mentioned by Aristotle.

Clangula Dim. from L. *clangere*, to resound; ref. whistling wings of the Long-tailed Duck or Oldsquaw *C. hyemalis* and the Common Goldeneye *Bucephala*.

clappertoni After Captain Hugh Clapperton (1788–1827) Scottish explorer and collector in tropical Africa, 1822–1827.

clara L. *clarus*, clear, brilliant, distinct.

clarae No expl. (Blasius (1890), *Journ. für Orn.*, **38**, 144–8); a female eponym.

Claravis L. *clarus*, distinct, brilliant; *avis*, a bird.

clarisse After Clarisse Parzudaki (fl. 1841) wife of French collector Charles Parzudaki.

clarki/clarkii After **1** J. H. Clark (b. 1830) US surveyor, boundary commissioner, naturalist, and collector (*Aechmophorus)*. **2** Harry M. Clark (fl. 1935) US naturalist (*Otus*).

clarus see **clara.**

clathratus L. *clathri*, a lattice (Gr. *klethra*, a lattice, a bar); *-atus*, possessing; ref. barred tail pattern of the Lattice-tailed Trogon *Trogon*.

cleaveri After Mr Cleaver (fl. 1874) who collected on the Gold Coast (Ghana).

clemenciae After Clémence Lesson (fl. 1825) wife of French ornithologist and author René Primavère Lesson.

Cleptornis Gr. *kleptes*, a robber (*klepto*, to steal); *ornis*, a bird; ref. the French name *Iles des Larrons*, Robbers Islands, for the Marianas in the Pacific; the Golden White-eye *C. marchei* is confined to the islands of Saipan and Agiguan in the Marianas.

Clibanornis Gr. *klibanos*, an oven; *ornis*, a bird; although the nest of the Canebrake Ground-creeper *C. dendrocolaptoides* is unknown this name reflects its affinity to the ovenbirds (v. *Furnarius)*.

climacocerca Gr. *klimax, klimakos*, a ladder; *kerkos*, the tail.

Climacteris Gr. *klimakter, klimakteros*, the step of a staircase or rung of a ladder; the Australian treecreepers are well adapted to tree climbing.

climacurus Gr. *klimax*, a ladder; *-ouros*, tailed .

clotbey After Antoine-Barthélmy Clot (1793–1868) French physician and author in Egypt, given the Ottoman honorary title *bey* by Muhammad Ali in 1832 for services to medicine.

clypeata L. *clypeatus*, shield-bearing (*clypeare*, to arm with a shield; *clypeus*, a shield); ref. the bill of the Northern Shoveler *Anas* 'much broader towards the tip than at the base, excavated like a buckler, of a round circumference' (Ray (1678), p. 370).

Clypicterus L. *clypeus*, a shield; genus *Icterus*, American oriole; the Casqued Oropendola *C. oseryi* has a knobbed frontal shield.

Clytoceyx Gr. *klutos*, splendid, noble; genus *Ceyx*, kingfisher.

Clytoctantes Gr. *klutos*, splendid, famous; *ktantes*, a murderer (i.e. a shrike); ref. the dagger-shaped bill and shrike-like appearance of the Recurve-billed Bushbird *C. alixii*.

Clytolaema Gr. *klutos*, splendid, beauteous; *laimos*, the throat.

Clytomyias Gr. *klutos*, splendid, noble; Mod. L. *myias*, flycatcher .

Clytorhynchus Gr. *klutos*, splendid, noble; *rhunkhos*, the bill.

Clytospiza Gr. *klutos*, splendid, beauteous; *spiza*, a finch.

Cnemarchus Gr. *knemos*, a mountain slope; *arkhos*, a ruler or chief.

Cnemophilus Gr. *knemos*, a mountain slope; *philos*, loving, fond of.

Cnemoscopus Gr. *knemos*, a mountain slope; *skopos*, a watcher, a seeker (*skeptomai*, to watch, to examine).

Cnemotriccus Gr. *knemos*, a mountain slope; Mod. L. *triccus*, tyrant flycatcher.

Cnipodectes Gr. *knips, knipos*, an insect; *dektes*, a biter (*dakno*, to bite).

coccinea /coccineus L. *coccineus*, scarlet-coloured (Gr. *kokkinos*, scarlet).

coccinicollaris L. *coccineus*, scarlet-coloured; *collaris*, collared.

coccinigastra L. *coccineus*, scarlet-coloured; *gaster, gasteris*, the belly.

Coccopygia Gr. *kokkos*, a scarlet berry, the kermes oak berry; *puge*, the rump.

Coccothraustes Gr. *kokkothraustes*, a kind of bird mentioned by Hesychius, not further identified (*kokkos*, a kernel; *thrauo*, to break or shatter).

Coccycolius Gr. *kokkux*, the cuckoo; Mod. L. *colius*, starling; the Emerald Starling *C. iris* was thought to closely resemble the bronze cuckoos.

Coccyzus Gr. *kokkuzo*, to cry cuckoo (*kokkux*, the cuckoo).

cochinchinensis After Cochin-China, a former province of French Indochina, now part of Vietnam.

Cochlearius L. *cochlearium*, a spoon; ref. bill shape of the Boat-billed Heron *C. cochlearius*.

Cochoa Nepalese name *cocho* for the Purple Cochoa *C. purpurea*.

cockerelli After J. Cockerell (fl. 1895) Australian collector in Samoa, 1874 and the Bismarck Archipelago, 1876.

cocoi Cayenne Indian names *cuca* or *cocoi* for the Cocoi Heron *Ardea* (cf. Tupí Indian name *socó* for a heron).

coelebs L. *caelebs*, single, unmarried; according to Linnaeus (1758), *Systema Naturae*, (Ed. 10), book 1, p. 179, the female Chaffinches *Fringilla* alone migrate, leaving the cocks to lead a bachelor existence. Here, and in the following entries, authors have been guilty of incorrectly separating the ligature *œ* as *oe* instead of *ae*, although the two were frequently regarded as interchangeable.

coelestis L. *caelestis*, celestial, heavenly, glorious.

coelicolor Mod. L. *caelicolor*, sky-blue (L. *caelum*, the heavens; *color*, colour).

Coeligena L. *caeligenus*, heaven-born, celestial.

Coenocorypha L. *coenum* (an incorrect form of *caenum)*, mud, mire; *coryphaeus*, a leader or chief.

Coereba Tupí (Brazilian) Indian name *güirá coereba* for some small blue, black and yellow bird.

coeruleicinctus L. *caeruleus*, azure blue; *cinctus*, girdled, banded (*cingere*, to encircle).

coeruleogularis L. *caeruleus*, azure blue; Mod. L. *gularis*, throated, of the throat (L. *gula*, the throat).

coerulescens see **caerulescens.**

cognita L. *cognita*, known (i.e. discovered) (*cognoscere*, to learn) (cf. *cognata*, related).

Colaptes Gr. *kolaptes*, a chiseller (i.e. a woodpecker) (*kolapto*, to chisel or pick away).

colchicus L. *Colchicus*, of Colchis, a country on the eastern shores of the Black Sea (the modern republic of Georgia, USSR) where the Argonauts found many pheasants.

colensoi After the Revd. William Colenso (1811–1899) naturalist, ethnologist, philologist, and missionary to New Zealand.

Colibri Spanish name *colibri* for a hummingbird (apparently originally from a Carib Indian name).

Colinus Nahuatl (Mexican) Indian names *colin* or *zolin* (variants include *acolin, ococolin* and *zonecolin)* for a quail.

Colius Gr. *koloios*, the Jackdaw *Corvus* (cf. syn. genus *Coloeus,* jackdaw); the relevance of this name for the Afrotropical mousebirds is unclear and probably arbitrary. Dowsett (1988) suggests it is from Gr. *koleos,* a scabbard or sheath, with ref. to the long tails of these birds. Elsewhere in ornithology *colius* is used in its original sense (e.g. *Psarocolius)* or to indicate 'starling' (v. *Hypocolius).*

Coliuspasser Genus *Colius*, mousebird (considered by Buffon (1770) to be intermediate between the whydahs *Vidua* and the bullfinches *Pyrrhula* !); genus *Passer*, sparrow; ref. the long tails of the sparrow-like whydahs.

collaria L. *collare*, a collar, a neckchain; *-ius,* resembling, possessing.

collaris L. *collaris*, pertaining to the neck (i.e. collared, -necked) (*collare*, a collar).

collectoris L. *collector, collectoris,* a collector, a student; the Ochre-bellied Kingfisher *Ceyx* was named for the collector Albert Meek (v. *meeki)* and his assistants.

colliei After A. Collie (1794–1835) naval surgeon on HMS *Blossom* 1826–1827.

Collocalia Gr. *kolla*, glue; *kalia*, a nest; the nests of cave swiftlets are made from a cement formed in the salivary glands of breeding birds, and form the base of the Chinese delicacy birds-nest soup.

Colluricincla Syn. genus *Collurio*, shrike (Gr. *kollurion,* a bird ; v. *collurio*); Mod. L. *cinclus*, thrush.

collurio Gr. *kollurion*, a thrush-sized bird mentioned by Aristotle and Hesychius, not further identified, later conjectured to be a shrike *Lanius* or a Fieldfare *Turdus pilaris*.

collurioides Specific name *Lanius collurio* (cf. syn. genus *Collurio*, shrike); Gr. *-oides*, resembling.

collybita L. *collybista*, a moneychanger (Gr. *kollubistes*, a moneychanger); from supposed resemblance of the Common Chiffchaff's *Phylloscopus* distinctive song to that of money being counted or clinked together.

colma French *le Colma*, the name given to the Rufous-capped Ant-thrush *Formicarius* by Buffon (1770) with ref. to the female's spotted throat (*collier*, necklace; *marque*, mark).

colombiana /colombianus /colombica After Colombia, named for Christopher Columbus or Cristóbal Colón (v. *columbiana).*

Colonia Spanish settlers' name *el Colon* given to the Long-tailed Tyrant *C. colonus* in Paraguay and Argentina (L. *colonus*, a farmer, a rustic).

colonus L. *colonus*, a farmer, a rustic.

Colopteryx Gr. *kolos*, short, stunted; *pterux*, a wing.

Coloramphus Gr. *kolos*, short, stunted; *rhamphos*, the bill.

coloratus L. *coloratus*, coloured (especially red-coloured) (*colorare*, to colour).

coloria L. *colorius*, coloured, variegated.

colubris Spanish name *colibri* for a hummingbird (v. *Colibri*).

Columba L. *columba*, a pigeon or dove.

columbarius L. *columba*, a pigeon or dove; *-arius*, pertaining to; the Merlin *Falco* was formerly known as the Pigeon Hawk, from its supposed prey (cf. *columbarius*, a dove-keeper).

columbiana /columbianus After **1** the Columbia River, western USA (*Nucifraga, Olor*) and **2** Colombia (v. *colombiana*) (*Crypturellus, Odontophorus, Porzana, Sicalis*). To this day the name of Colombia is misspelt in newspapers and mispronounced on radio and television.

Columbigallina L. *columba*, a dove or pigeon; *gallina*, a hen.

Columbina L. *columbinus*, pertaining to a dove (*columba*, a dove or pigeon).

columboides L. *columba*, a dove; Gr. *-oides*, resembling.

Colymbus Gr. *kolumbis*, a waterbird mentioned by Aristophanes, Aristotle, Dionysius, and other authors, not otherwise identified, but later associated with a grebe *Podiceps* or diver *Gavia*.

comata /comatus L. *comatus*, hairy (*coma*, hair; Gr. *kome*, hair).

comechingonus After the Sierras de Comechingones, Córdoba Province, Argentina.

comeri After George Comer (fl. 1887) US seaman and collector in the South Atlantic.

comitata L. *comitatus*, a following, accompanying (i.e. closeness); of the African Blue Flycatcher *Muscicapa* Cassin (1857), *Proc. Acad. Nat. Sci.*, 35, wrote, '…intimately allied to, and resembling the preceding. Much resembling *B. epulatus*, but larger and easily distinguished'.

communis L. *communis*, common, general.

comorensis After the Comoro Islands (French *Iles Comores*), in the Indian Ocean north-west of Madagascar.

Compsocoma Gr. *kompsos*, pretty, elegant; *kome*, hair .

Compsospiza Gr. *kompsos*, pretty; *spiza*, a finch.

Compsothlypis Gr. *kompsos*, pretty; Mod. L. *thlypis*, a warbler.

Compsothraupis Gr. *kompsos*, pretty; genus *Thraupis*, tanager.

comptus L. *comptus*, adorned, neat (*comere*, to adorn).

comrii After R. M. Comrie (1832–1882) British naval surgeon and collector.

concinens L. *concinens*, singing, harmonizing (*concinere*, to sing in chorus).

concinna /concinnus L. *concinnus*, beautiful, elegant.

concolor L. *concolor*, uniform, similar in colour, plain.

concretus Mod. L. *concretus*, large, strong (L. *concretus*, hard, solid; *concrescere*, to grow).

condamini After Charles Marie de la Condamine (1701–1774) French scientist and traveller on the Amazon, 1735–1743.

condita L. *conditus*, hidden, stored away (*condere*, to store); the Grey-winged Cotinga *Tijuca* was described from a museum specimen collected thirty years earlier.

conditicius L. *conditicius*, preserved (*condere*, to store); the Gilbert Wood Rail *Tricholimnas* is known only from a single specimen preserved in alcohol.

confinis L. *confinis*, similar, nearly allied.

congener L. *congener*, of the same race or kind (i.e. allied to, similar).

congensis After the Belgian Congo (Zaïre). The Congo received its name from the mediaeval kingdom of Kongo which, at its zenith, occupied much of modern Angola and western Zaïre.

congica After the River Congo (River Zaïre).

Conioptilon Gr. *konia,* a fine powder or dust; *ptilon,* a feather; ref. powder-down feathers on the body of the Black-faced Cotinga *C. mcilhennyi.*

conirostris L. *conus,* a cone; *-rostris,* billed .

Conirostrum L. *conus,* a cone; *rostrum,* the bill.

conjuncta L. *coniunctus,* connected, joined (*coniungere,* to unite); the Black-capped Triller *Lalage* was thought to connect and share the characteristics of four other groups or species of triller in Papuasia.

connivens L. *conivens,* winking, blinking (*conivere,* to wink).

Conopias Gr. *konops, konopos,* a gnat; *piazo,* to seize.

Conopoderas Gr. *konopotheras,* a gnat-hunting bird mentioned by Hesychius, not further identified (*konops, konopos,* a gnat; *-thera,* a hunter).

Conopophaga Gr. *konops,* a gnat; *phagos,* a glutton (*phagein,* to eat).

Conopophila Gr. *konops,* a gnat; *philos,* loving, fond of.

Conostoma Gr. *konos,* a cone; *stoma,* the mouth; ref. bill shape of the Great Parrotbill *C. oemodium.*

Conothraupis L. *conos,* a cone; genus *Thraupis,* tanager; ref. bill shape of the Black-and-white Tanager *C. speculigera.*

conoveri After H. B. Conover (1892–1950) US ornithologist and author.

consobrinorum L. *consobrinorum,* of the cousins (*consobrinus,* a cousin); ref. the cousins Paul Sarasin and Fritz Sarasin, Swiss zoologists and collectors (v. *sarasinorum).*

consobrinus L. *consobrinus,* a cousin; of the Chinese Penduline Tit *Remiz* Swinhoe (1879), *Proc. Zool. Soc. London,* 133, wrote,'I consider the discovery of this species most interesting, as affording a case analogous to that of *Cyanopica,* which appears restricted to Spain and Portugal in Europe and then turns up in China about the Yangtsze and northwards, extending to Japan, in a somewhat modified form. The Penduline Tit occurs only in South Europe; and we find it again rather changed on the banks of the Yangtsze 850 miles from the sea'.

conspicillata /conspicillatus Mod. L. *conspicillatus,* spectacled (L. *conspicillum,* a place to look from).

constanti After C. Constant (1820–1905) French taxidermist and collector.

contaminatus L. *contaminatus,* unclean, dirty (*contaminare,* to pollute); no expl. (Lichtenstein (1854), *Nomen. av. mus. zool. Berol.,* p. 64), but probably based on a poor specimen.

contempta L. *contemptus,* disdain, contempt ; of the Lord Howe Whistler *Pachycephala* Hartert (1898), *Bull. Brit. Orn. Cl.,* **8**, 15, explained, 'The yellow *Pachycephala* of Lord Howe Island has been treated the same as *P. gutturalis* in the 'Catalogue of Birds ', vol. viii, where at least three different species or subspecies are united under the title of *P. gutturalis'.*

Contopus Gr. *kontos,* a shaft or pole; *pous,* the foot; the pewees were distinguished by their short tarsi.

contra According to Albin (1738), *Natural history of birds,* Vol. **3**, pl. 21, the Pied Mynah *Sturnus* was called *contra* by the natives of Bengal (cf. Assamese name *kan kurika).*

Conuropsis Syn. genus *Conurus,* parakeet (Gr. *konos,* a peak; *oura,* the tail); Gr. *opsis,* appearance.

conversii After M. Convers (fl. 1846) French naturalist and collector in Colombia.

convexus L. *convexus,* arched, convex; ref. the large ivory-coloured casque of the Southern Pied Hornbill *Anthracoceros.*

cookii After Captain James Cook (1718–1779) English explorer and navigator.

cooperi /cooperii After **1** William Cooper (1798–1864) US zoologist, collector, conchologist and author *(Accipiter).* **2** J. G. Cooper (1830–1902) US ornithologist (*Otus).*

coprotheres Gr. *kopros,* filth, dung; *-theras,* a hunter; the Cape Vulture *Gyps* feeds on all sorts of carrion.

Copsychus Gr. *kopsukhos,* the Blackbird *Turdus merula.*

coquereli /coquerellii After Chevalier Coquerel (1822–1867) French entomologist and collector in Madagascar, 1847.

coqui Adopted from name *coqui* given to the Coqui Francolin *Francolinus* by Smith (1836), *Rep. exped. centr. Africa,* p. 55, in imitation of its distinctive call 'ko..kwee ko..kwee'.

Coquus L. *coquus,* a cook; the local French name for the Mauritius Cuckoo-shrike *Coracina typica* is *Le Cuisenier*, the cook.

cora After Cora, an Inca priestess of the sun in Jean François Marmontel's novel *Les Incas, ou la destruction de l'Empire du Pérou*, 1777, and the subject of an opera libretto *Cora och Alonzo* by Baron Adlerbeth, 1781.

Coracias Gr. *korakias,* a type of jackdaw (*korax,* the raven).

Coracina /coracinus Gr. *korakinos,* a little raven (dim. of *korax,* the raven) (cf. L. *coracinus,* raven-black).

Coracopsis Gr. *korax, korakos,* the raven; *opsis,* appearance.

Coracornis Gr. *korax, korakos,* the raven; *ornis,* a bird; a contrived eponym after H. C. Raven (1889–1944) US explorer and collector in the East Indies 1914–1918.

Coragyps Gr. *korax,* the raven; *gups,* a vulture.

coralensis After the small coral island of Aratika in the Tuamotus (L. *corallum,* coral).

corallirostris L. *corallinus,* coral-red; *-rostris,* billed .

Corapipo No expl. (Bonaparte (1854), *Ateneo Italiano*, **2** [11], 316); probably from Gr. *kore,* the pupil of the eye (i.e. black); Mod. L. *pipo,* manakin; ref. the glossy black plumage of the White-throated Manakin *C. gutturalis.*

corax Gr. *korax,* the raven.

coraya French name *Le Coraya* given to the Coraya Wren *Thryothorus* by Buffon (1770) as a phonetic rendering of *queue rayée,* striped tail (v. *bambla).*

Corcorax Gr. *korone,* the crow; *korax,* the raven (cf. *korkora,* a bird mentioned by Hesychius, not further identified).

cordofanica /cordofanicus After the province of Kordofan in the Sudan.

corensis After the town of Coro, formerly Venezuela ('Little Venice', after Indian huts on stilts in the waters of Lake Maracaibo), Falcón, Venezuela.

Cormobates Gr. *kormos,* a log; *-bates,* roaming.

corniculata /corniculatus L. *corniculatus,* horned (*corniculum,* a little horn) (cf. Med. L. *corniculus,* a small crow; L. *cornix, cornicis,* the crow).

cornix L. *cornix,* the crow.

cornuta /cornutus L. *cornutus,* horned (*cornu,* a horn).

coromanda /coromandelianus/ coromandelica /coromandelicus / coromandus After the Coromand or Coromandel coast north and south of Madras, India. Coromandel is a corruption of Cholamandalam, an old Dravidic kingdom of eastern India.

coronata /coronatus L. *coronatus,* crowned (*coronare,* to crown).

corone Gr. *korone,* the crow.

coronoides Specific name *Corvus corone*; Gr. *-oides*, resembling.

coronulatus L. *coronula,* a little crown; *-atus,* possessing.

correndera Local Aragonese (Spanish) name *correndera* for the Meadow Pipit *Anthus pratensis*, given to the Correndera Pipit *Anthus* by Azara (1805) because he considered the two spp. almost indistinguishable.

corrugatus Mod. L. *corrugatus,* wrinkled, corrugated (L. *corrugare,* to wrinkle); ref. marks on bill and casque of the Wrinkled Hornbill *Rhyticeros.*

corruscus L. *coruscus,* twinkling, flashing, brilliant (*coruscare,* to twinkle).

coruscans L. *coruscans,* twinkling (*coruscare,* to twinkle).

corvina L. *corvinus,* raven-like, corvine (*corvus,* the raven) (cf. Mod. L. *corvinus,* lustrous black).

Corvinella Dim. of L. *corvinus,* raven-like; of the Yellow-billed Shrike *C. corvina* Shaw (1809), *Gen. zool.*, Vol. **7**, p. 337, wrote, 'This bird, which Monsr. Levaillant considers as a species of Shrike, might perhaps with equal propriety be arranged under the genus *Corvus*...the general appearance of the bird indicates a strong alliance to the longtailed kind of Corvi'.

Corvultur Genus *Corvus,* raven; genus *Vultur,* condor (also formerly used for the Black Vulture *Aegypius).*

Corvus L. *corvus,* the raven.

Corydon Gr. *korudon,* a kind of crested lark mentioned by Aristotle and Aristophanes (variant form of *korudalos).*

coryi After Charles Barney Cory (1857–1921) US businessman, collector, ornithologist, and author (*The birds of Haiti and San Domingo*, 1885; *The birds of the West Indies*, 1889; *The birds of Illinois and Wisconsin*, 1909).

coryphaeus **1** French *le Coriphée,* the name given to the Karoo Scrub Robin by Levaillant (1802), ref. its beauty and voluptuous song (Gr. myth. Coryphaea, an epithet of the goddess Diana) (*Erythropygia).* **2** Gr. *koruphaia,* of the head, a tuft on the head (*koruphe,* the crown of the head) (*Pogoniulus).*

Coryphaspiza Gr. *koruphe,* the crown of the head; *spiza,* a finch.

Coryphistera Gr. *koruphe,* the crown of the head; *sterrhos,* stiff; ref. conspicuous upstanding crest of the Lark-like Brushrunner *C. alaudina.*

Coryphoenas Gr. *koruphe,* the crown of the head; *oinas,* a dove.

Coryphospingus Gr. *koruphe,* the crown of the head; *spingos,* a finch.

Coryphotriccus Gr. *koruphe,* the crown of the head; Mod. L. *triccus,* tyrant flycatcher.

Corythaeola Gr. *koruthaiolos,* with shaking plume (*korus, koruthos,* the head; *aiolos,* rapid, quick-moving).

corythaix Gr. *koruthaix,* with waving plumes (*korus, koruthos,* the head; *aisso,* to move rapidly).

Corythaixoides Specific name *Tauraco corythaix*; Gr. *-oides*, resembling.

Corythopis Gr. *koruthopis,* having a lark-like appearance.

Corythornis Gr. *korus, koruthos,* the head, a helmet (i.e. crested); *ornis,* a bird .

Coscoroba Local Chilean names *coscoroba, coscaroba* or *cascaroba* for the Coscoroba Swan *C. coscoroba.*

Cosmopsarus Gr. *kosmos,* an ornament; *psar,* the starling.

Cossypha Gr. *kossuphos,* the Blackbird *Turdus merula.*

costae After Louis Marie Pantaléon Costa Marquis de Beauregard (1806–1864) Savoyard statesman, amateur historian, and archaeologist.

costaricensis After Costa Rica (Spanish for 'The Rich Coast', in anticipation of the gold never found there), discovered by Columbus in 1502 and settled by the Spanish in 1563.

Cotinga Tupí (Brazilian) Indian name *cotinga* for these colourful birds.

cotta In describing the Jamaican Elaenia *Myiopagis* Gosse (1849), *Ann. Mag. Nat. Hist* ., Ser. 2, **3**, 257, wrote, 'This little Tyrant, for want of any obvious peculiarities to distinguish it from others of its genus, I have named from the locality where I first met with it, the Cotta-wood, a tangled coppice on Grand Vale Mountain, in the parish of St Elizabeth'.

Coturnicops L. *coturnix, coturnicis,* the quail; Gr. *ops,* the face (i.e. appearance).

Coturnix L. *coturnix,* the quail.

Coua Malagasy name *koa* for cuckoos and couas.

couchii After General Darius Nash Couch (1822–1897) US Army in New Mexico, 1853.

couesi After Elliott B. Coues (1842–1899) US ornithologist, collector and author (*Key to North American birds*, 1872; *Check list of North American birds*, 1882; *Hand-*

book of field and general ornithology, 1890).

couloni After Paul Louis de Coulon (1804–1894) Swiss naturalist.

coultasi After W. F. Coultas (b. 1899) collector in the Pacific, 1929–1937.

courseni After C. Blair Coursen (fl. 1970) US field ornithologist and author.

courtoisi After Rév. Père F. Courtois (1860–1928) French missionary to China 1901–1928, naturalist, and author (*Les oiseaux du musée de Zi-Ka-Wei*, 1912).

coxeni After Charles Coxen (1809–1876) English naturalist, settler, and parliamentarian in Australia.

Cracticus Gr. *kraktikos,* clamorous (*krazo,* to screech).

cranchii After John Cranch (1758–1816) British explorer and naturalist in tropical Africa, 1816.

Cranioleuca Gr. *kranion,* the skull, the head; *leukos,* white.

crassa L. *crassus,* dense, thick, heavy; ref. heavy bill (*Aplonis, Poicephalus)* or densely feathered rump (*Napothera).*

crassirostris L. *crassus,* thick, heavy; *-rostris,* billed .

crassus see **crassa.**

Crateroscelis Gr. *krateros,* strong, stout; *skeles,* the leg.

cratitius L. *cratitius,* wattled (strictly mud and wattled wickerwork, but used here in the sense of wattled or carunculated) (*cratis,* a joint, a rib).

craveri After Federico Craveri (1815–1890) Italian chemist, meteorologist, and collector in Mexico 1840–1859.

Crax No expl. (Linnaeus (1758), *Systema Naturae*, (Ed.10), book **1**, p. 157), but probably the Latin equivalent of Ulisse Aldrovandi's (1599) (*Ornithologia)*, 'Gallus curassavicus' or Curaçao Hen. However, perhaps based on Gr. *kras,* the head, with ref. to noticeable crest, or Gr. *keras,* a horn, with ref. to tubercules at base of bill.

Creadion Gr. *kreadion,* a morsel of meat (dim. of *kreas,* a piece of meat or flesh); the Saddleback *C. carunculatus* has orange gape wattles.

creagra /Creagrus L *creagra,* a flesh-hook (Gr *kreagra,* a meat-hook); ref the long hooked bill of the Swallow-tailed Gull *C. furcatus.* Specifically refers to the long, hook-tipped outer tail feathers of the Long-trained Nightjar *Macropsalis.*

Creatophora Gr. *kreas, kreatos,* flesh; *-phoros,* carrying, bearing; the male Wattled Starling *C. cinerea* carries fantastic head wattles in breeding plumage.

creatopus Gr. *kreas,* flesh (i.e. pink or flesh-coloured); *pous,* the foot.

crecca Swedish *kricka,* the Green-winged Teal *Anas* .

Crecopsis Genus *Crex,* corncrake (Gr. *krex, krekos)*; Gr. *opsis,* appearance.

crenatus Mod. L. *crenatus,* notched (Med. L. *crena,* a notch); ref. emarginated primaries of the Yellow-tufted Pipit *Anthus.*

crepitans L. *crepitans,* breaking wind (*crepitare,* to break wind); given to the Grey-winged Trumpeter *Psophia,* 'as the sound it made was supposed to be produced *per anum'* (Langton (1920)).

crestatus L. *cristatus,* crested (*crista,* a crest).

Creurgops Gr. *kreourgos,* a butcher (i.e. a butcherbird or shrike); *ops,* the face (i.e. appearance); ref. heavy, notched bill of the Rufous-crested Tanager *C. verticalis.*

Crex Gr. *krex,* a long-legged bird mentioned by Herodotus, Aristophanes, Aristotle, and other authors, since identified with a wide variety of birds including the Corncrake *C. crex* (on the assumption that the name was onomatopoeic), the Ruff *Philomachus,* and the Black-winged Stilt *Himantopus.*

Crinifer L. *crinis,* a hair; *- fer,* bearing, carrying ; the plantain-eaters have bushy crests of lanceolate feathers.

Criniferoides Genus *Crinifer,* plantain-eater; Gr. *-oides*, resembling.

crinifrons L. *crinis,* a hair; *frons,* the brow, the forehead.

Criniger L. *criniger,* long-haired.

crinitus L. *crinitus,* long-haired, hairy (*crinis,* a hair).

crispifrons L. *crispus,* curled, wrinkled (*crispare,* to curl); *frons,* the forehead, brow; the forehead of the Limestone Wren Babbler *Napothera* has a scaly appearance.

crispus L. *crispus,* curled, curly-headed (*crispare,* to curl).

crissale / crissalis Mod. L. *crissalis,* of the vent, crissum or lower tail coverts (L. *crissare,* to move the haunches).

cristata L. *cristatus,* crested (*crista,* a crest).

cristatella /cristatellus Dim. of L. *cristatus,* crested.

cristatum / cristatus see **cristata.**

crocea L. *croceus,* saffron-coloured, golden yellow (*crocus,* saffron; Gr. *krokos).*

Crocethia Gr. *kroke,* a beach pebble (*krokale,* a beach); *theio,* to run.

croceus see **crocea.**

Crocias Gr. *krokias,* an unknown saffron-coloured stone (*krokos,* saffron).

croconotus Gr. *krokos,* saffron; *-notos,* backed .

crossleyi /Crossleyia After A. Crossley (fl. 1875) collector in the Cameroons, Rhodesia, and Madagascar 1870–1873.

Crossoptilon Gr. *krossoi,* a fringe or tassel; *ptilon,* a feather; from appearance of decomposed tail-feathers of the eared pheasants.

Crotophaga Gr. *kroton,* a tick; *phagos,* a glutton (*phagein,* to eat).

crudigularis L. *crudus,* bleeding; Mod. L. *gularis,* throated (L. *gula,* the throat); ref. black and red spotted throat band of the White-throated Hill Partridge *Arborophila.*

cruenta L. *cruentus,* bloody (*cruor,* blood, gore).

cruentata /cruentatum /cruentatus L. *cruentatus,* bloody, blood-stained (*cruor,* blood, gore).

cruentus see **cruenta.**

crumeniferus L. *crumena,* a leather money-pouch carried around the neck; *-fera,* carrying, bearing.

cruralis L. *cruralis,* of the leg or shin (*crus, cruris,* the leg or shin).

cruziana Erroneous toponym after the city of Santa Cruz, Bolivia; the Croaking Ground Dove *Columbina* is found on the coasts of Ecuador, Peru, and northern Chile.

Crypsirina Gr. *krupsis,* hiding, concealment; *rhis, rhinos,* the nose; the nostrils of the Racquet-tailed Treepie *C. temia* are hidden by feathers.

crypta Gr. *kruptos,* hidden, secret (*krupto,* to hide, conceal).

cryptica L. *crypticus,* covered, concealed (Gr. *kruptikos).*

Cryptigata Syn. genus *Cryptolopha,* warbler (Gr. *kruptos,* hidden; *lophos,* a crest); specific name *Phylloscopus trivirgatus* .

cryptoleuca /cryptoleucus Gr. *kruptos,* hidden, secret; *leukos,* white; ref. hidden or obscured white patches on neck or wing.

cryptolophus Gr. *kruptos,* hidden, secret; *-lophos,* crested (*lophos,* a crest); ref. partially concealed black and white crest of Olivaceous Piha *Lipaugus.*

Cryptolybia Gr. *kruptos,* hidden, secret (i.e. newly described); genus *Lybius,* barbet.

Cryptophaps Gr. *kruptos,* hidden, secret; *phaps,* a pigeon; ref. dull plumage of the Sombre Pigeon *C. poecilorrhoa.*

Cryptoplectron Gr. *kruptos,* hidden, secret; *plectron,* a cock's spur; the cock Painted Bush Quail *Perdicula erythrorhyncha* lacks spurs.

Cryptospiza Gr. *kruptos,* hidden, secret; *spiza,* a finch; ref concealed crimson wing-coverts of the Red-faced Crimsonwing *C. reichenovii.*

cryptoxanthus Gr. *kruptos,* concealed, hidden; *xanthos,* golden-yellow, yellow.

Crypturellus Dim. of syn. genus *Crypturus,* tinamou (Gr. *kruptos,* hidden; *oura,* the tail); all tinamous have very short tails, often covered by the tail-coverts.

cryptus see **crypta.**

cubanensis /cubensis After the island of Cuba, in the Greater Antilles, discovered by Christopher Columbus in 1492, and probably named for the indigenous Ciboney Indians.

cubla French *Le Cubla,* the name given to the Southern Puffback *Dryoscopus* by Levaillant (1799) (*cou-blanc,* white rump) (cf. Singuni name *qubana).*

cucullata/cucullatus Late L. *cucullatus,* hooded (L. *cucullus,* a hood or cowl).

cuculoides L. *cuculus,* the cuckoo; Gr. *-oides*, resembling.

Cuculus L. *cuculus,* the cuckoo.

cujubi Tupí (Brazilian) Indian name *cujubini* for the piping guan.

Culicicapa L. *culex, culicis,* a midge or gnat; *capere,* to seize (cf. *Muscicapa).*

Culicivora L. *culex, culicis,* a midge or gnat; *-vorus,* eating.

culik Cayenne Creole name *koulik* for the Guiana Toucanet *Selenidera.*

culminatus Late L. *culminatus,* culminate, apical, having reached the highest point; ref. yellow ridged upper mandible of the Yellow-ridged Toucan *Ramphastos.*

cumanensis After the city of Cumaná, Sucre, Venezuela.

cumingi After Hugh Cuming (1791–1865) English naturalist, conchologist, and collector in tropical America 1822–1826, 1828–1830; Polynesia 1827–1828; East Indies 1836–1840.

Cuncuma Nepalese name *kankam* or *kunkum* for Pallas's Fish Eagle *Haliaeetus leucoryphus.*

cuneata L. *cuneatus,* wedge-shaped, pointed like a wedge (*cuneus,* a wedge).

cunicularia L. *cunicularius,* a miner (strictly a military engineer) (*cuniculus,* an underground passage, a mine).

cupido L. *cupido,* love, physical desire; the erectile neck feathers of the Greater Prairie Chicken *Tympanuchus* were thought to resemble the wings of Cupid, the god of love.

cuprea L. *cupreus,* coppery (*cuprum,* copper) (cf. Mod. L. *cupreus,* copper coloured).

cupreiceps L. *cupreus,* coppery; *-ceps,* crowned, headed.

cupreocauda L. *cupreus,* coppery; *cauda,* the tail.

cupreoventris L. *cupreus,* coppery; *venter, ventris,* the belly.

cupreus see **cuprea.**

cupripennis L. *cuprum, cupri,* copper; *pennis,* winged (*penna,* a feather).

Curaeus Araucano (Chilean) Indian names *cureu, kuren* or *kueren* for the Austral Blackbird *C. curaeus* (cf. Gr. *koureus,* a bird mentioned by Hesychius).

curruca L. *curruca,* a kind of bird mentioned by Juvenal, and latterly identified as a sort of warbler *Sylvia.*

currucoides Specific name *Sylvia curruca* ; Gr. *-oides*, resembling.

cursitans L. *cursitans,* running (*cursitare,* to run up and down, to run about).

cursor L. *cursor,* a runner (*currere,* to run).

Cursorius L. *cursorius,* addicted to running (*currere,* to run).

curtata L. *curtatus,* shortened, cut short (*curtare,* to shorten); according to Sclater (1870), *Proc. Zool. Soc. London* [1869], 636, the Ash-browed Spinetail *Cranioleuca* was, '...not unlike *S. erythrops,* but has the red colour entirely confined to the crown of the head'.

curucui Tupí (Brazilian) Indian name *surucuí* or *surucuá* for a smaller type of bird.

Curucujus Specific name *Trogon curucui.*

curvipennis L. *curvus,* curved; *pennis,* winged (*penna,* a wing).

curvirostra L. *curvus,* curved; *rostrum,* the bill.

curvirostre/curvirostris L. *curvus,* curved; *-rostris,* billed.

Cutia Nepalese name *khatya* or *khutya* for the Nepal Cutia *C. nipalensis.*

cuvieri/cuvierii After **1** Léopold Chrêtien Frédéric Dagobert Baron Cuvier (1769–1832) French anatomist and author better known by his *nom de plume* Georges Cuvier (*Tableau élémentaire de l''histoire naturelle des animaux,* 1798; *Le règne animal distribué d'après son* *organisation*, 1817; *Histoire des sciences naturelles*, 1841) (*Canirallus, Phaeochroa, Talegalla*) and his brother **2** Frédéric Georges Cuvier (1775–1838) French zoologist and author (*Dictionnaire des sciences naturelles*, 1816; *Histoire naturelle des mammifères*, 1824; *De l'histoire naturelle des cétacés*, 1836) (*Falco*).

cuyensis After the island of Cuyo, between Palawan and Panay, Philippine Islands.

cyana Gr. *kuanos*, blue, dark blue.

cyane /cyanea L. *cyaneus*, dark blue, sea blue (Gr. *kuaneos*, dark blue).

cyaneovirens L. *cyaneus*, dark blue; *virens*, green.

cyaneoviridis L. *cyaneus*, dark blue; *viridis*, green.

Cyanerpes Gr. *kuaneos*, dark blue; *herpes*, a creeper (*herpo*, to creep).

cyanescens Inferred L. *cyanescens*, dark bluish (*cyaneus*, dark blue; v. *caerulescens)*.

cyaneus see **cyanea.**

cyaniceps L. *cyaneus*, dark blue; *-ceps*, headed, crowned.

cyanicollis L. *cyaneus*, dark blue; Mod. L. *-collis*, necked.

Cyanicterus L. *cyaneus*, dark blue; Mod. L. *icterus*, yellow.

cyanifrons L. *cyaneus*, dark blue; *frons*, the forehead, the brow.

cyanirostris L. *cyaneus*, dark blue; *-rostris*, billed.

cyaniventer /cyaniventris L. *cyaneus*, dark blue; *venter, ventris*, the belly.

cyanocampter Gr. *kuanos*, dark blue; *kamptos*, a bend, an angle; ref. the blue lesser and median wing-coverts of the Blue-shouldered Robin Chat *Cossypha*.

cyanocephala /cyanocephalus Gr. *kuanos*, dark blue; *-kephalos*, headed.

Cyanochen Gr. *kuanos*, dark blue; *khen*, a goose.

Cyanocitta Gr. *kuanos*, dark blue; *kitta*, the jay (v. *Cissa)*.

Cyanocompsa Gr. *kuanos*, dark blue; *kompsos*, elegant, pretty.

Cyanocorax Gr. *kuanos*, dark blue; *korax*, the raven (i.e. of the crow family).

cyanogaster Gr. *kuanos*, dark blue; *gaster*, the belly.

cyanogenia Gr. *kuanos*, dark blue; *genus*, the cheek.

cyanoides Specific name *Passerina cyanea* ; *-oides*, resembling.

cyanolaema Gr. *kuanos*, dark blue; *laimos*, the throat.

cyanoleuca Gr. *kuanos*, dark blue; *leukos*, white.

Cyanolimnas Gr. *kuanos*, dark blue; Mod. L. *limnas*, a rail or crake.

Cyanoliseus Gr. *kuanos*, dark blue; *alusis*, a chain; ref. bluish white collar of the Burrowing Parrot *C. patagonus*.

Cyanoloxia Gr. *kuanos*, dark blue; genus *Loxia*, crossbill (used in ornithology for an assortment of heavy-billed finches or finch-like birds).

Cyanolyca Gr. *kuanos*, dark blue; *lukos*, a sort of crow mentioned by Aristotle and Hesychius, probably the Jackdaw *Corvus monedula*.

cyanomelaena /cyanomelas Gr. *kuanos*, dark blue; *melas, melanos*, black.

Cyanomitra Gr. *kuanos*, dark blue; *mitra*, a head band, a cap.

Cyanonympha Gr. *kuanos*, dark blue; *numphe*, a nymph (i.e. a woodnymph, a dryad).

cyanopectus Gr. *kuanos*, dark blue; L. *pectus*, the breast.

Cyanophaia Gr. *kuanos*, dark blue; *phaios*, dusky.

Cyanophasis Gr. *kuanos*, dark blue; Mod. L. *phasis*, pheasant.

cyanophrys Gr. *kuanos*, dark blue; *-ophrus*, browed (*ophrus*, the eyebrow).

Cyanopica Gr. *kuanos*, dark blue; genus *Pica*, magpie.

cyanopis Gr. *kuanopis*, dark-eyed, blue-eyed.

cyanopogon Gr. *kuanos*, dark blue; *pogon*, a beard.

Cyanopsitta Gr. *kuanos*, dark blue; *psittas*, a parrot.

cyanoptera/cyanopterus Gr. *kuanopteros,* with blue-black feathers, blue-winged.

Cyanoptila Gr. *kuanos,* dark blue; *ptilon,* a feather, plumage.

cyanopus Gr. *kuanopos,* of dark aspect.

cyanopygius Gr. *kuanos,* dark blue; Mod. L. *pygius,* rumped (Gr. *puge,* the rump).

Cyanoramphus Gr. *kuanos,* dark blue; *rhamphos,* the bill.

Cyanosylvia Gr. *kuanos,* dark blue; genus *Sylvia,* warbler.

cyanotis/cyanotus Gr. *kuanos,* dark blue; *-otis,* eared.

cyanouroptera Gr. *kuanos,* dark blue; *oura,* the tail; *-pteros,* winged; the Blue-winged Siva *Minla* has blue tail and wings.

cyanoventris Gr. *kuanos,* dark blue; L. *venter, ventris,* the belly.

cyanura/cyanurus Gr. *kuanos,* dark blue; *-ouros,* tailed.

cyanus see **cyana.**

Cyclarhis Gr. *kuklos,* a ring or circle; *rhis,* the nose; the Rufous-browed Peppershrike *C. gujanensis* is '...remarkable for its round, naked nostrils...' (Swainson (1824), *Zool. Journ.*, **1**, 294).

Cyclorrhynchus Gr. *kuklos,* a ring, circle, or shield; *rhunkhos,* the bill; ref. the stout, upturned, parrot-like bill of the Parakeet Auklet *C. psittacula*.

cygnoides L. *cygnus,* the swan; Gr. *-oides*, resembling.

Cygnus L. *cygnus,* the swan (Gr. *kuknos,* the swan).

cylindricus Mod. L. *cylindricus,* cylindrical (Gr. *kulindikos,* cylindrical); ref. casque shape of the Brown-cheeked Hornbill *Bycanistes*.

Cymbilaimus Gr. *kumbe,* a cup or boat; misspelt genus *Lanius,* shrike; ref. heavy hooked bill of the Fasciated Antshrike *C. lineatus*.

Cymbirhynchus Gr. *kumbe,* a boat or cup (cf. *kumbe,* probably an error for *komba,* a sort of crow *Corvus* mentioned by Hesychius); *rhunkhos,* the bill; ref. broad, boat-like bill of the Black-and-red Broadbill *C. macrorhynchos*.

Cymochorea Gr. *kuma,* a wave, swell or surge; *khoreia,* a dancing (*khoros,* a dance).

Cynanthus Gr. *kuanos,* dark blue; *anthos,* a flower or bloom.

Cyornis Gr. *kuanos,* dark blue; *ornis,* a bird.

cyornithopsis Genus *Cyornis,* blue flycatcher (Gr. *ornis, ornithos)*; Gr. *opsis,* appearance.

Cyphorhinus Gr. *kuphos,* bent, bowed (*kupto,* to stoop); *rhis, rhinos,* the nose; the Neotropical song wrens have thick bills with a high, ridged culmen.

cypriaca L. *Cypriacus,* Cyprian, of Cyprus.

Cypseloides Syn. genus *Cypselus,* a swift (L. *cypselus,* a swift; Gr. *kupselos,* a kind of swallow); Gr. *-oides*, resembling.

Cypsiurus Gr. *kupselos,* a kind of swallow; *-ouros,* tailed.

Cypsnagra Syn. genus *Cypselus,* a swift (Gr. *kupselos,* a kind of swallow); syn. genus *Tanagra,* tanager; ref. the odd characters of the White-rumped Tanager *C. hirundinacea*.

Cyrtonyx Gr. *kurtos,* curved, bent; *onux,* a claw, a nail; the Montezuma Quail *C. montezumae* was described as having the '...nails very long, much lengthened, curved and thick at the point' (Gould (1844), *Monograph of the Odontophorinae*, Pt. **1**, pl. 2, and text).

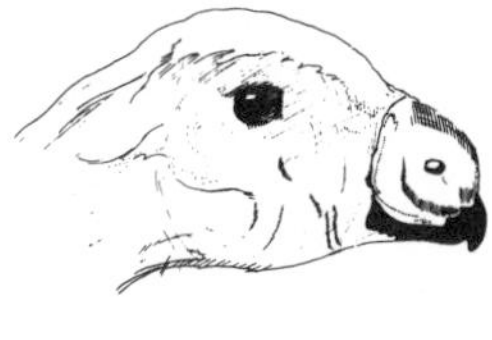

Cereopsis novaehollandiae Cape Barren Goose

D

dabbenei After Roberto Dabbene (1864–1938) Argentine ornithologist.

Dacelo Anagram of genus *Alcedo,* kingfisher, coined to reflect relationship.

Dacnis Gr. *daknis,* a type of bird from Egypt mentioned by Hesychius and the grammarian Pompeius Festus, not otherwise identified.

dacotiae Meade-Waldo (1889), *Ibis,* 504, gives the type locality of the Canarian Chat *Saxicola* as '*Ins[ula] Fuerteventura, Mauritanice Dacos'*. Hartert (1913), *Die vögel der Paläarktischen fauna*, Vol. **2**, p. 1086, also used the epithet for the Lanzarote race of the Common Kestrel *Falco tinnunculus,* and I believe *Dacos* to be a Roman adaptation of an ancient Mauretanian (Moroccan) name for one or more of the Canary Islands.

dactylatra Gr. *daktulos,* a finger (i.e. the primary wing feathers); L. *ater,* black.

Dactylortyx Gr. *daktulos,* a finger (i.e. a toe); *ortux,* the quail; the Singing Quail *D. thoracicus* has very long toes.

Dafila No expl. (Stephens (1824), *Gen. zool.*, Vol. **7**, Pt. 2, p. 126); said to be a Brazilian name for some sort of duck, but Coues (1882) scornfully rejects it as a nonsense-word.

daguae After the Río Dagua, Valle del Cauca, Colombia.

dahli After F. O. Dahl (1856–1919) German zoologist, collector in Melanesia 1896–1897, and author (*Das leben der vögel auf den Bismarckinseln*, 1899).

dalhousiae After Christina Countess of Dalhousie (1786–1839) wife of George Ramsay, 9th Earl of Dalhousie, Commander-in-Chief East Indies, 1829–1832.

damarensis After Damaraland, South West Africa (Namibia).

dambo Swahili name *dambo* for a peaty, often almost permanently wet, grassland.

damii After D. C. van Dam (fl. 1873) Dutch naturalist and collector in Madagascar, 1863.

dammermani After Karel Willem Dammerman (1885–1951) Dutch zoologist, botanist, collector, and author.

Damophila After Damophila (seventh century BC), a poetess of Lesbos and intimate of Sappho.

danae Gr. myth. Danae, daughter of King Acrisius of Argos, who cast her adrift.

danjoui After André Danjou (fl. 1919) French Consul at Singapore, 1919.

Daphoenositta Gr. *daphoinos,* blood-red; genus *Sitta,* nuthatch.

Daption Anagram of Portuguese *pintado,* pied, spotted (L. *pictus,* coloured, painted).

Daptrius Gr. *daptria,* an eater (*dapto,* to devour); ref. scavenging habits of the Neotropical caracaras.

darjellensis After the town of Darjeeling, West Bengal, India.

darnaudii After J. d'Arnaud (fl. 1843) French explorer and big-game hunter in the Sudan and Ethiopia, 1839–1843.

darwini/darwinii After Charles Robert Darwin (1809–1882) British naturalist, creator of Darwinism, and author (*Origin of species by means of natural selection*, 1859).

Dasylophus Gr. *dasus,* hairy, shaggy; *lophos,* a crest.

Dasyornis Gr. *dasus,* hairy, shaggy; *ornis,* a bird; ref. the strong rictal bristles from which the bristlebirds receive their name.

dasypus Gr. *dasupous,* rough-footed, hairy-footed (*dasus,* hairy; *pous,* a foot).

daubentoni After Edmé Louis d'Aubenton (1732–1788) French naturalist and author (*Les planches enluminéez d'histoire naturelle*, 1765).

dauma Bengali name *dama* for the Orange-headed Ground Thrush *Zoothera citrina*. According to Macleod (1954), p. 48, *dama* is the Bengali name for White's Thrush *Zoothera*, but no native names are recorded in Ali and Ripley (1973),Vol. **9**, for the latter, and there is no explanation in Latham (1787) *Gen. synop*., Supp. 1, p. 145 ('It is called *Cowal*, from its note imitating that word') or Latham (1790) *Index orn*., Vol. **1**, p. 362 (where called the 'Dauma Thrush').

daurica /dauuricae /dauuricus After Dauria, a region of south-eastern Siberia east of Lake Baykal bordering Mongolia and Manchuria.

davidi After **1** Abbé Père Armand David (1826–1900) French missionary to China 1858–1874, naturalist, and author (*Les oiseaux de la Chine*, 1877) (*Garrulax, Niltava, Parus, Stachyris, Strix)*. **2** André David-Beaulieu (b. 1896) French colonial administrator in Indochina, naturalist, and author (*Les oiseaux du Tranninh*, 1944) (*Arborophila)*.

davidiana /davidianus see **davidi 1**.

davisoni After W. R. Davison (d. 1893) curator of Raffles Museum, Singapore, 1887–1893.

dayi After Colonel Lee G. Day (fl. 1960) US financier, sponsor of expeditions to tropical America, 1915 and 1927.

dea L. *dea,* a goddess.

debilis L. *debilis,* feeble, weak (i.e. small).

decaocto L. *deca-,* ten; *octo,* eight; Gr. myth. a hard-worked maidservant, bemoaning her pay of a miserly 18 pieces a year, prayed for release and was changed by the gods into a dove that echoed her mournful cries.

decipiens L. *decipiens,* deceiving (*decipere,* to catch, cheat); the Mourning Collared Dove *Streptopelia* was previously confused with the Collared Dove *S. decaocto*.

deckeni After Baron Carl Claus von der Decken (1833–1865) German explorer and author (*Reisen in ost-Afrika*, 1869).

Deconychura Gr. *deka,* ten; *onux,* a claw; *oura,* the tail; of Cherrie's Woodcreeper *D. typica* Cherrie (1891), *Proc. US Nat. Mus*., **14**, 339, wrote, 'The generic name here chosen refers to the peculiarities of the tail, with ten rectrices having the stiffened stems extended beyond the webs and bent down…'.

decora L. *decora,* graceful, beautiful (*decor,* beauty).

decoratus L. *decoratus,* adorned, beautiful.

decumanus L. *decumanus,* a legionnaire of the tenth legion (*decimus,* tenth); ref. the dagger- or sword-like bill and shield-like casque of the Crested Oropendola *Psarocolius*.

decurtatus L. *decurtatus,* mutilated (*curtus,* shortened, mutilated); the Grey-headed Greenlet *Hylophilus* is a small sp. with a very short tail.

decussatus L. *decussatus,* decussate, marked with X-shaped crosses (*decussis,* the number ten (Roman X)).

dedemi After Baron F. K. van Dedem (fl. 1910) Dutch naturalist and collector in Sumatra.

defilippi /defilippiana After Filippo de' Defilippi (1814–1867) Italian naturalist, traveller, and collector.

deglandi After Côme Damien Degland (1787–1856) French naturalist and author (*Ornithologie Européene*, 1849).

degodiensis After the Degodi region, Sidamo Province, Ethiopia.

deiroleucus Gr. *deire,* the throat; *leukos,* white.

delalandei /delalandi After Pierre Antoine Delalande (1787–1823) French naturalist, explorer, and collector in Brazil, 1816 and tropical Africa, 1818–1822.

delatrii After Adolphe de Lattre (fl. 1847) French naturalist and collector in Mexico, 1838 and Colombia, 1847.

delattrei After Henri de Lattre (fl. 1838) French naturalist and collector in Mexico (with brother Adolphe), 1838.

delawarensis After the Delaware River, Pennsylvania, USA, named for Thomas West, 3rd Baron De La Warr (1557–1618) Governor and Captain-General of Virginia, 1610.

delegorguei After Adolphe Delegorgue (1814–1850) French traveller and collector in South Africa, 1839–1850.

delesserti After Adolphe Delessert (d. 1843) French naturalist and collector in India, 1834–1839.

delicata L. *delicatus,* dainty, nice.

delicatula Dim. from L. *delicatus,* dainty, nice.

Delichon Anagram of syn. genus *Chelidon,* a swallow (Gr. *khelidon,* a swallow), coined to reflect relationship.

deliciosus L. *deliciosus,* delightful, delicate.

Delothraupis Gr. *delos,* clear, visible (i.e. different); genus *Thraupis,* tanager.

delphinae No expl. (Lesson (1839), *Rev. Zool.* (Paris), 44), where the Brown Violetear *Colibri* is first referred to as *Oiseau-mouche Delphine.* In Gr. myth. Delphinios was an epithet of Apollo the sun god.

Deltarhynchus Gr. *delta,* the letter Δ; *rhunkhos,* the bill; ref. the short but wide bill of the Flammulated Flycatcher *D. flammulatus.*

demersus L. *demersus,* plunging, sinking (*demergere,* to sink).

demissa L. *demissus,* humble, modest (*demittere,* to let down).

Dendragapus Gr. *dendron,* a tree; *agapao,* to love, be fond of.

Dendrexetastes Gr. *dendron,* a tree; *exetastes,* a searcher (*exetazo,* to inspect).

Dendrobiastes Gr. *dendron,* a tree; *biastes,* a tyrant, a violent man.

Dendrocincla Gr. *dendron,* a tree; Mod. L. *cinclus,* a thrush.

Dendrocitta Gr. *dendron,* a tree; *kitta,* the magpie (v. *Cissa).*

Dendrocolaptes Gr. *dendrokolaptes,* a woodpecker (*dendron,* a tree; *kolapto,* to chisel).

dendrocolaptoides Genus *Dendrocolaptes,* woodcreeper; Gr. *-oides,* resembling.

Dendrocopos Gr. *dendron,* a tree; *kopos,* striking, beating (*kopto,* to cut).

Dendrocygna Gr. *dendron,* a tree; genus *Cygnus,* swan.

Dendroica Gr. *dendron,* a tree; *-oikos,* living, dwelling (*oikeo,* to inhabit).

Dendronanthus Gr. *dendron,* a tree; genus *Anthus,* pipit.

Dendropicos Gr. *dendron,* a tree; *pikos,* a woodpecker.

Dendroplex Gr. *dendron,* a tree; *plesso,* to strike.

Dendrortyx Gr. *dendron,* a tree; *ortux,* the quail.

denhami After Lieutenant-Colonel Dixon Denham (1786–1826) English soldier, explorer in tropical Africa, Lieutenant-Governor of Sierra Leone, 1826, and author (*Narrative of travels and discoveries in northern and central Africa,* 1826).

deningeri After K. Deninger (1878–1918) Dutch zoologist and collector in the Moluccas, 1911–1912.

dennistouni After John Dennistoun (fl. 1924) British naturalist and collector in the Philippines, 1894.

densus L. *densus,* thick, dense.

dentata L. *dentatus,* toothed (*dens, dentis,* a tooth) (i.e. with a notched bill).

denti After Captain R. E. Dent (b. 1882) British explorer and collector in tropical Africa, 1901 and 1906.

dentirostris L. *dens, dentis,* a tooth; *-rostris,* billed.

derbiana /derbianus /derbyana /derbyi After Edward Smith Stanley, 13th Earl of Derby (1775–1851) English zoologist and founder of Knowsley Menagerie and Museum.

Derbyomyia After the 13th Earl of Derby (v. *derbiana)*; Gr. *muia,* a fly (i.e. a small bird).

Deroptyus Gr. *deire,* the neck; *ptuon,* a fan; ref. distinctive ruff of red and blue feathers of the Red-fan Parrot *D. accipitrinus.*

deserti L. *desertum, deserti,* a desert (*deserere,* to forsake, to abandon).

deserticola L. *desertum,* a desert; *-cola,* a dweller.

desgodinsi After Abbé Auguste Desgodins (1826–1913) French missionary to Tibet.

desmaresti /desmarestii After Anselme Gaetan Desmarest (1784–1838) French zoologist and author (*Histoire naturelle des tangaras, des manakins et des todiers*, 1805).

desmursii After Marc Athanase Parfait Oeillet des Murs (1804–1878) French naturalist and author (*Iconographie ornithologique*, 1849; *La verité sur le coucou*, 1879).

desolata After Desolation Island (Kerguelen Island), so named by Captain Cook because of its isolation and barrenness (L. *desolatus,* forsaken, abandoned; *desolare,* to abandon).

Dessonornis Misspelt from Gr. *bessa,* a wooded valley; *ornis,* a bird.

deva After Deva, a good spirit or god of Hindu myth.

devillei After Émile Deville (1824–1853) French collector in tropical America, 1843–1847.

diabolicus L. *diabolicus,* devilish, diabolical; the natives of Sulawesi believe the call of the Satanic Nightjar *Eurostopodus* to be the sound of the bird taking out someones eyes.

diadema L. *diadema,* a diadem (Gr. *diadema,* a diadem, especially the blue and white band around the Persian kings tiara).

diademata /diadematum /diadematus L. *diadematus,* diademed, adorned with a diadem.

dialeucos Gr. *dialeukos,* marked with white.

diana Roman myth. Diana the huntress, virgin goddess of the moon and the chase.

diardi After Pierre Médard Diard (1795–1863) French collector in the East Indies, 1827–1848.

Diardigallus After Pierre Diard (v. *diardi)*; genus *Gallus,* fowl.

diardii see **diardi.**

diazi After Augustín Díaz (1829–1893) Mexican military engineer, geographer, and explorer.

Dicaeum Gr. *dikaion,* supposedly an Indian bird mentioned by Aelianus, but the name probably refers to the scarab beetle *Scarabaeus.*

dichroa Gr. *dikhrous,* two-coloured (*di-,* two; *khros,* colour).

dichrocephalus Gr. *dikhrous,* two-coloured; *-kephalos*, headed.

Dichromanassa Gr. *di-,* double; *khroma,* colour, complexion; *nassa,* a kind of waterbird, probably some sort of duck (form of *netta)*, but used in orn. to indicate a heron.

dichrous see **dichroa.**

Dichrozona G *dikhrous,* two-coloured; *zone,* a belt or girdle (i.e. banded).

dickeyi After Donald Ryder Dickey (1887–1932) US zoologist and author (*The birds of El Salvador*, 1938).

dickinsoni After J. Dickinson (1832–1863) British doctor and missionary to Nyasaland (Malawi).

dicolorus L. *di-,* two; *colorius,* coloured, variegated.

Dicrocercus Gr. *dikros,* forked; *kerkos,* the tail.

Dicrurus Gr. *dikros,* forked; *-ouros,* tailed.

didinus Syn. genus *Didus,* the Dodo; L. *-inus,* resembling.

Didunculus Dim. of syn. genus *Didus,* Dodo (Portuguese *doudo,* a simpleton); ref. stout, hooked, dodo-like bill of the Tooth-billed Pigeon *D. strigirostris.*

dieffenbachii After Ernst Dieffenbach (1811–1855) German naturalist, collector in New Zealand, 1840, and author (*Travels in New Zealand,* 1843).

diemenensis /diemenianus Erroneous toponyms after Van Diemen's Land, the name by which Tasmania was known until 1856, named for Antony van Diemen 1593–1645, Dutch colonial administrator and patron of Abel Tasman.

difficilis L. *difficilis,* difficult (i.e. of uncertain systematic position).

diffusus L. *diffusus,* diffuse, extensive (*diffundere,* to spread); ref. grey-washed head and neck of the Southern Grey-headed Sparrow *Passer*.

Diglossa Gr. *diglossos,* double-tongued (originally in the sense of being bilingual); the flower-piercers have specialized tongues for nectar feeding.

Diglossopis Genus *Diglossa,* flower-piercer; Gr. *ops,* the face (i.e. appearance).

dignissima L. *dignissima,* very worthy (super. of *dignus,* worthy).

dignus L. *dignus,* worthy, fitting.

dilectissima L. *dilectissimus,* very select, choice (super. of *dilectus,* select, choice; *deligere,* to choose).

diluta L. *dilutus,* weak, slight.

dilutior L. *dilutior,* paler, weaker, more diluted (*dilutus,* diluted, weak).

dimidiata /dimidiatus L. *dimidiatus,* halved, divided (i.e. two-coloured).

dimorpha Gr. *dimorphos,* dimorphic, of two forms; the Papuan Hawk Owl *Uroglaux* was considered to have underparts resembling the Short-eared Owl *Asio otus* and upperparts like the Rufous Owl *Ninox rufa.*

dinellianus After L. M. Dinelli (fl. 1939) naturalist and collector in Argentina 1905–1939.

dinemelli /Dinemellia After Dinemelli (fl. 1840) a collector in Ethiopia about whom nothing is known.

Dinopium Gr. *deinos,* huge, powerful; *ops, opos,* appearance.

Dinornis Gr. *deinos,* huge, terrible; *ornis,* a bird.

diodon Gr. *di-,* two, double; *odon,* a tooth; ref. notched bill of the Rufous-thighed Kite *Harpagus.*

Diomedea Gr. myth. Diomedes, a Greek warrior whose companions were turned into seabirds (cf. L. *aves Diomedias,* unknown seabirds, perhaps gannets *Sula).*

Diopezus Gr. *dios,* mighty, noble; *pezos,* on foot, pedestrian.

diophthalma Gr. *di-,* two, double; *ophthalmos,* the eye; ref. dark blue loral spots of the Double-eyed Fig Parrot *Opopsitta.*

diops Gr. *di-,* two, double; *ops,* the eye; ref. white loral spots.

Dioptrornis Gr. *diopter,* a spy; *ornis,* a bird; the White-eyed Slaty Flycatcher *D. fischeri* has typical wait and pounce flycatcher habits.

Diphogena Gr. *di-,* two, double; *phlogion,* a little flame (dim. of *phlogia* or *phlox,* a flame). This generic name is often spelled *Diphlogena* or *Diphlogaena.*

diphone Gr. *di -,* two, double; *phone,* the voice; the Japanese Bush Warbler *Cettia* is renowned for its distinctive voice, the inspiration for many poems and ballads.

Diphyllodes Gr. *di-,* two, double; *phullon,* a leaf; *-oides,* resembling; ref. the curved tail-wires of the Magnificent Bird of Paradise *D. magnificus.*

Diplochelidon Gr. *diploos,* double; *khelidon,* a swallow; the Black-collared Swallow *D. melanoleuca* was considered to resemble swallows of the genera *Hirundo* and *Orochelidon.*

Diplootocus Gr. *diploos,* double; *ootokos,* egg-laying (*oon,* an egg; *tokos,* birth; *tikto,* to hatch or lay); Moussier's Redstart *Phoenicurus moussieri* probably has two broods.

discolor L. *discolor,* of different colours.

discors L. *discors,* different, unlike, discordant.

Discosura Gr. *diskos,* a plate; *oura,* the tail; ref. racquetted outer tail feathers of the Racquet-tailed Coquette *D. longicauda.*

discurus Gr. *diskos,* a plate; *-ouros,* tailed; the central tail-feathers of the Blue-crowned Racquet-tailed Parrot *Prioniturus* terminate in spatules or discs.

disjuncta L. *disiunctus,* separated, distant (*disiungere,* to loosen); fifteen years separated the collection of specimens and the description of the Yapacana Antbird *Myrmeciza.*

dispar L. *dispar,* different, dissimilar; ref. the great dissimilarity in the sexes of the Allied Greybird *Coracina* (thought to resemble or be allied to two other spp.) and of the Orange-bellied Kingfisher *Ceyx* (totally different in appearance).

disposita L. *dispositus,* orderly, arranged; according to Ripley & Marshall (1967), *Proc. Biol. Soc. Washington,* **80**, 243, 'The name *disposita* refers to the furtive, secretive and skulking habits...' of the Olive-backed Flycatcher *Ficedula.*

Dissemurus L. *disseminare,* to spread; Gr *oura,* the tail.

dissimilis L. *dissimilis,* unlike, dissimilar; of the Black-breasted Thrush *Turdus* Blyth (1847), *Journ. Asiat. Soc. Bengal,* **16**, 144, wrote, 'I succeeded in getting a male, which, as I all along suspected, proved to be clad in not quite so homely a garb as his mate. The male is, indeed, rather a handsome Thrush'.

dissita L. *dissitus,* spread, scattered (*disserere,* to scatter); ref. isolated, insular distribution of the Coiba Spinetail *Cranioleuca.*

Dissoura Gr. *dissos,* divided, double; *oura,* the tail.

distans L. *distans,* different, distinct (*distare,* to be separate).

distincta L. *distincta,* distinct, separate (*distinguere,* to distinguish).

Diuca Araucano (Chilean) Indian names *diuca* or *siuca* for the Common Diuca Finch *D. diuca.*

divaricatus L. *divaricatus,* spread out (*divaricare,* to spread eagle); Raffles (1822), *Trans. Linn. Soc. London,* **13**, 305, described the Ashy Minivet *Pericrocotus* as having the '...tail as long as the body, and forked...'.

Dives L. *dives,* rich, costly; ref glossy silken plumage of the Melodious Blackbird *D. dives.*

dixoni After Charles Dixon (1858–1926) British oologist, journalist, and author.

dodsoni After W. Dodson (fl. 1895) traveller and collector in Arabia and tropical Africa.

dohertyi After William Doherty (1857–1901) US collector.

dohrni/dohrnii After **1** Karl Augustus Dohrn (1806–1901) German entomologist and collector (*Ramphodon*). **2** W. L. Dohrn (1838–1913) German collector on Principe, 1865 (*Horizorhinus*).

dolei After Sanford Ballard Dole (1844–1926) US/Hawaiian judge and politician, President of Hawaiian Republic, 1893–1898.

doliatus Mod. L. *doliatus,* banded, striped (perhaps from L. *dolere,* to suffer pain).

Dolichonyx Gr. *dolikhos,* long; *onux,* a claw; the Bobolink *D. oryzivorus* has a long, lark-like hind claw.

Doliornis Gr. *dolios,* crafty, wily; *ornis,* a bird; the Bay-vented Cotinga *D. sclateri* was described from a single male shot on 9 June 1871, no other specimens being collected.

Dolospingus Gr. *dolos,* false, deceitful; *spingos,* a finch; Elliot (1871), *Ibis,* 402, considered the White-naped Seedeater *D. fringilloides,* '...not unlike the members of the genus *Spermophila*...but the bird precludes such a disposition...there is no genus known to me under which it may be classified'.

domesticus L. *domesticus,* domestic (*domus,* a house).

Domicella Med. L. *domicella,* a damsel, a young lady (dim. of L. *domina,* a lady).

domicellus Med. L. *domicellus,* a young gentleman, a squire (dim. of L. *dominus,* a husband); an erroneous gender change of the feminine substantive *Psittacus Domicella* described by Linnaeus (1758), but subsequently placed in the masculine genus *Lorius.*

domicola L. *domi,* in the house (*domus,* a house); *-cola,* a dweller, inhabitant.

dominica After Santo Domingo, a former name for the island of Hispaniola, named for St Dominic, the Spanish prior Domingo de Guzmán (Spanish *Santo Domingo)* (1170–1221), founder of the Dominican friars.

dominicana/dominicanus Given to species with black and white plumage, in allusion to the black and white habits of the Dominican, Black Friars, or Jacobins. The Dominicans were prominent in the Inquisition and as missionaries to Latin America.

dominicensis/dominicus see **dominica.**

Donacobius Gr. *donax, donakos,* a reed; *bios,* manner of living (i.e. inhabiting).

Donacospiza Gr. *donax, donakos,* a reed; *spiza,* a finch.

donaldsoni After A. Donaldson-Smith (1864–1939) US zoologist and collector in Somaliland, 1894–1895.

dorae After Dora Philby (1888–1957) wife of British explorer and Arabist Harry St John Bridge Philby (v. *philbyi).*

dorbignyi/dorbygnianus After Alcide Dessalines d'Orbigny (1802–1857) French naturalist, explorer in tropical America, 1826–1833 and author (*Voyage dans l'Amérique méridionale*, 1835).

doriae After Marchese Giacomo Doria (1840–1913) Italian naturalist and explorer.

Doricha After Doricha, or Dorikha, a Greek courtesan or hetaira (seventh century BC).

doris Gr. myth. Doris, a sea goddess, daughter of Oceanus and Tethys and mother of the Nereides.

dorotheae After Dorothy White (fl. 1916) daughter of Australian naturalist Henry L. White.

dorsale/dorsalis Med. L. *dorsalis* or *dorsualis,* backed, of the back (L. *dorsum* or *dorsus,* the back).

dorsimaculatus/dorsomaculatus L. *dorsum, dorsi,* the back; *maculatus,* spotted, blotched (*maculare,* to stain).

dorsostriatus L. *dorsum, dorsi,* the back; Mod. L. *striatus,* striped, streaked (L. *stria,* a furrow).

Doryfera Gr. *doru,* a spear; L. *-fera,* carrying; ref. long, straight bills of the Neotropical lancebills (cf. Gr. *doruphoros,* the lance-bearer).

dougallii After Peter McDougall (1777–1814) Scottish physician and collector.

douglasii After David Douglas (1798–1834) British traveller, botanist, and collector in North America, 1823–1834 and Hawaii, 1834.

dowii After Captain J. M. Dow (1827–1892) US explorer, naturalist, and collector.

Drepanis Gr. *drepane,* a sickle or scythe; ref. bill shape of the extinct Hawaiian mamos (cf. Gr. *drepanis,* the sickle-wing, a bird mentioned by Aristotle and Hesychius, latterly identified as a swift *Apus* or a falcon *Falco*).

Drepanoplectes Gr. *drepanon,* a sickle; Mod. L. *plectes,* weaver-bird (Gr. *pleko,* to weave).

Drepanoptila Gr. *drepanon,* a sickle; *ptilon,* a feather, a wing; ref. shape of divided, curved wing feathers of the Cloven-feathered Dove *D. holosericea.*

Drepanorhynchus Gr. *drepanon,* a sickle; *rhunkhos,* the bill.

Drepanornis Gr. *drepanon,* a sickle; *ornis,* a bird; ref. bill shape of the New Guinea sicklebills.

Dreptes Gr. *drepton,* a kiss (*drepto,* to pluck); ref. nectar feeding habits of the Giant Sunbird *D. thomensis.*

dresseri After Henry Eeles Dresser (1838–1915) English traveller, ornithologist, oologist, and author (*A history of the birds of Europe*, 1871; *A monograph of the Meropidae*, 1884; *A monograph of the Coraciidae*, 1893).

Drioctistes Gr. *drios,* a thicket, a copse; *ktistes,* an occupier (*ktizo,* to occupy).

Dromaeocercus Genus *Dromaius,* emu; Gr. *kerkos,* the tail; ref. long, decomposed tail-feathers of the Brown Emu-tail *D. brunneus.*

Dromaius Gr. *dromaios,* fleet of foot, swift; the flightless Emu *D.*

novaehollandiae escapes all enemies by running away.

Dromas Gr. *dromas,* running, whirling (*trekho,* to run); ref. hunting habits of the Crab Plover *D. ardeola.*

Dromococcyx Gr. *dromos,* running; *kokkux,* the cuckoo.

drownei After Frederick P. Drowne (fl. 1930) US traveller and collector in the Galápagos Islands, 1897–1898 and the Solomon Islands, 1930.

dryas Gr. *druas,* a dryad, a tree nymph (*drus,* the oak tree).

Drymocichla Gr. *drumos,* a wood, an oak coppice; *kikhle,* a thrush.

Drymodes Gr. *drumodes,* woody (i.e. of the woods) (considered by some authors to be an error for *drumodos,* a woodland singer).

Drymophila Gr. *drumos,* a wood; *philos,* loving, fond of.

Drymornis Gr. *drumos,* a wood; *ornis,* a bird.

Dryobates Gr. *drus, druos,* the oak tree, but later used of any tree; *-bates,* inhabitant.

Dryocichloides Syn. genus *Dryocichla,* robin chat (Gr. *drus, druos,* the oak tree; *kikhle,* a thrush); Gr.*-oides,* resembling.

Dryocopus Gr. *druokopos,* a woodpecker (*drus, druos,* the oak tree; *kopos,* a cutter).

Dryolimnas Gr. *drus,* the oak tree; Mod. L. *limnas,* a rail or crake.

Dryoscopus Gr. *drus,* the oak tree; *skopos,* a look-out, a guardian.

Dryotriorchis Gr. *drus,* the oak tree; *triorkhes,* a bird of prey, probably the Common Buzzard *Buteo*, mentioned by Aristophanes, Aristotle, and others, so called because it was thought to have three testicles (*tri-,* three; *orkhis,* a testicle).

dubia /dubium /dubius L. *dubius,* doubtful, uncertain (i.e. as to relationships or systematic position).

Dubusia After Vincente Bernard Léonard Vicomte du Bus de Ghisignies (1808–1874) Belgian ornithologist and author (*Esquisses ornithologiques*, 1845).

duchaillui After Paul Belloni du Chaillu (1838–1903) French/US explorer and anthropologist.

ducorps After Adjutant L. J. Ducorps (fl. 1858) French naval administrator and explorer.

Ducula Nepalese name *dukul* for the Mountain Imperial Pigeon *D. badia.*

dufresniana After L. Dufresne (1752–1832) French conchologist.

dugandi After A. Dugand (b. 1906) Colombian zoologist.

duidae After Mount Duida (Cerro Duida), Amazonas, Venezuela.

duivenbodei After Maarten Dirk van Renesse van Duyvenbode (fl. 1860) Dutch merchant and planter in the Moluccas.

dulcis L. *dulcis,* charming, delightful.

Dulus Gr. *doulos,* a slave; according to Brisson (1760) the Palm Chat *D. dominicus* was called *Esclave,* the slave, by the people of Santo Domingo because of its dark colour and subservient behaviour compared to the native tyrant flycatcher or kingbird *Tyrannus.*

dumasi After M. J. Dumas (d. 1917) traveller and collector in the Moluccas.

dumetaria L. *dumetum,* a thicket; *-arius,* relating to, pertaining to.

Dumetella Dim. from L. *dumetum,* a thicket; the Grey Catbird *D. carolinensis* inhabits dense thickets.

Dumetia L. *dumetum, dumeti,* a thicket; *-ius,* indicating characteristic of.

dumetoria see **dumetaria.**

dumetorum L. *dumetorum,* of the thickets.

dumicola L. *dumicola,* a thicket-dweller (*dumus, dumi,* a thorn bush; *-cola,* a dweller).

dumontii No expl. (Lesson (1827), *Bull Sci. Nat.*, **10**, 159.) After either Charles Henri Frédéric Dumont de Sainte Croix (d.1830) French ornithologist and father in law of Lesson, or Admiral Jules Sebastien César Vicomte Dumont d'Urville (d. 1842) French explorer and surveyor.

dunni After Colonel Henry Nason Dunn (1864–1952) British big-game hunter.

dupetithouarsii After Admiral Abel Aubert Dupetit-Thouars (1793–1864) French explorer in the Pacific, 1836–1839.

Dupetor Gr. *doupetor,* a clatterer; ref. drumming or booming calls of the Black Bittern *D. flavicollis.*

duponti /dupontii After **1** Léonard Puech Dupont (1795–1828) French naturalist, explorer, and collector *(Chersophilus).* **2** Monsieur Dupont (d.1873) French natural history dealer *(Tilmatura).*

dussumierii After J. J. Dussumier (1792–1883) French traveller and collector.

duvaucelii After Alfred Duvaucel (1796–1824) French naturalist and collector in Sumatra.

duyvenbodei see **duivenbodei.**

dwighti After J. Dwight (1858–1929) US ornithologist.

Dyaphorophyia Gr. *diaphoros,* different; *phuo,* to be; the different sexes of the Chestnut Wattle-eye *D. castanea* had formerly been considered two separate spp.

dybowskii After Jan Dybowski (1856–1928) Polish traveller in tropical Africa, 1892.

Dysithamnus Gr. *duo,* to enter; *thamnos,* a bush.

Dysmorodrepanis Gr. *dusmoros,* ill-fated; genus *Drepanis,* mamo; the extinct, curve-billed Lanai Ou *D. munroi* is known from only one specimen.

Dacelo novaeguineae Laughing Kookaburra

E

earlei After Willis Earle (fl. 1855) English merchant in India, 1844.

eatoni After the Revd. Alfred Edmond Eaton (1845–1929) English naturalist and explorer.

eburnea L. *eburneus,* ivory-coloured, ivory white (*ebur,* ivory).

ecaudata /ecaudatus L. *ex-* (*e-* before *c),* lacking; *caudatus,* tailed (*cauda,* a tail); indicates a very short tail rather than the lack of one.

echo L. *echo,* an echo, a woodnymph (Gr. *ekho,* a woodnymph).

Eclectus Gr. *eklektos,* chosen, select.

Ectopistes Gr. *ektopistes,* a wanderer (i.e. a migrant).

edithae After Edith Cole (1859–1949) British botanist and entomologist in Somaliland, 1895.

Edithornis After Edith Baker (fl. 1937) wife of US financier George Fisher Baker; Gr. *ornis,* a bird.

edolioides Syn. genus *Edolius,* drongo (French name *edolio* for the drongo *Dicrurus*, said to be from an African native name, in imitation of its call, but Hesychius refers to Gr. *eidalis,* a sort of bird, not further identified, and to *edolios,* a sort of heron (? an error for *erodios*)); Gr. *-oides*, resembling.

edouardi After Édouard Verreaux (1810–1868) French natural history dealer (v. *verreauxi).*

eduardi After Edward Newton (fl. 1862) '...a gentleman who has recently visited Madagascar, and whose zealous efforts have very materially forwarded our knowledge of the ornithology of the East African archipelago' (Hartlaub (1862), *Proc. Zool. Soc. London*, 152).

edward After Edward Wilson (fl. 1857) British naturalist and collector.

edwardsi /edwardsii After **1** Henri M. Milne-Edwards (1800–1885) French zoologist and author (*Cours élémentaire de zoologie*, 1834) (*Bangsia, Calonectris, Carpodacus*). **2** Alphonse Milne-Edwards (1835–1900) French zoologist and author (*Histoire naturelle des oiseaux de Madagascar*, 1876) (*Lophura, Psittaculirostris).*

egertoni After Sir Philip de Malpas Grey-Egerton (1806–1881) English palaeontologist.

egregia L. *egregius,* admirable, excellent, extraordinary.

Egretta French *aigrette,* an egret or little heron (Anglo-Norman *aigret).*

eichhorni After Albert F. Eichhorn (d. 1913) Australian farmer and collector in New Guinea, 1900 and the Solomon Islands, 1903, 1907.

eidos Gr. *eidos,* known (*eido,* to know).

eisenmanni After Eugene Eisenmann (1906–1981) US ornithologist and author (*The species of Middle American birds*, 1955).

eisentrauti After M. Eisentraut (b. 1902) German zoologist and collector.

ekmani After Eric Ekman (1885–1939) botanist in the West Indies, 1928.

elachus Gr. *elakhus,* small.

Elaenia Gr. *elainios,* oleaginous, oily green (cf. *elaina,* a small bird that frequented the olive groves mentioned by Alexandrios, perhaps some kind of tit *Parus* or a warbler *Sylvia* or *Hippolais* (*elaion,* the olive grove)).

Elainopsis Genus *Elaina* (a former spelling of *Elaenia)*, elaenia; Gr. *opsis*, appearance.

Elanoides Genus *Elanus*, kite; Gr. *-oides*, resembling.

Elanus Gr. *elanos*, a kite (also known as *iktinos)*.

elaphrus Gr. *elaphros*, small, nimble.

elata/elatus L. *elatus*, exalted, proud (*efferre*, to raise up).

Electron Gr. *elektron*, amber.

elegans L. *elegans*, fine, choice.

elegantior L. *elegantior*, more fine, neater (comp. of *elegans)*.

elegantissima L. *elegantissima*, very fine (super. of *elegans)*.

elegantula Dim. from L. *elegans*, fine, choice.

eleonorae After Giudicessa Eleonora d'Arborea (1350–1404) heroine of Sardinia's struggle for independence against Aragon, who made laws protecting nesting birds of prey.

Eleothreptus Gr. *heleothreptus*, marsh-bred (*helos*, a marsh; *trepho*, to thrive); ref. swamp habitat of the Sickle-winged Nightjar *E. anomalus*.

elgini After James Bruce, 8th Earl of Elgin (1811–1863) Governor-General of India, 1860–1863.

eliciae After Elicia Alain (fl. 1846) about whom nothing else is known.

elisabeth After Queen Isabel I of Castile and Spain (1451–1504: reigned 1474–1504) 'la Reina inmortal' (Lembeye (1850), *Aves de la isla de Cuba*, p. 39). It is curious that Lembeye used the English equivalent, Elisabeth, of the Spanish name Isabel.

eliza After Eliza Lefèvre (fl. 1839) wife of French zoologist Amadée Lefèvre.

ellenbecki After Dr Ellenbeck (fl.1904) German collector in Ethiopia.

ellioti/elliotii After Daniel Giraud Elliot (1835–1915) US ornithologist and author (*A monograph of the Pittidae*, 1861; *A monograph of the Phasianidae*, 1870; *A monograph of the Paradiseidae*, 1873; *A monograph of the Bucerotidae*, 1877)

ellisianus After William Ellis (fl. 1780) British surgeon's mate on Cook's third voyage, artist, and collector.

Elminia After the town and port of Elmina, Gold Coast (Ghana).

elpenor 'After Elpenor, a member of Odysseus' crew who, like the rail, was stranded on an island and fell to his death' (Olson in Ripley (1977), p. 354).

elphinstonii After the Hon. Mountstuart Elphinstone (1779–1859) British diplomat and Governor of Bombay, 1819–1827.

Elseyornis After Joseph R. Elsey (1834–1858) English explorer and naturalist; Gr. *ornis*, a bird.

eludens L. *eludens*, evading, mocking (*eludere*, to defeat).

Elvira No expl. (Mulsant Verreaux and Verreaux (1866), Essai d'une Classification Méthodique des Trochilidés, in *Mém. Soc. Imp. Sci. Nat. Cher.*, **12**, 176), but probably after Elvire 'the name by which the poet Lamartine idealised Mme. Julie Charles...with whom he was at one time deeply in love' (Harvey and Heseltine (1959), p. 243).

Emarginata L. *emarginatus*, emarginated, marginned; ref. indented second primary of the Sickle-winged Chat *Cercomela sinuata*.

Emberiza Old German names *emmeritz* and *embritz* for a bunting.

Emberizoides Genus *Emberiza*, bunting; Gr. *-oides*, resembling.

Embernagra French name *Embernagre* given to the Great Pampa Finch *E. platensis* by Lesson (1831), *Traité d'ornithologie*, p. 465 (genus *Emberiza*, bunting; syn. genus *Tanagra*, tanager); these finches were considered to be intermediate between buntings and tanagers, hence their alternative name *tangaras-bruants* or tanager-buntings.

Emblema L. *emblema*, inlaid or mosaic work; ref. complicated plumage patterns of the Painted Firetail *E. picta*.

emiliae After **1** Emilie Hose (fl. 1888) wife of Charles Hose (v. *hosii)*

(*Chlorocharis*). **2** Emilia Snethlage (1868–1929) German ornithologist and author (*Microrhopias*).

emiliana After Émile Parzudaki (d. 1862) French natural history dealer, traveller, and collector.

eminentissima L. *eminentissimus*, very conspicuous, outstanding (super. of *eminens*, prominent).

emini /Eminia/eminibey After Emin Bey, later Emin Pasha, the adopted name of Eduard Schnitzler (1840–1892) German administrator in the Ottoman service, Pasha (Governor) of Equatorial Province, Egyptian Sudan, 1878–1889, naturalist, and collector. The title bey is of lower rank than pasha.

emmae After Emma Gräfin von Berlepsch (fl. 1891) wife of German ornithologist Hans Graf von Berlepsch (v. *berlepschi*).

emphanum Gr. *emphanes*, reflecting, shining back.

Empidonax Gr. *empis, empidos*, a gnat or mosquito; *anax*, a lord or king.

Empidonomus Gr. *empis*, a gnat or mosquito; *nemo*, to rule over.

Empidornis Gr. *empis*, a gnat or mosquito; *ornis*, a bird.

enarratus L. *enarratus*, explained in detail (*enarrare*, to explain); Gray had a difference of opinion with Sharpe regarding the provenance of the Collared Nightjar *Caprimulgus* and sought to explain the correct situation and history with his description (Gray (1871), *Ann. & Mag. Nat. Hist.*, [4] **8**, 428).

enca Javanese (Malayan) names *enca* or *engkak* for a crow.

Endomychura Gr. *endomukhos*, secret, lurking within; *oura*, the tail; these small Nearctic murrelets have very short tails almost hidden by tail-coverts.

enganensis After the island of Engano (Enggano), Sumatra.

Enicognathus Gr. *henikos*, unique, strange; *gnathos*, the lower jaw; ref. slender, curved bill of the Slender-billed Conure *E. leptorhynchus*.

enicura /Enicurus Gr. *henikos*, unique, strange; *oura*, the tail; ref. the long, graduated, deeply-forked tails of the male Slender Sheartail *Doricha* and the Indomalayan forktails.

enigma see **aenigma.**

Enneoctonus Gr. *ennea*, nine; *kteino*, to kill; in Germany the Great Grey Shrike *Lanius excubitor* was traditionally known as *neuntöter*, nine killer, because it was supposed not to rest until it had killed nine birds every day.

Enodes L. *enodis*, smooth, plain (cf. Gr. *enodas*, whiskered).

Ensifera L. *ensifera*, sword-bearing (*ensis*, a sword; *-fera*, carrying, bearing).

ensipennis L. *ensis*, a sword; *pennis*, winged (*penna*, a feather); ref. sabre-like primary shafts of the White-tailed Sabrewing *Campylopterus*.

Entomodestes Gr. *entoma*, insects; *edestes*, an eater, a devourer (*edo*, to eat).

Entomyzon Gr. *entoma*, insects; *muzo*, to suck (i.e. a honeyeater).

Entotriccus Gr. *entomos*, cut up; Mod. L. *triccus*, tyrant flycatcher; ref. the narrow, falcated primaries of the Cinereous Tyrant *E. striaticeps*.

enucleator L. *enucleare*, to remove the kernel.

Eolophus Gr. *eos*, the dawn; *lophos*, a crest; ref. pale rose-pink crest of the familiar Australian Galah *E. roseicapillus*.

Eopsaltria Gr. *eos*, the dawn; *psaltria*, a singer.

Eos Gr. *eos*, the dawn, the east; ref. the bright red plumage and East Indian home of the Red-and-blue Lory *E. histrio*, and the deep pink plumage of the Pink-rumped Rosefinch *Carpodacus*.

epauletta French *épaulette*, a shoulder ornament (*épaule*, the shoulder); from appearance of white scapular stripe of the Gold-naped Finch *Pyrrhoplectes*.

Ephippiorhynchus Gr. *ephippios*, a saddle; *rhunkhos*, the bill; ref. bright yellow frontal shield or saddle on bill of the Saddle-billed Stork *E. senegalensis*.

Ephthianura Gr. *e-*, for euphony; *phthinas,* decreasing (cf. *ephthos,* refined); *oura,* the tail; Gould (1838), *Synop. Bds. Austr.*, Pt. **4**, App.3, described the White-faced Chat *E. albifrons* as having the '...tail short and truncate'.

epichlora Gr. *epi-*, excessively; *khloros,* green.

epilepidota Gr. *epi-*, excessively; *lepidotos,* scaly (*lepis,* a scale).

Epimachus Gr. *epimakhos,* a fighter; ref. scimitar bills of the New Guinea sicklebills.

episcopus Med. L. *episcopus,* a bishop; **1** a rendering of *L' Évesque*, the bishop, Brisson's name for the Blue-grey Tanager *Thraupis* on account of its episcopal blue plumage. **2** ref. the episcopal black skullcap of the Woolly-necked Stork *Ciconia.*

epomidis Gr. *epomidios,* on the shoulder (*epomis, epomidos,* the shoulder).

epomophora Gr. *epomis,* the shoulder; *-phoros,* carrying; ref. large, white, lozenge-shaped patches on the shoulders of the Royal Albatross *Diomedea.*

epops Gr. *epops,* the Hoopoe *Upupa.*

epulata L. *epulatus,* satiated, feasted (*epulari,* to feast); Cassin (1855), *Proc. Acad. Nat. Sci.*, **7**, 326, considered the Little Grey Flycatcher *Muscicapa,* 'A small but robust species... Its general form is much as though it had been accustomed to good cheer on the banks of the river Moonda, or wherever else it may have found sufficient entertainment'.

eques L. *eques,* a cavalryman; ref. military red and black plumage of the Red-spot Myzomela *Myzomela.*

Erannornis Gr. *erannos,* lovely; *ornis,* a bird.

Erator L. *errator,* a wanderer; ref. uncertain generic status of this group of tityras, originally proposed as a subgenus.

erckelii After Theodor Erckel (1811–1897) German taxidermist and collector in Ethiopia, 1830–1834.

Eremalauda Gr. *eremos,* a desert, a wilderness; genus *Alauda,* lark.

Eremialector Gr. *eremia,* a desert; *alektor,* a domestic chicken or fowl.

Eremiornis Gr. *eremia,* a desert; *ornis,* a bird.

eremita L. *eremita,* a hermit (*eremus,* a desert); ref. solitary habitat or distribution.

Eremobius Gr. *eremos,* a desert; *bios,* manner of living.

Eremomela Gr. *eremos,* a desert; *melos,* a song (i.e. a songster).

Eremophila Gr. *eremophiles,* desert-loving, solitude loving.

Eremopterix Gr. *eremos,* a desert; *pterux,* a bird (*pteron,* a feather).

Ereunetes Gr. *ereunetes,* a searcher (*ereunao,* to seek).

Ergaticus Gr. *ergatikos,* hard-working, busy.

ericetorum Med. L. *ericetorum,* of the heaths (Med. L. *ericetum,* a heath; L. *erice,* heather; Gr. *ereike,* heather, broom).

Eriocnemis Gr. *erion,* wool, cotton; *knemis,* a boot, legging (*kneme,* the shin); from large white or black leg-puffs of the Neotropical pufflegs.

Erismatura Gr. *ereisma, ereismatos,* a prop or stay; *oura,* the tail.

erithachorides Adopted from the pre-Linnaean name *Chloris erithachorides* given to the Mangrove Warbler *Dendroica* by Louis Feuillée (1714), *Observations physiques, mathématiques et botaniques, faites sur les côtes de l'Amérique Méridionale* (Mod. L. *erithacus*, robin; Gr. *-ides*, resembling).

Erithacus L. *erithacus,* a winter bird that changed into the redstart *Phoenicurus* in summer, mentioned by Pliny (Gr. *erithakos,* a bird mentioned by Aristotle, Hesychius, and others, usually identified as the robin, but also as some sort of mimetic bird). Used specifically for species with red breasts like the robin (*Ceyx, Pyrrhura)* or for the mimetic African Grey Parrot (*Psittacus).*

erlangeri After Baron Carlo von Erlanger (1872–1904) German naturalist and collector in Tunisia, 1893 and 1897, and tropical Africa, 1900–1901.

Erolia No expl. (Vieillot (1816), *Analyse*, p. 55); perhaps coined arbitrarily, or an error for *Acrolia* (Gr. *akros*, edge of the tide or beach).

erythaca see **Erithacus.**

erythrauchen Gr. *eruthros*, red; *aukhen*, the neck, the throat.

Erythrinus L. *erythros*, red (Gr. *eruthros*, red); *-inus*, possessing.

erythrocephala/erythrocephalus Gr. *eruthros*, red; *-kephalos*, headed.

erythrocerca/Erythrocercus Gr. *eruthros*, red; *kerkos*, the tail.

erythrochlamys Gr. *eruthros*, red; *khlamus*, a mantle, a cloak (i.e. upperparts).

erythrocnemis Gr. *eruthros*, red; *kneme*, the leg, the shin.

Erythrocnus Gr. *eruthros*, red; *oknos*, a kind of heron.

erythrogaster Gr. *eruthros*, red; *gaster*, the belly.

erythrogenys Gr. *eruthros*, red; *genus*, the cheek.

Erythrogonys Gr. *eruthros*, red; *gonu*, the knee.

erythroleuca Gr. *eruthroleukos*, red and white, reddish-white.

erythrolophus Gr. *eruthros*, red; *lophos*, a crest.

erythromelas Gr. *eruthromelas*, red and black, blackish-red.

erythronemius see **erythrocnemis.**

erythronota/erythronotos/erythronotus Gr. *eruthros*, red; *-notos*, backed.

erythrophris/erythrophrys Gr. *eruthros*, red; *-ophrus*, browed (*ophrus*, the eyebrow).

erythrophthalma/erythrophthalmus Gr. *eruthros*, red; *ophthalmos*, the eye.

erythropleura Gr. *eruthros*, red; *pleura*, a rib (i.e. the side).

erythrops Gr. *eruthros*, red; *ops*, the eye, the face; ref. red or rufous face (*Cisticola, Cranioleuca, Myiagra, Odontophorus, Quelea*), spectacles (*Climacteris*), or iris (*Neocrex*).

erythroptera/erythropterus Gr. *eruthros*, red; *-pteros*, winged.

erythropthalmus see **erythrophthalma.**

erythropus Gr. *eruthropous*, red-footed (*eruthros*, red; *pous*, the foot).

Erythropygia/erythropygius Gr. *eruthros*, red; Mod. L. *pygius*, rumped (Gr. *-pugos*, rumped; *puge*, the rump).

erythropygus Gr. *eruthros*, red; *-pugos*, rumped (*puge*, the rump).

erythrorhyncha/erythrorhynchos/erythrorhynchus Gr. *eruthros*, red; *rhunkhos*, the bill.

Erythrosterna Gr. *eruthros*, red; *sternon*, the breast.

erythrostictus Gr. *eruthros*, red; *stiktos*, spotted, dappled.

Erythrothlypis Gr. *eruthros*, red; Mod. L. *thlypis*, warbler or thin-billed tanager.

erythrothorax Gr. *eruthros*, red; *thorax*, the breast.

erythrotis Gr. *eruthros*, red; *-otis*, eared.

Erythrotriorchis Gr. *eruthros*, red; *triorkhes*, a bird of prey (v. *Dryotriorchis*).

Erythrura/erythrurus Gr. *eruthros*, red; *oura*, the tail.

Esacus Gr. *aisakos*, a shorebird, not further identified. In myth. Esacus or Aesakos, son of Priam, was metamorphosed into a shorebird or seabird.

esculenta L. *esculentus*, edible (*esca*, food).

estella No expl. (d'Orbigny and Lafresnaye (1838), *Synopsis Avium* in *Mag. Zool.*, Cl. 2, 36); perhaps based on a personal name Estelle (L. *stella*, a star).

estherae After Esther Finsch (fl. 1957) daughter of German ornithologist Otto Finsch.

Estrilda Adapted from *Loxia astrild* (q.v.) the name originally given to the African Waxbill *Estrilda*. The generic name has been amended to *Astrilda* or *Estrelda*. It has been suggested that the name derives from a German or Dutch avicultural term for a waxbill. However, Reichenbach (1850), *Av. syst. nat.*, wrote that the name is not German (nor anything else!) and castigated the linguistically proficient authors who should have known better than to foist it on the scientific community.

etchecopari After Robert Daniel Etchécopar (b. 1906) French ornithologist and author (*Les oiseaux du nord de l'Afrique*, 1964; *Les oiseaux du proche et du moyen Orient*, 1970; *Les oiseaux de Chine, de Mongolie et de Corée*, 1978).

ethologus L. *ethologus,* a mimic (i.e. closely resembling) (Gr. *ethologos,* a mimic).

Eubucco Gr. *eu-*, good (in orn. *eu-* implies abundance, greatness, or a true relationship); Mod. L. *bucco,* barbet.

Eucephala Gr. *eu-,* good; *kephale,* the head.

euchloris/euchlorus Gr. *eukhloros,* greenish.

euchrysea Gr. *eukhrusos,* rich in gold.

Eucometis Gr. *eukomes,* lovely-haired.

eucosma Gr. *eukosmos,* decorous, well adorned.

Eudocimus Gr. *eudokimos,* famous, glorious.

Eudosia No expl. (Mulsant (1876), *Annales de la Société Linnéenne de Lyon,* (sér. nouv.) **22**, 228); probably after the Empress Eudosia (d. AD 460) philosopher and wife of the Byzantine Emperor Theodosius II (reigned 402–450).

Eudromia/Eudromias Gr. *eudromia,* swiftness.

Eudynamys Gr. *eu-,* good; syn. genus *Dynamene,* koel (Gr. myth Dunamene, one of the Nereids) (cf. Gr. *eudunamos,* mighty).

Eudyptes Gr. *eu-,* good; *duptes,* a diver.

Eudyptula Dim. of genus *Eudyptes,* penguin.

Eugenes Gr. *eugenes,* noble, proud.

Eugenia/eugeniae After Empress Eugénie of the French (1826–1920: reigned 1853–1870) wife of Napoléon III.

Eugerygone Gr. *eu-,* good; genus *Gerygone,* gerygone, warbler.

Eugralla Gr. *eu-,* good; L. *grallae,* stilts; ref. long legs and cursorial habits of the Ochre-flanked Churrín *E. paradoxa.*

Eulabeornis Gr. *eulabes,* cautious, wary; *ornis,* a bird.

Eulabes Gr. *eulabes,* pious, devout (v. *religiosa).*

Eulacestoma Gr. *eulaka,* a ploughshare; *stoma,* the mouth; ref. strange plough-shaped bill of the Wattled Ploughbill *E. nigropectus.*

Eulampis Gr. *eulampes,* bright-shining.

euleri After Carl Euler (1834–1901) Swiss Vice-Consul in Rio de Janeiro, Brazil, 1867–1901.

Eulidia French *Eulidie,* the name given to the Chilean Woodstar *E. yarrellii* by Mulsant and Verreaux (1877), *Histoire naturelle des oiseaux-mouches*, Vol. **4**, p. 114. Perhaps from Gr. *eu-*, good, and Lydia, a mistress of the poet Horatius Flaccus, or Lydia, the wealthy and celebrated ancient kingdom of Asia Minor.

Eulipoa Gr. *eu-,* good; syn. genus *Lipoa,* megapode (v. *Leipoa*).

eulophotes Gr. *eulophos,* well crested; *-otes,* featuring (*eidos,* form, appearance).

Eumomota Gr. *eu-,* good; genus *Momotus,* motmot.

Eumyias Gr. *eu-,* good; Mod. L. *myias,* flycatcher.

Euneornis Gr. *eu-,* good; *neos,* new; *ornis,* a bird.

eunomus French *Merle eunome,* the name given to the Dusky Thrush *Turdus* by Temminck (1831), *Planches coloriées,* Vol. **87**, pl. 514, apparently in reference to its large organized winter flocks (Gr. *eunomos,* orderly).

Eunymphicus Gr. *eu-,* good; genus *Nymphicus,* parakeet.

Euodice Gr. *eu-,* good; *odikos,* musical, singing (*ode,* a song).

euophrys Gr. *euophrus,* fine-browed, with fine eyebrows.

euops Gr. *euopsia,* good looks.

eupatria Gr. *eupatereia,* a noble daughter.

Eupelia Gr. *eu-,* good; *peleia,* a dove or pigeon.

Eupetes Gr. *eupetes,* favourable, fortunate.

Eupetomena Gr. *eu-,* good; *petomenos,* flying.

Euphagus Gr. *eu-,* good; *phagos,* a glutton.

Eupherusa Gr. *eu-,* good; *pherousa,* bearing (*phero,* to carry).

Euphonia Gr. *euphonia,* fine-voiced.

Euplectes Gr. *eu-,* good; *plektos,* woven, plaited (*pleko,* to weave); ref. complex, woven nests of the Afrotropical bishops.

Eupoda Gr. *eupous, eupodos,* with good feet, swift of foot.

Eupodotis Gr. *eupodia,* swift, speedy; genus *Otis,* bustard.

eupogon Gr. *eu-,* good; *pogon,* a beard.

Euptilotis Gr. *eu-,* good; *ptilon,* a feather; *-otis,* eared; ref. unique ear-tufts of the Eared Trogon *E. neoxena.*

eurhythmus Gr. *euruthmos,* orderly, graceful, well proportioned; both adult and immature Schrenk's Bittern *Ixobrychus* were considered equally handsome.

Eurillas Gr. *eurus,* broad, wide; *illas,* a thrush (in orn. *illas* is used for an assortment of unrelated, supposedly thrush-like birds); ref. broad bill of the Little Greenbul *E. virens.*

eurizonoides Syn. genus *Euryzona,* rail (Gr. *eurus,* broad; *zone,* a band or girdle); Gr. *-oides,* resembling.

Eurocephalus Gr. *eurus,* broad, wide; *-kephalos,* headed; from large-headed appearance of the white-crowned shrikes.

europaea /europaeus L. *Europaeus,* European, belonging to Europe.

Eurostopodus Gr. *eurostos,* strong, powerful; *pous, podos,* the foot.

Euryapteryx Gr. *eurus,* broad, wide; genus *Apteryx,* kiwi.

Euryceros Gr. *eurus,* broad, wide; *keros,* a horn; ref. massive casqued bill of the Madagascar Helmetbird *E. prevostii.*

eurygnatha Gr. *eurus,* broad, wide; *gnathos,* the lower jaw (i.e. the bill).

Eurylaimus Gr. *eurus,* broad, wide; *laimos,* the throat; Horsfield (1821), *Trans. Linn. Soc. London,* **13**, 170, wrote of the Banded Broadbill *E. javanicus,* 'The distinguishing characters of this genus are a broad, depressed, cuneated beak (passing backwards into a narrow incurved margin, which extends beneath the eyes and forms an excessively wide throat)…'.

eurynome Gr. myth. Eurynome, daughter of Oceanus and Tethys or Thetis, mother of Leucothoe.

Eurynorhynchus Gr. *euruno,* to widen, make broad; *rhunkhos,* the bill; ref. spatulate bill tip of the Spoon-billed Sandpiper *E. pygmaeus.*

euryptera Gr. *eurus,* broad, wide; *-pteros,* winged.

Euryptila Gr. *eurus,* wide, broad; *ptilon,* a feather; ref. broad primary feathers of the Kopje Warbler *E. subcinnamomea.*

Eurypyga Gr. *eurus,* wide, broad; *puge,* the rump; ref. long, broad tail of the Sun Bittern *E. helias.*

eurystomina Gr. *eurustomos,* wide-mouthed; *-inos,* possessing.

Eurystomus Gr. *eurustomos,* wide-mouthed (i.e. broad-billed).

euryura Gr. *eurus,* broad, wide; *-ouros,* tailed.

euryzona Gr. *eurus,* broad, wide; *zone,* a girdle (i.e. a band).

Euscarthmus Gr. *euskarthmos,* high leaping; ref. active feeding habits of the pygmy tyrants.

Euschistospiza Gr. *eu-,* good; syn. genus *Schistospiza,* grey-crested finch (Late. L. *schistus,* slate; Gr. *spiza,* a finch).

euteles Gr. *euteles,* shabby, mean.

Euthlypis Gr. *eu-,* good; Mod. L. *thlypis,* warbler.

eutilotus Gr. *eu-,* good; *ptilotos,* feathered; ref. thickly feathered lower back of the Puff-backed Bulbul *Pycnonotus*.

Eutoxeres Gr. *eu-*, good; *toxeres*, an archer, armed with a bow (*toxon*, a bow); ref. sickle- or bow-shaped bills of the sicklebill hummingbirds.

Eutrichomyias Gr. *eu-*, good; *thrix, trikhos*, the hair; Mod. L. *myias*, flycatcher; the Cerulean Paradise Flycatcher *E. rowleyi* is a rare, perhaps extinct, species remarkable for its very long rictal bristles or 'hairs'.

Eutriorchis Gr. *eu-*, good; *triorkhes*, a type of hawk (v. *Dryotriorchis*).

Euxenura Gr. *eu-*, good; *xenos*, strange; *oura*, the tail.

evelynae No expl. (Bourcier (1847), *Proc. Zool. Soc. London*, **15**, 44), where the Bahama Woodstar *Calliphlox* is first referred to as *Troch. d'Evelyn;* a female eponym.

everetti After Alfred Hart Everett (1848–1898) English administrator and collector in the East Indies.

eversmanni After Eduard Friedrich Eversmann (1794–1860) German naturalist, collector in Kirghizia and Siberia, and author (*Addenda ad celeberrimi Pallasii Zoographiam Rosso-Asiaticam, Aves*, 1835).

ewingi After the Revd. Thomas J. Ewing (d. 1876) Australian teacher, naturalist, and collector.

exarhatus L. *exaratus*, ploughed up, furrowed (*exarare*, to plough up); ref. ridge-like markings on bill and casque of the White-faced Hornbill *Penelopides*.

Excalfactoria L. *excalfactorius*, warming, heating; Jardine (1838), *The Naturalist's Library*, wrote of these small quails, 'They are here [China] domesticated and trained to fight...They are also used by the Chinese to warm their hands in cold weather, their bodies being thought to contain a large proportion of animal heat from the pugnacious disposition of their tempers'.

excellens L. *excellens*, eminent, remarkable (*excellere*, to be distinguished).

excelsa L. *excelsus*, distinguished, eminent, lofty.

excelsior L. *excelsior*, more distinguished, higher (comp. of *excelsus*, distinguished).

excubitor L. *excubitor*, a sentinel, a guard; the Great Grey Shrike *Lanius* 'derives its trivial designation from the use made of it as a sentinel by falconers when catching wild hawks' (Newton and Gadow (1896), p. 66).

excubitoroides Specific name *Lanius excubitor;* Gr. *-oides*, resembling.

exilis L. *exilis*, little, slender.

eximia /eximium /eximius L. *eximius*, exceptional, select, distiguished (*eximere*, to take away).

exortis L. *exortus*, the east (*exoriri*, to rise); the large rosy throat spot of the Tourmaline Sunangel *Heliangelus* resembles the rising sun.

explorator L. *explorator*, an explorer or scout (*explorare*, to search out). **1** Based on the French name *L'Espionneur*, the lookout, given to the Sentinel Rock Thrush *Monticola* by Levaillant (1802) because of its habit of perching prominently in an erect stance. **2** After the HMS *Challenger* Deep Sea Exploratory Expedition to the South Seas, 1873–1876 (*Zosterops*).

exquisita L. *exquisitus*, choice, exquisite (*exquiere*, to seek out).

exsul L. *exsul*, an exile, a banished person; ref. isolated range of the Chestnut-backed Antbird *Myrmeciza*.

externa L. *externus*, a stranger, a foreigner.

exulans L. *exsulans*, wandering (*exsulare*, to be exiled, banished) (i.e. wide-ranging).

exustus L. *exustus*, burnt, scorched (*exurere*, to burn up); ref. dark brown coloration of the Chestnut-bellied Sandgrouse *Pterocles*.

eytoni After Thomas Chalmers Eyton (1809–1880) English ornithologist and author (*Monograph of the Anatidae or duck tribe*, 1838).

F

fabalis L. *fabalis,* of the beans (*faba,* the broad bean); ref. diet of the Bean Goose *Anser*.

fagani After Charles Edward Fagan (1855–1921) British ornithologist and collector in tropical Africa, 1905–1906 and New Guinea, 1910–1912.

faiostricta Gr. *phaios,* grey, dusky; *stiktos,* dappled, spotted.

falcata /falcatus L. *falcatus,* sickle-shaped, furnished with sickles (*falx, falcis,* a sickle); from shape of scapular (*Anas*), rump (*Ptilocichla*), or primary feathers (*Campylopterus*).

falcinellus Dim. from L. *falx, falcis,* a sickle; ref. shape of bill (*Plegadis*) or bill-tip (*Limicola*).

Falcipennis L. *falx,* a sickle; *pennis,* winged (*penna,* a feather); ref. very narrow outer primary tips of the Sickle-winged Grouse *F. falcipennis*.

falcirostris L *falx,* a sickle; *-rostris,* billed.

falcklandii After the Falkland Islands in the South Atlantic Ocean, named for Anthony Cary, 5th Viscount Falkland (1656–1694), Navy Treasurer, 1690.

Falco Late L. *falco,* a falcon (L. *falx,* a sickle; ref. the hooked talons).

Falculea L. *falcula,* a small sickle; from bill shape of the Sickle-billed Vanga *F. palliata*.

falcularius L. *falcula,* a small sickle; *-ius,* resembling.

Falcunculus Dim. from L. *falcula,* a small sickle; ref. robust bill and high arched culmen of the Crested Shriketit *F. frontatus*.

falkensteini After Johann Falkenstein (1842–1917) German surgeon and collector in tropical Africa, 1873–1876.

falklandicus see **falcklandii.**

fallax L. *fallax,* deceitful, false (*fallere,* to deceive) (i.e. previously undescribed, confused with other species, or thought to be in the wrong genus); for example, Schlegel (1866), *Ned. Tjdschr. Diek*., **3**, 187, considered the Sulawesi Kingfisher *Ceyx* to be falsely allied to *Ceyx* by reason of its small size and tridactyl toes and described it in *Dacelo*.

familiare /familiaris L. *familiaris,* friendly, familiar (i.e. common or confiding).

famosa L. *famosus,* renowned (*fama,* talk).

fanny /fannyi After Frances 'Fanny' Wilson (fl. 1846) wife of British collector Edward Wilson.

fanovanae After the Fanovana Forest in eastern Madagascar.

farinosa L. *farinosus,* sprinkled with flour (*farina,* meal, flour dust); ref. 'dusted' appearance of the Mealy Amazon *Amazona*.

farquhari After Admiral Sir Arthur Murray Farquhar (1855–1937) British mariner.

fasciata Late L. *fasciatus,* banded, striped (L. *fascia,* a band, a stripe).

fasciatoventris Late L. *fasciatus,* banded, striped; L. *venter, ventris,* the belly.

fasciatum /fasciatus see **fasciata.**

fasciicauda L. *fascia,* a band, a streak; *cauda,* the tail.

fasciinucha L. *fascia,* a band, a streak; Med. L. *nucha,* the nape (Arabic *nukha,* spinal marrow).

fasciiventer L. *fascia,* a band, a streak; *venter,* the belly.

fasciogularis L. *fascia,* a band, a streak; Mod. L. *gularis,* throated, of the throat (L. *gula,* the throat).

fasciolata /fasciolatus Mod. L. *fasciolatus,* marked with small bars (Med. L. *fasciolatus,* bandaged).

fastosus L. *fastosus,* proud (*fastus,* pride).

fastuosa L. *fastuosus,* proud, haughty (*fastus,* pride).

feadensis After the island of Fead, or Abgarris, Nuguria Islands, Papua New Guinea.

feae After Leonardo Fea (1852–1903) Italian ornithologist and collector.

featherstoni After Isaac Earl Featherston (1813–1876) Scottish/New Zealand statesman.

fedoa Mod. L. *fedoa,* a godwit (said to be based on a now lost Old English name for a godwit) (cf. local Venetian (Italian) name *vetola* for a godwit).

feldegg After Colonel Baron Christoph von Feldegg (1780–1845) Austrian naturalist and collector in the Balkans.

felix L. *felix,* happy, fortunate.

femoralis Late L. *femorale,* a covering for the thigh (L. *femur,* the thigh).

ferdinandi After Tzar Ferdinand of Bulgaria (1861–1948: reigned 1908–1918), ornithologist and collector.

fernandensis /fernandezianus After the volcanic Juan Fernández Islands, isolated in the eastern Pacific Ocean, named for the Spanish navigator Juan Fernández, who discovered them in 1563.

fernandinae After Conde de la Fernandina (fl. 1850) Cuban landowner.

ferina L. *ferinus,* wild, game; the Grey Pochard *Aythya* was formerly a popular gamebird whose meat was much sought after.

Ferminia After Fermin Z. Cervera (fl. 1926) Cuban landowner.

ferox L. *ferox,* fierce, brave.

ferrea Mod. L. *ferreus,* iron-grey (L. *ferreus,* made of iron; *ferrum,* iron).

ferreorostris Mod. L. *ferreus,* iron-grey; L. *-rostris,* billed.

ferrocyanea Mod. L. *ferrocyaneus,* steely-blue (L. *ferrum,* iron; *cyaneus,* dark blue).

ferruginea L. *ferrugineus,* rusty, ferruginous (*ferrugo, ferruginis,* iron rust).

ferrugineifrons L. *ferrugineus,* rusty, ferruginous; *frons,* the forehead, the brow.

ferrugineipectus L. *ferrugineus,* rusty, ferruginous; *pectus,* the breast.

ferrugineiventre L. *ferrugineus,* rusty, ferruginous; *venter, ventris,* the belly.

ferrugineus see **ferruginea.**

ferruginosa /ferruginosus L. *ferruginosus,* rusty, ferruginous (*ferrugo,* iron rust).

festiva /festivus L. *festivus,* gay, festive, cheerful (*festus,* festive).

Ficedula L. *ficedula,* a small fig-eating bird that changed into the Blackcap *Sylvia atricapilla* in winter, mentioned by Pliny, Terentius Varro, the poet Junius Juvenalis, and other authors.

ficedulina L. *ficedula,* a small bird; *-inus,* resembling.

figulus L. *figulus,* a potter, a worker in clay (*fingere,* to mould or fashion); the Wing-banded Hornero *Furnarius* builds a distinctive nest of clay.

filicauda L. *filum, fili,* a thread; *cauda,* the tail.

fimbriata /fimbriatum L. *fimbriatus,* fringed (*fimbriae,* fringe, border).

finlaysoni After George Finlayson (1790–1823) Scottish naturalist and surgeon in India and Indochina.

finschi /Finschia /finschii After Friedrich Hermann Otto Finsch (1839–1917) German diplomat, administrator, ornithologist, collector, and author (*Die papageien*, 1867; *Die vögel ost Afrikas*, 1870).

fischeri After **1** Johann Gotthelf Fischer von Waldheim (1771–1853) German palaeontologist, zoologist, and author (*Entomographia Imperii Rossici*, 1820; *Jubilaeum semisaecularem*, 1847) (*Somateria*). **2** George Fischer (fl. 1896) Dutch surgeon and collector in the East Indies (*Ptilinopus*). **3** Gustav Adolf Fischer (1848–1886) German explorer and collector in tropical Africa, 1876–1886 (*Agapornis, Dioptrornis, Phyllastrephus, Spreo, Vidua*).

Fiscus L. *fiscus*, tax, public revenues. The Fiscal Shrike *Lanius collaris* was so called because its rapacious habits were considered unequalled even by revenue-officers or tax-collectors.

fissirostris L. *fissus*, split, cloven (*findere*, to split); *-rostris*, billed.

fistulator L. *fistulator*, a piper.

flabelliformis L. *flabellum*, a small fan (*flabra*, breezes); *forma*, shape, form.

flagrans L. *flagrans*, flaming, glittering (*flagrare*, to burn).

flammea L. *flammeus*, fiery, flame-coloured, flaming (*flamma*, a flame).

flammeolus Mod. L. *flammeolus*, flame-red, red.

flammeus see **flammea.**

flammiceps L. *flamma*, a flame; *-ceps*, headed, crowned.

flammigerus L. *flamma*, a flame; *-gera*, carrying (*gerere*, to carry or wear).

flammula L. *flammula*, a little flame (dim. of *flamma*, a flame).

flammulata/flammulatus Mod. L. *flammulatus*, ruddy, with flame-like markings.

flava L. *flavus*, golden-yellow, yellow.

flavalus L. *flavus*, golden-yellow; *ala*, a wing.

flaveola/flaveolus Dim. from L. *flavus*, golden-yellow.

flavescens L. *flavescens*, golden-yellowish (*flavescere*, to become golden yellow).

flavibuccalis L. *flavus*, golden-yellow; *bucca*, the cheek; *-alis*, pertaining to.

flavicans Mod. L. *flavicans*, yellowish, golden-yellowish (L. *flavus*, golden-yellow; *-icans*, closely resembling).

flavicapilla L. *flavus*, golden-yellow; *-capillus*, capped.

flavicaudata L. *flavus*, golden-yellow; *caudatus*, tailed (*cauda*, the tail).

flaviceps L. *flavus*, golden-yellow; *-ceps*, headed.

flavicollis L. *flavus*, golden-yellow; Mod. L. *-collis*, necked.

flavida L. *flavidus*, golden, yellowish.

flavifrons L. *flavus*, golden-yellow; *frons*, the forehead, the brow.

flavigaster L. *flavus*, golden-yellow; *gaster*, the belly.

flavigula L. *flavus*, golden-yellow; *gula*, the throat.

flavigularis L. *flavus*, golden-yellow; Mod. L. *gularis*, throated, of the throat (L. *gula*, the throat).

flavinucha/flavinuchus L. *flavus*, golden-yellow; Med. L. *nucha*, the nape (Arabic *nukha*, spinal marrow).

flavipectus L. *flavus*, golden-yellow; *pectus*, the breast.

flavipennis L. *flavus*, golden-yellow; *pennis*, winged.

flavipes L. *flavus*, golden-yellow; *pes*, the foot.

flaviprymna L. *flavus*, golden-yellow; Gr. *prumnos*, the hindmost part (i.e. rump and tail).

flavirictus L. *flavus*, golden-yellow; *rictus*, the open mouth, gaping jaws; ref. bright yellow rictal streak of the Yellow-gaped Meliphaga *Meliphaga*.

flavirostra L. *flavus*, golden-yellow; *rostrum*, the bill.

flavirostris L. *flavus*, golden-yellow; *-rostris*, billed.

flaviscapis L. *flavus*, golden-yellow; *scapulae*, the shoulders, the back.

flaviventer/flaviventris L. *flavus*, golden-yellow; *venter*, *ventris*, the belly.

flavivertex L. *flavus*, golden-yellow; *vertex*, the crown of the head.

flavocincta/flavocinctus L. *flavus*, golden-yellow; *cinctus*, encircled, girdled, wreathed, crowned (*cingere*, to encircle).

flavocrissalis L. *flavus,* golden-yellow; Mod. L. *crissalis,* crissal, of the vent.

flavogaster see **flavigaster.**

flavogrisea L. *flavus,* golden-yellow; Med. L. *griseus,* grey.

flavolateralis L. *flavus,* golden-yellow; *lateralis,* lateral, of the flanks (*latus,* the side, the flank).

flavolivacea L. *flavus,* golden-yellow; Mod. L. *olivaceus,* olive-green (L. *oliva,* the olive).

flavopectus see **flavipectus.**

flavostriatus L. *flavus,* golden-yellow; Mod. L. *striatus,* striped, streaked (L. *stria,* a furrow).

flavotincta L. *flavus,* golden-yellow; *tinctus,* coloured, tinged, dyed (*tingere,* to dye).

flavovelata L. *flavus,* golden-yellow; *velatus,* veiled (i.e. of the crown and neck) (*velare,* to cover, to veil).

flavovirens Mod. L. *flavovirens,* yellowish-green (L. *flavus,* golden-yellow; *virens,* green).

flavovirescens L. *flavus,* golden-yellow; *virescens,* greenish (*virescere,* to become green).

flavoviridis L. *flavus,* golden-yellow; *viridis,* green.

flavus see **flava.**

floccosus Late L. *floccosus,* full of flocks of wool (L. *floccus,* a flock of wool); the Australian Pilotbird *Pycnoptilus* has '...plumage dense, lengthened and silky, especially on the flanks' (Gould (1851), *Proc. Zool. Soc. London,* 95).

florensis After the island of Flores, Lesser Sunda Islands, Indonesia.

floriceps L. *flos, floris,* a flower; *-ceps,* headed, crowned.

Florida After **1** the state of Florida, USA (Spanish *Pascua Florida,* Easter, the time of year when the Spanish explorer Juan Ponce de León discovered the area in 1513). **2** L. *floridus,* flowery, blooming (*Tangara).*

floris L. *flos, floris,* a flower; after the island of Flores, Indonesia (Spanish *flores,* a flower).

Florisuga L. *flos,* a flower; *sugere,* to suck.

flosculus L. *flosculus,* a little flower, an ornament.

fluminea L. *flumineus,* riverine, of a river (*flumen, fluminis,* a stream or river).

fluminensis Mod. L. *Fluminensis,* of Rio de Janeiro, Brazil (*Flumen Januarii)* (L. *flumen,* a stream or river; Portuguese *rio,* a river).

fluviatilis L. *fluviatilis,* riverine, of a river (*fluvius,* a river).

Fluvicola L. *fluvius, fluvii,* a river, a stream; *-cola,* a dweller, inhabitant.

foersteri After F. Foerster (1865–1918) German botanist and collector.

foetidus L. *foetidus,* stinking, foetid; supposedly from the evil-smelling flesh of the Bare- necked Fruitcrow *Gymnoderus.*

foghaensis After Mount Fogha (Gunung Madang or Mount Madang) on the island of Buru in the Moluccas.

forbesi After **1** Henry Ogg Forbes (1851–1932) Scottish naturalist and collector (*Lonchura, Rallicula).* **2** William Alexander Forbes (1855–1883) British zoologist and collector in tropical Africa and Brazil (*Charadrius, Curaeus, Emberiza).* **3** Sir Victor Courtenay Walter Forbes (1889–1958) British diplomat in Mexico, Spain, and Peru (*Atlapetes).*

forficata /forficatus L. *forfex, forficis,* a pair of scissors; *-atus,* possessing; ref. a forked tail.

Formicarius L. *formicarius,* of an ant (*formica,* an ant); the antbirds received their name from trailing ant armies and feeding on the insects disturbed or put to flight.

Formicivora /formicivorus L. *formica,* an ant; *-vorus,* eating.

formosa L. *formosus,* beautiful.

formosae After the island of Formosa (Taiwan).

formosus see **formosa.**

fornsi After Ramón M. Forns (fl. 1858) resident on Cuba.

Forpus No expl. (Boie (1858), *Journ. für Orn.*, **6**, 363). Probably arbitrary, but perhaps from Gr. *phoreo,* to possess, and *pous,* the foot, ref. the dextrous toes of the Neotropical parrotlets. Prestwich (1963): App. C, suggests an anagram of L. *profus,* spread out, extended (*profundere,* to stretch at full length) with ref. to their short wedge-shaped tails.

forsteni After Eltio Alegondas Forsten (1811–1843) Dutch collector in the East Indies, 1838–1843.

forsteri After Johann Reinhold Forster (1729–1798) and his son Georg Forster (1754–1794) naturalist and artist respectively aboard Cook's *Resolution* voyage of discovery, 1772–1775.

fortipes L. *fortis,* stout, robust; *pes,* the foot.

fortis L. *fortis,* stout, robust, strong.

fossii After W. Fosse (fl. 1856) German collector in the Gabon.

Foudia Malagasy name *fody* or *foudi* for the small fodies or fody-weavers.

Foulehaio Tongan native names *foulehaoi, fulehau* or *fuleheu* for the Wattled Honeyeater *F. carunculata*.

fraenatus L. *frenatus,* bridled (*frenare,* to bridle); ref. face pattern.

francescae After Frances 'Fanny' Wilson (fl. 1846) wife of English collector Edward Wilson.

francesii After Lady Frances Cole (d. 1847) wife of General Sir Galbraith Lowry Cole, Governor of Cape Colony, 1828–1833.

franciae After Francia Bourcier (fl. 1846) daughter of Jules Bourcier, French Consul to Ecuador,1849–1850 and naturalist.

francicus French name *Ile de France* for Mauritius.

franciscanus Med. L. *Franciscus,* Francis; L. *-anus,* relating to; ref. the scarlet robes worn by the Franciscan order (founded by St Francis of Assisi, 1182–1226).

Francolinus French name *francolin* for the francolins (Italian *francolino,* a little hen).

Franklinia /franklinii After Major James Franklin (1783–1834) British Army in India, geologist, ornithologist, and author.

frantzii After Alexander von Frantzius (1821–1877) German naturalist and collector in Brazil, 1849–1853 and Costa Rica, 1853–1868.

fraseri /Fraseria After Louis Fraser (d. 1866) English natural history dealer, collector in Nigeria, 1841–1842, Ecuador, 1859 and California, 1860, and author (*Zoologica typica,* 1849).

frater L. *frater,* a brother (i.e. closely related); Sclater and Hartlaub (1881), *Proc. Zool. Soc. London,* 171, wrote of the Sokotra Chestnut-wing Starling *Onychognathus* and the Somali Chestnut-wing Starling *O. blythii,* 'The occurrence of two typical species of *Amydrus* in so small an island as Socotra is very curious; but there seems to be no doubt about the fact'.

Fratercula Med. L. *fraterculus,* a friar; ref. the hooded appearance and black and white plumage of the Atlantic Puffin *F. arctica*.

fratrum L. *fratrum,* brotherly (i.e. closely related); Shelley (1900), *Ibis,* 522, considered Woodward's Batis *Batis,* '...most nearly allied to *P. mixta* and ... agrees well with the plumage of the adult male of that species'.

Frederickena After Frederick Vavasour McConnell (1868–1914) English traveller and collector (v. *Mackenziaena).*

freethi After Captain G. Freeth (fl. 1892) US Governor of Laysan, Hawaiian Islands.

Fregata French mariner's name *frégate,* for the swift, rapacious frigatebirds, in allusion to the light swift warships bearing the same name.

Fregetta Dim. of genus *Fregata,* frigatebird.

Fregilupus Syn. genus *Fregilus,* chough (Med. L. *frigella,* the Rook *Corvus)*; French name *huppe,* given to the crested Bourbon Starling *F. varius* by early settlers on Réunion (cf. *huppe fasciée*, the Hoopoe *Upupa*).

fremantlii After Major Guy Fremantle (b. 1867) British Army and collector in Somaliland, 1896.

frenata /frenatus see **fraenatus.**

freycinet /freycineti After Captain Louis Claude Desaulses de Freycinet (1779–1842) French navigator and explorer in the Pacific, 1817–1820.

Fringilla L. *fringilla*, a small bird mentioned by Terentius Varro, the poet Valerius Martialis, and the grammarian Pompeius Festus, identified by some later authors with the Robin *Erithacus* and by others with the Chaffinch *F. coelebs* (Gr. *phrugilos*, a bird mentioned by Aristophanes, thought to be a sparrow *Passer*).

Fringillaria L. *fringilla*, a small bird, probably a finch; *-arius*, relating to.

fringillaris L. *fringilla*, a small bird, probably a finch; *-aris*, relating to.

fringillarius see **Fringillaria.**

fringillinus L. *fringilla*, a small bird, identified by some as the Robin *Erithacus*; *-inus*, resembling; ref. rich isabelline face and throat of the Red-throated Tit *Parus*.

fringilloides Genus *Fringilla*, finch; Gr. *-oides*, resembling.

frontale /frontalis Mod. L. *frontalis*, fronted, browed (L. *frons, frontis*, the brow, the forehead).

frontatum /frontatus Mod. L. *frontatus*, fronted, browed (L. *frons, frontis*, the brow, the forehead).

frugilegus L. *frugilegus*, crop-picking, fruit-gathering (*frux, frugis*, fruit; *legere*, to pick) (cf. Med. L. *frigella*, the Rook *Corvus*).

frugivorus L. *frux, frugis*, fruit; *-vorus*, eating.

fruticeti L. *fruticetum, fruticeti*, a thicket (*frutex*, a shrub).

fucata L. *fucatus*, painted, coloured (*fucare*, to paint).

fuciphagus Gr. *phukos*, seaweed; *phagos*, a glutton (*phagein*, to eat); in the mistaken belief that the edible nests of the swiftlets *Aerodramus* were composed of seaweed or other plant matter collected by the birds.

fucosa L. *fucosus*, painted, coloured, beautiful (*fucare*, to paint).

fuelleborni After Friedrich Fülleborn (1866–1933) German naturalist and academic in German East Africa (Tanzania), 1896–1900.

fuertesi After Louis Agassiz Fuertes (1874–1927) US bird artist, explorer, and collector.

fugax L. *fugax*, fleeing, cowardly (i.e. skulking).

fulgens L. *fulgens*, shining, glittering (*fulgere*, to flash).

fulgida /fulgidus L. *fulgidus*, shining, glittering (*fulgere*, to flash).

Fulica L. *fulica*, a coot. Specifically the name refers to the lobed, coot-like toes of the Sungrebe *Heliornis*.

fulicarius L. *fulica*, a coot; *-arius*, relating to; ref. lobed, coot-like toes of the Grey Phalarope *Phalaropus*.

fulicata L. *fulica*, a coot; *-atus*, resembling; ref. pied plumage of the Indian Robin *Saxicoloides*.

fuliginiceps L. *fuligo, fuliginis*, soot; *-ceps*, headed.

fuliginosa /fuliginosus Late L. *fuliginosus*, sooty (L. *fuligo, fuliginis*, soot).

fuligiventer L. *fuligo, fuliginis*, soot; *venter*, the belly.

fuligula L. *fuligo*, soot; *gula*, the throat.

Fulmarus Old Norse name *fulmar*, the foul mew or gull, for the Northern Fulmar *F. glacialis*.

fulva L. *fulvus*, tawny, yellowish-brown.

fulvescens Mod. L. *fulvescens*, buffy, tawny (L. *fulvus*, tawny).

Fulvetta Dim. of L. *fulvus*, tawny, yellowish-brown.

fulvicapilla L. *fulvus*, tawny; *-capillus*, capped.

fulviceps L. *fulvus*, tawny; *-ceps*, headed.

fulvicervix L. *fulvus*, tawny; *cervix*, the nape, the neck.

fulvicollis L. *fulvus,* tawny; Mod. L. *-collis,* necked.

fulvicrissa L. *fulvus,* tawny; Mod. L. *crissum,* the vent.

fulvifrons L. *fulvus,* tawny; *frons,* the forehead or brow.

fulvigula L. *fulvus,* tawny; *gula,* the throat.

fulvipectus L. *fulvus,* tawny; *pectus,* the breast.

fulviventris L. *fulvus,* tawny; *venter, ventris,* the belly.

fulvogularis L. *fulvus,* tawny; Mod. L. *gularis,* throated, of the throat (L. *gula,* the throat).

fulvus see **fulva.**

fumicolor L. *fumus, fumi,* smoke; *color,* colour, tint.

fumifrons L. *fumus,* smoke; *frons,* the forehead or brow.

fumigatus L. *fumigatus,* smoked (i.e. dusky, sooty-coloured) (*fumigavi,* to smoke).

funebris L. *funebris,* funereal (i.e. dark-coloured) (*funus, funeris,* a funeral).

funerea /funereus L. *funereus,* funereal (*funus,* a funeral).

furcata /furcatus Med. L. *furcatus,* forked, cloven (L. *furca,* a two-pronged fork).

furcifer L. *furcifer,* a fork-carrier (originally applied to slaves punished by wearing a pronged pole) (*furca,* a two-pronged fork; *ferre,* to carry or bear).

Furnarius L. *furnaria,* a baker; ref. the distinctive oven-shaped nests of the Neotropical horneros (Spanish *hornero,* a baker).

fusca L. *fuscus,* dark, dusky, brown, black. Despite its classical meaning this epithet is used in orn. to indicate a very wide spectrum of colours from black, greys, and browns to a range of dark tints including slate-blue and dusky-orange.

fuscans Mod. L. *fuscans,* blackish (L. *fuscus,* dark; *-icans,* resembling).

fuscata L. *fuscatus,* dark (*fuscare,* to darken).

fuscater L. *fuscus,* dark, brown; *ater,* black.

fuscatus see **fuscata.**

fuscescens Mod. L. *fuscescens,* brownish, blackish (L. *fuscus,* dark, dusky).

fuscicapilla /fuscicapillus L. *fuscus,* dark, dusky; *-capillus,* crowned, capped.

fuscicauda L. *fuscus,* dark, dusky; *cauda,* the tail.

fuscicollis L. *fuscus,* dark, dusky; Mod. L. *-collis,* necked.

fuscigularis L. *fuscus,* dark, dusky; Mod. L. *gularis,* of the throat (L. *gula,* the throat).

fuscipennis L. *fuscus,* dark, dusky; *pennis,* winged.

fuscirostris L. *fuscus,* dark, dusky; *-rostris,* billed.

fuscobrunneus L. *fuscus,* dark, dusky; Mod. L. *brunneus,* brown.

fuscocaerulescens L. *fuscus,* dusky, brown; *caerulescens,* bluish (v. *caerulescens).*

fuscocapillum see **fuscicapilla.**

fuscocinereus L. *fuscus,* dark, dusky; *cinereus,* ash-coloured.

fuscoflava L. *fuscus,* dark, dusky; *flavus,* golden-yellow.

fusconota L. *fuscus,* dark, dusky; Gr. *-notos,* backed.

fuscoolivaceus L. *fuscus,* dark, dusky; Mod. L. *olivaceus,* olive-green, olivaceous (L. *oliva,* the olive).

fuscorufa /fuscorufus L. *fuscus,* dark, dusky; *rufus,* red, ruddy.

fuscus see **fusca.**

fytchii After Major-General Albert Fytche (1820–1892) Chief Commissioner in British Burma, 1867–1871.

Falco vespertinus Red-footed Falcon

G

gabar French *Le Gabar* the name given to the Gabar Goshawk *Micronisus* by Levaillant (1799) (probably from *garde,* a watchman, guard; *barré,* barred).

gabela After the town of Gabela, Cuanza Sul, Angola.

Gabianus Local Provençal (French) name *gabian* for a gull *Larus* (Italian *gabbiano,* a gull; Spanish *gaviota,* a gull; L. *gavia,* a seabird).

gabonensis /gabonicus After Gabon, formerly Gaboon, probably named from Portuguese *gabão,* a hooded cloak, ref. the shape of the Gabon estuary.

gaimardi /gaimardii After Paul Gaimard (1793–1858) French surgeon-naturalist and explorer.

Galachrysia Gr. *gala, galaktos,* milk; *khruseos,* golden, gilt.

galactotes Gr. *gala,* milk; *-otes,* resembling.

galapagoensis After the Galapagos Islands in the eastern Pacific Ocean (Spanish *galápagos,* tortoises) annexed by Ecuador in 1832.

galatea Gr. myth. Galatea, a sea-nymph or a rustic maiden.

Galbalcyrhynchus Genus *Galbula,* jacamar; syn. genus *Alcyone,* kingfisher (Gr. *alkuon,* the kingfisher); *rhunkhos,* the bill; ref. dagger bill and kingfisher-like appearance of the White-eared Jacamar *G. leucotis.*

galbanus L. *galbanus,* greenish-yellow (*galbanum,* a yellowish sap).

Galbula L. *galbula,* a small yellowish bird (dim. of *galbina,* a small yellow bird; *galbus,* yellow); identified by subsequent authors as the Golden Oriole *Oriolus* or a kind of 'woodpecker', the witwall, which builds a pendulous nest.

galeata /galeatus L. *galeatus,* helmeted (i.e. crested) (*galea,* a helmet).

Galeopsar L. *galea,* a helmet; Gr. *psar,* the starling; ref. the cushion of soft feathers on the forehead of the Bristle-crowned Starling *G. salvadorii.*

galericulata /galericulatus Mod. L. *galericulatus,* hooded (L. *galericulum,* a small wig or cap; *galerus,* a skull-cap).

Galerida L. *galeritus,* the Crested Lark *G. cristata.*

galerita /galeritus L. *galeritus,* hooded, capped (*galerus,* a skull-cap).

galgulus L. *galgulus,* a small bird that slept hanging suspended by its feet mentioned by Pliny (perhaps confused with *galbula,* a bird which builds a pendulous nest); the Indomalayan hanging parrots *Loriculus* roost by hanging upside down.

galinieri After Captain Joseph Germain Galinier (fl. 1852) French explorer in Ethiopia, 1839–1843 and author (*Voyage en Abyssinie,*1847).

gallardoi After Angel Gallardo (fl. 1974) and José Maria Gallardo (fl. 1974) Argentinian naturalists and landowners.

Gallicolumba L. *gallus,* the farmyard cock; *columba,* a pigeon; ref. ground-dwelling habits of the bleeding-hearts and ground doves.

Gallicrex L. *gallus,* the farmyard cock; genus *Crex,* crake; ref. scarlet horn or wattle on forehead of the Watercock *G. cinerea.*

gallicus L. *Gallicus,* Gallic, of Gaul (i.e. of France).

gallinacea L. *gallinaceus,* of poultry (*gallina,* a hen); ref. head wattle of the Comb-crested Jacana *Irediparra.*

Gallinago Mod. L. *gallinago,* the snipe (L. *gallina,* a hen; probably with ref. to speckled brown plumage).

Gallinula L. *gallinula,* a little hen or chicken; used by Gesner (1555) for the Common Moorhen or Waterhen *G. chloropus* with ref. to its cocked tail and halting progress.

Gallirallus L. *gallus,* the farmyard cock; genus *Rallus,* rail; ref. brown speckled plumage and ground loving habits of the Weka *G. australis.*

Gallirex L. *gallus,* the farmyard cock; *rex,* a king; ref. bare facial wattles and large crest of the Purple-crested Turaco *G. porphyreolophus.*

gallopavo Mod. L. *gallopavus,* the wild turkey (Gesner (1555)).

Galloperdix L. *gallus,* the farmyard cock; *perdix,* a partridge; ref. the leg-spurs of the Indian spurfowl.

Gallus L. *gallus,* the farmyard cock; the Red Junglefowl *G. gallus* is the wild ancestor from which all domestic breeds are derived.

gambagae After the town of Gambaga, Northern Province, Ghana.

gambeli /gambelii After William Gambel (1819–1849) US explorer, naturalist, and collector.

gambensis After The Gambia, former British colony extending along both banks of the River Gambia to the Atlantic.

gambieri After the Gambier Islands, a group of the Tuamotu Archipelago, French Polynesia, named for Admiral James Gambier (1756–1833).

Gampsonyx Gr. *gampsonux,* hooked-claw, used as an epithet of the vulture (*gampsos,* curved; *onux,* a claw).

Gampsorhynchus Gr. *gampsos,* curved; *rhunkhos,* the bill; the White-hooded Babbler *G. rufulus* has the bill 'outline curving towards the tip, which is strongly emarginated, and forms a distinct hook' (Blyth (1844), *Journ. Asiat. Soc. Bengal,* **13**, 370).

garbei After P. E. W. Garbe (1853–1925) zoologist at Museu Paulista, Brazil.

garleppi After Gustav Garlepp (1862–1907) German collector in tropical America, 1883–1897.

garnotii After Prosper Garnot (1794–1838) French naturalist and collector.

Garritornis L. *garrire,* to babble, to chatter; Gr. *ornis,* a bird.

Garrodia After Alfred Henry Garrod (1846–1879) English zoologist and anatomist.

garrula see **Garrulus.**

Garrulax L. *garrulus,* babbling, chattering; *-ax,* inclining towards.

Garrulus L. *garrulus,* babbling, chattering (*garrire,* to chatter).

garzetta Italian names *garzetta* or *sgarzetta* for the little white heron or egret *Egretta.*

gaudichaud After Charles Gaudichaud-Beaupré (1780–1854) French botanist, pharmacologist, and author.

gaumeri After G. F. Gaumer (1850–1929) collector in Mexico, 1885–1893.

Gavia L. *gavia,* a seabird, perhaps a gull *Larus,* mentioned by Pliny and the philosopher Lucius Appuleius.

gayaquilensis After the city of Santiago de Guayaquil, Guayas, Ecuador.

gayi After Claude Gay (1800–1873) French botanist in Chile, 1824–1841.

Gecinulus Dim. of syn. genus *Gecinus,* woodpecker (Gr. *ge,* earth; *kineo,* to move).

geelvinkiana /geelvinkianum After Geelvink Bay, Irian Jaya (Indonesian New Guinea).

geislerorum After B. Geisler (1857–1945) and his brother H. G. Geisler (fl. 1892) German taxidermists and collectors in New Guinea, 1884–1892.

Gelochelidon Gr. *gelao,* to laugh; *khelidon,* a swallow (i.e. a sea swallow or tern).

geminus L. *geminus,* twin, double (i.e. closely resembling); the Rusty-patched Fruit Dove *Ptilinopus* was considered closely related to the Lilac-capped Fruit Dove *P. coronulatus*.

genei After Giuseppe Gené (1800–1847) Italian naturalist and author.

genibarbis L. *genis*, the cheeks; *barba*, the beard.

Gennaeus Gr. *gennaios*, noble, thoroughbred.

gentianus L. *gentiana*, the gentian; the Gentian-blue Kingfisher *Ceyx* exhibits three shades of brilliant blue found also in the gentians *Gentiana* (Gentius, a king of Illyria).

gentilis L. *gentilis*, noble (cf. Med. L. *falco gentilis*, the gentle falcon, the goshawk).

Geobates Gr. *geo-*, earth (*ge*, the ground, earth); *-bates*, roaming, dwelling.

Geococcyx Gr. *geo-*, earth; *kokkux*, the cuckoo.

Geocolaptes Gr. *geo-*, earth; genus *Colaptes*, woodpecker.

geoffroyi /Geoffroyus After Étienne Geoffroy Saint-Hilaire (1772–1844) French zoologist and author (*Histoire naturelle des mammifères*, 1824; *Études progressives d'une naturaliste*, 1835).

Geokichla Gr. *geo-*, earth; *kikhle*, a thrush.

Geomalia Gr. *geo-*, earth; genus *Malia*, malia, a babbler.

Geopelia Gr. *geo-*, earth; *peleia*, a pigeon or dove.

Geophaps Gr. *geo-*, earth; *phaps*, a pigeon.

Geopsittacus Gr. *geo-*, earth; *psittakos*, a parrot.

georgiana After **1** King George Sound, Western Australia (*Eopsaltria*). **2** the state of Georgia, USA, named for George II (1683–1760: reigned 1727–1760) King of England (*Melospiza*).

georgianus /georgica /georgicus After South Georgia Island, South Atlantic Ocean, named for George III (1738–1820: reigned 1760–1820) King of England.

Geositta Gr. *geo-*, earth; genus *Sitta*, nuthatch; ref. burrowing activities of the Neotropical miners.

Geospiza Gr. *geo-*, earth; *spiza*, finch.

Geothlypis Gr. *geo-*, earth; Mod. L. *thlypis*, warbler.

Geotrygon Gr. *geo-*, earth; *trugon*, the Turtle Dove *Streptopelia*.

Geranoaetus Gr. *geranos*, the Crane *Grus; aetos*, an eagle; ref. grey plumage, loud cries, and dorsal plumes of the Grey Buzzard Eagle *G. melanoleucus*.

Geranospiza Gr. *geranos*, the Crane *Grus*; *spizias*, a hawk. The author Johann von Kaup made frequent and erroneous use of *spiza* (v. *Spiza)* instead of *spizias*.

germainii /germani After L. R. Germain (1827–1927) French Army in Indochina, 1866–1878.

Geronticus Gr. *gerontos*, an old man; *-ikos*, belonging to; alluding to bald head of the Bald Ibis *G. calvus*.

gertrudae After Gertrude Whitney (1877–1942) US socialite and philanthropist.

Gerygone Gr. *gerugonos*, echoes (*geruo*, to sing; *gone*, a child); ref. fine plaintive songs of the Australasian gerygones or warblers.

gestroi After Raffallo Gestro (1845–1935) Italian zoologist.

ghiesbreghti After Augustus B. Ghiesbreght (1810–1862) Belgian collector in tropical America, 1836–1862.

gibberifrons L. *gibber*, *gibberis*, a hump; *frons*, the forehead; ref. bulging forehead of the Grey Teal *Anas*.

gibsoni After Mr Gibson (fl. 1840) collector in Colombia for the taxidermist and natural history dealer Loddiges.

gierowii After H. Gierow (fl. 1878) Swedish collector in Angola.

gigantea /giganteus L. *giganteus*, gigantic, giant (*gigas, gigantis*, a giant).

gigas L. *gigas*, a giant.

gilberti After Gilbert Nkwocha (fl.1955) African skinner and collector.

gilletti After Major F. A. Gillett (1872–1944) British big-game hunter in Somaliland, 1894.

giloloensis After the island of Gilolo or Jailolo (Halmahera) in the Moluccas, Indonesia.

gilvicollis L. *gilvus,* pale yellow; Mod. L. *-collis,* necked.

gilvigaster L. *gilvus,* pale yellow; *gaster,* the belly.

gilviventris L. *gilvus,* pale yellow; *venter, ventris,* the belly.

gilvus L. *gilvus,* pale yellow.

gingica Erroneous toponym after the town of Gingi (Gingee), Coromandel (Tamil Nadu), India; the Collared Hill Partridge *Arborophila* is confined to the mountains of south-eastern China.

ginginianus After the town of Gingi (Gingee), Tamil Nadu, India.

githagineus Syn. Botanical genus *Githago,* corn cockle (L. *gith,* Roman coriander; *-ago,* resembling); L. *-ineus,* resembling; ref. the reddish colouring of the ribbed calyces of the plant and the rose-tinged plumage of the Trumpeter Finch *Bucanetes.*

glabricollis L. *glaber,* bald; Mod. L. *-collis,* necked.

glabrirostris L. *glaber,* bald, smooth, without hair; *-rostris,* billed; the Black Catbird *Melanoptila* lacks rictal bristles.

glacialis L. *glacialis,* icy (*glacies,* ice).

glacialoides Specific name *Fulmarus glacialis;* Gr. *-oides,* resembling.

gladiator L. *gladiator,* a gladiator, a brigand (i.e. a shrike).

glandarius L. *glandarius,* producing acorns (*glans, glandis,* an acorn). The name refers to the acorn-eating and acorn-collecting habits of the Acorn Jay *Garrulus,* and the noisy, jay-like behaviour of the Great Spotted Cuckoo *Clamator.*

Glareola Dim. from L. *glarea,* gravel; from conjectured habitat of the Collared Pratincole *G. pratincola.*

glauca L. *glaucus,* glaucous, bluish-grey (Gr. *glaukos,* bluish-green).

glaucescens Mod. L. *glaucescens,* sea-green.

Glaucidium Dim. of Gr. *glaux,* the Little Owl *Athene* (cf. *glaukidion,* a little grey fish; dim. of *glaukos,* a grey fish).

glaucinus Gr. *glaukinos,* bluish-grey.

Glaucionetta Gr. *glaukion,* with glaring eyes; *netta,* a duck (cf. *glaukion,* a kind of duck with yellow eyes).

Glaucis Gr. *glaukos,* pale green, glaucous.

glaucocaerulea L. *glaucus,* glaucous; *caeruleus,* dark blue.

glaucocolpa L. *glaucus,* glaucous; Gr. *kolpos,* the belly or bosom.

glaucogularis L. *glaucus,* glaucous; Mod. L. *gularis,* throated (L. *gula,* the throat).

glaucoides Syn. specific name *Larus glaucus* (a former name for the Glaucous Gull *L. hyperboreus);* Gr. *-oides,* resembling.

glaucopis Gr. *glaukopis,* with gleaming eyes.

glaucopoides Specific name *Thalurania glaucopis;* Gr. *-oides,* resembling

glaucops Gr. *glaukops,* with gleaming eyes.

glaucurus Gr. *glaukos,* glaucous, bluish-green; *-ouros,* tailed.

glaucus see **glauca.**

Gliciphila Gr. *glukus,* sweet, sweet-tasting; *philos,* loving, fond of.

globulosa L. *globulus,* a globule, a little ball (*globus,* a ball); *-osus,* abundance of; ref. shape of red wattles and tubercules of the Wattled Curassow *Crax.*

gloriosa L. *gloriosus,* glorious, famous.

gloriosissima L. *gloriosissimus,* most glorious (super. of *gloriosus,* glorious).

Glossopsitta Gr. *glossa,* the tongue; Mod. L. *psitta,* a parrot (Gr. *psittakos,* a parrot); the brush-tipped tongues of the lorikeets are an adaptation to feeding on nectar and pollen.

glottis Gr. *glottis,* a migratory bird mentioned by Aristotle, not further identified. Subsequently associated, on flimsy evidence, with the flamingo *Phoenicopterus* or the Greenshank *Tringa nebularia* (cf. Swedish names *glutt* or *gluttsnäppa* for the Greenshank).

glyceria Gr. *glukerios,* sweet, delightful.

Glycichaera Gr. *glukus,* sweet, sweet tasting; *khairo,* to delight in.

Glyphorhynchus Gr. *gluphis,* a knife; *rhunkhos,* the bill; the Wedge-billed Woodcreeper *G. spirurus* has a short dagger-shaped bill.

Gnathospiza Gr. *gnathos,* the jaw; *spiza,* finch; the Sulphur-throated Finch *G. taczanowskii* has a heavy grosbeak-like bill.

gnoma Mod. L. *gnomus,* a gnome, a dwarf.

Gnorimopsar Gr. *gnorimos,* notable, well-known; *psar,* the starling (in orn. *psar* is also used to indicate the black, iridescent, or uniformly coloured members of the American Icteridae, once considered united with the starlings (Sturnidae)).

godeffroyi After J. C. Godeffroy (1813–1885) German ornithologist and founder of Godeffroy Museum, Hamburg, 1860.

godefrida Late L. *Godefrida,* Geoffrey; after Étienne Geoffroy Saint-Hilaire (1772–1844) French zoologist (v. *Geoffroyus).*

godini After Louis Godin (1704–1760) French scientist in Peru, 1735.

godlewskii After Wiktor Witold Godlewski (1831–1900) Polish farmer and field naturalist exiled to Siberia.

godmani After Frederick Du Cane Godman (1834–1919) English ornithologist and author (*Biologia centrali Americana*, 1879; *A monograph of the petrels*, 1907).

goeldii After Emil August Goeldi (1859–1917) German zoologist resident in Brazil and author (*Aves do Brasil*, 1894; *Die vögelwelt des Amazonensstromes*, 1901).

goeringi After Anton Goering (1836–1905) German naturalist and artist in Venezuela, 1866–1872.

goertae Senegalese native name *goertan* for the Grey Woodpecker *Mesopicos*.

Goethalsia After George Washington Goethals (1858–1928) US chief engineer Panama Canal, 1907–1914 and Civil Governor Panama Canal Zone, 1914–1916.

goffini After Lieutenant Andrea Leopold Goffin (d. 1863) Dutch naval officer.

goiavier No expl. (Scopoli (1786), *Deliciae florae et faunae insubricae*, fasc. **2**, p. 96); perhaps a corruption of the Tamil (Sri Lanka) name *kuruvi* used in combination for various small birds, including bulbuls and babblers.

goisagi Japanese name *goi-sagi* for the Black-crowned Night Heron *Nycticorax* (*goi,* a bittern; *sagi,* a heron).

golandi After Brigadier-General Goland Clark (1875–1944) British field ornithologist and collector in tropical Africa and the Middle East.

goldiei After Andrew Goldie (fl. 1882) Scottish botanical collector.

goldmani /Goldmania After Major Edward Alphonso Goldman (1873–1944) US mammalogist, collector, and author (*Mammals of Panama*, 1920).

golensis After the Gola Forest, Sierra Leone.

goliath After Goliath of Gath, biblical giant champion of the Philistines slain by the boy David (Samuel I, xvii).

gongonensis After the village of Gongoni, near Mombasa, Kenya.

goodenovi After Bishop Samuel Goodenough (1743–1829) naturalist and vice-president of Linnean Society of London.

goodfellowi /Goodfellowia After Walter Goodfellow (1866–1953) English explorer, ornithologist, and collector.

goodsoni After Arthur T. Goodson (1873–1931) assistant ornithologist at Rothschild's Tring Museum, 1893–1931.

gordoni No expl. (Mathews (1912), *Austral Avian Record*, **1**, 102); a male eponym.

Gorsachius Japanese name *goi-sagi* for the Black-crowned Night Heron (v. *goisagi).*

goslingi After Captain G. B. Gosling (1872–1906) of the Niger-Nile expedition, 1904.

gossei After Philip Henry Gosse (1810–1888) English zoologist in Jamaica, 1844–1846, and author (*The birds of Jamaica*, 1847).

goudoti/goudotii After J. Goudot (fl. 1843) French zoologist in Colombia, 1822–1843.

goughensis After Gough Island, Tristan da Cunha group, in the South Atlantic Ocean.

gouldi After John Gould (1804–1881) English publisher, naturalist, and artist (*A century of birds from the Himalaya Mountains*, 1832; *The birds of Europe*, 1832; *A monograph of the toucans*, 1833; *A monograph of the Macropodidae*, 1841; *A monograph of the Odontophorinae*, 1844; *The mammals of Australia*, 1845; *A monograph of the Trochilidae*, 1849; *The birds of Asia*, 1850; *The birds of Great Britain*, 1862; *The birds of New Guinea*, 1875; *A monograph of the Pittidae*, 1880).

gouldiae After Elizabeth Gould (1804–1841) artist and wife of John Gould (v. *gouldi*).

gouldii see **gouldi.**

gounellei After Edmond Gounelle (1850–1914) French naturalist and collector in Brazil, 1887.

Goura New Guinea aboriginal name *goura* for the crowned pigeons.

govinda Hindu myth. Govinda, the cow-finder, another name for Krishna the dark one.

goyderi After George Woodroffe Goyder (1826–1898) Surveyor-General of South Australia and explorer.

graceannae After Grace Anna Lewis (1821–1912) US botanist and teacher.

graciae After Grace Darling Page (b. 1847) sister of US ornithologist Elliott Coues.

gracilipes L. *gracilis*, slender; *pes*, the foot.

Gracula L. *graculus*, the Jackdaw *Corvus monedula* (cf. syn. genus *Graculus*, chough *Pyrrhocorax*).

graculina L. *graculus*, the Jackdaw; *-inus*, resembling.

graculus see **Gracula.**

graduacauda L. *gradus, graduus*, a step; *cauda*, the tail; ref. graduated tail of Audubon's Oriole *Icterus*.

graeca L. *Graecus*, Grecian, Greek (*Graecia*, Greece).

graeffii After Eduard Graeffe (1833–1916) Austrian ornithologist and author.

Grafisia No expl. (Bates (1926), *Bull. Brit. Orn. Cl.*, **46**, 105); perhaps ref. the German nobiliary title *Graf*, count (the genus has been amended to *Grafsia*), or Gr. *grapis*, a bird mentioned by Hesychius, or Mod. L. *gravus*, a kind of porphyry or purple stone.

Grallaria L. *grallae*, stilts; *-arius*, relating to; ref. long legs and cursorial habits of the Neotropical antpittas.

Grallaricula Dim. of genus *Grallaria*, antpitta.

Grallina Mod. L. *grallinus*, with stilts, stilted (L. *grallae*, stilts); the Magpie-lark *G. cyanoleuca* feeds mainly on the ground.

gramineus **1** L. *gramineus*, grassy (*gramen, graminis*, grass); ref. habitat (*Megalurus, Pooecetes*). **2** Mod. L. *gramineus*, grass-green; ref. colour (*Tanygnathus*).

Graminicola L. *gramen, graminis*, grass; *-cola*, inhabitant, dweller.

grammacus Gr. *grammikos*, linear, striped (*gramme*, a line); ref. striped head pattern of the Lark Sparrow *Chondestes*.

grammiceps L. *grammicus*, of lines, lettered; *-ceps*, headed.

grammicus L. *grammicus*, of lines, lettered (Gr. *grammikos*, linear).

granadense/granadensis After New Granada, a former republic of South America variously incorporating Panama, Colombia, Venezuela and Ecuador (Republic of New Granada 1830–1858; Granadine Confederation 1858–1863). Of those spp. with this epithet *Myiozetetes* was, in fact, described from Panama during one of the intermittent periods when that country was independent of the Confederation, and *Idioptilon* was described from a 'Bogota' trade skin. In early orn. 'Bogota' was used in much the same way as Cayenne (i.e. designating a sp. of unknown provenance from the general area of north or north-western tropical America).

Granatellus Dim. of Med. L. *granatus,* a garnet; ref. bright red breast patches of the Mexican Red-breasted Chat *G. venustus.*

Granatina **1** Med. L. *granatus,* a garnet; L. *-inus,* resembling; ref. the violet-purple cheeks and ear-coverts of the Violet-eared Grenadier *Uraeginthus granatina.* **2** French *breve grenadin,* the name given to the Garnet or Grenadier Pitta *Pitta* by Temminck ref. resemblance to a grenadiers uniform.

Grandala No expl. (Hodgson (1843), *Journ. Asiat. Soc. Bengal,* **12**, 447); perhaps from L. *grandis,* grand, and *ala,* a wing.

grandidieri After Alfred Grandidier (1836–1921) French ornithologist, collector in Madagascar, 1865 and author (*Histoire naturelle des oiseaux de Madagascar*, 1876).

grandis L. *grandis,* large, great.

granti After **1** William Robert Ogilvie Grant (1863–1924) English ornithologist (*Penelope).* **2** Captain Claude H. B. Grant (1878–1958) British ornithologist, collector, and author (*African handbook of birds,* 1952) (*Phoeniculus).*

grantia No expl. (McClelland (1840), *Proc. Zool. Soc. London*, **7**, 165).

Grantiella Dim. after W. R. Ogilvie Grant (v. *granti* **1**).

granulifrons L. *granulum, granuli,* a granule; *frons,* the forehead; ref. the knobbed and wattled cere of the Carunculated Fruit Dove *Ptilinopus.*

Graphephasianus Gr. *graphe,* drawing, engraving; *phasianos,* a pheasant; ref. the heavily barred tail of the Copper Pheasant *Syrmaticus soemmerringii.*

grata L. *gratus,* pleasing, welcome.

Graucalus Gr. *graukalos,* an ash-coloured bird mentioned by Hesychius, not further identified.

graueri /Graueria After Rudolf Grauer (1871–1927) Austrian ornithologist and collector in tropical Africa, 1904–1911.

gravis L. *gravis,* heavy (i.e. large).

Graydidascalus After J. E. Gray (v. *grayi* **1**); Gr. *didaskalos,* a teacher or master; an acknowledgement of Gray's influence on ornithology during his lifetime.

grayi /grayii After **1** John Edward Gray (1800–1875) English ornithologist and author (*Gleanings from the menagerie at Knowsley Hall,* 1846) (*Ammomanes, Ardeola, Hylocharis, Zosterops)* and his brother **2** George Robert Gray (1808–1872) ornithologist, first assistant keeper British Museum (Department of Zoology) 1869, and author (*Genera of birds*, 1844) (*Chenorhamphus, Pipile, Turdus).*

graysoni /graysonii After A. J. Grayson (1819–1867) US ornithologist and collector in Mexico, 1859–1860.

gregalis L. *gregalis,* sociable (*grex, gregis,* a flock or herd).

gregarius L. *gregarius,* gregarious (*grex, gregis,* a flock).

greyii After Sir George Grey (1812–1898) Governor of South Australia, 1841–1845, New Zealand, 1845–1853, 1861–1867.

grillii After J. W. Grill (1815–1864) Swedish zoologist.

grimwoodi After Major I. R. Grimwood (b. 1912) chief game warden in Kenya, 1959–1964.

grisea Med. L. *griseus,* grey.

grisegena Med. L. *griseus,* grey; L. *gena,* the cheek.

griseicapilla /griseicapillus Med. L. *griseus,* grey; L. *-capillus,* capped.

griseicauda Med. L. *griseus,* grey; L. *cauda,* the tail.

griseiceps Med. L. *griseus,* grey; L *-ceps,* crowned, headed.

griseicollis Med. L. *griseus,* grey; Mod. L. *-collis,* necked.

griseigula Med. L. *griseus,* grey; L. *gula,* the throat.

griseigularis Med. L. *griseus,* grey; Mod. L. *gularis,* throated (L. *gula,* the throat).

griseipectus Med. L. *griseus,* grey; L. *pectus,* the breast.

griseisticta Med. L. *griseus,* grey; Gr. *stiktos,* spotted, dappled.

griseiventris Med. L. *griseus,* grey; L. *venter, ventris,* the belly.

griseldis No expl. (Hartlaub (1891), *Abh. natur. Ver. Bremen,* **12**, 7); perhaps a variant form from Med. L. *griseus,* grey, or an eponym after an otherwise unknown Griselda.

griseocapillus see **griseicapillus.**

griseocephalus Med. L. *griseus,* grey; Gr. *-kephalos,* headed.

griseoceps see **griseiceps.**

griseocristatus Med. L. *griseus,* grey; L. *cristatus,* crested (*crista,* a crest).

griseogularis see **griseigularis.**

griseolus Mod. L. *griseolus,* grey.

griseomurina Med. L. *griseus,* grey; L. *murinus,* of mice (*mus, muris,* a mouse) (i.e. colour).

griseonota Med. L. *griseus,* grey; Gr. *-notos,* backed.

griseonucha Med. L. *griseus,* grey; *nucha,* the nape (Arabic *nukha,* spinal marrow).

griseopyga Med. L. *griseus,* grey; Gr. *-pugos,* rumped (*puge,* the rump).

griseostriatus Med. L. *griseus,* grey; Mod. L. *striatus,* streaked, striped (L. *stria,* a furrow).

griseotincta Med. L. *griseus,* grey; L. *tinctus,* coloured, dyed (*tingere,* to dye).

Griseotyrannus Med. L. *griseus,* grey; genus *Tyrannus,* tyrant flycatcher, kingbird.

griseovirescens Med. L. *griseus,* grey; L. *virescens,* greenish (*virescere,* to become green).

griseus see **grisea.**

grisola Mod. L. *griseolus,* grey.

griveaudi After Paul Griveaud (fl. 1959) French entomologist and collector in Madagascar.

grossus L. *grossus,* thick; ref. thick, heavy bill of the Slate-coloured Grosbeak *Pitylus.*

grosvenori After Gilbert Hovey Grosvenor (1875–1966) US philanthropist and President of the National Geographic Society, 1920–1954.

Grus L. *grus,* a crane.

grylle Swedish *grissla,* a guillemot (cf. German *gryllteiste,* the Black Guillemot *Cepphus).*

gryphus L. *gryphus,* a griffin; ref. hooked bill of the Andean Condor *Vultur.*

grzimeki After Bernhard Grzimek (fl. 1973) German zoologist and author.

Guadalcanaria After Guadalcanar (Guadalcanal) Island, Solomon Islands.

guadeloupensis After the island of Guadeloupe in the West Indies.

guainumbi Tupí (Brazilian) Indian name *guainumbi* or *guinambi* for a type of hummingbird.

guajana After the Guianas (i.e. tropical America); an erroneous toponym, the Banded Pitta *Pitta* ranging from Thailand to Borneo and Bali.

gualaquizae After the town of Gualaquiza, Morona-Santiago, Ecuador.

guarauna Tupí (Brazilian) Indian name *guarana* for some sort of marshbird.

guarayanus After the Río Guarayos, Santa Cruz, Bolivia.

guarouba Tupí (Brazilian) Indian name *guarajuba* for a type of parakeet.

guatemalae /guatemalensis /guatimalensis After Guatemala.

guatimozinus After Guatemozin (d. 1520) nephew of Montezuma and last Emperor of the Aztecs, executed on the orders of Cortes.

gubernator L. *gubernator,* a governor; after Emin Pasha, Pasha (governor) of Equatorial Province, Egyptian Sudan, 1878–1889 (v. *emini).*

Gubernatrix L. *gubernatrix,* a governess; from *le Bruant Commandeur,* the French name given to the Yellow Cardinal *G. cristata* by Temminck on account of its imposing crest (like the feathered and plumed hats worn by military commanders of the day).

guerinii After Félix Édouard Guérin-Méneville (1799–1874) French entomologist.

guianensis After the Guianas, deriving their name from a Carib Indian word, meaning 'land of water', for the swampy lowlands.

guifsobalito Abyssinian (Ethiopian) native name *guifso batito dimmo-won jerck* for the Black-billed Barbet *Lybius*.

guildingii After the Revd. Lansdown Guilding (1797–1831) painter, naturalist, and collector resident in the West Indies.

guilielmi Med. L. *Guilelmus,* William (and European equivalents); after Wilhelm II Emperor of Germany and King of Prussia (1859–1941: reigned 1888–1918).

guillemardi After Francis Henry Hill Guillemard (1853–1933) British geographer, traveller, and author.

guimeti After Jean Baptiste Guimet (1795–1871) French chemist.

guinea After Guinea, a name applied to tropical West Africa from the Gambia to the Congo.

Guira Güaraní (Paraguayan) Indian name *güirá,* a bird.

Guiraca Güaraní (Paraguayan) Indian name *güirá,* a bird.

guirahuro Güaraní (Paraguayan) Indian name *güirahúro,* black and yellow bird, for the Yellow-rumped Marshbird *Pseudoleistes.*

gujanensis see **guianensis.**

gularis Mod. L. *gularis,* throated, of the throat (L. *gula,* the throat).

gulgula No expl. (Franklin (1831), *Proc. Zool. Soc. London,* **I**, 119); probably ref. the fine song of the Small Skylark *Alauda* (L. *gula,* the throat), although *gulgul* is a name for the Indian Mynah *Acridotheres tristis* in Madhya Pradesh, India, and Langton (1920) records that *gulgula* in Punjabi means 'globular'.

gulielmi Med. L. *Guilelmus,* William (and European equivalents); after Captain William Jardine (1834–1869) son of British naturalist Sir William Jardine.

gulielmitertii Med. L. *Guilelmus,* William (and European equivalents); L. *tertius,* third; after Willem III King of Holland (1817–1890: reigned 1849–1890).

gundlachii After Juan Gundlach (1810–1896) Cuban ornithologist and author (*Catalogo de los aves Cubanas*, 1873).

gunningi After Jan Willem Bowdewyn Gunning (1860–1912) South African zoologist, director of Pretoria Museum, 1896–1912.

gurneyi After John Henry Gurney (1819–1890) English banker, collector, and author.

gustavi After Gustaaf Schlegel (1840–1903) Dutch ornithologist and Sinologist.

guttata L. *guttatus,* speckled, spotted (*gutta,* a drop).

guttaticollis L. *guttatus,* spotted; Mod. L. *-collis,* necked.

guttatum /guttatus see **guttata.**

Guttera L. *gutta,* a drop, a spot; *-fera,* bearing.

guttifer L. *gutta,* a spot, a drop; *-fera,* bearing.

guttulatus L. *guttula,* a little drop, a spot; *-atus,* possessing.

guttuligera L. *guttula,* a little drop, a spot; *-gera,* bearing, a carrier.

guttulus L. *guttula,* a little drop, a spot.

gutturalis Mod. L. *gutturalis,* throated, of the throat (L. *guttur,* the throat) (cf. Med. L. *gutturalis,* speaking, speech).

gutturata L. *guttur,* the throat; *-atus,* possessing; ref. yellow chin spot of the Speckled Spinetail *Cranioleuca.*

guy After J. Guy (fl. 1833) French naturalist.

Gyalophylax Gr. *gualon,* a valley; *phulax,* a watcher, a sentinel (*phulasso,* to watch).

Gygis Gr. *guges,* a mythical bird mentioned by Dionysius.

Gymnobucco Gr. *gumnos,* bare, naked; Mod. L. *bucco,* a barbet; ref. bare skin on face and crown of the Naked-faced Barbet *G. calvus.*

gymnocephala /gymnocephalus Gr. *gumnos,* bare, naked; *-kephalos,* headed.

Gymnocichla Gr. *gumnos,* bare, naked; *kikhle,* a thrush; the male Bare-crowned Antbird *G. nudiceps* has a bare, bright blue crown.

Gymnocrex Gr. *gumnos,* bare, naked; genus *Crex,* crake; the Bald-faced Rail *G. rosenbergii* has bare, light blue facial skin.

Gymnoderus Gr. *gumnos,* bare, naked; *deras,* the skin; the Bare-necked Fruit-crow *G. foetidus* has extensive cobalt-blue folds of bare skin on the neck.

gymnogenys Gr. *gumnos,* bare, naked; *genus,* the cheek.

Gymnoglaux Gr. *gumnos,* bare, naked; *glaux,* the Little Owl *Athene;* the Bare-legged Owl *G. lawrencii* has long, bare, unfeathered legs.

Gymnogyps Gr. *gumnos,* bare, naked; *gups,* a vulture; the bare-headed California Condor *G. californianus* is now (1990) represented only by captive specimens.

Gymnomystax Gr. *gumnos,* bare, naked; *mustax,* a moustache (*mastax,* the jaws); the Oriole Blackbird *G. mexicanus* has a bare, black ocular area and moustachial streak.

Gymnomyza Gr. *gumnos,* bare, naked; Mod. L. *myza,* honeyeater (Gr. *muzo,* to suck); ref. the naked, orange-yellow skin around the eyes and cheeks of the Crow Honeyeater *G. aubryana.*

Gymnopelia Gr. *gumnos,* bare, naked; *peleia,* a dove or pigeon.

Gymnophaps Gr. *gumnos,* bare, naked; *phaps,* a pigeon; the Papuan Mountain Pigeon *G. albertisii* has an extensive area of naked, red skin on the face.

Gymnopithys Gr. *gumnos,* bare, naked; genus *Pithys,* antbird; the Rufous-throated Antbird *G. rufigula* has a bare, bluish-white eye-patch.

gymnops Gr. *gumnos,* bare, naked; *ops,* the eye; ref. bare skin surrounding eye.

Gymnorhina /Gymnorhinus Gr. *gumnos,* bare, naked; *rhis, rhinos,* the nostril; ref. the exposed nostrils of the Australian Magpie *G. tibicen* and the Piñon Jay *G. cyanocephalus.*

Gymnoschizorhis Gr. *gumnos,* bare, naked; syn. genus *Schizorhis,* go-away bird (Gr. *skhiza,* an arrow; *rhis,* the nostril; ref. slit-shaped nostrils); the Bare-faced Go-away Bird *Corythaixoides personata* has a naked black face.

Gymnostinops Gr. *gumnos,* bare, naked; syn. genus *Ostinops,* oropendola; the Montezuma Oropendola *G. montezuma* has extensive bare lavender, pink, or blue skin at the base of the bill.

Gypaetus Gr. *gupaietos* (a corrupt version of *hupaietos*), an eagle or vulture.

Gypohierax Gr. *gups,* a vulture; *hierax,* a hawk.

Gypopsitta Gr. *gups,* a vulture; Mod. L. *psitta,* a parrot (Gr. *psittakos,* a parrot).

Gyps Gr. *gups,* a vulture.

gyrola Dim. from L. *gyrus,* a ring, a circle (Gr. *guros*); ref. yellow nuchal collar of the Bay-headed Tanager *Tangara.*

gyroloides Specific name *Tangara gyrola;* Gr. *-oides,* resembling.

Gavia immer Great Northern Diver

H

haastii After Sir John Francis Julius von Haast (1824–1887) New Zealand explorer, geologist, and author.

habessinica After Abyssinia (Ethiopia) (v. *abyssinica*).

Habia Güaraní (Paraguayan) Indian name *habia* for various sorts of finches and tanagers.

Habropteryx Gr. *habros*, soft, delicate; *pterux*, the wing; the Pink-legged Rail *H. insignis* has soft, reduced wing feathers.

Habroptila/habroptilus Gr. *habros*, soft, delicate; *ptilon*, a feather; ref. the soft, decomposed wing feathers of the Invisible Rail *H. wallacii* and the soft body plumage of the flightless Kakapo *Strigops*.

haemacephala Gr. *haima*, blood (i.e. crimson red); *-kephalos*, headed.

haemastica Gr. *haimatikos*, bloody (*haima, haimatos*, blood).

haematina L. *haematinus*, blood-coloured (Gr. *haimatinos*, bloody).

Haematoderus Gr. *haima, haimatos*, blood; *deras*, the skin.

haematodus Gr. *haimatodes*, blood-red.

haematogaster Gr. *haima*, blood; *gaster*, the belly.

haematonota/haematonotus Gr. *haima*, blood; *-notos*, backed.

Haematopus L. *haematopus*, blood-foot (Gr. *haima*, blood; *pous*, the foot).

haematopygus Gr. *haima*, blood; *-pugos*, rumped (*puge*, the rump).

Haematortyx Gr. *haima*, blood; *ortux*, the quail.

Haematospiza Gr. *haima*, blood; *spiza*, finch.

haematostictum Gr. *haima*, blood; *stiktos*, spotted, dappled.

haematotis Gr. *haima*, blood; *-otis*, eared.

haematuropygia Gr. *haima*, blood; *ouropugion*, the rump; the Red-vented Cockatoo *Cacatua* has deep orange undertail-coverts.

haemorrhous Gr. *haima*, blood; *orrhos*, the rump.

haesitata Mod. L. *haesitatus*, hesitant, faltering (L. *haesitare*, to stick fast).

hagedash/Hagedashia Afrikaans *hagedash* (also *hadelde)*, an onomatopoeic name given to the Hadada Ibis *H. hagedash* by early settlers in South Africa.

hainana After the island of Hainan, China.

Halcyon Gr. *halkuon*, a mythical bird, long associated with the kingfisher, which nested on the sea. It was beloved of the gods, who calmed the waves whilst it incubated and raised its young. Such days of peace and calm became known as 'halcyon days'.

Haliaeetus/haliaetus L. *haliaeetos*, the sea-eagle *H. albicilla* or Osprey *Pandion* (Gr. *haliaetos*, the osprey; *hals*, the sea; *aetos*, an eagle).

Haliastur Gr. *hali-*, sea (*hals*, the sea); L. *astur*, a hawk.

Halietor Gr. *halieos*, a fisherman.

hallae After Beryl Patricia Hall (b. 1917) British ornithologist and author (*An atlas of speciation in African passerine birds*, 1970).

halli After **1** Robert Hall (1867–1949) Australian ornithologist and President of the Royal Australasian Ornithologists Union, 1912–1913 (*Macronectes*). **2** Major Harold Wesley Hall (b. 1888) Australian zoologist, explorer, and collector (*Pomatostomus*).

Halobaena Gr. *hals, halos,* the sea; *baino,* to go, to walk.

Halocyptena Gr. *hals,* the sea; *okus,* speedy; *ptenos,* winged.

hamatus L. *hamatus,* hooked, curved (*hamus, hami,* a hook).

hambroeki After A. Hambroek (d. 1661) Dutch missionary to Taiwan.

hamertoni After Colonel Albert Ernest Hamerton (1873–1955) English explorer in Somaliland, 1904–1906.

hamiltoni After J. E. Hamilton (1891–1957) British scientist in the Falkland Islands, 1919–1957.

Hamirostra L. *hamus, hami,* a hook; *rostrum,* the bill.

hamlini After Hannibal Hamlin (fl. 1929) collector in the Pacific.

hammondii After Surgeon-General William Alexander Hammond (1828–1900) US neurologist and collector.

Hapalopsittaca Gr. *hapalos,* delicate; *psittakos,* a parrot; ref. narrow, untoothed bill of the Rusty-faced Parrot *H. amazonina*.

Hapaloptila Gr. *hapalos,* delicate; *ptilon,* a feather; ref. soft, loose-webbed plumage of the White-faced Nunbird *H. castanea*.

Haplochelidon Gr. *haploos,* simple, plain; *khelidon,* a swallow.

haplochrous Gr. *haploos,* simple, plain; *khroos,* the body, the skin; ref. plain uniform plumage (*Turdus*) or upperparts (*Accipiter*).

haplonota Gr. *haploos,* simple, plain; *-notos,* backed.

Haplophaedia Gr. *haploos,* sheer, plain; *phaidros,* radiant, brilliant (*phao,* to shine).

Haplospiza Gr. *haploos,* simple, plain; *spiza,* a finch.

hardwickei/hardwickii After **1** Major General Thomas Hardwicke (1756–1835) Indian Army and naturalist (*Chloropsis*). **2** Charles Hardwicke (fl. 1831) collector in Tasmania (*Gallinago*).

harlani After Richard Harlan (1796–1843) US surgeon and naturalist.

harmani After Lieutenant H. J. Harman (1850–1883) Indian Army and surveyor.

harmonica L. *harmonicus,* harmonious.

Harpactes Gr. *harpaktes,* a robber (i.e. of fruit).

Harpagus L. *harpago,* a rapacious person (Gr. *harpazo,* to seize, to plunder).

Harpia Gr. *harpe,* a bird of prey mentioned by Aristotle, Dionysius, Hesychius, and Pliny, not further identified. In Gr. myth. the harpies (*Harpuiai*) were winged monsters with sharp claws having the face of a woman and the body of a vulture, '…they emitted an infectious smell and spoiled whatever they touched by their filth and excrements' (Lemprière (1984), p. 293).

Harpiprion Gr. *harpe,* a sickle or scimitar; *prion,* a sawyer (*prio,* to saw); ref. long, narrow crest feathers and scimitar-shaped bill of the Plumbeous Ibis *H. caerulescens*.

Harpyhaliaetus Genus *Harpia,* harpy eagle; genus *Haliaeetus,* sea-eagle.

harpyja see **Harpia.**

Harpyopsis Genus *Harpia,* harpy eagle; Gr. *opsis,* appearance.

harrisi/harrisii After **1** Edward Harris (1799–1863) US farmer, naturalist, and sponsor of Audubon (*Aegolius*). **2** Charles Miller Harris (fl. 1898) English collector and taxidermist for Tring Museum (*Nannopterum*).

harterti After Ernst Johann Otto Hartert (1859–1933) German ornithologist and curator of Rothschild collection, Tring.

Hartertula Dim. after Ernst Hartert (v. *harterti*).

hartlaubi/hartlaubii/Hartlaubius After Karl Johann Gustav Hartlaub (1814–1900) German ornithologist and author (*Die vögel Madagaskars*, 1877).

harwoodi After Leonard Harwood (fl. 1899) English naturalist and taxidermist.

hasitata see **haesitata.**

hastatus L. *hastatus,* armed with a spear, hastate; the Sinaloa Screech Owl *Otus* has '...the markings...on the pileum and hind neck in the form of irregularly rhomboid or hastate spots' (Ridgway (1887), *Proc. US Nat. Mus.*, **10**, 268).

hattamensis After the village of Hatam, Arfak, Irian Jaya (Indonesian New Guinea).

hauxwellii After J. Hauxwell (fl. 1854) collector resident in Peru.

hebetior L. *hebetior,* dull, sluggish (*hebes,* dull).

Hedydipna Gr. *hedudeipnos,* sweet-feeding, sweet-eating.

Hedymeles Gr. *hedumeles,* sweet-singing.

heermanni After Adolphus Lewis Heermann (1818–1865) US explorer and field naturalist.

heilprini After Angelo Heilprin (1853–1907) Hungarian/US scientist, explorer, and author (*The geographical and geological distribution of animals*, 1907).

heinei After Ferdinand Heine (1809–1894) German ornithologist and founder of Museum Heineanum, Halberstadt.

heinrichi After **1** Gerd Heinrich (b. 1896) German field ornithologist, collector in the East Indies, and author (*Der vögel schnarch. Zwei jahre rallenfang und urwaldforschung in Celebes*, 1932) (*Cacomantis, Geomalia)* and his wife **2** Hildegard Heinrich (fl. 1954) ornithologist and collector in Sulawesi, 1930, 1932 and Angola, 1954 (*Cossypha).*

Heinrichia see **heinrichi 1.**

heinrothi After Oskar Heinroth (1871–1945) German zoologist, collector in the Pacific, 1900–1901 and author (*Die vögel Mitteleuropas*,1924).

Heleia Gr. *eleia,* a small bird mentioned by Callimachus, conjectured to be a warbler.

helenae After **1** Gr. myth. Helen, wife of Menelaus and the most beautiful woman of her age (*Hypothymis).* **2** Princesse Hélène d'Orléans (1814–1858) wife of patron of natural history Duc d'Orléans (*Lophornis).* **3** Helen Gundlach (fl. 1850) wife of Cuban ornithologist Juan Gundlach (*Mellisuga).* **4** Helena Forde (d. 1910) daughter of Australian oologist A. Scott (*Parotia).* **5** Helen Kelsall (fl. 1913) wife of Colonel H. Kelsall collector in tropical Africa, 1910–1913 (*Laniarius).*

heliaca Gr. *heliakos,* solar, of the sun (*helios,* the sun); ref. golden nape patch of the Imperial Eagle *Aquila.*

Heliactin Gr. *helios,* the sun; *aktis,* a sunbeam.

Heliangelus Gr. *helios,* the sun; *angelos,* an angel or messenger.

helianthea Gr. *helios,* the sun; *antheion,* a blossom.

helias Gr. *helias,* solar, of the sun

Helicolestes Gr. *helix, helikos,* anything twisted, snail-shell; *lestes,* a robber; the Slender-billed Kite *H. hamatus* preys exclusively on freshwater snails.

heliobates Gr. *heliobatos,* sun-scorched.

Heliobletus Gr. *heliobletos,* sun-scorched.

Heliocorys Gr. *helios,* the sun; Mod. L. *corys,* a lark.

heliodor No expl. (Bourcier (1840), *Rev. Zool.* [Paris], 275), where the Gorgetted Woodstar *Acestrura* is first referred to as *Oiseau-mouche Héliodore* (Gr. *helios,* the sun; *doree,* a gift (*doreo,* to give)).

Heliodoxa Gr. *helios,* the sun; *doxa,* splendour, glory.

Heliolais Gr. *helios,* the sun; Mod. L. *lais,* a warbler (Gr. *leios,* a type of warbler; *laios,* a thrush); ref. red patched wings of the Red-wing Warbler *H. erythroptera.*

Heliomaster Gr. *helios,* the sun; *master,* a searcher.

Heliopais Gr. *helios,* the sun; *pais,* a child; reflecting relationship of the Masked Finfoot *H. personata* with the Sungrebe *Heliornis.*

Heliornis Gr. *helios,* the sun; *ornis,* a bird; '...this name seems to have arisen from a mistake of Latham's who in 1785 supposed the *Oiseau de Soleil,* so translated by Fermin in 1769 from the Dutch *Sonne-vogel,* to be the present species, whereas it is obviously the *Eurypyga* (Sun Bittern)' (Newton and Gadow (1896), p. 252, footnote 1).

heliosylus Gr. *helios,* the sun; *sulao,* to carry off; ref. buffy underparts of the New Guinea Forest Bittern *Zonerodius.*

Heliothryx Gr. *helios,* the sun; *thrix,* the hair; ref. glittering crown of the Black-eared Fairy *H. aurita.*

helleri After Edmund Heller (1875–1939) US naturalist and collector in tropical Africa.

Hellmayrea/hellmayri After Carl Eduard Hellmayr (1878–1944) German/US ornithologist and author (*Catalogue of birds of the Americas*, 1924).

Helminthophila Gr. *helmins, helminthos,* a worm; *philos,* fond of, loving.

Helmitheros Gr. *helmins,* a worm; *-thera,* hunting (*therao,* to hunt).

heloisa No expl. (Lesson and Delattre (1839), *Rev. Zool.* [Paris], 15), where the Bumblebee Hummingbird *Atthis* is first referred to as *Oiseau-mouche Héloïse,* but, after the fashion of the time, probably for Héloïse (1101–1164), niece of the Canon of Notre Dame, who had a tragic love affair with Abelard, by whom she had a child.

helvola L. *helvolus* (a form of *helveolus),* pale yellow, yellowish.

hemichrysus Gr. *hemikhrusos,* half-golden.

Hemicircus Gr. *hemi,* half (i.e. small, short); *kerkos,* the tail; ref. short stumpy tail of the Grey-and-buff Woodpecker *H. concretus.*

Hemignathus Gr. *hemi,* half (i.e. small, short); *gnathos,* the lower jaw; ref. thinner, shorter lower mandible of the Lesser Akialoa *H. obscurus.*

hemilasius Gr. *hemi,* half; *lasios,* shaggy, hairy; ref. partially feathered legs of the Upland Buzzard *Buteo.*

hemileucurus Gr. *hemi,* half; *leukouros,* white-tailed.

hemileucus Gr. *hemi,* half; *leukos,* white.

Hemimacronyx Gr. *hemi,* half (i.e. akin to, similar); genus *Macronyx,* long-claw.

hemimelaena Gr. *hemi,* half; *melania,* blackness.

Hemiparra Gr. *hemi,* half (i.e. akin to, similar); syn. genus *Parra,* jacana (*parra,* a bird of ill omen); ref. long, jacana-like toes of the Long-toed Lapwing *Hoplopterus crassirostris.*

Hemiphaga Gr. *hemi,* half (i.e. between); syn. genus *Carpophaga,* fruit pigeon (Gr. *karpos,* fruit; *phagos,* a glutton); the New Zealand Pigeon *H. novaeseelandiae* was regarded as intermediate between *Carpophaga* and *Megaloprepia.*

Hemipipo Gr. *hemi,* half (i.e. small); Mod. L. *pipo,* manakin.

Hemiprocne Gr. *hemi,* half (i.e. small, resembling); L. *progne,* a swallow.

Hemipus Gr. *hemi,* small; *pous,* the foot; the pygmy trillers have very small legs and feet.

Hemispingus Gr. *hemi,* half (i.e. small); *spingos,* a finch; ref. small size of the hemispinguses.

Hemitesia Gr. *hemi,* half (i.e. small, related); genus *Tesia,* ground warbler.

Hemithraupis Gr. *hemi,* half (i.e. small); genus *Thraupis,* tanager.

Hemitriccus Gr. *hemi,* half (i.e. small); Mod. L. *triccus,* tyrant flycatcher.

hemixantha Gr. *hemi,* half; *xanthos,* yellow, golden.

hemprichii After Friedrich Wilhelm Hemprich (1796–1825) German naturalist and explorer in Egypt and Arabia, 1820–1825.

hendersoni After Lieutenant Colonel George Henderson (1837–1929) British soldier, explorer, and author (*Lahore to Yarkand*, 1873).

henicogrammus Gr. *henikos,* unique, single; *gramme,* a line; ref. concealed white nuchal bar of the Moluccan Goshawk *Accipiter.*

Henicopernis Gr. *henikos,* unique (i.e. related to); genus *Pernis,* honey-buzzard.

Henicophaps Gr. *henikos,* unique (i.e. related to); genus *Phaps,* bronzewing pigeon.

Henicorhina Gr. *henikos,* unique; *rhis, rhinos,* the nose; ref. long bill and partially concealed slit-like nostrils of the White-breasted Wood Wren *H. leucosticta.*

henrici/henricii After **1** Lieutenant Colonel A. H. Henrici (d. 1836) Dutch Army in the East Indies (*Megalaima*). **2** Henri Prince d'Orléans (1867–1901) French explorer and collector in Tibet and China, 1889–1890 (*Garrulax*).

henslowii After Revd. John Stevens Henslow (1796–1861) English botanist, mentor of Charles Darwin, and geologist.

henstii After G. van der Henst (fl. 1870) collector in Madagascar.

hepatica L. *hepaticus,* hepatic, of the liver (i.e. colour) (*hepar, hepatis,* the liver; Gr. *hepar, hepatos,* the liver) (cf. Mod. L. *hepaticus,* liver-coloured).

heraldica After HMS *Herald*, British warship that conducted explorations in the Pacific, 1845–1851.

herberti After **1** E. G. Herbert (1870–1951) English naturalist and collector (*Stachyris*). **2** Herbert Alexander (1874–1946) English artist and author, brother of explorer Captain Boyd Alexander (*Seicercus*).

herbicola L. *herba,* grass; *-cola,* inhabitant, dweller.

Herbivocula Dim. of syn. genus *Herbivox,* warbler (L. *herba,* grass, herbage; *vox,* voice).

hercules Gr. myth. Heracles or Hercules, a celebrated hero later deified; ref. the heroic proportions of Blyth's Kingfisher *Alcedo.*

herero After the Herero, a native people of Namibia, formerly South West Africa, who rose in revolt against the German colonists 1904–1908 and were almost exterminated.

herioti After F. Maitland-Heriot (1852–1925) English collector in the Philippines, 1884.

herminieri After Félix Louis l'Herminier (1779–1833) French naturalist in Guadeloupe, West Indies.

herodias Gr. *erodios,* a heron.

Herpetotheres Gr. *herpeton,* a reptile; *theras,* a hunter.

Herpsilochmus Gr. *herpo,* to creep about; *lokhme,* a thicket or copse.

herrani After Pedro Alcántara Herran (1800–1872) President of the Republic of New Granada (v. *granadense*), 1841–1844.

hershkovitzi After Philip Hershkovitz (b. 1909) US zoologist.

Hesperiphona Gr. *hesperos,* the evening; *phone,* a voice; the Evening Grosbeak *H. vespertina* was believed to sing only during the evening.

Heteralocha Gr. *heteros,* different; *alokhos,* a wife; ref. striking differences in shape and length of bill in the two sexes of the extinct Huia *H. acutirostris.*

Heteranax Gr. *heteros,* different, another; *anax,* a monarch, a king; ref. the wedge-shaped, shrike-like bill of the Black-bibbed Monarch *H. mundus.*

Heterhyphantes Gr. *heteros,* different, another; *huphantes,* a weaver.

Heterocercus Gr. *heteros,* different; *kerkos,* the tail; ref. very rounded tail of the Flame-crowned Manakin *H. linteatus.*

heteroclitus L. *heteroclitus,* irregular, abnormal, varying; ref. unusually varied callnotes of the Singing Parrot *Geoffroyus.*

Heterocnus Gr. *heteros,* different, unusual; *oknos,* a kind of heron.

Heteromirafra Gr. *heteros,* different, another; genus *Mirafra,* bush-lark.

Heteromunia Gr. *heteros,* different, another; syn. genus *Munia,* mannikin (Hindi name *munia* for the Black-headed Munia *Lonchura malacca* and other small waxbills).

Heteromyias Gr. *heteros,* different, another; Mod. L. *myias,* a flycatcher.

Heteronetta Gr. *heteros,* different, another; *netta,* a duck; the Black-headed

Duck *H. atricapilla* is the only parasitic duck.

Heterophasia Gr. *heteros,* different, another; *phasis,* appearance; of the Long-tailed Sibia *H. picaoides* Blyth (1842), *Journ. Asiat. Soc. Bengal,* **11**, 186, wrote, '...a curious meruline form, exhibiting affinity for various distinct genera, but which cannot be immediately approximated to any with which I am acquainted'.

heteropogon Gr. *heteros,* different (i.e. variegated); *pogon,* the beard.

Heteroptilorhis Gr. *heteros,* different; genus *Ptiloris,* riflebird.

Heterorhynchus Gr. *heteros,* different; *rhunkhos,* the bill; the Nukupuu *H. lucidus* has a long sickle-shaped upper mandible and much shorter decurved lower mandible.

Heteroscelus Gr. *heteroskeles,* with uneven (i.e. different) legs; ref. the overlapping, scutellated tarsal scales of the Grey-tailed Tattler *H. brevipes*.

Heterospingus Gr. *heteros,* different, another; *spingos,* a finch.

Heterospizias Gr. *heteros,* different, another; *spizias,* a hawk.

Heterotrogon Gr *heteros,* different, another; genus *Trogon,* trogon.

heterura/heterurus Gr. *heteros,* different; *-ouros,* tailed.

heudei After Pierre Marie Heude (1836–1902) French conchologist in China.

heuglini After Theodor von Heuglin (1824–1876) German explorer, ornithologist, and author (*Ornithologie nordost-Afrika's, der Nilquellen- und küstengebiete des Rothen Meeres und des nördlichen Somal-Landes,* 1869; *Reise in nordorst-Afrika,* 1877).

heyi After Michael Hey (1798–1832) German surgeon and collector.

hiaticula L. *hiatus,* a cleft; *-cola,* a dweller (equivalent to Gr. *kharadrios* (v. *Charadrius)*; *kharadra,* a ravine); the Ringed Plover *Charadrius* breeds amongst pebbles and rocks.

Hieraaetus Gr. *hierax,* a hawk; *aetos,* an eagle.

Hierococcyx Gr. *hierax,* a hawk; *kokkux,* the cuckoo; ref. the hawk-like appearance of the Indomalayan hawk cuckoos.

Hierofalco Gr. *hieros,* sacred; genus *Falco,* falcon (v. *augur)* (cf. *Falco sacer*, a former name for the Saker Falcon *F. cherrug*, is derived from the Arabic name *sakr* for a falcon and not from L. *sacer*, sacred).

Hierophasis Gr. *hieros,* sacred; Mod. L. *phasis,* pheasant.

hildebrandti After Johann Maria Hildebrandt (1847–1881) German zoologist and collector in tropical Africa, 1871–1879 and Madagascar, 1879–1881.

himalayana/himalayensis After the Himalaya Mountains (Sanskrit 'abode of snow'), the great mountain chain of central Asia.

Himantopus Gr. *himantopous,* a wading bird, now usually identified as the stilt; 'The old name of *himantopus* is taken from Pliny: and, by an aukward metaphor, implies that the legs are as slender and pliant as if cut out of a thong of leather' (White (1789), letter, 7 May 1779) (Gr. *himas, himantos,* a strap; *pous,* a foot).

Himantornis Genus *Himantopus,* stilt; Gr. *ornis,* a bird; ref. long red legs of the Nkulenga Rail *H. haematopus*.

Himatione Gr. *himation,* the Spartan military cloak, always blood red to conceal wounds which might give encouragement to the enemy; the Apapane *H. sanguinea* is deep crimson with black wings and tail.

hindei After Sidney Langford Hinde (1863–1931) naturalist and collector in Kenya, 1896–1901.

hindwoodi After Keith Alfred Hindwood (1904–1971) Australian businessman, ornithologist, and President of the Royal Australasian Ornithologists Union, 1944–1946.

hiogaster Gr. *io-,* dark, violet (*ion,* the violet); *gaster,* the belly.

Hippolais Gr. *hupolais,* a small ground-nesting bird mentioned by

Aristotle, Theophrastus, and Hesychius, not further identified, but thought by some later authors to be the Northern Wheatear *Oenanthe*.

hirsuta /hirsutum L. *hirsutus*, hairy, rough, bristled.

Hirundapus Genus *Hirundo*, swallow; genus *Apus*, swift.

hirundinacea /hirundinaceum /hirundinaceus L. *hirundo*, a swallow; *-aceus*, resembling.

Hirundinea /hirundineus Med. L. *hirundineus*, swallow-like (L. *hirundo, hirundinis*, a swallow); the long-winged Cliff Flycatcher *H. ferruginea* appropriates the nests of swallows and martins for its own use, having ejected the original owners. Specifically the epithet refers to the long, forked tail of the Swallow-tailed Bee-eater *Dicrocercus*.

Hirundo L. *hirundo*, a swallow.

hispanica L. *Hispanicus*, Spanish (*Hispania*, Spain).

hispaniolensis After **1** L. *Hispania*, Spain (*Passer*). **2** erroneously the island of Hispaniola (Spanish *Isla Española*) in the West Indies (*Poospiza*).

hispidus L. *hispidus*, rough, hairy; from grizzled appearance of throat of the White-bearded Hermit *Phaethornis*.

histrio L. *histrio*, an actor, a harlequin; ref. brightly splashed plumage of the Red-and-blue Lory *Eos*.

histrionica /Histrionicus L. *histrionicus*, theatrical, like a harlequin; ref. bright marked plumage of the drake Harlequin Duck *H. histrionicus* and the pied face pattern of the Flock Pigeon *Phaps*.

Histriophaps L. *histrio*, an actor, a harlequin; genus *Phaps*, pigeon (v. *histrionica*).

Histurgops Gr. *histurgeo*, to weave (i.e. a weaver bird); *ops*, the face (i.e. appearance).

hoatzin Nahuatl (Mexican) Indian name *uatsin* for some kind of large bird, probably not the Hoatzin *Opisthocomus*, which does not occur in Middle America, but a gamebird. The name is said to be onomatopoeic in origin and has been variously transcribed as *hoactzin, hoactli, hoatzin, hoacton, hoazin, heatotl* and *tobactli*. Hernandez (1615), gives *hoactli* and *hoacton* as Mexican Indian names for the adult and young Black-crowned Night Heron *Nycticorax* respectively.

hodgsoni see **hodgsonii**

hodgsoniae After Anne Hodgson (d. 1868) wife of English naturalist B. H. Hodgson (v. *hodgsonii*).

hodgsonii /Hodgsonius After Brian Houghton Hodgson (1800–1894) English diplomat, Resident to Nepal, 1833–1844, ethnologist, naturalist, and author.

hoedtii After D. S. Hoedt (1813–1893) Dutch civil servant and collector in the Moluccas, 1853–1867.

hoematotis see **haematotis.**

hoeschi After Walter Hoesch (1896–1961) German zoologist, collector in Namibia, 1930–1961 and author (*Die vögelwelt Deutsch-Südwestafrikas namentlich des Damara- und Namalandes*, 1940).

hoevelli After Baron G. W. C. van Hoevell (b. 1848) Dutch Governor of the Celebes (Sulawesi), 1902.

hoffmanni /hoffmannii After Carl Hoffmann (1823–1859) German explorer and naturalist in Costa Rica, 1854–1859.

hoffmannsi /hoffmansi After W. Hoffmanns (1865–1909) German collector in Peru, 1903 and Brazil, 1904–1908.

holerythra Gr. *holos*, complete; *eruthros*, red.

hollandicus After New Holland, the name by which the eastern half of continental Australia was known to Europeans during the two centuries after its discovery. In orn. generally synonymous with what is now the state of New South Wales.

holochlora /holochlorus Gr. *holokhloros*, all green.

holomelaena Gr. *holomelas*, black all over.

holopolia Gr. *holos,* complete, entire; *polios,* grey, hoary.

Holoquiscalus Gr. *holos,* complete, entire; genus *Quiscalus,* grackle.

holosericea/holosericeus Gr. *holoserikos,* silken.

holospilus Gr. *holos,* complete, entire; *spilos,* a spot.

holostictus Gr. *holos,* complete, entire; *stiktos,* spotted, blotched.

holstii After P. A. Holst (fl. 1889) Swedish naturalist in Japan and Taiwan, 1889.

hombroni After Bernard Hombron (1800–1853) French surgeon naturalist in the Pacific 1837–1840, and author (*Aventures les plus curieuses des voyageurs*, 1847).

homeyeri After Eugen Ferdinand von Homeyer (1809–1889) German ornithologist, opponent of Darwinism, and author (*Verzeichniss der vögel Deutschlands*, 1885).

homochroa/homochrous Gr. *homokhrous,* uniform, of one colour.

hondae After the town of Honda, Tolima, Colombia.

hoogerwerfi After A. Hoogerwerf (b. 1907) Dutch botanist and author.

hopkei After Gustav Hopke (fl. 1897) German collector.

Hoplopterus Gr. *hoplon,* a weapon; *-pteros,* winged; ref. sharp, curved carpal spurs of the Spur-winged Plover *H. spinosus.*

Hoploxypterus Gr. *hoplon,* a weapon; *oxus,* sharp; *-pteros,* winged; ref. sharp carpal spurs of the Pied Plover *H. cayanus.*

hordeaceus L. *hordeaceus,* of barley (*hordeum,* barley); ref. diet of the Fire-crowned Bishop *Euplectes.*

Horeites Gr. *oreiotes,* a mountaineer.

Horizorhinus Gr. *orizo,* to enclose, to divide; *rhis, rhinos,* the nose.

hornbyi After Admiral Sir Phipps Hornby (1785–1867) British Commander-in-Chief, Pacific station, 1847–1850.

hornemanni After Jens Wilken Hornemann (1770–1841) Danish botanist and author (*Flora Danica*, 1806).

horsfieldii After Thomas Horsfield (1773–1859) US naturalist, collector in Sumatra and Java, 1796–1818 and author (*Zoological researches in Java and the neighbouring islands*, 1824).

hortensis L. *hortensis,* of a garden (*hortus,* a garden).

hortulana Italian name *hortulane* for the Ortolan Bunting *Emberiza.*

hortulorum L. *hortulorum,* of little gardens (*hortulus,* a little garden).

horus Egyptian myth. Horus, the sun god.

hosii After Charles Hose (1863–1929) English naturalist in Sarawak, 1884–1907, collector, and author (*Fifty years of romance and research*, 1927; *The field book of a jungle wallah*, 1929).

hottentotta/hottentottus After the Hottentots, a nomad pastoral race of Namibia and South Africa, so called by the Dutch settlers who considered their staccato tongue unintelligible. In the case of *Dicrurus* the toponym is erroneous.

Houbaropsis Syn. genus *Houbara,* bustard (Arabic *hubara,* a bustard); Gr. *opsis,* appearance.

Houppifer French *houppe,* a tuft or crest; L. *-fer,* carrying, bearing.

housei After Frederick Maurice House (1865–1936) Australian sheep-farmer in Western Australia.

hova/hovarum After the Hova, a native clan of Madagascar.

huallagae After the Río Huallaga, La Libertad, Peru.

huancavelicae After the department of Huancavelica, south-central Peru.

hudsoni After William Henry Hudson (1842–1922) British naturalist resident in Argentina, 1842–1869, and author (*Argentine ornithology*, 1888; *Green mansions*, 1904; *Birds of La Plata*, 1920).

hudsonicus After Hudson Bay, Canada, named for the English navigator Henry Hudson who was cast adrift there in 1611.

huetii After Nicolas Huet (1770–1830) French painter and naturalist.

huhula French *Le Huhul,* the name given to the Black-banded Owl *Ciccaba* by Levaillant (1799) (cf. *hulotte,* an owl; L. *ulula,* a screech owl).

humbloti/Humblotia After Léon Humblot (fl. 1904) French Governor of Grand Comoro Island, 1884–1897 and collector in Madagascar.

humboldti After Friedrich Wilhelm Karl Heinrich Alexander Baron von Humboldt (1769–1859) German scientist and explorer.

humei After Allan Octavian Hume (1829–1912) British civil servant in India, ornithologist, collector, founder of *Stray Feathers - a journal of ornithology for India and dependancies,* and author (*Nest and eggs of Indian birds,* 1873; *Game birds of India, Burma and Ceylon,* 1879).

humeralis Late L. *humeralis,* of the shoulder or upper arm (L. *umerus,* the shoulder).

humiae After Mary Hume (d. 1890) wife of British ornithologist Allan Hume (v. *humei).*

humicola L. *humus, humi,* the earth, the ground; *-cola,* dweller.

humilis L. *humilis,* on the ground (can also mean lowly, poor, small).

hunsteini After C. Hunstein (1843–1888) German administrator in German New Guinea, 1885–1888.

hunteri After H. C. V. Hunter (1861–1934) big-game hunter in East Africa, 1886–1887.

hutchinsii After T. Hutchins (1730–1790) employee of Hudson's Bay Company, Canada.

huttoni After **1** William Hutton (d. 1851) US naturalist and collector (*Vireo).* **2** Frederick Wollaston Hutton (1836–1905) English/New Zealand geologist, zoologist, and author (*Catalogue of the birds of New Zealand,* 1871) (*Ptilinopus, Puffinus).*

hyacinthina/hyacinthinus L. *hyacinthinus,* hyacinthine, of the hyacinth (i.e. in colour) (Gr. *huakinthos,* the hyacinth).

hybrida L. *hybrida,* a hybrid, a mongrel; ref. doubtful status as a full species.

Hydranassa Gr. *hudro-,* water (*hudor,* water); *nassa,* a duck or some other waterbird.

Hydrobates Gr. *hudro-,* water; *-bates,* a dweller (*baino,* to go or tread).

hydrocharis Gr. *hudro-,* water; *kharis,* grace, loveliness.

Hydrochous Gr. myth. Hydrochous or Hudrokhoos, the water-pourer, the constellation Aquarius; the Waterfall Swift *H. gigas* nests under waterfalls.

hydrocorax Gr. *hudro-,* water; *korax,* the raven; the cries of these large hornbills *Buceros,* formerly called horned ravens or water ravens (Clusius (1605)), are thought to presage rain.

Hydrophasianus Gr. *hudro-,* water; genus *Phasianus,* pheasant; ref. long tail of the Pheasant-tailed Jacana *H. chirurgus.*

Hydroprogne Gr. *hudro-,* water; genus *Progne,* swallow; the terns were formerly known as sea-swallows.

Hydropsalis Gr. *hudro-,* water; *psalis,* a pair of shears or scissors; from supposed preferred habitat of the Scissor-tailed Nightjar *H. brasiliana*

hyemalis L. *hiemalis,* wintry, of winter (*hiems, hiemis,* winter); ref. migratory habits and winter distribution of the Long-tailed Duck *Clangula* and the Slate-coloured Junco *Junco.*

Hyetornis Gr. *huetos,* rain; *ornis,* a bird; the calls of the Chestnut-bellied Cuckoo *H. pluvialis* were thought to presage rain, hence its local name 'Rainbird' in Jamaica.

Hylacola Gr. *hule,* a woodland; *-cola,* a dweller.

Hylexetastes Gr. *hule,* a woodland; *exetastes,* a searcher (*exetazo,* to examine).

Hylia Gr. *hule,* a woodland; the Green Hylia *H. prasina* is a common forest species.

Hyliota Gr. *hule,* a woodland; *-otus,* possessing (i.e. inhabiting).

hylobius Gr. *hulobios,* wood-living.

Hylocharis Gr. *hule,* a woodland; *kharis,* grace, loveliness.

Hylocichla Gr. *hule,* a woodland; *kikhle,* a thrush.

Hylocitrea Gr. *hule,* a woodland; syn. genus *Muscitrea,* whistler (genus *Muscicapa,* flycatcher; syn. genus *Tchitrea,* paradise flycatcher).

Hylocryptus Gr. *hule,* a woodland; *kruptos,* hidden, secret (*krupto,* to hide).

Hyloctistes Gr. *hule,* a woodland; *ktistes,* an inhabitant.

Hylomanes Gr. *hulomanes,* passionately fond of the woods (*mainomai,* to adore).

Hylonympha Gr. *hule,* a woodland; *numphe,* a nymph or spirit.

Hylopezus Gr. *hule,* a woodland; *pezos,* walking (*peza,* the foot).

hylophila /Hylophilus Gr. *hule,* a woodland; *philos,* loving, fond of.

Hylophylax Gr. *hule,* a woodland; *phulax,* a watcher, a guardian.

Hylorchilus Gr. *hule,* a woodland; Mod. L. *orchilus,* wren (Gr. *orkhilos,* a small bird, usually identified with the wren *Troglodytes,* mentioned by Aristophanes, Aristotle, Pliny, Hesychius, and other authors) (cf. syn. genus *Orchilus,* kinglet *Regulus* (and also, later, pygmy-tyrant *Myiornis*)).

Hyloterpe Gr. *hule,* a woodland; *terpo,* to enjoy, delight in.

Hymenolaimus Gr. *humen, humenos,* a skin or membrane; *laimos,* the throat; ref. the membraneous upper chin of the Blue Duck *H. malacorhynchos.*

Hymenops Gr. *humen,* a skin or membrane; *ops,* the eye; ref. fleshy yellow membrane surrounding the eye of the Spectacled Tyrant *H. perspicillata.*

hyogastra Gr. *io-,* dark, violet (*ion,* the violet); *gaster,* the belly.

Hypargos Gr. *hupo-,* beneath; *argos,* bright, glistening; ref. the heavily spotted underparts of the Rosy Twinspot *H. margaritatus.*

hyperboreus Gr. *huperboreos,* in the extreme north, arctic.

Hypergerus Gr. *hupergeros,* extremely old, ancient; ref. the silver-haired appearance of the Oriole Warbler *H. atriceps.*

hypermetra Gr. *hupermetros,* excessive, beyond all measure; ref. large size of the Red-winged Bush Lark *Mirafra.*

hyperythra /hyperythrus Gr. *hupo-,* beneath; *eruthros,* red (cf. *huperuthros,* rather red, somewhat red).

Hyphantornis Gr. *huphantes,* a weaver (*huphaino,* to weave); *ornis,* a bird.

Hyphanturgus Gr. *huphantourgos,* a weaver (*huphaino,* to weave; *-ourgos,* working); ref. nestbuilding activities of the Spectacled Weaver *Ploceus ocularis.*

Hypnelus Gr. *hupnelos,* sleepy, drowsy; ref. lethargic behaviour of puffbirds.

Hypochera Gr. *hupo-,* somewhat like; syn. genus *Chera,* widowbird (Gr. *kheros,* a widow).

hypocherina Genus *Hypochera,* widowfinch; Gr. *-inos,* resembling.

hypochlora Gr. *hupokhloros,* greenish-yellow, pallid.

hypochloris Gr. *hupo-,* beneath; *khloris,* the greenfinch (i.e. green-coloured).

hypochondria L. *hypochondria,* the abdomen; ref. chestnut flanks and buff undertail coverts of the Rufous-sided Warbling Finch *Poospiza.*

hypochondriacus Gr. *hupokhondriakos,* of the abdomen.

hypochroma Gr. *hupo-,* beneath; *khroma,* colour; the Rufous-rumped Seedeater *Sporophila* has grey upperparts and chestnut underparts.

hypochryseus Gr. *hupokhrusos,* gilt, laden with gold.

Hypocnemis Gr. *hupoknemis,* somewhat booted; ref. partially booted tarsi of the Warbling Antbird *H. cantator.*

Hypocnemoides Genus *Hypocnemis,* antbird; Gr. *-oides,* resembling.

Hypocolius Gr. *hupo-,* somewhat like; genus *Colius,* mousebird; Bonaparte considered the Grey Hypocolius *H.*

ampelinus to be intermediate between the mousebirds and cotingas, but its relationships are still unclear. In orn. *colius* was also used to indicate 'starling' (e.g. *Coccycolius, Lamprocolius)*, and as late as 1930 Sclater tentatively placed the Hypocolius in the Sturnidae under the name 'Pink-billed Starling'.

Hypocryptadius Gr. *hupo-*, somewhat; *kruptadios,* secret, obscure, hidden.

Hypoedaleus Gr. *hupoidaleos,* somewhat swollen; ref. heavy bill of the Spot-backed Antshrike *H. guttatus.*

hypoglauca Gr. *hupo-,* beneath; *glaukos,* bluish green, glaucous.

Hypogramma Gr. *hupogramma,* inscribed underneath (originally of pottery); ref. boldly streaked underparts of the Purple-naped Sunbird *H. hypogrammicum.*

hypogrammica /hypogrammicum Gr. *hupo-*, beneath; *grammikos,* linear, of lines (*gramme,* a line).

hypoinchrous Gr. *hupo-,* beneath; *oinos,* wine; *khros,* skin, complexion.

hypolais see **Hippolais.**

hypoleuca /hypoleucos /hypoleucum /hypoleucus Gr. *hupo-,* beneath; *leukos,* white (cf. *hupoleukos,* whitish).

Hypomorphnus Gr. *hupo-*, somewhat like; genus *Morphnus,* eagle.

hypophaeus Gr. *hupo-,* beneath; *phaios,* dusky, brown, grey-brown.

Hypopicus Gr. *hupo-*, related to; genus *Picus,* woodpecker.

hypopolius Gr. *hupo-*, beneath; *polios,* grey (cf. *hupopolios,* somewhat grey).

hypopyrra / hypopyrrha / Hypopyrrhus Gr. *hupo-,* beneath; *purrhos,* flame-coloured (*pur,* a flame) (cf. *hupopurrhos*, reddish; *hupopuros*, with secret fire, with fire underneath).

Hypositta Gr. *hupo-*, somewhat like; genus *Sitta,* nuthatch.

hypospodia /hypospodium Gr. *hupo-,* beneath; *spodios,* ash-coloured, grey (*hupospodizo,* to be somewhat ash-coloured).

hypostictus Gr. *hupo-*, beneath; *stiktos,* spotted, dappled.

Hypotaenidia Gr. *hupo,* beneath; *tainidion,* a small band (*tainia,* a band).

Hypothymis Gr. *hupothumis,* a bird mentioned by Aristophanes, not otherwise identified.

hypoxantha /hypoxanthus Gr. *hupo-*, beneath; *xanthos,* golden, yellow.

hyrcanus L. *Hyrcanus,* Hyrcanian; Hyrcania was a wild mountainous country in what is now Kurdistan and northern Iran.

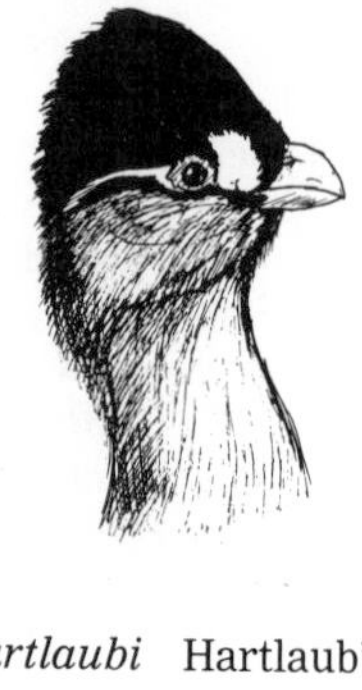

Tauraco hartlaubi Hartlaub's Turaco

I

iagoensis After the island of São Tiago or Santiago, Cape Verde Islands.

ianthinogaster L. *ianthinus*, violet-blue; *gaster*, the belly.

ibadanensis After the city of Ibadan, Oyo, Nigeria.

Ibidorhyncha Gr. *ibis, ibidos*, an ibis; *rhunkhos*, the bill.

ibis L. *ibis*, an ibis (Gr. *ibis*, an ibis; the Greek authors recognised three different types of ibis; the Glossy Ibis *Plegadis falcinellus*, the Sacred Ibis *Threskiornis aethiopicus*, and the Bald Ibis *Geronticus calvus)*.

Ibycter Gr. *ibukter*, a chanter of war-songs, a shouter (*ibuo*, to shout).

ichthyaetus Gr. *ikhthus*, a fish; *aetos*, an eagle.

Ichthyophaga Gr. *ikhthuophagos*, fish-eating (*ikhthus*, a fish; *phagos*, a glutton).

Icteria Gr. *ikteros*, a yellow bird, perhaps the Golden Oriole *Oriolus*, the sight of which was said to cure jaundice; L. *-ius*, resembling.

icterina/icterinus Mod. L. *icterinus*, yellow, yellowish (v. *Icterus)*.

icterioides Genus *Icterus*, oriole; Gr. *-oides*, resembling; ref. black and yellow plumage of the Black-and-yellow Grosbeak *Mycerobas*.

icterocephala/icterocephalus Mod. L. *icterus*, yellow; Gr. *-kephalos*, headed.

icteronotus Mod. L. *icterus*, yellow; Gr. *-notos*, backed.

icterophrys Mod. L. *icterus*, yellow; Gr. *ophrus*, the eyebrow.

Icteropsis Genus *Icterus*, oriole; Gr. *opsis*, appearance.

icteropygialis Mod. L. *icterus*, yellow; *pygialis*, rumped (Gr. *puge*, the rump).

icterorhynchus Mod. L. *icterus*, yellow; Gr. *rhunkhos*, the bill.

icterotis Mod. L. *icterus*, yellow; Gr. *-otis*, eared.

Icterus Gr. *ikteros*, a yellow bird, perhaps the Golden Oriole *Oriolus*, the sight of which was supposed to cure jaundice (cf. Mod. L. *icterus*, yellow).

Ictinaetus Gr. *iktinos*, a kite; *aetos*, an eagle.

Ictinia Gr. *iktinos*, a kite.

idae After Ida Laura Pfeiffer (1797–1858) Austrian traveller and author (*Reise einer Weinerin in das heilige land*, 1846; *Reise nach Madagaskar*, 1861).

idaliae Gr. myth. Idalia, an epithet of the goddess Venus.

Idiopsar Gr. *idios*, strange, distinct; *psar*, the starling (v. *Gnorimopsar)*; with its grey plumage, short tail, and long bill the Short-tailed Finch *I. brachyurus* was originally considered to be an icterid allied to the American grackles and blackbirds.

Idioptilon Gr. *idios*, strange, distinct; *ptilon*, a wing; of the White-eyed Tody Tyrant *I. zosterops* Berlepsch (1907), *Ornis*, **14**, 356, wrote, 'The wing formula is a rather curious one, some of the secondaries, viz., the 14th, 15th, and 16th, being as long as the longest primaries, while the 9th, 10th, and 11th are much abbreviated and of nearly equal length'.

idoneus L. *idoneus*, suitable, appropriate.

Ieracidea Gr. *hierakideus*, a little falcon (dim. of *hierax, hierakos*, a hawk or falcon).

Ifrita Arabic *ifrit,* a spirit, a djinn; ref. the uncertain affinities of the Blue-wreathed Ifrit *I. kowaldi.*

igata Maori name *igata* for the New Zealand Gerygone *Gerygone*.

igneus L. *igneus,* fiery, flaming.

ignicapillus L. *ignis,* fire; *-capillus,* capped, headed.

ignicauda L. *ignis,* fire; *cauda,* the tail.

igniferum L. *ignifer,* fiery, fire-bearing (*ignis,* fire; *ferre,* to carry).

ignipectus L. *ignis,* fire; *pectus,* the breast.

ignita L. *ignitus,* fiery, glowing (*ignire,* to set on fire).

igniventris L. *ignis,* fire; *venter, ventris,* the belly.

ignobilis L. *ignobilis,* obscure, unknown, undistinguished.

ignota L. *ignotus,* unknown, overlooked (*ignoscere,* to overlook).

ignotincta L. *ignis,* fire; *tinctus,* painted (*tingere,* to dye).

ignotus see **ignota.**

iheringi After Hermann von Ihering (or Jhering) (1850–1930) German zoologist, director of São Paulo Museum, Brazil, and author.

ijimae After I. Ijima (1861–1921) Japanese ornithologist and first President of Ornithological Society of Japan.

iliaca /iliacus L. *iliacus,* of the flanks (*ile,* the groin or flanks).

Ilicura No expl. (Reichenbach (1850), *Av. syst. nat.*, pl. 63), but probably from Gr. *elix, elikos,* a curl or lock of hair, a tendril; *oura,* the tail; ref. the thin, pointed, central tail feathers of the Pin-tailed Manakin *I. militaris.*

iliolophum L. *ilium,* the flank; Gr. *lophos,* a crest, a plume; ref. long-plumed flanks of the Dwarf Honeyeater *Oedistoma.*

Illadopsis Gr. *illas* (a variant form of *ilias),* a thrush; *opsis,* appearance.

imberbe /imberbis L. *imberbis,* beardless (i.e. youthful, small).

imerinus After the Imerina, a native clan and kingdom of central Madagascar, to which the last kings and queens of Madagascar belonged.

imitans L. *imitans,* imitating (i.e. resembling) (*imitari,* to copy).

imitator L. *imitator,* an imitator (i.e. resembling).

immaculata /immaculatus L. *immaculatus,* spotless, immaculate.

immer Icelandic name *himbrimi,* the 'surf-roarer', for the Great Northern Diver *Gavia.*

immunda L. *immundus,* unclean, dirty; ref. drab, dusky coloration of the Pale-bellied Mourner *Rhytipterna.*

immutabilis L. *immutabilis,* unchangeable; the adult and juvenile plumages of the Laysan Albatross *Diomedea* are scarcely separable.

imparatus L. *imparatus,* unprepared; Peters (1929), *Proc. Biol. Soc. Washington,* **42**, 123, censured previous authors by untangling the nomenclatural nightmare of *Corvus mexicanus* (properly the name of the cowbird *Scaphidura,* not the Mexican Crow *Corvus).*

impejanus After Lady Mary Impey (fl. 1784) early aviculturalist who kept monal pheasants *Lophophorus* in captivity, and wife of Sir Elijah Impey (d. 1809) Chief Justice of Bengal.

impennis L. *im-,* without; *pennis,* winged (*penna,* a feather).

imperatrix L. *imperatrix,* an empress; after Eugénie, Empress of the French (1826–1920: reigned 1853–1870), wife of Napoléon III.

imperialis L. *imperialis,* imperial (*imperium,* sovereignty); **1** ref. large size or splendid plumage (*Amazona, Ardea, Campephilus, Gallinago).* **2** after Khai Dinh Emperor of Annam (1885–1925: reigned 1916–1925) (*Lophura).*

impetuani Zulu (South African) native name *im' tiyane* for a waxbill or finch.

impiger L. *impiger,* active, busy.

implicata L. *implicatus,* united, entwined; of the Volcano Whistler *Pachycephala* Hartert (1929), *Am. Mus. Novit.*, **364**, 13, stated, 'This bird, quite overlooked hitherto in spite of former collections made on Guadalcanar, is a very aberrant *Pachycephala*'.

importunus L. *importunus*, troublesome, insolent (v. *Andropadus*).

imthurni After Sir Everard im Thurn (1852–1932) British diplomat and explorer in British Guiana (Guyana), 1877–1899, Lieutenant-Governor of Ceylon (Sri Lanka), 1901–1904, Governor of Fiji, 1904–1910.

inca After the Incas, the preconquest Indians of Peru, whose ruler, the Sapa Inca Atahuallpa, was executed by the Spanish conquistadores in 1533. The epithet refers to the Peruvian distribution of the species named, but is erroneous in the case of *Scardafella* (where Aztec (Mexican) would be more appropriate).

Incana /incanus L. *incanus*, light grey, hoary, quite grey.

Incaspiza After the Incas of Peru (v. *inca)*; Gr. *spiza*, a finch; the inca finches are found only in the Andes of Peru.

incerta /incertus L. *incertus*, uncertain, doubtful; ref. doubtful status or uncertain systematic position.

incognita L. *incognitus*, unknown (i.e. newly discovered).

inda L. *Indus*, Indian; an erroneous toponym in the case of *Chloroceryle*.

indica L. *Indicus*, Indian; an erroneous toponym in the case of *Urocolius*.

Indicator L. *indicator*, an indicator, a guide; the Greater Honeyguide *I. indicator* is well known for leading mammals (especially the Ratel *Mellivora capensis* and man) to bees' nests, enabling the stronger mammal to break open the nests and benefit from the honey, whilst the honeyguide eats the grubs, bees, and wax. As its specific and substantive names imply the Honeyguide Greenbul *Baeopogon* bears a remarkable resemblance to a honeyguide.

indicus see **indica.**

indigo Mod. L. *indigo*, indigo-coloured (Spanish *indico*, indigo).

indigotica /indigoticus Mod. L. *indigoticus*, indigo blue, deepest blue.

indistincta L. *indistinctus*, indistinct, obscure; ref. plain plumage.

indus see **inda.**

ineptus L. *ineptus*, silly, foolish (i.e. easily caught and killed, having no fear of man).

inerme L. *inermis*, helpless, defenceless (i.e. small).

inexpectata /inexpectatum L. *inexspectatus*, unexpected.

Inezia After Enriqueta Iñez Cherrie (b. 1907) daughter of US ornithologist George Cherrie.

infaustus L. *infaustus*, unlucky, unfortunate; Newton and Gadow (1896), p. 469, wrote of the Siberian Jay *Perisoreus*, 'Linnaeus seems to have been under a misapprehension when he applied to it the trivial epithet it bears; for by none of his countrymen is it deemed an unlucky bird, but rather the reverse. In fact, no one can list to the cheery sound of its ordinary calls with any but a hopeful feeling'. However, Temminck (1820), *Manuel d'ornithologie*, (Ed. 2), records that in Norway the species is known as a bird of ill omen.

infelix L. *infelix*, miserable, unhappy; of the type specimen of the Admiralty Pied Monarch *Monarcha* Sclater (1877), *Proc. Zool. Soc. London*, 552, bemoaned, 'Unfortunately it is in a very imperfect state, the bill and greater part of the tail feathers having been shot away…'.

infuscata /infuscatus L. *infuscatus*, dusky, clouded (*infuscus*, dusky, blackish).

ingens L. *ingens*, enormous.

ingoufi After Captain Jules A. Ingouf (b. 1846) French explorer and collector in Argentina, 1882.

innominatus L. *innominatus*, unnamed.

innotata L. *innotatus*, remarkable, noticeable (*innotavi*, to mark, to notice).

inopinatum L. *inopinatus*, unexpected.

inornata /inornatum /inornatus L. *inornatus*, unadorned, plain.

inquieta L. *inquietus,* restless, not quiet.

inquisitor L. *inquisitor,* an investigator, a searcher.

inscriptus L. *inscriptus,* marked.

insignibarbis L. *insignis,* notable, extraordinary; *barba,* the beard.

insignis L. *insignis,* notable, extraordinary, remarkable.

insolitus L. *insolitus,* unusual, strange.

insperatus L. *insperatus,* unexpected.

insulana L. *insulanus,* an islander, belonging to an island (*insula,* an island).

insularis L. *insularis,* insular, belonging to an island (*insula,* an island).

insulata L. *insulatus,* insulated, alone, isolated.

intermedia/intermedius L. *intermedius,* intermediate (i.e. sharing characteristics with or acting as a link between two other species).

internigrans L. *internigrans,* being black here and there.

interpres L. *interpres,* a messenger; **1** Newton and Gadow (1896), p. 997, footnote, wrote of the Ruddy Turnstone *Arenaria,* 'Linnaeus…who first met with this bird on the island of Gottland…was under the mistaken belief that it was there called Tolk (= *interpres*). But that name properly belongs to the Redshank, from the cry of warning to other animals that it utters on the approach of danger'. **2** French name *Merle messager* given to the Chestnut-capped Ground Thrush *Zoothera* by Temminck (1826), *Planches coloriées*, Vol.**78**, pl. 458.

involucris L. *involucre,* a wrap (*involvere,* to envelop); ref. heavily streaked upperparts of the Stripe-backed Bittern *Ixobrychus.*

Iodopleura Gr. *ioeides,* violet-coloured (*ion,* the violet); *pleura,* the side, the ribs; the purpletufts are distinguished by violet-coloured tufts or plumes at the sides of the breast.

iohannis L. *Iohannis,* John (and European equivalents); after an otherwise unknown collector for the Museu Goeldi expedition to the Rio Purús region of Brazil in 1904.

iolaima Gr. *io-,* violet; *laimos,* the throat.

Iole Gr. myth. Iole, daughter of King Eurytus of Oechalia and promised in marriage to Hercules.

Ionolaima Gr. *ion,* the violet; *laimos,* the throat.

iouschistos Gr. *ioeis,* dark; Mod. L. *schistus,* slate-coloured.

iozonus Gr. *iozonos,* with purple girdle, purple-banded.

iphis Gr. *iphis,* rapid, fleet.

Ipophilus Gr. *ipos,* a piece of wood; *philos,* loving, fond of.

iracunda L. *iracundus,* irascible, wrathful (*irasci,* to grow angry).

Irania After Iran, formerly Persia, the genus being described from a specimen of the White-throated Robin *I. gutturalis* supplied by Baron Auguste Finot, French Consul to Persia, 1863.

iredalei After Tom Iredale (b. 1880) English artist, naturalist, and secretary to Australian ornithologist Gregory MacAlister Mathews.

Irediparra After Tom Iredale (v. *iredalei*); syn. genus *Parra,* jacana (L. *parra,* a bird of ill omen).

Irena Gr. myth. Eirene, the goddess of peace (*eirene,* peace).

ireneae After Irene Morden (fl. 1967) US sponsor and collector in Kenya, 1965.

iriditorques Gr. *iris, iridos,* the rainbow; L. *torques,* a collar or necklace.

Iridophanes Gr. *iris,* the rainbow; *-phanes,* shining, appearing (*phaino,* to shine forth).

Iridoprocne Gr. *iris,* the rainbow; genus *Procne,* swallow.

Iridosornis Gr. *iris,* the rainbow; *ornis,* a bird.

iris Gr. *iris,* the rainbow.

irrorata L. *irroratus,* bedewed, sprinkled (*inrorare* or *irrorare,* to bedew); ref. fine vermiculations on belly and upperparts of the Waved Albatross *Diomedea.*

irupero Güaraní (Paraguayan) Indian name *iruperó* for the White Monjita *Xolmis*.

isaacsonii After J. P. Isaacson (fl. 1845) English naturalist.

isabella see **isabellina.**

isabellae After **1** Isabelle Thirion (fl. 1846) wife of French naturalist Eugène Thirion (*Iodopleura*). **2** Lady Isabel Burton (1831–1896) traveller, author, and wife of Sir Richard Burton (*Cossypha*).

isabellina/isabellinus Mod. L. *isabellinus,* isabelline, greyish-yellow.

Ischnosceles Gr. *iskhnos,* thin, meagre; *skelos,* the leg.

isidorei/isidori After Isidore Geoffroy Saint-Hilaire (1805–1861) French zoologist and author.

islandica Med. L. *Islandicus,* Icelandic.

Ispidina L. *hispida,* the kingfisher; *-inus,* resembling.

isura Gr. *isos,* equal; *-ouros,* tailed; ref. squarish tail of the Square-tailed Kite *Lophoictinia.*

italiae L. *Italiae,* of Italy (*Italia,* Italy).

Ithaginis Gr. *ithagenes,* true-born, genuine, original.

ituriensis After the Ituri District, Haut-Congo, Belgian Congo (Zaïre).

Ixobrychus Gr. *ixos,* the mistletoe (confused with *ixias,* a plant with reed-like foliage); *brukhao,* to roar (i.e. a reed roarer or boomer).

ixoides Genus *Ixos,* bulbul; Gr. *-oides,* resembling; the Plain Honeyeater *Pycnopygius* was described as having the form and colour of the Olive-winged Bulbul *Ixos* (now *Pycnonotus*) *plumosus.*

Ixonotus Genus *Ixos,* bulbul; Gr. *-notos,* backed; ref. lax, spotted rump feathers of the Spotted Greenbul *I. guttatus.*

Ixoreus Gr. *ixos,* the mistletoe; L. *-orius,* belonging to; the Varied Thrush *I. naevius* resembles the Mistle Thrush *Turdus viscivorus.*

Ixos Gr. *ixos,* the mistletoe; ref. supposed favourite food of the bulbuls.

Irania gutturalis White-throated Robin

J

Jabiru Tupí (Brazilian) Indian name *jabirú guasú*, or great jabiru, for the Jabiru *J. mycteria*.

Jabouilleia After Pierre Jabouille (1875–1947) French colonial administrator in Indochina, 1905–1933, ornithologist, and editor of *L'Oiseau*.

Jacamaralcyon French *jacamar*, jacamar (Tupí [Brazilian] Indian names *jacameri* or *jacamaciri*); syn. genus *Alcyone*, three-toed kingfisher (L. *alcyon*, the kingfisher); ref. small size and three toes of the Three-toed Jacamar *J. tridactyla*.

Jacamerops French *jacamar*, jacamar (v. *Jacamaralcyon*); genus *Merops*, bee-eater; ref. stout curved bill of the Great Jacamar *J. aurea*.

Jacana Tupí (Brazilian) Indian name *jasaná* for the Wattled Jacana *J. jacana* (Portuguese spelling *jaçaná*).

jacarina Tupí (Brazilian) Indian name *jacarini* for a type of finch.

jacksoni After Sir Frederick Jackson (1860–1929) Governor of Uganda, 1911–1917, naturalist, collector, and author (*The birds of Kenya Colony and the Uganda Protectorate*, 1938).

jacobinus After the Jacobin or Dominican friars (v. *dominicana*).

jacquacu Güaraní (Paraguayan) Indian name *yacú* for a chachalaca or curassow.

jacquinoti After Vice Admiral Charles Hector Jacquinot (1796–1879) French explorer in the Pacific, 1837–1840.

jacucaca see **jacquacu.**

jacula L. *iaculus*, darting (*iacere*, to hurl).

jacutinga Güaraní (Paraguayan) Indian name *yacútinga* for a chachalaca.

jamacaii Tupí (Brazilian) Indian name *jamacaii* for some small black and yellow bird.

jamaica/jamaicensis After the island of Jamaica in the Greater Antilles.

jambu Malay name *punai jambu* for a fruit dove *Ptilinopus* (Sanskrit *jambu*, the rose-apple tree).

jamesi After **1** H. Berkeley James (1846–1892) British businessman (*Phoenicoparrus*). **2** Frank Linsly James (1851–1890) British explorer in the Sudan, Somaliland, India, and Mexico (*Tchagra*).

jamesoni After James Sligo Jameson (1856–1888) Scottish explorer, hunter, naturalist, and author (*Story of the rear column of the Emin Pasha relief expedition*, 1890).

jandaya Tupí (Brazilian) Indian name *jendaya* for a sort of parrot (v. *Nandayus*).

jankowskii After Michel Jankowski (1840–1903) Polish naturalist exiled to Siberia, 1863–1903.

janthina L. *ianthinus*, violet-coloured (Gr. *ianthinos*, violet-coloured).

Janthothorax Gr. *ianthis*, the violet; *thorax*, the breast.

japonensis/japonica After Japan (French *Japon*).

jaraguana After the town of Jaragua, Goiás, Brazil.

jardineii/jardini/jardinii After Sir William Jardine, Bt. (1800–1874) English ornithologist and author (*Illustrations of ornithology*, 1826; *The natural history of* *hummingbirds*, 1833; *The natural history of the Nectariniidae*, 1834).

javanense /javanica /javanicus /javensis After the island of Java, Indonesia (Malay *Djawa* or *Jawa)*.

jefferyi After Jeffrey Whitehead (d. 1909) English stockbroker, father of explorer John Whitehead.

jelskii After Konstanty Jelski (1838–1896) Polish zoologist, curator of Lima Museum, Peru, 1874–1878.

jerdoni After Surgeon Major Thomas Claverhill Jerdon (1811–1872) Scottish field ornithologist, collector, and author (*Illustrations of Indian ornithology*, 1847; *The birds of India*, 1862; *The game birds and wildfowl of India*, 1864).

jobiensis After the island of Jobi (Japen), Geelvink Bay, Irian Jaya (Indonesian New Guinea).

jocosus L. *iocosus,* merry, humorous; ref. red cheeks of the Red-whiskered Bulbul *Pycnonotus*.

johannae After **1** Johanna Verreaux (fl. 1851) wife of French natural history dealer Jules Verreaux (*Nectarinia)*. **2** no expl. (Bourcier (1847), *Proc. Zool. Soc. London*, **15**, 45), where the Blue-fronted Lancebill *Doryfera* is first referred to as *Troch. de Jeanne.* **3** no expl. (de Dalmas (1900), *Bull. Brit. Orn. Cl.*, **11**, 36) (*Tangara)*.

johannis L. *Iohannis,* John; after John P. Clarke (b. 1896) British colonial policeman in British Somaliland, 1918–1920.

johni After John Whitehead (1860–1899) English explorer, naturalist, and collector.

johnstoni After Sir Harry Hamilton Johnston (1858–1927) English explorer, collector, colonial administrator, and author (*The story of my life*, 1923).

johnstoniae After Marion Johnstone (fl. 1920) English aviculturalist and author.

jonquillaceus French *jonquille,* the narcissus or jonquil (Spanish *junquillo;* L. *iuncus,* a rush); L. *-aceus,* resembling; ref. colouring of the Jonquil Parrot *Aprosmictus*.

josefinae After Josephine Finsch (fl. 1882) wife of German ornithologist F. H. Finsch.

josephinae After Josephine V. McConnell (b. 1906) daughter of British ornithologist and collector F. V. McConnell.

jourdanii After Mr Jourdan (fl. 1839) collector in Trinidad.

jouyi After P. L. Jouy (1856–1894) ornithologist and collector in Japan, 1881.

jubata /jubatus L. *iubatus,* crested, maned.

Jubula Dim. from L. *iuba,* a crest, a mane; ref. distinctive shaggy crest of the Maned Owl *J. lettii*.

jucunda L. *iucundus,* delightful, agreeable.

jugger Hindi name *jaggar* for the male or tiercel Laggar Falcon *Falco* (the female being known as *laggar)*.

jugularis Med. L. *jugularis,* jugular, of the throat (L. *iugulum,* the throat).

juliae After **1** Julia Marchioness of Tweeddale (1846–1937) wife of British collector Arthur, 9th Marquis of Tweeddale (*Arachnothera)*. **2** no expl. (Heine (1860), *Journ. für Orn.*, **8**, 184) (*Halcyon)*.

julianae After Queen Juliana of the Netherlands (b. 1909: reigned since 1948).

julie No expl. (Bourcier (1842), *Rev. Zool.*, 373); probably after Julie Mulsant (fl. 1842) wife of French naturalist Martial Mulsant or Julie Duquarie his granddaughter, or Zénaïde Laetitia Julie Princesse Bonaparte (v. *Zenaida*).

julieni After Alexis A. Julien (fl. 1863) collector resident on Sombrero Island, Virgin Islands.

julius L. *Iulius,* Julius; after Jules Pierre Verreaux (1808–1873) French natural history dealer.

jumana After the Rio Juma, Amazonas, Brazil.

juncidis Dim. from L. *iuncus,* a reed (i.e. a small reed dweller).

Junco Med. L. *junco,* the Reed Bunting *Emberiza* (L. *iuncus,* a reed).

juninensis After the town of Junín, Junín Department, Peru.

junoniae Hartert (1916), *Nov. Zool.*, **23**, 86, records that the White-tailed Laurel Pigeon *Columba* was so named, 'From its habitat: La Palma and Gomera, or, as they were formerly called, Junonia Mayor and Junonia Menor'.

juruanus After the Rio Juruá, Amazonas, Brazil.

Jynx Gr. *iunx*, the Wryneck *J. torquilla*, a bird that figured largely in ancient superstition.

Pithecophaga jefferyi Monkey-eating Eagle

K

kaempferi After Emil Kaempfer (fl. 1953) German naturalist in Brazil, 1926–1931.

Kakamega /kakamegae After the Kakamega Forest, Western Province, Kenya.

Kakatoe French name *kakatoès* for a cockatoo.

kalinowskii After J. Kalinowski (1860–1942) Polish collector in Peru, 1889–1942.

Kalochelidon Gr. *kalos,* beautiful; *khelidon,* a swallow.

kampalili After Mount Kampalili, Davao Province, Mindanao, Philippines.

kamtschatschensis After the Kamchatka Peninsula, north-eastern Siberia, USSR.

kandavensis After Kandavu Island (Kadavu Island), Fiji.

karamojae After Karamoja Province, eastern Uganda.

kasumba Malay name *burong kesumba* for a trogon (Sanskrit *kesumba,* a generic name for trees yielding yellowish-red dyes).

katangae After Katanga Province, Belgian Congo (Shaba Province, Zaïre).

katherina After Katherine De Vis (fl. 1905) wife of English naturalist the Revd. Charles De Vis (d. 1915), curator of the Queensland Museum, 1892–1905.

kaupi After Johann Jakob von Kaup (1803–1873) German zoologist and author (*Classification der säugethiere und vögel,* 1844).

Kaupifalco After Johann von Kaup (v. *kaupi*); genus *Falco,* falcon.

keartlandi After George Arthur Keartland (1848–1926) Australian field naturalist.

keayi After W. A. Keay (fl. 1900) sugar planter on Negros, Philippines.

keiensis After the Kei Islands, Moluccas, Indonesia (Kai Islands, Maluku, Indonesia).

kelaarti /Kelaartia After Lieutenant Colonel Edward Frederick Kelaart (1818–1860) British zoologist in Ceylon (Sri Lanka) and author (*Prodromus fauna Zeylanica*, 1852).

kelleyi After W. V. Kelley (1861–1932) US sponsor and philanthropist.

kempi After Robin Kemp (b. 1871) naturalist and collector in Sierra Leone, 1902–1904, Australia, 1912–1914 and Argentina, 1916–1917.

kennicottii After Robert Kennicott (1835–1866) US naturalist and explorer.

Kenopia Syn. genus *Keropia*, piopio (Maori names *keropia* or *koropio* for the Piopio *Turnagra)*; ref. supposed affinities of the Striped Wren Babbler *K. striata*.

kenricki After Major Kenrick (fl. 1889) British Army in Kenya.

keraudrenii After Pierre F. Keraudren (fl. 1844) Inspector General of French naval medical services.

kerearako Mangaia (Cook Islands) native name *kerearako* for the Kerearako Warbler *Conopoderas*.

keri After the Bellenden Ker Range, Queensland, Australia.

kerriae After Mrs E. L. Kerr (fl. 1915) US traveller in Colombia, 1905.

kessleri After Karl Thedorovich Kessler (1815–1881) German zoologist and collector.

Ketupa /ketupu Malay names *ketupok* or *burong ketupok* for the Malaysian Fish Owl *K. ketupu* in Sumatra and Java.

kibalensis After the Kibale Forest, Bunyoro Province, Uganda.

kieneri /kienerii After Louis Charles Kiener (1799–1881) French zoologist and conchologist.

kikuyuensis After the town of Kikuyu, Central Province, Kenya.

kilimensis After Kilima N'jaro, an alternative spelling of Mount Kilimanjaro, Tanganyika (Tanzania), the highest peak in Africa.

kinabaluensis After Mount Kinabalu, North Borneo (Sabah).

kingi /kingii After **1** Captain James King (1750–1784) in the Pacific with Captain Cook, 1776–1779 (*Conopoderas*). **2** Rear Admiral Philip Parker King (1791–1856) British marine surveyor, collector, and traveller in tropical America, 1825–1830 (*Aglaiocercus*).

kioloides French name *le Kiolo*, given to the Clapper Rail *Rallus longirostris* by Buffon (1770) (onomatopoeic Cayenne Indian name *kiolo*); Gr. *-oides*, resembling.

kirhocephalus Gr. *kirrhos*, tawny; *-kephalos*, headed.

kirkii After John Kirk (fl. 1845) collector on Tobago.

kirtlandii After Jared Potter Kirtland (1793–1877) US physician and zoologist.

Kittacincla Mod. L. *kitta*, magpie (v. *Cissa*); *cinclus*, a thrush.

kizuki French name *kisuki* given to the Japanese Pied Woodpecker *Dendrocopos* by Temminck (1836), *Planches coloriées*, Vol. **99**, apparently based on a Japanese name for a woodpecker.

klaas After Klaas (fl. 1784) African native servant to French explorer François Levaillant.

klagesi After S. M. Klages (b. 1875) collector in Venezuela, 1898–1913 and Brazil, 1922.

Klais After Kleis, daughter of Lesbian lyric poetess Sappho (v. *Sappho*).

kleinschmidti After T. Kleinschmidt (1834–1881) German explorer and collector in the Pacific.

klossi After Cecil Boden Kloss (1877–1949) director of Raffles Museum, Singapore, 1921–1936.

Knipolegus Gr. *knips, knipos*, an insect; *lego*, to gather or pick (cf. *knipologos*, the treecreeper *Certhia*).

knudseni After V. Knudsen (1822–1898) Norwegian ornithologist and collector on Hawaii.

kochi After Gottlieb von Koch (1849–1914) German taxidermist and collector.

koeniswaldiana After G. Koeniswald (fl. 1901) German ornithologist and author.

koepckeae After Maria Koepcke (d. 1977) German ornithologist resident in Peru and author (*Corte ecológico transversal en los Andes del Perú central*, 1954; *Die vögel des waldes von Zárate*, 1958; *Las aves del departamento de Lima*, 1964).

koliensis After the Koli River, Lango Province, Uganda.

kollari After Vincent Kollar (1797–1860) German entomologist and collector.

komadori Japanese name *koma dori*, meaning 'little toy bird' (*koma*, little toy; *tori* or *dori*, a bird) for the Japanese Robin *Erithacus akahige*, erroneously applied to the Ryukyu Robin *Erithacus* by Temminck (v. *akahige*).

kona After the Kona District of Hawaii, Hawaiian Islands.

kori Chuana (Botswana) native name *khori* for the Kori Bustard *Ardeotis*.

koslowi see **Kozlowia.**

kowaldi After C. Kowald (fl. 1890) government agent in British New Guinea.

Kozlowia After General Pyotr Kuzmich Kozlov (1863–1935) Russian explorer of central Asia and author (*Mongoliya i Kam*, 1905).

krameri After W. H. Kramer (d. 1765) Austrian naturalist and author.

kreffti After Johann Ludwig Gerhard Krefft (1830–1881) German artist, adventurer, and naturalist in Australia.

kretschmeri After Eugen Kretschmer (1868–1894) German naturalist and collector in Kenya.

kreyenborgi After H. Kreyenborg (1889–1963) German falconer.

krueperi After Theobald Johannes Krüper (1829–1921) German ornithologist, collector, and author (*Ornithologische notizen über Griechenland*, 1862; *Beitrag zur ornithologie Klein Asiens*, 1875).

kubaryi After Jan S. Kubary (1846–1896) Polish collector in the Pacific, 1868–1879 and New Guinea, 1885–1895.

kuehni After Heinrich Kühn (1860–1906) German naturalist and collector in the Celebes (Sulawesi), 1882–1885 and the Kai islands, 1898.

kuhli/kuhlii After Heinrich Kuhl (1796–1821) German ornithologist and author (*Conspectus psittacorum*, 1820).

kumlieni After Thure Ludwig Theodor Kumlien (1819–1888) Swedish/US naturalist.

kupeensis After Kupé Mountain, Kumba, British Cameroons (Cameroun).

Kupeornis After Kupé Mountain (v. *kupeensis*); Gr. *ornis*, a bird.

Canirallus kioloides Grey-masked Rail

L

labradorides French *pierre de Labrador,* feldspath, Labrador stone; Gr. *-oides,* resembling; ref. the shining silver and blue reflections in the green plumage of the Metallic-green Tanager *Tangara.*

labradorius After the Labrador coast, eastern Canada.

Lacedo Anagram of genus *Alcedo,* kingfisher, coined to reflect relationship.

lacernulata L. *lacernula,* a little cloak (*lacerna,* a cloak or mantle); *-atus,* possessing; ref. distinctive contrasting dark grey mantles of the Black-backed Imperial Pigeon *Ducula* and the White-necked Hawk *Leucopternis.*

lacertosa L. *lacertosus,* powerful (i.e. large).

lachrymosa/lachrymosus L. *lacrimosus,* tearful (*lacrima,* a tear); the Fan-tailed Warbler *Euthlypis* has a tear-shaped spot in front of each eye, and the Black-streaked Woodcreeper *Xiphorhynchus* has copious, tear-shaped streaks on the head, mantle, and underparts.

lacrymiger L. *lacrima,* a tear; *-ger,* carrying, bearing; ref. bold, white, tear-shaped spots on head and underparts of the Montane Woodcreeper *Lepidocolaptes.*

lacrymosum/lacrymosus L. *lacrimosus,* tearful (*lacrima,* a tear); the Lachrymose Mountain-Tanager *Anisognathus* has a yellow, tear-shaped spot beneath each eye, and the Spot-flanked Barbet *Tricholaema* has black, tear-shaped spots along each flank.

lactea/lacteus L. *lacteus,* milky, milk-white (*lac, lactis,* milk).

Lacustroica L. *lacus,* a lake (Gr. *lakos,* a lake); Gr. *-oikos,* living (*oikeo,* to inhabit); after Lake Way, East Murchison, Western Australia.

laemosticta Gr. *laimos,* the throat; *stiktos,* spotted, dappled.

laeta L. *laetus,* bright, glad.

laetior L. *laetior,* more bright (comp. of *laete,* brightly, gladly).

laetissima L. *laetissimus,* very bright (super. of *laete,* brightly, gladly).

laetus see **laeta.**

lafargei After Ensign A. A. T. Pavin de La Farge (1812–1839) aboard French exploratory vessel *La Zelée* in the Pacific, 1837–1839.

lafayettii After Marie Joseph Paul Yves Roch Gilbert Motier Marquis de Lafayette (1757–1834) French general and reformist.

Lafresnaya/lafresnayanus/lafresnayei/lafresnayi/lafresnayii After Baron Noël Frédéric Armand André de Lafresnaye (1783–1861) French ornithologist and collector.

lagdeni After Sir Godfrey Yeatman Lagden (1851–1934) English diplomat on the Gold Coast (Ghana), 1883 and Basutoland (Lesotho), 1893–1901.

Lagonosticta Gr. *lagon, lagonos,* the flanks; *stiktos,* spotted, dappled.

Lagopus L. *lagopus,* an alpine gamebird (Gr. *lagopous,* a gamebird, perhaps the Ptarmigan; *lagos,* the hare; *pous,* the foot).

lagrandieri After Vice Admiral P. P. de Lagrandière (1807–1876) French Governor of Cochin-China (Vietnam).

lahtora Hindi name *latora* for a shrike *Lanius*.

lais After Lais (fl. 340 BC), a celebrated Sicilian courtesan, murdered by the jealous wives of Alcibiades.

Lalage Gr. *lalages*, a bird mentioned by Hesychius, not further identified (cf. Lalage, a beautiful Roman woman beloved by the poet Horatius Flaccus).

lalandi After Pierre Antoine Delalande (1787–1823) French explorer (v. *delalandei)*.

lamberti After Aylmer Bourke Lambert (1761–1842) English botanist.

lamelligerus L. *lamella*, a small thin plate of metal; *-gera*, carrying; the African Openbill Stork *Anastomus* is 'so called from the curious flattening and broadening into shining horny-plates of its feather shafts, especially on the lower parts' (Newton and Gadow (1896), p. 655).

lamellipennis L. *lamella*, a small thin plate of metal; *pennis*, winged; the White-tailed Cotinga *Xipholena* has shining black wing coverts.

laminirostris L. *lamina*, a thin plate of metal; *-rostris*, billed; the Plate-billed Mountain Toucan *Andigena* has raised ivory white plates near the base of the upper mandible.

Lampornis Gr. *lampos*, bright; *ornis*, a bird.

Lampraster Gr. *lampros*, brilliant; *aster*, a star.

Lampribis Gr. *lampros*, brilliant; *ibis*, an ibis.

Lamprocolius Gr. *lampros*, brilliant; Mod. L. *colius*, a starling.

Lamprolaima Gr. *lampros*, brilliant; *laimos*, the throat.

Lamprolia Dim. of Gr. *lampros*, brilliant, splendid.

Lampronetta Gr. *lampros*, brilliant; *netta*, a duck.

Lampropsar Gr. *lampros*, brilliant; *psar*, the starling (v. *Gnorimopsar)*.

Lamprospiza Gr. *lampros*, brilliant; *spiza*, a finch.

Lamprothorax Gr. *lampros*, brilliant, splendid; *thorax*, the breast.

Lamprotornis Gr. *lamprotes*, splendour, brilliance; *ornis*, a bird.

lanaiensis After the island of Lanai, Hawaiian Islands.

lanceolata/lanceolatus L. *lanceolatus*, lanceolate, spear-shaped (*lancea*, a spear, lance).

landanae After the village of Landana, Cabinda, Angola.

langbianis After the Langbian Plateau, Annam (Vietnam).

langsdorffi/langsdorffii After Baron George Heinrich von Langsdorff (1744–1852) German naturalist and collector, Russian Consul to Brazil, 1813–1830.

languida L. *languidus*, sluggish, weak, inactive (*languere*, to be faint).

Laniarius Genus *Lanius*, shrike; L. *-arius*, relating to.

laniirostris Genus *Lanius*, shrike; L. *-rostris*, billed.

Laniisoma Genus *Lanius*, shrike; Gr. *soma*, the body; from supposed shrike-like appearance of the Shrike-like Cotinga *L. elegans*.

Lanio French name *lanion* for the shrike-tanagers (cf. genus *Lanius*, shrike).

Laniocera Genus *Lanius*, shrike; Gr. *keras*, a horn (i.e. the bill); ref. heavy, hooked bill of the Cinereous Mourner *L. hypopyrra*.

lanioides Genus *Lanius*, shrike; Gr. *-oides*, resembling.

Lanioturdus Genus *Lanius*, shrike; genus *Turdus*, thrush.

Lanius L. *lanius*, a butcher; the shrikes were formerly known as butcher-birds, from their habit of storing prey by impaling it on thorns and sharp twigs, giving the resemblance to a butcher's slaughterhouse.

lansbergei After J. W. van Lansberge (1830–1905) Governor-General of the Dutch East Indies (Indonesia), 1875–1881.

lansbergi After R. F. van Lansberge (1804–1873) Dutch administrator in the West Indies.

lanyoni After Wesley E. Lanyon (fl. 1988) US ornithologist.

laperouse After Captain Jean François de Galaup Comte de la Pérouse (1741–1788) French explorer in the Pacific, 1785–1788.

lapponica/lapponicus Mod. L. *Lapponicus*, of Lapland, northern Scandinavia (*Lapponia*).

largipennis L. *largus*, ample; *pennis*, wings (*penna*, a feather); ref. thickened outer primary shafts of the Grey-breasted Sabrewing *Campylopterus*.

Larosterna Genus *Larus*, gull; genus *Sterna*, tern.

Larus L. *larus*, a rapacious seabird, probably a gull (Gr. *laros*, a seabird).

larvata/larvatus L. *larvatus*, masked (*larva*, a mask, a spectre).

Larvivora Mod. L. *larva*, a caterpillar (L. *larva*, a ghost); L. *-vorus*, eating.

latebricola L. *latebricola*, a skulker, one that lives in hiding.

lateralis L. *lateralis*, lateral, of the side (*latus*, the side).

Laterallus Contracted anagram of subspecific name *Rallus lateralis* (now regarded as a subsp. of *L. melanophaius*).

lathami/Lathamus After John Latham (1740–1837) English ornithologist and author (*General synopsis of birds*, 1781; *Index ornithologicus*, 1790; *A general history of birds*, 1821).

Lathrotriccus Gr. *lathros*, secret, hidden; Mod. L. *triccus*, tyrant flycatcher; the distinctiveness of these flycatchers was formerly obscured within the genus *Empidonomus*.

Laticilla L. *latus*, broad, wide; Mod. L. *cilla*, the tail (v. *Motacilla*).

laticincta L. *latus*, broad, wide; *cinctus*, girdled (*cingere*, to encircle).

latifrons L. *latus*, broad, wide; *frons*, the forehead, the brow; ref. broad white patch on forehead of the Bornean Falconet *Microhierax*.

latimeri After G. Latimer (fl. 1866) US Consul to Puerto Rico.

latirostre/latirostris L. *latus*, broad, wide; *-rostris*, billed.

latistriatus L. *latus*, broad, wide; Mod. L. *striatus*, striped, streaked (L. *stria*, a furrow).

Latoucheornis After John David Digues La Touche (1861–1935) Inspector of Customs, China, 1882–1921, and author (*A handbook of the birds of eastern China*, 1925); Gr. *ornis*, a bird.

latrans L. *latrans*, barking (*latrare*, to bark).

laudabilis L. *laudabilis*, praiseworthy, laudable.

laurae After Laura Boulton (fl. 1934) wife of US ornithologist W. Rudyerd Boulton (b. 1900).

lauterbachi After Carl Lauterbach (1864–1937) German botanist and collector in New Guinea, 1896.

lavinia After Lavinia Bowen (fl. 1858) US natural history painter.

lavongai After the island of Lavongai (New Hanover), Bismarck Archipelago, Papua New Guinea.

lawesii After Revd. William George Lawes (1839–1907) British missionary to New Guinea.

lawrencei/lawrencii After George Newbold Lawrence (1806–1895) US businessman, collector, and ornithologist.

layardi After Edgar Leopold Layard (1824–1900) English diplomat, civil servant, naturalist, and author (*The birds of South Africa*, 1867).

laysanensis After the island of Laysan, Hawaiian Islands.

lazuli Med. L. *lazulus*, azure, deep blue (Persian *lazhward*, lapis lazuli).

leachii After William Elford Leach (1790–1836) English zoologist and author (*The zoological miscellany*, 1814).

leadbeateri After Benjamin Leadbeater (d. 1837) English natural history dealer.

leari After Edward Lear (1821–1888) English author, traveller, and artist (*Illustrations of the family of Psittacidae or parrots*, 1830).

lebruni After Édouard A. Lebrun (1852–1904) French taxidermist and collector in Argentina, 1882.

leclancheri/leclancherii After C. R. A. Léclancher (1804–1857) French surgeon and explorer.

lecontei/leconteii After **1** John Le Conte (1818–1891) US physician and naturalist (*Ammodramus*). **2** John Lawrence Le Conte (1825–1883) US entomologist, biologist, and author (*Myioceyx, Toxostoma*).

ledanti After Jean-Paul Ledant (b. 1951) Belgian ecologist.

Legatus L. *legatus,* a commander, a governor.

leguati After François Leguat (1637–1735) French explorer, naturalist on Rodrigues, 1691 and author (*Voyage et avantures en deux îles désertes des Indes Orientales*, 1708).

Leiothrix Gr. *leios,* smooth; *thrix,* hair; ref. sleek appearance of the Red-billed Leiothrix *L. lutea.*

Leipoa Gr. *leipo,* to abandon; *oon,* an egg; the megapodes are remarkable for their specialised incubation techniques, the eggs being left in mounds of rotting vegetation, hot sand, or volcanic ash to incubate on their own.

Leistes Gr. *leistes,* a robber, a plunderer; the Red-breasted Blackbird *L. militaris* sometimes takes eggs and small young from other birds nests.

lembeyi After J. Lembey (fl. 1850) Cuban ornithologist and author (*Aves de Cuba*, 1850).

lemosi After Antonio J. Lemos-Gúzman (fl. 1962) Governor of Cauca, Colombia.

lempiji Javanese name *lempi - ji* for the Collared Scops Owl *Otus.*

lemprieri After E. G. Lemprière (fl. 1884) collector on Palawan, Philippines.

lendu After the Lendu Plateau, Ituri District, Belgian Congo (Zaïre).

lentiginosus L. *lentiginosus,* freckled, dotted (*lentigo,* a freckle, a lentil-shaped spot).

Leonardina After Major General Leonard Wood (1860–1927) US Governor-General of the Philippines, 1906–1908.

leontica L. *leonticus,* leonine, belonging to a lion (*leo,* the lion); after Sierra Leone (Spanish 'Lion Range').

lepe After Lepe (Lepi) Mission Station, Benguella, Angola.

lepida L. *lepidus,* charming, elegant.

Lepidocolaptes Gr. *lepis, lepidos,* a scale; genus *Colaptes,* woodpecker; the scaly plumaged woodcreepers inhabit tropical America.

Lepidogrammus Gr. *lepis,* a scale; *gramma,* that which is drawn or inscribed; ref. black spangled head feathers of the Scale-feathered Malkoha *L. cumingi.*

Lepidopyga Gr *lepis,* a scale; *puge,* the rump; ref. scaled undertail coverts of the Shining-green Hummingbird *L. coeruleigularis.*

Lepidopygia Gr *lepis,* a scale; Mod. L. *pygius,* rumped; ref. yellow fringed uppertail coverts of the Bibbed Mannikin *L. nana.*

lepidus see **lepida.**

Leptasthenura Gr. *leptos,* thin; *asthenes,* weak; *oura,* the tail; ref. lengthened, decomposed tail feathers of the tit spinetails.

Leptodon Gr. *leptos,* thin, slender; *odon,* a tooth; ref. sharp, curved tip of upper mandible of the Grey-headed Kite *L. cayanensis.*

leptogrammica Gr. *leptos,* thin, fine; *grammikos,* linear (*gramme,* a line); ref. narrowly barred underparts of the Brown Wood Owl *Strix.*

Leptolophus Gr. *leptos,* slender, narrow; *lophos,* a crest.

Leptophaps Gr. *leptos,* delicate, slender; *phaps,* a pigeon.

Leptopoecile Gr. *leptos,* delicate, fine; syn. genus *Poecile,* tit (Gr. *poikilis,* a small bird mentioned by Aristotle, identified by subsequent authors as some sort of finch).

Leptopogon Gr. *leptos,* fine, slender; *pogon,* a beard; ref. fine rictal bristles of the Slaty-capped Flycatcher *L. superciliaris.*

Leptopterus Gr. *leptos,* fine, delicate; *-pteros,* feathered; the Black-and-white Vanga *L. chabert* was considered allied to the syn. genus *Leptopteryx,* woodswallow (Gr. *pterux,* a wing).

Leptoptilos Gr. *leptos,* fine, delicate; *ptilon,* a feather; ref. specialized, soft, fluffy feathers on undertail coverts of the Marabou Stork *L. crumeniferus.*

leptorhynchus Gr. *leptos,* delicate, slender; *rhunkhos,* the bill.

Leptosittaca Gr. *leptos,* slender, thin; *sittakos* (variant of *psittakos),* a parrot; ref. delicate proportions and long, pointed tail of the Golden-plumed Conure *L. branickii.*

Leptosomus Gr. *leptos,* delicate, fine; *soma,* the body; the Cuckoo Roller *L. discolor* has a pair of powder down patches, one on either side of the rump, which impart a delicate greyish bloom to the plumage. Specifically the name refers to the soft, lax plumage and Cuckoo Roller-like appearance of the Short-legged Ground-Roller *Brachypteracias.*

Leptotila Gr. *leptos,* slender, thin; *ptilon,* a feather; from shape of outermost primary of the doves in this genus.

Leptotriccus Gr. *leptos,* small, slight; Mod. L. *triccus,* tyrant flycatcher.

lepturus Gr. *leptos,* slender, small; *-ouros,* tailed.

Lerwa Bhutea (Bhutan) name *lerwa* for the Snow Partridge *L. lerwa.*

Lesbia Gr. *Lesbias,* a woman of Lesbos.

leschenaulti /leschenaultii After Jean Baptiste Louis Claude Théodore Leschenault de la Tour (1773–1826) French botanist and collector in Australia, 1801–1802, Java, 1803–1806 and India, 1816–1822.

lessoni /Lessonia/lessonii After René Primavère Lesson (1794–1849) French naval surgeon, naturalist, explorer, and author (*Manuel d'ornithologie,* 1828; *Histoire naturelle des oiseaux-mouches,* 1829; *Histoire naturelle des colibris,* 1830; *Centurie zoologique,* 1830; *Traité d'ornithologie,* 1831; *Les trochilidées,* 1831; *Illustrations de zoologie,* 1831; *Histoire naturelle des oiseaux de paradis et des epimaques,* 1834).

letitiae After Laetitia del Gallo (b.1850), daughter of Julie Désirée Marchesa di Roccagiovine and grand-daughter of French ornithologist Prince Charles Bonaparte.

lettii After Mr Lett (fl. 1889) resident in Liberia.

leucaspis L. *leucaspis,* white-shielded (Gr. *leukaspis,* white-shielded); ref. white throat patch of White-cheeked Antbird *Gymnopithys.*

Leucippus Gr. myth. Leucippus, son of Oenomaus, who developed a passion for Daphne and disguised himself as a woman companion to be near her. When his deception was revealed he was killed by the outraged attendants of the goddess. Both sexes of the Buffy Hummingbird *L. fallax* share the same plumage.

leucoblephara Gr. *leukos,* white; *blepharon,* eyelid; the White-rimmed Warbler *Phaeothlypis* has white eyelids and narrow eyebrow.

Leucocarbo Gr. *leukos,* white; syn. genus *Carbo,* cormorant (specific name *Phalacrocorax carbo).*

leucocephala /leucocephalus Gr. *leukokephalos,* white-headed.

Leucochloris Gr. *leukos,* white; *khloros,* green.

leucogaster /leucogastra Gr. *leukos,* white; *gaster,* belly.

leucogastroides Specific name *Lonchura leucogastra;* Gr. *-oides,* resembling.

leucogenis /leucogenys Gr. *leukos,* white; *genus,* the cheek.

leucogeranus Gr. *leukos,* white; *geranos,* the crane.

leucognaphalus Gr. *leukos,* white; *gnaphos,* the mouth or jaw; the White-necked Crow *Corvus* has a white patch at base of upper mandible.

leucogrammica /leucogrammicus Gr. *leukos,* white; *grammikos,* linear, lined; ref. white streaked plumage.

leucolaema /leucolaemus /leucolaima Gr. *leukos,* white; *laimos,* the throat.

leucolepis Gr. *leukos,* white; *lepis,* a scale or shell.

leucolophus Gr. *leukolophos,* white-crested.

leucomela Gr. *leukos,* white; *melas,* black (cf. *leukomelas,* whitish-black).

leucomelaena Gr. *leukos,* white; *melais, melaina,* black.

leucomelan Gr. *leukos,* white; *melan,* black.

leucomelana Gr. *leukos,* white; *melas, melanos,* black.

leucomelas see **leucomela.**

leucomystax Gr. *leukos,* white; *mustax,* moustache (*mastax,* the mouth).

Leuconerpes Gr. *leukos,* white; *herpes,* a creeper (i.e. a tree-creeper or woodpecker).

leuconota /leuconotos /leuconotus Gr. *leukos,* white; *-notos,* backed.

leucoparaeus /leucopareia Gr. *leukopareios,* white-cheeked.

Leucopeza Gr. *leukos,* white; *peza,* the foot.

leucophaea **1** Gr. *leukos,* white; *phaea,* the eyes (*Collocalia, Microeca).* **2** Gr. *leukos,* white; *phaios,* dusky, brown (*Cormobates).*

Leucophaeus Gr. *leukophaes,* shining white, gleaming.

leucophaius Gr. *leukos,* white; *phaios,* dusky, brown.

leucophoea Gr. *leukophaes,* shining white, gleaming.

Leucophoyx Gr. *leukos,* white; *phoux,* a heron.

leucophrus /leucophrys Gr. *leukophrus,* white-browed.

leucophthalma /leucophthalmus Gr. *leukos,* white; *ophthalmos,* the eye.

leucopis Gr. *leukos,* white; *ops,* the eye; ref. white eye-ring of White-rimmed Brush Finch *Atlapetes.*

leucopleura /leucopleurus Gr. *leukos,* white; *pleura,* the side, the rib.

leucopodus Gr. *leukopous, leukopodos,* white-footed.

leucopogon Gr. *leukos,* white; *pogon,* the beard.

Leucopolius Gr. *leukos,* white; *polios,* grey, hoary.

leucops Gr. *leukos,* white; *ops,* the eye.

Leucopsar Gr. *leukos,* white; *psar,* the starling.

leucopsis Gr. *leukos,* white; *opsis,* the face, appearance.

leucoptera Gr. *leukopteros,* white-winged.

Leucopternis Gr. *leukos,* white; *pternis,* a type of hawk.

leucopterus see **leucoptera.**

leucopteryx Gr. *leukopterux,* white-winged.

leucopus Gr. *leukopous,* white-footed.

leucopyga Gr. *leukos,* white; *-pugos,* rumped (*puge,* the rump).

leucopygia Gr. *leukos,* white; Mod. L. *pygius,* rumped (Gr. *puge,* the rump).

leucopygialis Gr. *leukos,* white; Mod. L. *pygialis,* rumped (Gr. *puge,* the rump).

leucopygius see **leucopygia.**

leucopyrrhus Gr. *leukos,* white; *purrhos,* red, flame-coloured (*pur,* fire).

leucoramphus Gr. *leukos,* white; *rhamphos,* the bill.

leucorhoa Gr. *leukos,* white; *orrhos,* the rump.

leucorhynchus Gr. *leukorhunkhos,* white-nosed (i.e. white-billed).

leucorodia Gr. *leukorodios,* the spoonbill (*leukos,* white; *erodios,* a heron).

leucorrhoa /leucorrhous see **leucorhoa.**

leucoryphus Gr. *leukos,* white; *-koruphos,* crowned (*koruphe,* the crown of the head).

Leucosarcia Gr. *leukos,* white; *sarx, sarkos,* the flesh, the body.

leucoscepus Gr. *leukos,* white; *skepas,* a covering; ref. pale buff or white streaked mantle of the Yellow-necked Spurfowl *Pternistis.*

leucosoma/leucosomus Gr. *leukos,* white; *soma,* the body.

leucospila Gr. *leukos,* white; *spilos,* a spot.

leucospodia Gr. *leukos,* white; *spodios,* ash-coloured, grey.

leucostephes Gr. *leukostephes,* white-wreathed; ref. white forehead, moustaches and cheek-plumes of the White-fronted Melidectes *Melidectes.*

leucosterna Gr. *leukos,* white; *sternon,* the breast.

leucosticta/Leucosticte/leucostictus Gr. *leukostiktos,* white-speckled, grizzled.

leucostigma Gr. *leukos,* white; *stigma,* a mark, a spot.

leucothorax Gr. *leukothorax,* white-chested, with white cuirass.

leucotis/leucotos Gr. *leukos,* white; *-otis,* eared.

Leucotreron Gr. *leukos,* white; genus *Treron,* green pigeon.

leucura/leucurum/leucurus Gr. *leukouros,* white-tailed.

leuphotes Gr. *lophos,* a crest; *-otes,* featuring; the tall spiky crest is common to all Indomalayan bazas.

levaillanti/levaillantii After François Levaillant (1753–1824) French ornithologist, explorer, collector, and author (*Histoire naturelle des oiseaux d'Afrique,* 1796; *Histoire naturelle d'une partie d'oiseaux nouveaux et rares,* 1801; *Histoire naturelle des oiseaux de paradis et des rolliers,* 1801).

levaillantoides Specific name *Francolinus levaillantii;* Gr. *-oides,* resembling.

leveriana After Sir Ashton Lever (1729–1788) English naturalist, collector, and founder of Leverian Museum, 1760.

levigaster L. *laevum,* favourable (in augury); *gaster,* the belly; the Mangrove Gerygone *Gerygone* has bright white underparts.

levraudi After M. Levraud (fl. 1857) French Consul to Venezuela, 1856–1857.

lewinii After John William Lewin (1770–1819) English naturalist, artist, collector in Australia, 1800–1819 and author (*A natural history of the birds of New South Wales,* 1838).

lewis After Meriwether Lewis (1774–1809) US soldier and explorer.

lewisii After T. H. Lewis (1854–1917) British naval surgeon and collector in the Pacific.

leytensis After the island of Leyte in the Philippines.

lherminieri After Félix Louis l'Herminier (1779–1833) French naturalist exiled to Guadeloupe, West Indies.

lhuysii After Édouard Drouyn de Lhuys (1805–1881) French statesman and diplomat.

libonyanus No expl. (Smith (1836), *Rep. exped. central Africa,* p. 45); probably based on an unrecorded Chuana native name.

Lichenops Gr. *likhen, likhenos,* scurvy, lichen; *ops,* the eye; ref. fleshy yellow membrane surrounding eye of the Spectacled Tyrant *Hymenops perspicillata.*

Lichenostomus Gr. *likhen,* moss, lichen; *stoma,* the mouth; ref. brush-tipped tongue of the Purple-gaped Honeyeater *L. cratitius.*

Lichmera Gr. *likhmeres,* to dart the tongue; the Brown Honeyeater *L. indistincta* feeds on nectar and honeydew.

lichtensteini/lichtensteinii After Martin Hinrich Carl Lichtenstein (1780–1857) German ornithologist, collector, and author (*Reisen in südlichen Africa,* 1810).

Licmetis Gr. *likmetes,* a winnower of wheat, a field worker.

lictor L. *lictor,* a lictor, a magistrate's attendant who bore the rods of office and carried out judgements and executions; a tyrannical epithet undeserved by the quiet Lesser Kiskadee *Philohydor.*

lidthi After Theodor Gerard van Lidth de Jeude (1788–1863) Dutch zoologist.

lignarius L. *lignarius,* a carpenter (*lignum,* wood).

lilianae After Lilian Sclater (fl. 1909) British naturalist and traveller in East Africa, 1894.

lilliae After Lillie Stone (1872–1940) wife of US ornithologist and author Witmer Stone.

limae After J. L. Lima (1874–1936) Brazilian zoologist.

Limicola L. *limicola,* a mud-dweller (*limus, limi,* mud, mire; *-cola,* dweller).

limnaeetus Gr. *limne,* a marsh or mere; *aetos,* an eagle.

Limnocorax Gr. *limne,* a marsh; *korax,* the raven; ref. black, crow-like plumage of the marsh-dwelling Black Crake *L. flavirostris.*

Limnoctites Gr. *limne,* a marsh or mere; *ktites,* an inhabitant (*ktizo,* to occupy).

Limnodromus Gr. *limne,* a marsh or mere; *dromos,* running (i.e. inhabiting).

Limnodytes Gr. *limne,* a marsh or mere; *dutes,* a diver, a plunger (*duo,* to plunge) (cf. syn. genus *Dytes*, grebe).

Limnornis Gr. *limne,* a marsh or mere; *ornis,* a bird.

Limnothlypis Gr. *limne,* a marsh or mere; Mod. L. *thlypis,* a warbler.

Limosa L. *limosus,* muddy (*limus,* mud); the non-breeding habitats of godwits include muddy estuaries, saltmarshes, and flooded fields.

linchi Javanese name *linchi* for a cave swiftlet *Collocalia.*

lincolnii After Thomas Lincoln (1812–1883) US naturalist.

lindsayi After Hugh Hamilton Lindsay (1802–1881) English naturalist and collector.

linearis L. *linearis,* linear, consisting of lines, lined.

lineata /lineatum /lineatus L. *lineatus,* lined, marked with lines (*linea,* a line).

lineifrons L. *linea,* a line; *frons,* the forehead, the brow.

lineiventris L. *linea,* a line; *venter, ventris,* the belly.

lineola L. *lineola,* a little line (dim. of *linea,* a line).

linteatus L. *linteatus,* clothed in linen (*linteum,* linen cloth); the Flame-crowned Manakin *Heterocercus* has a silky, white throat ruff.

lintoni After Morris Albert Linton (b.1887) US businessman.

Linurgus Gr. *linon,* flax, linen; *ourgos,* working, weaving (*ergon,* woven handiwork); no expl.(Reichenbach (1850), *Av. syst. nat.*, pl. 78) but probably ref. assumed relationship of the Oriole Finch *L. olivaceus* to the weaverbirds.

Liocichla Syn. genus *Liothrix,* leiothrix (v. *Leiothrix)*; Gr. *kikhle,* a thrush; Swinhoe (1877), *Ibis,* 473, considered the Yellow-spectacled Liocichla *L. steerii,* 'In general character a *Liothrix,* but with the stronger legs and shorter wings of a *Garrulax* [laughing thrush]'.

Lioptilus Gr. *leios,* smooth; *ptilon,* plumage, feather; ref. sleek appearance of the Bush Blackcap *L. nigricapillus.*

Liosceles Gr. *leios,* smooth; *skele,* legs (*skelos,* the leg); ref. booted, unscutellate legs of the Rusty-belted Tapaculo *L. thoracicus.*

Lipaugus Gr. *lipauges,* dull, lacking in brightness (*leipo,* lacking; *auge,* a bright light).

Lissotis Gr. *lissos,* smooth; genus *Otis,* bustard; these Afrotropical bustards lack crests.

Lithiophanes Gr. *litheos,* marbled (*litheos,* stone); *-phanes,* shining, appearing.

litsipsirupa Onomatopoeic Chuana (Botswana) native name *letshutshuroopoo,* meaning 'ground scraper', for the Groundscraper Thrush *Turdus.*

littoralis L. *littoralis,* littoral, of the shore (*litus, litoris,* seashore, beach).

livens L. *livens,* bluish, black and blue (*livere,* to be bluish).

liventer L. *liventer,* bluish, lead-coloured (*livere,* to be bluish).

livia Mod. L. *liveus,* blue-grey (L. *livere,* to be bluish).

livida L. *lividus,* bluish, lead-grey (L. *livere,* to be bluish).

livingstonei /livingstonii After David Livingstone (1813–1873) Scottish missionary to tropical Africa, anti-slaver, explorer, and author.

llaneae After Llane Hadden (fl. 1982) wife of New Zealand ornithologist Don Hadden.

lobata /lobatus Mod. L. *lobatus,* lobate, lobed (Gr. *lobos,* a lobe).

Lobibyx Gr. *lobos,* a lobe; Mod. L. *ibyx,* a lapwing or plover; ref. large yellow face wattles of the Masked Lapwing *Hoplopterus miles.*

Lobiophasis Gr. *lobos,* a lobe; Mod. L. *phasis,* a pheasant; ref. extravagant blue face wattles of the Wattled Pheasant *Lophura bulweri.*

Lobipes Gr. *lobos,* a lobe; L. *pes,* the foot; ref. lobed, coot-like toes of the Red-necked Phalarope *Phalaropus lobatus.*

Lobipluvia Gr. *lobos,* a lobe; contracted genus *Pluvialis,* plover; ref. yellow face wattles of the Yellow-wattled Lapwing *Hoplopterus malarbaricus.*

Lobivanellus Gr. *lobos,* a lobe; genus *Vanellus,* lapwing; ref. red eye wattle of the Red-wattled Lapwing *Hoplopterus indicus.*

Loboparadisea Gr. *lobos,* a lobe; genus *Paradisaea,* bird of paradise; ref. green wattles at base of bill of the male Yellow-breasted Bird of Paradise *L. sericea.*

Loboptiloris Gr. *lobos,* a lobe; genus *Ptiloris,* riflebird.

Loborhamphus Gr. *lobos,* a lobe; *rhamphos,* the bill.

Lobotos Mod. L. *lobatus,* lobate (Gr. *lobos,* a lobe); the Wattled Cuckoo-shrike *Campephaga lobata* has large orange gape wattles.

Lochmias Gr. *lokhmios,* thicket-dwelling (*lokhme,* a thicket; *lokhos,* a lair).

Locustella Dim. of L. *locusta,* a grasshopper; ref. whirring, insect-like songs of the grasshopper warblers.

Loddigesia After G. Loddiges (1784–1846) British taxidermist and natural history dealer, specializing in hummingbirds.

loehkeni After W. Löhken (fl. 1969) collector in Brazil.

loennbergi After Axel Johan Einar Lönnberg (1865–1942) Swedish zoologist.

lombokia After the island of Lombok, between Bali and Sumbawa, Indonesia.

lomvia Swedish name *lomvia* for an auk or diver (cf. Norwegian *lomvi).*

Lonchura Gr. *lonkhe,* a spear-head; *oura,* the tail; ref. pointed central tail-feathers of the Spice Finch *L. punctulata* and other munias.

longicauda L. *longus,* long; *cauda,* the tail.

longicaudatus L. *longus,* long; *caudatus,* tailed.

longicaudus see **longicauda.**

longicornis L. *longus,* long; *cornu,* a horn; ref. long, feathered ear-tufts of the Slender-billed Scops Owl *Otus.*

longimembris L. *longus,* long; *membris,* with the legs (*membrum,* a limb).

longipennis L. *longus,* long; *pennis,* winged (*penna,* a feather).

longipes L. *longipes,* long-footed.

longirostra L. *longus,* long; *rostrum,* the bill.

longirostre /longirostris L. *longus,* long; *-rostris,* billed.

longuemarei /longuemareus After G. de Longuemare (fl. 1841) French amateur collector.

Loomelania After Leverett M. Loomis (1857–1928) US ornithologist (cf. Gr. *loon,* better, desirable); specific name *Procellaria melania.*

lopesi After José Lopes (fl. 1903) Portuguese collector in tropical Africa.

Lophaetus Gr. *lophos,* a crest; *aetos,* an eagle.

Lophodytes Gr. *lophos,* a crest; *dutes,* a diver or plunger (*duo,* to dive, to plunge into).

Lophoictinia Gr. *lophos,* a crest; genus *Ictinia,* kite; the Square-tailed Kite *L. isura* has only a slight crest.

Lopholaimus Gr. *lophos,* a crest; *laimos,* the throat; ref. stiffened, elongated neck feathers of the Topknot Pigeon *L. antarcticus*.

Lophonetta Gr. *lophos*, a crest; *netta*, a duck.

Lophophaps Gr. *lophos*, a crest; *phaps*, a pigeon.

Lophophorus Gr. *lophos*, a crest; *-phoros*, carrying, wearing (*phero*, to carry).

Lophopsittacus Gr. *lophos*, a crest; *psittakos*, a parrot.

Lophorina Gr. *lophos*, a crest; *rhis*, the nose; the Superb Bird of Paradise *L. superba* has a tuft of glossy feathers covering the nostrils.

Lophornis Gr. *lophos*, a crest; *ornis*, a bird.

Lophortyx Gr. *lophos*, a crest; *ortux*, the quail (cf. syn. genus *Ortyx*, quail).

Lophospingus Gr. *lophos*, a crest; *spingos*, a finch.

Lophostrix Gr. *lophos*, a crest; genus *Strix*, owl.

lophotes Gr. *lophos*, a crest; *-otes*, featuring.

Lophotibis Gr. *lophos*, a crest; *ibis*, an ibis.

Lophotis Gr. *lophos*, a crest; genus *Otis*, bustard.

Lophotriccus Gr. *lophos*, a crest; Mod. L. *triccus*, tyrant flycatcher.

Lophozosterops Gr. *lophos*, a crest; genus *Zosterops*, white-eye.

Lophura Gr. *lophos*, a crest; *oura*, the tail; ref. arched, crested appearance of tail feathers of the Indomalayan gallopheasants.

lorata L. *loratus*, bound with thongs, strapped (*lorum*, a strap or thong); ref. black loral stripe of the Peruvian Tern *Sterna*.

lorealis Mod. L. *lorealis*, loral, of the lores (L. *loreus*, of thongs, bridled).

lorentzi After Hendrik Antoon Lorentz (1871–1944) Dutch explorer, collector, and diplomat.

lorenzi/lorenzii After **1** Ludwig Ritter Lorenz von Liburnau (1856–1943) Austrian zoologist (*Phyllastrephus*). **2** Theodor K. Lorenz (1842–1909) German zoologist and collector in Russia (*Phylloscopus*).

Loria/loriae After Lamberto Loria (1855–1913) Italian ethnologist and collector in New Guinea.

loricata/loricatus L. *loricatus*, clothed in mail, breast-plated (*lorica*, a cuirass).

Loriculus Dim. of genus *Lorius*, lory.

Lorius/lory Malay names *luri* or *lori* for the lories.

lotenia After Johan Gideon Loten (1710–1789) Dutch naturalist and Governor of Ceylon (Sri Lanka), 1752–1757.

louisiadensis After the Louisiade Archipelago, Papua New Guinea.

lovensis After the settlement of Love's Creek, near Alice Springs, Northern Territory, Australia.

loveridgei After Arthur Loveridge (b. 1891) US herpetologist, collector, and author (*Many happy days I've squandered*, 1949; *I drank the Zambesi*, 1954).

lowei After Percy Roycroft Lowe (1870–1948) British ornithologist and President of British Ornithologists' Union, 1938–1943.

loweryi After George H. Lowery, Jr. (1915–1978) US zoologist and author (*Louisiana birds*, 1955).

Loxia Gr. *loxos*, crosswise; the crossbills have distinctive crossed mandibles; in orn. the name is used for a variety of finch-like birds with or without curved or heavy bills.

Loxigilla Genus *Loxia*, crossbill; genus *Fringilla*, finch; ref. heavy bill of the Lesser Antillean Bullfinch *L. noctis*.

Loximitris Genus *Loxia*, crossbill; syn. genus *Chrysomitris*, goldfinch (Gr. *khrusometris*, the Goldfinch *Carduelis)*.

Loxioides Genus *Loxia*, crossbill; Gr. *-oides*, resembling; the Hawaiian Palila *L. bailleui* is distinguished by its finch-like appearance and thick, stubby bill.

Loxipasser Genus *Loxia*, crossbill; genus *Passer*, sparrow; ref. thick, stubby bill of the Yellow-shouldered Grassquit *L. anoxantha*.

Loxops Genus *Loxia,* crossbill; Gr. *ops,* the face (i.e. appearance); alluding to red, crossbill-like plumage of the Hawaiian Akepa *L. coccineus.*

loycae Local Chilean names *loica* or *lloica* for the Red-breasted Blackbird *Leistes militaris.*

lubomirskii After Prince Ladislas Lubomirski (1824–1882) Polish conchologist.

luciae After **1** Lucy Brewer (fl. 1867) daughter of US ornithologist T. M. Brewer (*Amazilia*). **2** Lucy Hunter Baird (1848–1913) daughter of US ornithologist Spencer Fullerton Baird (*Vermivora*).

luciani After J. Lucian Bucquet (1807–1889) French entomologist.

lucida/lucidus L. *lucidus,* clear, bright.

lucifer L. *lucifer,* light-bearing (*lux, lucis,* light; *-fer,* bearing; *-ferre,* to carry).

lucionensis/luconensis After the island of Luzon, largest of the Philippines.

luctuosa/luctuosus L. *luctuosus,* mournful (*luctus,* mourning); ref. dark, funereal plumage.

ludoviciae Late L. *Ludovicia,* Louise; after **1** Louise Geoffroy Saint-Hilaire (fl. 1847) wife of French zoologist Isidore Geoffroy Saint-Hilaire (*Doryfera*). **2** Louisa Lort Phillips (d. 1946) wife of big-game hunter E. Lort Phillips (*Turdus*).

ludoviciana/ludovicianus Late L. *Ludovicius,* Louis; L. *-anus,* belonging to; **1** after the Louisiana Territory or Colony, the Mississippi basin claimed by France and named for Louis XIV King of France (1638–1715: reigned 1643–1715). Covering a far larger area than the present state, it was a Spanish possession from 1763, before being repossessed by the French and sold by Napoléon Bonaparte to the USA in 1803 (*Lanius, Pheucticus, Thryothorus*). **2** after the Louisiana Territory, USA (*Piranga*). **3** after Louis C. G. Clarke (1881–1960) British naturalist, collector in Somaliland, 1913 and curator of the Fitzwilliam Museum, Cambridge (*Caprimulgus*).

ludwigii After Baron Carl Friedrich von Ludwig (1784–1847) German botanist in South Africa, 1805–1847.

luehderi After W. Lühder (1847–1873) German collector in the Cameroons, 1872.

lugens/Lugensa L *lugens,* mourning (*lugere,* to mourn); ref. sombre, funereal plumage.

lugentoides Specific name *Oenanthe lugens;* Gr. *-oides,* resembling.

lugubris L. *lugubris,* mournful; ref. sombre plumage.

Lullula Dim. from onomatopoeic name *Lulu* given to the Wood Lark *L. arborea* by Buffon (1770).

lumachellus Italian *lumachella,* fire marble, a dark limestone containing shells (*lumacha,* a snail) which emit fiery reflections; ref. fiery red spots on forehead and throat of the Hooded Visorbearer *Augastes.*

luminosa L. *luminosus,* bright, glorious (*lumen, luminis,* light, glory).

lunata/lunatus L. *lunatus,* sickle-shaped, crescent-shaped (*luna,* the moon).

Lunda Faeroese name *lunda* for the Atlantic Puffin *Fratercula* (cf. Norwegian *lunde*).

lunulata/lunulatus L. *lunulatus,* crescent-shaped, falcated (*lunula,* a little moon).

lurida L. *luridus,* ghastly, lurid, pale yellow or reddish-brown.

lurio No expl. (Vincent (1933), *Bull. Brit. Orn. Cl.,* **53**, 173); after the Rio Lurio, northern Moçambique.

Lurocalis Corrupted from Gr. *oura,* the tail, and *kolos,* stunted; ref. short, square tail of the Short-tailed Nighthawk *L. semitorquatus.* Cabanis (1863), *Museum Heineanum,* amended the name to *Urocolus.*

Luscinia L. *luscinia,* the nightingale.

luscinioides L. *luscinia,* the nightingale; Gr. *-oides,* resembling; Savi's Warbler *Locustella* was thought to resemble the Nightingale *Luscinia* in coloration.

Lusciniola L. *lusciniola,* a little nightingale (dim. of *luscinia,* the nightingale); the song of the Moustached Warbler *L. melanopogon* recalls that of the Nightingale *Luscinia.*

lutea L. *luteus,* saffron-yellow (*lutum,* yellow dye).

luteicapilla L. *luteus,* saffron-yellow; *-capillus,* capped.

luteifrons L. *luteus,* saffron-yellow; *frons,* the forehead, the brow.

luteirostris L. *luteus,* saffron-yellow; *-rostris,* billed.

luteiventris L. *luteus,* saffron-yellow; *venter, ventris,* the belly.

luteocephala L. *luteus,* saffron-yellow; Gr. *-kephalos,* headed.

luteola/luteolus L. *luteolus,* yellowish, pale yellow (*luteus,* saffron-yellow).

luteoschistaceus L. *luteus,* saffron-yellow; Mod. L. *schistaceus,* slate-grey.

luteoventris see **luteiventris.**

luteovirens L. *luteus,* saffron-yellow; *virens,* green.

luteoviridis Mod. L. *luteoviridis,* uranium green.

lutescens Mod. L. *lutescens,* somewhat yellowish (L. *luteus,* saffron-yellow) (cf. L. *lutescens,* muddy-coloured; *lutum, luti,* mud).

lutetiae L. *Lutetia,* Paris; after Louis Philippe Comte de Paris (1834–1894), grandson of Louis-Philippe (King of the French 1830–1848), and from 1883 head of the House of Bourbon.

luteus see **lutea.**

lutosus L. *lutosus,* muddy, dirty (*lutum,* mud).

luxuosa L. *luxus,* luxury, extravagance; *-osus,* abundance (cf. *luxuriosus,* luxurious).

luzonica/luzoniensis see **lucionensis.**

lyalli After D. Lyall (fl. 1894) lighthouse keeper at Stephens Island, New Zealand, whose cat killed the entire population of the extinct Stephens Island Wren *Xenicus.*

Lybius Gr. *libuos* or *lebios,* a bird mentioned by Aristotle and Aristophanes, possibly allied to the woodpeckers, but not further identified.

Lycocorax Gr. *luke,* light; *korax,* the raven; ref. glossy plumage and crow-like appearance of the Paradise Crow *L. pyrrhopterus.*

Lymnocryptes Gr. *limne,* a marsh or mere; *krupto,* to hide (i.e. dwelling).

lynesi After Rear Admiral Hubert Lynes (1874–1942) British ornithologist, collector in China, 1910–1913 and tropical Africa, 1926–1927, and author (*Review of the genus* Cisticola, 1930).

lyra L. *lyra,* the lyre (Gr. *lura,* the lyre); ref. tail shape of the Lyre-tailed Nightjar *Uropsalis.*

Lyrurus Gr. *lura,* the lyre; *-ouros,* tailed.

Lysurus Gr. *lusis,* loosing, releasing; *-ouros,* tailed; Ridgway (1898), *Auk*, **15**, 225, described the Sooty-faced Finch *L. crassirostris* as having the '...tail rounded, almost graduated, with the rectrices broad, though pointed at the tips, the webs semi-decomposed terminally'.

Bartramia longicauda Upland Sandpiper

M

Macabra French *macabre,* gruesome, ghoulish.

macao Tupí (Brazilian) Indian names *macao, maca,* and *macavuana* for the macaws.

macclounii After J. MacClounie (fl. 1938) British naturalist and collector in Nyasaland (Malawi), 1895–1906.

maccoa After Maccoa, an alternative spelling of Macao, China; an erroneous toponym.

macconnelli After Frederick Vavasour McConnell (1868–1914) English traveller and collector.

maccormicki After Robert McCormick (1800–1890) British naval surgeon, explorer, and naturalist.

macdonaldi After Colonel Max Macdonald (fl. 1890) US naturalist with US Fisheries Dept.

macei After M. Macé (fl. 1750) French traveller in India.

macgillivrayi After **1** John MacGillivray (1821–1867) British naturalist and collector (*Pseudobulweria*). **2** Alexander Sykes Macgillivray (1853–1907) Australian cattle rancher and field naturalist (*Barnardius*).

Macgregoria /macgregoriae After Lady Mary McGregor (fl. 1897) wife of British diplomat and naturalist Sir William McGregor.

macgregorii After Sir William McGregor (1847–1919) British colonial governor, ornithologist, and collector.

macgrigoriae After Jane G. McGrigor (fl. 1836) daughter of British zoologist Sir John McGrigor.

Machaerhamphus Gr. *makhaira,* a knife or razor; *rhamphos,* the bill; ref. hooked bill of the Black Bat Hawk *M. alcinus.*

Machaerirhynchus Gr. *makhaira,* a knife or razor; *rhunkhos,* the bill; ref. broad, sharp-tipped bill of the Yellow-breasted Boatbill *M. flaviventer.*

Machaeropterus Gr. *makhaira,* a knife or razor; *-pteros,* winged; ref. stiffened, enlarged inner remiges of the Striped Manakin *M. regulus.*

Machetornis Gr. *makhetes,* warlike, bellicose; *ornis,* a bird; the Cattle Tyrant *M. rixosus* has pugnacious habits, typical of the tyrant flycatchers.

machiki After Surgeon Captain József Machik (fl. 1883) Hungarian conchologist and collector in the Moluccas.

Mackenziaena After Helen Mackenzie McConnell (1871–1954) wife of collector Frederick McConnell (v. *macconnelli).*

mackinlayi After Archibald Mackinlay (1850–1924) Scottish explorer and naturalist.

mackinderi After Sir Halford John Mackinder (1861–1947), British geographer who made the first ascent of Mount Kenya in 1899.

mackinnoni After Archibald Donald MacKinnon (1864–1937) Scottish medical officer in Africa.

macleayana After Sir William John Macleay (1820–1891) Scottish/Australian politician and naturalist.

macleayii After Alexander Macleay (1767–1848) British entomologist, collector, and Colonial Secretary to New South Wales, Australia.

macloviana French name *Iles Malouines* (after fishermen from St Malo) for the Falkland Islands; L. *-anus,* belonging to.

macquariensis After the Macquarie Islands, south-east of Tasmania, named for Major-General Lachlan Macquarie (d. 1824), Governor of New South Wales, 1761–1824.

Macroagelaius Gr. *makros,* long, great; genus *Agelaius,* blackbird.

macrocephala /Macrocephalon Gr. *makrocephalos,* long-headed, great-headed.

macrocerca /macrocercus Gr. *makrokerkos,* long-tailed.

macrocerus Gr. *makros,* long; *oura,* the tail (Peters (1964), Vol. **X**, p. 238, footnote 1).

macrodactyla /macrodactylus Gr. *makrodaktulos,* long-toed.

Macrodipteryx Gr. *makros,* long; *dis,* double; *pterux,* wing; ref. the distinctive flag-tipped wing of the male Standard-winged Nightjar *M. longipennis* in the breeding season.

macrolopha Gr. *makros,* long; *lophos,* a crest.

Macronectes Gr. *makros,* long, great; *nektes,* a swimmer (*nekho,* to swim) (cf. syn. genus *Nectris,* shearwater).

Macronous Gr. *makros,* long; *onux,* a claw; ref. long hind-claw of the Fluffy-backed Tit Babbler *M. ptilosus.*

Macronyx Gr. *makros,* long; *onux,* a claw; ref. very long hind-claw from which the Afrotropical longclaws get their name.

Macropsalis Gr. *makros,* long; *psalis,* a pair of shears or scissors; ref. long outer tail feathers of the Long-trained Nightjar *M. creagra.*

macroptera /macropterus Gr. *makropteros,* long-winged.

macropus Gr. *makros,* long; *pous,* the foot.

Macropygia Gr. *makros,* long; Mod. L. *pygius,* rumped (Gr. *puge,* the rump); ref. long tails of the cuckoo doves.

macrorhyncha /macrorhynchos /macrorhynchus Gr. *macrorhunkhos,* long-billed.

macrorrhina Gr. *makros,* long; *rhis, rhinos,* the nose (i.e. the bill).

Macrosphenus Gr. *makros,* long; *sphenos,* a wedge; ref. long, heavy, wide-based bill of the Yellow Longbill *M. flavicans.*

macrotis Gr. *makrotes,* long-eared (*makros,* long; *ous, otos,* the ear).

macroura /macrourus /macrurus Gr. *makros,* long; *-ouros,* tailed.

macularia /macularius L. *macula,* a spot, a mark; *-arius,* possessing.

maculata /maculatum /maculatus L. *maculatus,* spotted, blotched (*maculare,* to spot).

maculialatus L. *macula,* a spot; *alatus,* winged (*ala,* a wing; originally the wing of an army in battle array).

maculicauda /maculicaudus L. *macula,* a spot; *cauda,* the tail.

maculicoronatus L. *macula,* a spot; *coronatus,* crowned (*coronare,* to crown).

maculifer L. *macula,* a spot; *-fer,* carrying.

maculifrons L. *macula,* a spot; *frons,* the forehead, the brow.

maculipectus L. *macula,* a spot; *pectus,* the breast.

maculipennis L. *macula,* a spot; *pennis,* winged (*penna,* a feather).

maculirostris L. *macula,* a spot; *-rostris,* billed.

maculosa /maculosus L. *maculosus,* spotted, speckled.

mada After Mount Mada on the island of Buru, Moluccas, Indonesia,

madagarensis Corrupted from *madagascariensis* (q.v.).

madagascariensis /madagascarinus After the island of Madagascar. The island first appeared on a Portuguese map of 1502, and early accounts speak of a coastal people, the Matacassi, a name subsequently corrupted to Madacassa and Malagasy and applied to all the native clans and, finally, the whole island. An erroneous toponym in the case of *Numenius,* where *macassarensis* was doubtless intended (after the port of Makassar, Celebes (Ujung Pandang, Sulawesi)).

Madanga After the Madanga Mountains on the island of Buru in the Moluccas, Indonesia.

madaraspatensis After the city and seaport of Madras, Tamil Nadu, India (Mod. L. *Maderaspata).*

madaraszi After Gyula von Madarász (1858–1931) Hungarian ornithologist and author.

madeira After the island of Madeira in the North Atlantic Ocean.

maderaspatana Erroneous toponym after Madras (v. *madaraspatensis),* when Linnaeus meant to write *madagascariensis.*

maesi After Albert Maës (d. 1914) French aviculturalist.

maforensis After the island of Mafor or Mafoor (Numfor), Geelvink Bay, Irian Jaya (Indonesian New Guinea).

magellani/magellanica/magellanicus After the Straits of Magellan, Chile, joining the South Atlantic Ocean to the Pacific Ocean, named for the Portuguese explorer Ferdinão Magelhaes. The toponym refers to the general area of Tierra del Fuego, but is erroneous in the case of *Carduelis.*

magentae After the Italian warship *Magenta* which circumnavigated the globe in 1865, commanded by Arminjon (v. *arminjoniana).*

magicus L. *magicus,* mysterious, magical (Gr. *magikos).*

magister L. *magister,* a master, a teacher; **1** after Spencer Fullerton Baird (v. *bairdi)* in recognition of his influence on US ornithology (*Vireo).* **2** ref. the large size and strong bill of Wied's Flycatcher *Myiarchus.*

magna L. *magnus,* large, great.

magnifica L. *magnificus,* grand, splendid.

magnificens L. *magnificens,* magnificently, splendidly.

magnificus see **magnifica.**

magnirostre/magnirostris L. *magnus,* great; *-rostris,* billed.

magnolia Botanical genus *Magnolia,* magnolia (after Pierre Magnol (1638–1715) French botanist); ref. supposed habitat of the Magnolia Warbler *Dendroica.*

magnum/magnus see **magna.**

maguari Tupí (Brazilian) Indian name *maguari* for the Maguari Stork *Euxenura* (cf. Güaraní (Paraguayan) Indian names *baguari* or *mbaguari).*

mahali No expl. (Smith (1836), *Rep. exped. centr. Africa,* p. 51), but probably based on an African native name.

mahrattensis After the Mahratta country of India; the Mahrattas or Marathas are a martial race of central India.

maillardi After L. Maillard (fl. 1862) French botanist, engineer, and author (*Ile Réunion,* 1862).

maior L. *maior,* greater (comp. of *magnus,* great).

maja Adapted from the name *maja* or *maia* by which an unknown finch, the 'ricebird', was known in Mexico (Hernandez (1615), *Historia avium novae Hispaniae*) (Gr. myth. Maia, the brightest of the Pleiades, daughter of Atlas and mother of Mercury).

major see **maior.**

makawai After Jali Makawa (fl. 1932) African native collector and skinner in Northern Rhodesia (Zambia).

makirensis After Makira Bay, San Cristóbal Island (Makira), Solomon Islands.

malabarica/malabaricus After the Malabar coast and district (Kerala), the first part of India known to Europeans.

malacca/malaccense/malaccensis/malacense/malacensis After the town and district of Malacca (Melaka), Malaya.

malachitacea Mod. L. *malachiteus,* malachite-green (Gr. *malakhe,* the mallow); L. *-aceus,* resembling.

malachurus Gr. *malakos,* soft, weak; -*ouros,* tailed; the Southern Emu-wren *Stipiturus* has a long, thin tail composed of only six feathers.

Malacocincla Gr. *malakos,* soft; Mod. L. *cinclus,* thrush; ref. the full, lax plumage of Abbott's Babbler *M. abbotti.*

Malaconotus Gr. *malakos,* soft; -*notos,* backed; ref. soft, elongated rump feathers which distinguish the bush shrikes from the true shrikes.

Malacopteron Gr. *malakos,* soft; *pteron,* a feather; ref. soft, lax rump feathers, typical of many babbler genera.

Malacoptila /malacoptilus Gr. *malakos,* soft; *ptilon,* plumage.

Malacorhynchus Gr. *malakos,* soft; *rhunkhos,* the bill; ref. soft flaps on upper mandible of the Pink-eared Duck *M. membranaceus.*

Malacothraupis Gr. *malakos,* soft; genus *Thraupis,* tanager; ref. soft, lax plumage of the Slaty Tanager *Creurgops dentata.*

malagae After Malaga 'high plateau locality, probably in La Paz area', Bolivia (Paynter Traylor and Winter (1975), p. 36).

malaitae After Malaita Island (Mala), Solomon Islands.

malarbaricus Misspelt version of *malabaricus* (q.v.).

malaris Mod. L. *malaris,* malar, of the cheek (L. *mala,* the cheek).

malayanus /malayensis After the Malayan Peninsula (*Anthracoceros, Chalcites)* or Malaysian archipelago (Sumatra, Java, Borneo, and associated smaller islands) (*Ictinaetus).*

malcolmi After Major General Sir John Malcolm (1769–1833) Governor of Bombay, 1827.

Malcorus Corrupted from Gr. *malakos,* soft, delicate; -*ouros,* tailed.

maldivarum After the Maldive Islands (Divehi Raajje), Indian Ocean.

maleo Malayan names *maleo* or *moleo* for a scrubfowl *Megapodius* on the island of Amboina in the Moluccas (cf. Tongan *malao,* megapode).

malherbi /malherbii After Alfred Malherbe (1804–1866) French ornithologist and author (*Monograph des Picidées,* 1859).

Malia Contraction of genus *Timalia,* babbler; Schlegel (1880), *Notes Leyden Mus.*, **2**, 165, questioned the relationships of the Malia *M. grata,* 'The bird in question belongs evidently to the group of the Timaliae, strongly characterized...by the vaulded form of the tail. In vain have I tried to find for the apparently unknown species a place in one of the numerous genera established in favour of the Timalia group'.

malimbica /malimbicus /Malimbus After the town of Malimbe, Portuguese Congo (Cabinda), Angola

mallee Aboriginal (Australian) native name *mallee* for a type of eucalyptus scrub, the habitat of the Mallee Emu-wren *Stipiturus.*

malouinus French name *Iles Malouines* (after fishermen from St Malo, France) for the Falkland Islands (cf. Spanish *Islas Malvinas).*

malura see **Malurus.**

maluroides Genus *Malurus,* fairy wren; Gr. -*oides*, resembling.

Malurus Gr. *malos* (a form of *amalos*), soft, weak; *oura,* the tail.

mana No expl.(Wilson (1891), *Ann. Mag. Nat. Hist.*, Ser. 6, **7**, 460); probably an Hawaiian native name.

Manacus English name *manakin* for these small, brilliantly coloured birds (cf. Dutch *manneken,* pretty little thing).

manadensis After the town of Manado, Celebes (Sulawesi). An erroneous toponym in the case of *Monarcha.*

mandellii After L. Mandelli (d.1881) tea planter in Assam, India.

Mandingoa After the Mandingo, a native people of the Niger valley in tropical West Africa.

mangle Local Brazilian name *saracura do mangue,* mangrove rail (Tupí Indian name *saracura,* wood rail), for the Little Wood Rail *Aramides.*

mango English colonists' name *mango-bird* for the Jamaican Mango *Anthracothorax* because it searches the flowers of mango trees (Malay *manga,* the mango tree).

manilata Probably a misspelling from L. *manicatus,* long sleeved (*manicae,* sleeves); ref. the long tail and broad wings of the Red-bellied Macaw *Ara.*

manillensis/manilloe After the city of Manilla (Manila), Luzon, Philippines.

manipurensis After the state of Manipur, north-eastern India.

manoensis After the Mano district, German East Africa (Tanzania).

Manorina Gr. *manos,* thin; *rhin* (a form of *rhis),* the nostril; ref. long, arrow-shaped nostrils of the Australian miners.

mantananensis After Mantanani Island (Mantanan), British North Borneo (Sabah).

mantchuricum After the territory of Manchuria, north-eastern China (now divided into various provinces including Heilongjiang, Jilin, and Liaoning). Manchuria was the home of the Manchu or Ch'ing dynasty which ruled China from 1644 until 1912.

mantelli After Walter B. Mantell (1820–1895) British amateur geologist and naturalist resident in New Zealand.

mantoui After M. Mantou (fl. 1890) French plume dealer.

manu After the Manú National Park (Río Manú), Madre de Dios, south-eastern Peru.

Manucodia Contracted form of *Manucodiata,* the name assigned to a group of birds of paradise by Brisson (1760), based on Old Javanese (Malay) *manuk dewata,* bird of the gods (v. *apoda).*

manusi After Manus Island, Admiralty Islands, Papua New Guinea.

manyar Tamil (Sri Lankan) name *manja* for various weaverbirds.

maracana Tupí (Brazilian) Indian name *maracaná* for a macaw.

marail Carib (Cayenne) Indian name *marai* for a guan.

maranhaoensis After the state of Maranhão, Brazil.

maranonica/maranonicus After the Río Marañón, headstream of the River Amazon, in northern Peru.

marcapatae After the town of Marcapata, Cuzco, Peru.

marchei After Alfred A. Marche (1844–1898) French explorer and author (*Trois voyages dans l'Afrique occidentale,* 1879; *Luçon et Palaouan,* 1887).

Mareca Brazilian Portuguese name *marréco* for the smaller kinds of duck *Anas* (cf. Roman myth. Marica, a river nymph or water nymph).

margaretae After Margarete Mayr (d. 1989) wife of the US/German ornithologist Ernst Mayr (v. *mayri).*

margarethae After **1** Princess Louise Margaret Alexandra Victoria Agnes (1860–1917) Duchess of Connaught (*Charmosyna).* **2** Margarethe Platen (fl. 1890) wife of German collector Carl Platen (*Ceyx).* **3** no expl. (Heine (1859), *Journ. für Orn.*, **7**, 406) (*Tanysiptera).*

margaritaceiventer L. *margarita,* a pearl; *-aceus,* resembling; *venter,* the belly.

margaritae After Margaret Lander Holt (fl. 1931) wife of US ornithologist E. Holt.

margaritatus L. *margaritatus,* adorned with pearls (*margarita,* a pearl; Gr. *margarites,* a pearl).

Margaroperdix L. *margarita,* a pearl; *perdix,* a partridge; ref. the heavily spotted belly of the cock Madagascar Partridge *M. madagarensis.*

Margarops Gr. *margarites,* a pearl; *ops,* the eye; ref. glaring white eye of the Pearly-eyed Thrasher *M. fuscatus.*

Margarornis Gr. *margarites,* a pearl; *ornis,* a bird; ref. pearl-shaped spots on underparts of the Pearled Treerunner *M. squamiger.*

marginalis Med. L. *marginalis,* marginal, pertaining to the edge or border (L. *margo, marginis,* the edge or border); the Striped Crake *Porzana* has the feathers of the upperparts edged with white.

marginata/marginatus L. *marginatus,* bordered, edged (*marginare,* to emarginate).

mariae After **1** Grand Duchess Maria (1819–1876), daughter of Tzar Nicholas I of Russia, wife of Maximilian Prince of Eichstadt (*Pteroglossus*). **2** Maria von Madarász (fl. 1905) wife of Hungarian ornithologist Gyula von Madarász (*Bradypterus*). **3** Florence Mary Benson (fl. 1960) wife of English ornithologist Constantine Benson (*Nesillas*).

mariei After E. A. Marie (1835–1889) French naturalist and collector in New Caledonia, 1869; Guadeloupe, 1874; and Madagascar, 1878.

marila Gr. *marile,* charcoal embers; ref. grey back of the Great Scaup *Aythya*.

marina/marinus L. *marinus,* marine, of the sea (*mare, maris,* the sea).

mariquensis After the Mariqua (Marico) River, Transvaal, South Africa/ Botswana.

maritima/maritimus L. *maritimus,* maritime, of the sea (*mare, maris,* the sea).

markhami After Sir Clements Robert Markham (1830–1916) English geographer and traveller.

Marmaronetta Gr. *marmaros,* marble; *netta,* a duck; ref. the Marbled Teal's *M. angustirostris* dappled plumage.

marmorata/marmoratus L. *marmoratus,* marbled (*marmor,* marble).

marshali/marshalli After **1** Joe T. Marshall (fl. 1979) US field ornithologist (*Otus*). **2** Alan John Marshall (1911–1967) Australian zoologist and collector (*Opopsitta*).

martii After Carl Friedrich Phillip von Martius (1794–1868) German botanist and collector in Brazil, 1817–1820.

martinica/martinicensis After the island of Martinique in the West Indies.

martius L. *picus Martius,* a type of woodpecker with a tufted crest mentioned by Pliny (*martius,* martial, war-like).

masafuerae After the island of Más a Fuera, Juan Fernández Islands.

Mascarinus French name *le Mascarin,* given to the Mascarene Parrot *M. mascarinus* by Buffon (1770). The Mascarene islands (Mauritius, Réunion, Rodrigues) were named for the Portuguese navigator Pedro Mascarenhas who came upon them in 1505.

Masius Bonaparte (1850), *Consp. gen. avium*, Vol.**1**, p. 175, wrote of the Golden-winged Manakin *M. chrysopterus,* 'My genus is dedicated to Aloysius Masi, the very essence of Italy, exiled yet dear, talented, intellectual, energetic, and distinguished! Would that Italy or France could always produce such sons!' The only candidate for this eulogy (here freely translated from the original Latin) would seem to be the republican Colonel Masi, a supporter of Garibaldi, but no further details are provided.

massena After François Victor Masséna Prince d'Essling and Duc de Rivoli (1795–1863) French ornithologist and collector.

Massornis Gr. *masson,* larger (comp. of *makros,* long); *ornis,* a bird; ref. size of the Greater Manakin *M. major* compared to genus *Schiffornis* in which originally placed.

masukuensis After the Masuku Mountains, northern Nyasaland (Malawi).

mathewsi After Gregory MacAlister Mathews (1876–1949) Australian ornithologist and author (*The birds of Australia*, 1910; *Systema avium Australasianarum*, 1927).

matsudairae After Viscount Yorikatsu Matsudaira (1876–1945) Japanese ornithologist.

matthewsii After Andrew Matthews (d. 1841) English botanist and collector in Peru and Chile, 1830–1841.

matthiae After St Matthias Island (Mussau), Bismarck Archipelago, Papua New Guinea.

maugaeus /maugei After R. Maugé de Cely (d. 1802) French zoologist and collector in the West Indies, 1796–1798 and Australia, 1800–1802.

maura L. *Maurus,* Moorish, African (i.e. black).

mauri After Ernesto Mauri (1791–1836) Italian botanist.

mauritianus After the island of Mauritius, Indian Ocean, named for Stadhouder Maurits, Prince of Orange, 1567–1625.

maurus see **maura.**

mavornata Misspelt coining of *inornata* (q.v.).

mavors L. *Mavors,* Mars, god of war.

maxillosus L. *maxilla,* the jaw; *-osus,* abundance; ref. short, thick bill of the Thick-billed Saltator *Saltator.*

maximiliani After Maximilian Alexander Philip Prince zu Wied-Neuwied (1782–1867) German explorer and collector in Brazil, 1815–1817.

maxima /maximus L. *maximus,* greatest (super. of *magnus,* great).

mayeri After **1** August Franz Joseph Carl Mayer (1787–1865) German anatomist and collector (*Nesoenas).* **2** Fred W. Shaw Mayer (b. 1899) Australian naturalist and manager of Nondugl Aviaries, Wahgi, Papua New Guinea, 1954–1963 (*Astrapia).*

maynana After the Mayna Indians of Peru, in whose country the Plum-throated Cotinga *Cotinga* was discovered.

mayottensis After the island of Mayotte, Comoro islands.

mayri After Ernst Walter Mayr (b. 1904) German/US ornithologist, collector, and author (*List of New Guinea birds*, 1941; *A classification of recent birds,* 1951).

Mayrornis After Ernst Mayr (v. *mayri);* Gr. *ornis,* a bird.

mccallii After Brigadier General G. A. McCall (1802–1868) US zoologist and collector.

mcclellandii After John MacClelland (1805–1875) British zoologist and geologist.

mccownii After Major General John Porter McCown (1815–1879) US soldier, naturalist, and collector.

mcgregori After Richard C. MacGregor (1871–1936) US zoologist, collector, and author (*A manual of Philippine birds*, 1909).

mcilhennyi After John S. McIlhenny (fl. 1965) US sponsor.

mcleannani After James McLeannan (fl. 1860) US engineer on Panama railway.

mcleodii After Richard Randall McLeod (fl. 1899) US collector in Mexico, 1883–1888 and Arizona, 1898.

meadewaldoi After Edmund Gustavus Bloomfield Meade-Waldo (1855–1934) English explorer, ornithologist, and collector.

mearnsi /Mearnsia After Edgar Alexander Mearns (1858–1916) US soldier in Mexico, 1892–1894, Philippines, 1903–1907, and collector in tropical Africa, 1909, 1911.

mechowi After Major A. von Mechow (fl. 1881) Austrian explorer and collector in Angola.

Mecocerculus Dim from Gr. *mekos,* long, and *kerkos,* the tail; ref. longish tail of the atypical White- throated Tyrannulet *M. leucophrys.*

media L. *medius,* intermediate, middle (i.e. intermediate in size between two other species).

mediocris L. *mediocris,* middling; Shelley (1885), *Proc. Zool. Soc. London,* 228, considered the Eastern Double-collared Sunbird *Nectarinia,* 'Intermediate between *C(innyris) chalybeus* and *C(innyris) chloropygius'.*

medius see **media.**

meeki /meekiana After Albert S. Meek (1871–1943) English explorer, collector in New Guinea, the Solomon Islands, and Australia, and author (*A naturalist in cannibal land*, 1913).

Megabyas Gr. *megas,* great; genus *Bias,* vanga flycatcher.

megacephala /megacephalus Gr. *megas,* great; *-kephalos,* headed (cf. *megalokephalos,* large-headed).

Megaceryle Gr. *megas,* great; genus *Ceryle,* kingfisher.

Megacrex Gr. *megas,* great; genus *Crex,* crake.

Megadyptes Gr. *megas,* great; *duptes,* a diving bird.

megaensis After the town of Mega, Sidamo Province, Ethiopia.

megala Gr. *megalos,* great, large (*megas,* great).

Megalaima Gr. *megas,* great; *laimos,* the throat; from puffed cheeks and full-throated appearance of the Indomalayan barbets when calling.

Megalapteryx Gr. *megalos,* great, large; genus *Apteryx,* kiwi.

Megaloprepia Gr. *megaloprepeia,* splendour, magnificence.

megalopterus Gr. *megalos,* great, large; *-pteros,* winged.

megalorynchos Gr. *megalos,* great, large; *rhunkhos,* the bill.

megalotis Gr. *megalos,* great, large; *-otis,* eared (cf. *megalotes,* greatness, magnitude).

megalura Gr. *megalos,* great, large; *oura,* the tail.

Megalurulus Dim. of genus *Megalurus,* grassbird.

Megalurus Gr. *megalos,* great, large; *-ouros,* tailed.

Megapodius Gr. *megas,* great, large; *pous, podos,* the foot; the megapodes are chicken-like birds with large feet.

megarhyncha /megarhynchos /megarhynchus Gr. *megas,* great, large; *rhunkhos,* the bill.

Megastictus Gr. *megas,* great, large; *stiktos,* spotted, dappled.

Megatriorchis Gr. *megas,* great, large; *triorkhes,* a hawk (v. *Dryotriorchis).*

Megaxenops Gr. *megas,* great, large; genus *Xenops,* xenops.

Megazosterops Gr. *megas,* great, large; genus *Zosterops,* white-eye.

meiffrenii After Guillaume Michel Jérome Meiffren Baron Laugier de Chartrouse (fl. 1829) French collector and author (*Nouveau recueil des planches coloriées d'oiseaux*, 1820).

Meiglyptes Gr. *meion,* smaller (comp. of *mikros,* small); *gluptes,* a sculptor or carver (i.e. a woodpecker).

melacoryphus Gr. *melankoruphos,* a small black-headed bird mentioned by Aristotle and Aristophanes, not further identified, but associated with a tit *Parus* or a warbler *Sylvia* by subsequent authors (*melas,* black; *koruphe,* the crown of the head).

melaena Gr. *melaino,* to turn black.

Melaenornis Gr. *melaino,* to turn black; *ornis,* a bird.

melambrotus Gr. *melambrotos,* the land of Negroes (i.e. black).

Melampitta Gr. *melam-,* black (*melas,* black); genus *Pitta,* pitta.

Melamprosops Gr. *melam-,* black, dark (*melas,* black); *pros,* near; *ops,* the eye; the Po'o-uli *M. phaeosoma* of Maui is also known as the Black-faced Honeycreeper.

melanaria Gr. *melas, melanos,* black; L. *-arius,* possessing.

melancholicus L. *melancholicus,* melancholy (Gr. *melankholikos,* melancholy); ref. irritable disposition of the Tropical Kingbird *Tyrannus* (the ancients believed that gloomy melancholy was caused by excessive amounts of splenetic black bile).

Melanerpes Gr. *melas, melanos,* black; *herpes,* a creeper (*herpo,* to creep) (i.e. a tree- creeper or woodpecker).

melania Gr. *melania,* blackness.

melanicterus Gr. *melas,* black; Mod. L. *icterus,* yellow (v. *Icterus).*

Melanitta Gr. *melas,* black; *netta,* a duck.

melanocephala /melanocephalum / melanocephalus Gr. *melas,* black; *-kephalos,* headed.

melanoceps Gr. *melas,* black; L. *-ceps,* headed.

Melanocharis Gr. *melas,* black; *kharis,* grace, loveliness.

melanochlamys Gr. *melas,* black; *khlamus,* a cloak or mantle; ref. black upperparts.

Melanochlora/melanochloros Gr. *melas,* black; *khloros,* green or yellow.

melanochroa/melanochrous Gr. *melanokhros,* black, swarthy.

Melanocorypha/melanocoryphus **1** Gr. *melas, melanos,* black; syn. genus *Corypha*, Lark (*koruphos,* a bird mentioned by Hesychius, not further identified, but here confused with *koruthos,* a lark) (cf. *melankoruphos,* a black-capped songbird mentioned by Aristotle, Aristophanes, and other authors, later identified with the Blackcap *Sylvia* or the Marsh Tit *Parus).* **2** Gr. *melas,* black; *koruphe,* the head (*Cygnus).*

melanocorys Gr. *melas,* black; Mod. L. *corys,* a lark.

melanocyanea Gr. *melas,* black; L. *cyaneus,* dark blue (Gr. *kuaneos,* dark blue).

Melanodera Gr. *melas,* black; *dere,* the throat.

Melanodryas Gr. *melas,* black; *druas,* a dryad or wood-nymph (*drus,* the oak).

melanogaster Gr. *melas,* black; *gaster,* the belly.

melanogenia/melanogenys Gr. *melas,* black; *genus,* the cheek.

melanolaimus Gr. *melas,* black; *laimos,* the throat.

melanoleuca/melanoleucos/melanoleucus Gr. *melas,* black; *leukos,* white.

melanolophus Gr. *melas,* black; *lophos,* a crest.

melanonota/melanonotus Gr. *melas,* black; *-notos,* backed.

Melanopareia Gr. *melas,* black; *pareion,* the cheek.

Melanoperdix Gr. *melas,* black; *perdix,* a partridge.

melanopezus Gr. *melas,* black; *peza,* the foot.

melanophaius Gr. *melanophaios,* dark grey.

Melanophoyx Gr. *melas,* black; *phoux,* a kind of heron.

melanophrys Gr. *melanophrus,* black-browed.

melanopis Gr. *melanopis,* black-looking.

Melanoploceus Gr. *melas,* black; genus *Ploceus,* weaver.

melanopogon Gr. *melas,* black; *pogon,* the beard.

melanops Gr. *melas,* black; *ops,* the face.

melanopsis Gr. *melas,* black; *opsis,* appearance.

melanoptera/melanopterus Gr. *melanopteros,* black-winged.

Melanopteryx Gr. *melas,* black; *pterux,* the wing.

Melanoptila Gr. *melas,* black; *ptilon,* plumage, a feather.

melanopygia Gr. *melas,* black; Mod. L. *pygius,* rumped (Gr. *puge,* the rump) (cf. *melampugos,* black-bottomed).

melanorhamphos Gr. *melas,* black; *rhamphos,* the bill.

melanorhyncha/melanorhynchus Gr. *melas,* black; *rhunkhos,* the bill.

melanorrhoa Gr. *melas,* black; *orrhos,* the rump or tail.

melanospila Gr. *melas,* black; *spilos,* a spot; ref. black patch on nape of the Black-naped Fruit Dove *Ptilinopus*.

Melanospiza Gr. *melas,* black; *spiza,* a finch.

melanosternon Gr. *melas,* black; *sternon,* the breast.

melanosticta Gr. *melanostiktos,* black-spotted.

melanota Gr. *melas,* black; *-notos,* backed.

melanothorax Gr. *melas,* black; *thorax,* the breast.

Melanotis Gr. *melas,* black; *-otis,* eared.

melanotos see **melanota.**

Melanotrochilus Gr. *melas,* black; genus *Trochilus,* hummingbird.

melanotus see **melanota.**

melanoxantha/melanoxanthos/melanoxanthum/melanozanthos Gr. *melas, melanos,* black; *xanthos,* yellow, golden.

melanura/melanurus Gr. *melanouros,* black-tailed.

melaschistos Gr. *melas,* black; Mod. L. *schistos,* slate-coloured.

melba No expl. (Linnaeus (1758), *Systema Naturae,* (Ed. 10), book **1**, p. 180 [*Pytilia*], p. 192 [*Apus*]). According to Albertus Magnus (1478) (*De Avibus*), *melba* is an Old German name for a gull *Larus;* there may be a tenuous connection between the Alpine Swift *Apus* (long thought to be a kind of swallow) and a gull (related to the terns or sea swallows)(see also Macleod (1954), p. 20), but the relevance of the name to the Melba Finch *Pytilia* is unclear.

meleagrides L. *meleagrides,* guineafowl (Gr. *meleagrides).*

Meleagris Gr. *meleagris,* the guineafowl; the ancients seem to have distinguished two types of guineafowl, but the American wild turkey was unknown to them. The name 'turkey' was used for the guineafowl in the sixteenth and early seventeenth centuries and by this confusion Linnaeus used the name *Meleagris* for the wild turkey.

Meliarchus Gr. *meli,* honey; *arkhos,* a king or chief; ref. large size of the Moustached Honeyeater *M. sclateri.*

Melichneutes Gr. *meli,* honey; *ikhneutes,* a tracker or hunter (v. *Indicator).*

Melidectes Gr. *meli,* honey; *dektes,* a beggar or receiver.

Melidora French name *Mélidore d'Euphrosine* given to the Hook-billed Kingfisher *M. macrorrhina* by Lesson (1830), *Traité d'Ornithologie,* Vol. **4**, p. 249; perhaps ref. the heavy yellow bill (Gr. *meli,* honey; *doru,* a spear; *doree,* a gift).

Melierax Gr. *melos,* a song; *hierax,* a hawk; the chanting goshawks have melodious whistling calls.

Melignomon Gr. *meli,* honey; *gnomon,* a guide or interpreter (v. *Indicator).*

Melilestes Gr. *meli,* honey; *lestes,* a robber.

melindae After the town of Melinda (Malindi), Coast Province, Kenya.

Melionyx Genus *Meliphaga,* tufted honeyeater; genus *Certhionyx,* pied honeyeater.

Meliornis Gr. *meli,* honey; *ornis,* a bird.

Meliphaga Gr. *meli,* honey; *phagos,* a glutton, an eater (*phagein,* to eat).

meliphilus Gr. *meli,* honey; *philos,* loving, fond of.

Melipotes Gr. *meli,* honey; *potes,* a drinker, a sucker (*pino,* to drink).

Melithreptus Gr. *melithreptos,* honey-fed.

Melitograis Gr. *meli, melitos,* honey; *grais,* an old woman; ref. the bare neck and white-streaked crown of the White-streaked Friarbird M. *gilolensis.*

melitophrys Gr. *meli,* honey; *-ophrus,* eyebrowed.

Melittophagus Gr. *melitta,* a bee; *phagos,* a glutton, an eater.

Melizophilus Gr. *melizo,* to sing, to warble; *philos,* loving, fond of.

melleri After C. J. Meller (1836–1869) botanist on Mauritius, 1865.

mellianus After R. E. Mell (b. 1878) German biologist in China.

Mellisuga /mellisugus L. *mel, mellis,* honey; *sugere,* to suck; the hummingbirds feed on nectar and minute insects.

mellivora L. *mel,* honey; *-vorus,* eating.

mellori After Joseph William Mellor (1869–1938) English chemist.

Melocichla Gr. *melos,* a song; *kikhle,* thrush.

melodia L. *melodia,* a pleasant song (i.e. a songster).

melodus L. *melodus,* melodious (*meloda,* a song).

Melophus Gr. *melas,* black; *lophos,* a crest.

Melopsittacus Gr. *melos,* a song; *psittakos,* a parrot; ref. warbling song of the familiar Budgerigar *M. undulatus.*

Melopyrrha Gr. *melas,* black; genus *Pyrrhula,* bullfinch; ref. chunky appearance and heavy bill of the black Cuban Bullfinch *M. nigra.*

meloryphus Gr. *melo-,* ruddy (*melon,* an apple); *koruphe,* the crown of the head.

Melospiza Gr. *melos,* a song; *spiza,* a finch.

Melozone Gr. *melon,* the cheek; *zone,* a band or girdle; ref. crescentic black and chestnut earpatch of the White-faced Ground-Sparrow *M. biarcuatum.*

melpoda Gr. *melpodos,* singing songs (*melpo,* to sing).

membranaceus L. *membranaceus,* of skin, membraneous (*membrana,* a thin skin); the Pink-eared Duck *Malacorhynchus* has a huge bill tipped with prominent membraneous flaps.

menachensis After the town of Menacha (Manakhah), Yemen.

menagei After Louis F. Menage (1850–1924) US philanthropist.

menbeki Papuan (New Guinea) name *menebiki* for the Great Black Coucal *Centropus.*

menckei After B. Mencke (d. 1901) German zoologist, leader of 1st German South Seas Expedition, 1900–1901.

mendanae After the Mendaña Islands (southern Marquesas) in the Pacific Ocean; the southern Marquesas were discovered in 1595 by the Spanish explorer Álvaro de Mendaña de Neira.

mendeni After J. J. Menden (fl. 1939) Dutch naturalist and collector in the East Indies, 1935–1939.

mendiculus L. *mendiculus,* beggarly, of a beggar (i.e. small) (*mendicus,* a beggar).

mendozae After A. H. Mendoza (d.1606) Portuguese explorer in the Pacific.

menetriesii After Édouard Ménétries (1802–1861) French zoologist, collector in Brazil, 1822–1824, and Russia 1829–1830 and author (*Catalogue raisonée des objets de zoologie recueillis dans un voyage au Caucase,* 1832).

meninting Javanese (Malay) name *meninting* for the Deep-blue Kingfisher *Alcedo.*

mennelli After Frederic Philip Mennell (b. 1880) mineralogist and curator of Bulawayo Museum, Southern Rhodesia (Zimbabwe).

menstruus L. *menstruus,* monthly, menstrual; the Blue-headed Conure *Pionus* has blood red undertail coverts.

mentalis Mod. L. *mentalis,* chinned, pertaining to the chin (L. *mentum, menti,* the chin).

mentawi After the Mentawi Islands (Mentawai Islands or Kepulauan Mentawai), Sumatra.

Mentocrex L. *mentum,* the chin (cf. Gr. *menos,* strength, fierceness); genus *Crex,* crake; the Grey-masked Rail *Canirallus kioloides* has a white chin and throat.

Menura Gr. *mene,* the moon; *oura,* the tail; the Australian lyrebirds have long, crescent-shaped tail plumes. According to Wolstenholme (1926) the name is derived from Gr. *menos,* might, and *oura,* the tail, and means 'mighty tail'.

mercenaria L. *mercenarius,* mercenary; from plated, armoured appearance of head and breast feathers of the Scaly-naped Amazon *Amazona.*

mercierii After M. Mercier (fl. 1848) French botanist.

Merganetta Genus *Mergus,* merganser; Gr. *netta,* a duck.

merganser L. *mergus,* a kind of waterbird; *anser,* a goose; the Goosander *Mergus* is the largest of the mergansers.

Mergellus Dim. of genus *Mergus,* merganser.

Mergus L. *mergus,* a kind of waterbird mentioned by Pliny, Terentius Varro, and the poet Horatius Flaccus, not further identified.

meridae/meridanus After the state of Mérida, Venezuela.

meridonalis L. *meridionalis,* southern (*meridies,* the south, noon).

merlini After **1** Condesa Mercedes Jaruco de Merlin (1788–1852) Spanish writer and musician (*Saurothera*). **2** Martial H. Merlin (b. 1860) French Governor-General of Indochina, 1924 (*Arborophila*).

Meropogon Genus *Merops,* bee-eater; Gr. *pogon,* a beard; ref. distinctive elongated throat feathers of the Purple-bearded Bee-eater *M. forsteni.*

Merops Gr. *merops,* the bee-eater.

merrilli After Elmer Drew Merrill (1876–1956) US botanist and collector in the Philippines, 1902–1929.

merula L. *merula,* the Blackbird *Turdus* (the female blackbird is various shades of brown, and this plumage is alluded to in *Dendrocincla).*

Merulaxis French name *Mérulaxe* given to the Slaty Bristlefront *M. ater* by Lesson (1830), *Traité d'ornithologie,* Vol. **5,** p. 397 (syn. genus *Merula,* blackbird *Turdus;* genus *Synallaxis,* spinetail).

merulinus Med. L. *merulinus,* blackbird-like, of a blackbird (v. *merula).*

meruloides Specific name *Dendrocincla merula;* Gr. *-oides,* resembling.

Mesembrinibis Gr. *mesembrinos,* southern (*mesembria,* the south); *ibis,* an ibis; the Cayenne Ibis *M. cayennensis* has a wide range in tropical South America.

Mesia No expl. (Hodgson (1837), *India Review,* **2** [1], 34); probably based on a Nepalese native name.

Mesitornis Syn. genus *Mesites,* mesite (French name *mésite,* mesite, said to be from a Malagasy native name; cf. Gr. *mesites,* a mediator); Gr. *ornis,* a bird.

mesochrysa Gr. *mesos,* middle; *khrusos,* gold; ref. deep orange-yellow belly of the Bronze-green Euphonia *Euphonia.*

Mesoenas Gr. *meso-,* half (i.e. resembling); *oinas,* a dove; ref. the pigeon-like gait and uncertain systematic position of the White-breasted Mesite *Mesitornis variegata.*

mesoleuca Gr. *mesoleukos,* half white; ref. white centre of underparts.

mesomelas Gr. *meso-,* half (*mesos,* middle); *melas,* black.

Mesophoyx Gr. *mesos,* middle, intermediate; *phoux,* a kind of heron.

Mesopicos Gr. *meso-,* half (*mesos,* middle); *pikos,* a woodpecker; ref. small size of these Afrotropical woodpeckers.

metabates Gr. *metabasis,* change, alteration (i.e. different from other species).

Metabolus Gr. *metabolos,* changeable; ref. polymorphic plumages of the Truk Monarch *M. rugensis.*

metallica/metallicus L. *metallicus,* metallic, of metal (*metallum, metalli,* metal).

Metallura L. *metallum, metalli,* metal; Gr. *oura,* the tail; ref. glittering purple and coppery tails of the metaltail hummingbirds.

metcalfii After P. H. Metcalfe (d. 1913) British naturalist and collector in the Solomons.

metopias Gr. *metopias,* beetle-browed.

Metopidius Gr. *metopidios,* on the forehead (*metopon,* the forehead); the Bronze-winged Jaçana *M. indicus* has a red shield on the forehead.

Metoponia Gr. *metopon,* the forehead; ref. the bright orange forehead of the Red-fronted Serin *Serinus pusillus.*

Metopothrix Gr. *metopon,* the forehead; *thrix,* the hair; ref. plush-like feathers on forehead of the Orange-fronted Plushcrown *M. aurantiacus.*

Metriopelia Gr. *metrios,* temperate, moderate; *peleia,* a dove; these small doves are adapted to life in the temperate zones of the Andes.

mevesii After F. W. Meves (1814–1891) Swedish zoologist.

mexicana After Mexico (named after Mexitli, the Aztec god of war). Erroneous toponyms in the cases of *Tangara* and *Todus.*

mexicanoides Specific name *Colaptes mexicanus* (now regarded as a subsp. of *C. cafer);* Gr. *-oides,* resembling.

mexicanum/mexicanus see **mexicana.**

meyeni After Franz Julius Ferdinand Meyen (1804–1840) German surgeon, botanist, and collector.

meyerdeschauenseei After Rodolphe Meyer de Schauensee (1901–1986) US ornithologist and author (v. *rodolphei).*

meyeri After **1** Bernhard Meyer (1767–1836) German naturalist, collector, and author (*Naturgeschichte der vögel*

Deutschlands, 1805; *Taschenbuch der deutschen vögelkunde*, 1810) (*Poicephalus*). **2** Adolf Bernhard Meyer (1840–1911) German ornithologist, collector in the East Indies, anthropologist, and author (*The birds of the Celebes and neighbouring islands*, 1898) (*Chalcites, Epimachus, Pachycephala, Philemon*).

meyerianus see **meyeri 2.**

michleri After Brigadier General Nathaniel Michler (1827–1881) US Army in Panama.

micraster Gr. *mikros*, small; *aster*, a star.

Micrastur Gr. *mikroa*, small; syn. genus *Astur*, goshawk (L. *astur*, a hawk).

Micrathene Gr. *mikros*, small; genus *Athene*, owl.

Microbates Gr. *mikros*, small; *-bates*, roamer, walker.

Microcerculus Dim. from Gr. *mikros*, small; *kerkos*, the tail.

Microchera Gr. *mikros*, small; *khera*, a widow; ref. black plumage of the tiny Snowcap *M. albocoronata*.

Microcochlearius Gr. *mikros*, small; genus *Cochlearius*, boat-billed heron; ref. long, broad bill of the Boat-billed Tody Tyrant *M. josephinae*.

Microdynamis Gr. *mikros*, small; syn. genus *Dynamene*, koel, cuckoo (Gr. *dunamis*, strength, power).

Microeca Gr. *mikros*, small; *oika*, to be like (*eioka*, to be like).

Microgoura Gr. *mikros*, small; genus *Goura*, crowned pigeon.

Microhierax Gr. *mikros*, small; *hierax*, a hawk or falcon.

Microligea Gr. *mikros*, small; syn. genus *Ligea*, warbler (Gr. myth. Ligea, a Nereid; *ligeia*, the nightingale).

Micromacronus Gr. *mikros*, small; genus *Macronous*, tit-babbler.

micromegas Gr. *mikros*, small; *megas*, large; from large number of small spots on neck and underparts of the Antillean Piculet *Nesoctites*.

Micromonacha Gr. *mikros*, small; syn. genus *Monacha*, puffbird (Gr. *monakhos*, a monk).

Micronisus Gr. *mikros*, small; syn. genus *Nisus*, sparrowhawk.

Micropalama Gr. *mikros*, small; *palame*, the palm (of the hand); the Stilt Sandpiper *M. himantopus* has small palm-shaped lobes between the toes.

Micropanyptila Gr. *mikros*, small; genus *Panyptila*, swallow-tailed swift.

Microparra Gr. *mikros*, small; syn. genus *Parra*, jacana (L. *parra*, a bird of ill omen).

Micropsitta Gr. *mikros*, small; Mod. L. *psitta*, a parrot (Gr. *psittake* or *psittakos*, a parrot).

microptera Gr. *mikroptera*, small-winged.

Micropternus Gr. *mikros*, small; *pterna*, the heel; ref. very short first toe and claw of the Rufous Woodpecker *M. brachyurus*.

micropterum /micropterus see **microptera.**

Micropygia Gr. *mikros*, small; Mod. L. *pygius*, rumped (Gr. *puge*, the rump); ref. short tail of the tiny Ocellated Crake *M. schomburgkii*.

Microrhopias Gr. *mikros*, small; syn. genus *Rhopias*, antbird (Gr. *rhops, rhopis*, a shrub, a bush).

microrhyncha /microrhynchum / microrhynchus Gr. *mikros*, small; *rhunkhos*, the bill.

Microsarcops Gr. *mikros*, small; *sarx, sarkos*, flesh; *ops*, the face; ref. small yellow wattles at base of bill of the Grey-headed Lapwing *Hoplopterus cinereus*.

Microscelis Gr. *mikros*, small; *skelos*, the leg.

Microsittace Gr. *mikros*, small; *sittakos* (a variant of *psittakos*), a parrot.

microsoma Gr. *mikros*, small; *soma*, the body; the Least Storm Petrel *Halocyptena* is the smallest species of storm-petrel.

Microspingus Gr. *mikros*, small; genus *Hemispingus*, tanager, hemispingus.

Microstilbon Gr. *mikros*, small; *stilbon*, glistening, glittering, a name for the planet Mercury.

Microxenops Gr. *mikros,* small; genus *Xenops,* xenops.

micrura Gr. *mikros,* small; *oura,* the tail.

migrans L. *migrans,* migrating (*migrare,* to move from one place to another).

migratorius L. *migrator,* a migrant, a wanderer; *-orius,* belonging to.

mikado Japanese *mikado,* illustrious gate, a name given to the emperor (*mi,* august; *kado,* a door); after the Emperor Meiji or Mutsuhito (1852–1912: reigned 1867–1912).

milanjensis After the Milanji Hills, Nyasaland (Mlanje Mountains, Malawi).

miles L. *miles,* a soldier; the Masked Lapwing *Hoplopterus* has wings armed with sharp carpal spurs.

Miliaria L. *miliaria,* a kind of bird, perhaps a finch or bunting, fattened for the table (*milium,* millet, on which the birds were fed).

militaris L. *militaris,* military, soldierly (*miles, militis,* a soldier); applied to spp. with scarlet foreheads (*Ara, Ilicura*), bright red breasts (*Leistes*), or crimson plumage (*Haematoderus*), in allusion to the military caps and jackets of the day.

milleri After **1** Waldron de Witt Miller (1879–1929) US zoologist, curator of American Museum of Natural History (*Waldronia*). **2** Leo E. Miller (b. 1887) US collector in tropical America, 1910–1917 (*Grallaria, Xenops*).

milleti After F. Millet (fl. 1919) French superindent of forests in Annam (Vietnam).

milnei After Alphonse Milne-Edwards (1835–1900) French zoologist and author (*Histoire naturelle des oiseaux de Madagascar*, 1876).

milo After Milo (fl. 511 BC), a giant Greek athlete; ref. the large size of the Buff-headed Coucal *Centropus*.

Milvago L. *milvago,* a kite (*milvus,* a kite; *-ago,* resembling).

Milvus L. *milvus,* a kite.

Mimeta Gr. *mimetes,* a mimic, an imitator (*mimeomai,* to copy); the Papuo-Moluccan orioles *Oriolus* closely resemble the more aggressive friarbirds *Philemon.*

mimikae After the Mimika River, Irian Jaya (Indonesian New Guinea).

Mimizuku Japanese *mimizuku,* eared owl.

Mimodes Genus *Mimus,* mockingbird; Gr. *-oides,* resembling.

Mimus L. *mimus,* a mimic actor; ref. the mimetic abilities of the Northern Mockingbird *M. polyglottos.*

mindanensis After the island of Mindanao, Philippines.

mindorensis After the island of Mindoro, Philippines.

miniaceus Med. L. *miniaceus,* scarlet (L. *minium,* red lead).

miniatus L. *miniatus,* painted vermilion, coloured with red lead (*minium,* red lead).

minima /minimus L. *minimus,* smallest, least (super. of *parvus,* small).

Minla Nepalese name *minla* for the Red-tailed Minla *M. ignotincta.*

minlosi After Emilio Minlos (fl. 1884) collector in Colombia.

Mino Hindi name *maina* for starlings and mynahs.

minor L. *minor,* smaller, less (comp. of *parvus,* small).

minula /minulla /minullus Med. L. *minulus,* very small (dim. of L. *minus,* less).

minuta L. *minutus,* small, little (*minuere,* to make smaller).

minutilla /minutillus Med. L. *minutellus,* very small (L. *minutalis,* insignificant).

minutissimum /minutissimus L. *minutissimus,* extremely small (super. of *minutus,* small).

minutus see **minuta.**

Mionectes Gr. *meionektes,* of damaged appearance; ref. attenuated 9th primary of the Streak-necked Flycatcher *M. striaticollis.*

mira L. *mirus,* wonderful, marvellous.

mirabilis L. *mirabilis,* marvellous, wonderful (*mirari,* to wonder).

Mirafra No expl. (Horsfield (1821), *Trans. Linn. Soc. London,* **13** [1], 159). Horsfield made great use of Javanese native names, but this does not appear to be one of them. Gotch (1981), p. 232, writes that the name is from L. *mirus,* wonderful, and *Afra,* African; the first part of this assumed etymology may be correct, but the type species, *M. javanica,* is one of only three members of *Mirafra* that does not occur in the Afrotropics.

miranda L. *mirandus,* wonderful (*mirari,* to wonder).

mirandae After Alipio de Miranda-Ribeiro (1874–1939) Brazilian zoologist.

mirandollei After M. Mirandolle (fl. 1862) resident in Suriname.

mirificus L. *mirificus,* astonishing, wonderful.

Miro Maori name *miro-miro* for the New Zealand Tit *M. macrocephala.*

mirus see **mira.**

Misocalius Gr. *miso-,* hating (*misos,* hatred); *kalia,* a nest; ref. parasitic habits of the Black-eared Cuckoo *M. osculans.*

mississippiensis After the Mississippi Territory, admitted to the USA in 1817, named for the Mississippi River (Algonquin Indian *missisipi,* great river).

mitchellii After David William Mitchell (d. 1859) English zoologist and Secretary of the Zoological Society of London.

mitrata /mitratus L. *mitratus,* crowned, mitred (L./Gr. *mitra,* a head-dress).

Mitrephanes Gr. *mitra,* a cap or head-dress; *-phanes,* appearing (*phaino,* to shine forth); ref. conspicuous pointed crest of the Tufted Flycatcher *M. phaeocercus.*

Mitrospingus Gr. *mitra,* a cap or head-dress; *spingos,* a finch; ref. yellowish cap of the Dusky-faced Tanager *M. cassinii.*

Mitu Tupí (Brazilian) Indian names *mitú, mutú* or *mutum* for the curassows, in imitation of their hooting calls.

mixta /mixtus L. *mixtus,* mixed, mingled (*miscere,* to mix); ref. birds which share characteristics with two or more other spp.

miyakoensis After the island of Miyako shima, Ryukyu Islands, Japan.

mlokosiewiczi After Ludwik Franciszek Mlokosiewicz (1831–1901) Polish naturalist, botanist, Russian army, and forester in the Caucasus.

Mniotilta Gr. *mnion,* moss; *tilo,* to pluck or pull; ref. supposed nesting material of the Black-and-white Warbler *M. varia.*

moabiticus After the mountainous country of Moab, east of the Dead Sea, incorporated into the ancient kingdom of Israel under King Solomon (cf. L. *Moabitis,* a Moabite).

mocinno After José Mariano Mociño (1757–1819) Mexican botanist and naturalist.

mocino An unwarranted emendation of *mocinno* (q.v.). Eisenmann (1959), *Auk,* **76**, 108, wrote, 'Latin words with "nn" generally softened in Spanish into "ñ", and the "ñ" sound was formerly written "nn". The tilde represents the suppressed extra "n" of the older orthography. Transliteration into Latin of the Spanish "ñ" by "nn" was standard scientific usage – recommended by the A. O. U. Code of Nomenclature: 65, 1892'.

modesta /modestus L. *modestus,* modest, unassuming (i.e. plain).

modularis L. *modulari,* to sing, to warble (cf. Med. L. *modulus,* tuneful).

Modulatrix L. *modulatrix,* a female musician.

modulator L. *modulator,* a musician (*modulari,* to sing, to play an instrument).

moesta L. *moestus,* sad,sorrowful; ref funereal plumage.

Moho Hawaiian native name *'o 'o* for these beautiful honeyeaters (cf. *moho* for the extinct rail *Pennula).*

Mohoua Maori name *mohua* for these New Zealand warblers.

molinae After the Abate Giovanni Ignazio Molina (1740–1829) Chilean naturalist and author (*Saggio sulla storia naturale del Chili*, 1782).

molitor L. *molitor*, a builder; ref. finely-built nest of the Chin-spot Batis *Batis*.

molleri After Adolphe F. Moller (1842–1920) Portuguese botanist and collector on São Tomé, 1887.

mollis L. *mollis*, soft, tender; ref. soft, lax plumage of the Soft-plumaged Petrel *Pterodroma*.

mollissima L. *mollissimus*, very soft (super. of *mollis*, soft); ref. dense, soft textured plumage.

moloneyanus After Sir Cornelius Alfred Moloney (1848–1913) British colonial administrator in Nigeria, 1887–1890.

Molothrus Gr. *molobros*, a glutton or greedy beggar. According to Coues (1882), the name is derived from Gr. *molos*, a quarrel or struggle, and *throsko*, to assault, to rape; ref. parasitic habits of the American cowbirds.

Molpastes Gr. *molpastes*, a minstrel, a singer (*molpazo*, to sing of; *melpo*, to sing).

molucca/moluccensis After the Moluccas or Spice Islands, Indonesia (Malay *Maluku*). An erroneous toponym in the case of *Dendrocopos*, where *malaccensis* was intended (v. *malacca*).

mombassica After the port of Mombasa in Kenya.

momota see **Momotus.**

momotula Dim. of genus *Momotus*, motmot.

Momotus Mod. corruption *momot* of supposed Mexican name *motmot* for the Neotropical motmots

monacha Mod. L. *monachus*, hooded (i.e. like a monk; Gr. *monakhos*, a monk; *monos*, alone).

Monachella Dim. of syn. genus *Monachus*, monarch flycatcher (error for *Monarchus*).

monachus see **monacha.**

Monarcha L. *monarcha*, a monarch (Gr. *monarkhes*, a monarch).

Monasa Gr. *monases*, a solitary, a monk (*monas*, alone, solitary); ref. plain plumage and quiet behaviour of the black nunbirds.

mondetoura After Pauline Mondétour (fl. 1856) widow of French zoologist Étienne Geoffroy Saint-Hilaire.

monedula L. *monedula*, the jackdaw.

moneduloides Specific name *Corvus monedula;* Gr. *-oides*, resembling.

mongolica/mongolus After Mongolia, a large area of central Asia formerly part of the Chinese empire, and now divided into Mongolia proper, or Outer Mongolia, and Inner Mongolia, or Nei Mongol Zizhiqu, a province of China.

Monias French name *monias* given to the Monias *M. benschi* by Oustalet and Grandidier (Gr. *monias*, solitary).

monileger/moniliger L. *monile*, a necklace or collar; *-ger*, carrying.

monocerata Gr. *monos*, single, only; *keras, keratos*, a horn; the Rhinoceros Auklet *Cerorhinca* has a distinctive horn at the base of the upper mandible during the breeding season.

monogrammicus Gr. *monos*, single, alone; *grammikos*, lined, linear; ref. black median throat stripe of the Lizard Buzzard *Kaupifalco*.

monorhis Gr. *monos*, single, alone; *rhis*, the nostril; ref. united tubular nostrils of Swinhoe's Storm Petrel *Oceanodroma*.

monorthonyx Gr. *monos*, single, alone; *orthos*, straight; *onux*, a claw; ref. straight claw on inner toe of the Snow Mountain Quail *Anurophasis*.

montagnii After Jean François Camille Montagne (1784–1866) French botanist.

montana L. *montanus*, a mountaineer, of the mountain (*mons, montis*, a mountain).

montanella Dim. of L. *montanus*, a mountaineer.

montani After J. Montano (fl. 1881) French anthropologist and author (*Voyages aux Philippines et en Malaise*, 1886).

montanus see **montana.**

monteiri After Joachim John Monteiro (d. 1878) collector in Angola, 1860–1875, and author (*Angola and the River Congo*, 1875).

montezuma /montezumae After Montezuma or Moctezuma Xocoyotzin (1480–1520) Emperor of the Aztecs, or people of the Mexica, who died of wounds inflicted by his own subjects, enraged at his support for the Spanish conquistadores under Cortes.

Monticola /monticolum / monticolus L. *monticola,* a highlander, a mountain dweller (*mons, montis,* a mountain; *-cola,* a dweller).

Montifringilla L. *mons, montis,* a mountain; *fringilla,* a finch.

montis L. *montis,* of a mountain.

montium L. *montium,* of mountains.

montivagus L. *montivagus,* mountain roaming, wandering over the mountains.

moquini After Horace Benedict Alfred Moquin-Tandon (1804–1863) French collector, ornithologist, and author (*Ornithologie Canarienne*, 1842).

moreaui After Reginald E. Moreau (1897–1970) British ornithologist, collector, and author (*The bird faunas of Africa and its islands*, 1966; *The Palaearctic-African bird migration systems*, 1972).

moreirae After C. Moreira (1869–1946) Brazilian zoologist.

morelleti After P. M. A. Morellet (1809–1892) Spanish zoologist and collector in the Canary Islands and Guatemala.

morenoi After Francisco Josue Pascasio Moreno (1852–1919) Argentine naturalist and founding director of La Plata Museum, 1884.

morinellus Dim. from Gr. *moros,* foolish, silly; of the Mountain Dotterel *Eudromias* Ray (1678), quoting Dr Key to Gesner, says, 'I call it Morinellus for two reasons, first, because it is frequent among the Morini (*Flemmings) and, next, because it is a foolish bird, even to a proverb, we calling a foolish dull person a Dotterel'.

morio L. *morio* (contraction of *mormorion),* a dark brown stone; ref. the sooty-brown plumage of the Brown Jay *Psilorhinus,* the dark cinnamon underparts of the female Sulawesi Cicadabird *Coracina,* and the chestnut primaries of the Red-wing Starling *Onychognathus* (cf. *Maurus,* Moorish (i.e. black)).

Morococcyx Gr. *moros,* silly, foolish; Mod. L. *coccyx,* cuckoo (Gr. *kokkux);* ref. clownish face pattern of the Lesser Ground Cuckoo *M. erythropygus.*

morphnoides Genus *Morphnus,* eagle; *-oides*, resembling.

Morphnus Gr. *morphnos,* a kind of eagle or vulture.

morphoeus Gr. myth. Morpheus, son of Sleep and god of dreams; ref. drowsy, lethargic behaviour of the White-fronted Nunbird *Monasa.*

morrisonia /morrisoniana / morrisonianus After Mount Morrison (Hsin Kao Shan), Taiwan.

mortieri After Bartholomé Charles Comte Dumortier (b. 1797) Belgian botanist and politician.

Morus Gr. *moros,* silly, foolish; ref the tame and confiding behaviour of gannets and boobies, allowing themselves to be easily caught for food or sport.

moschata Late L. *moschatus,* musky; the drake Muscovy Duck *Cairina,* formerly Musk Duck, has a musky smell.

mosquera After General Tomás Cipriano de Mosquera (1798–1878) Dictator of New Granada, 1845–1849; President of Granadine Confederation, 1862–1863; President of Colombia, 1863–1864, 1866–1867 (v. *granadense).*

mosquitus Spanish *mosquito,* a little fly, a mosquito (dim. of *mosca,* a fly); ref. small size and buzzing flight of the Ruby-topaz Hummingbird *Chrysolampis.*

Motacilla L. *motacilla,* a wagtail (Gr. *muttex,* a kind of bird mentioned by Hesychius). The mistaken use of *-cilla* for 'tail' in orn. goes back to mediaeval writers who misread *motacilla,* Varro's name for the wagtail and a dim. from L. *motare,* to move about or shake, as 'shaketail'.

motacilloides Genus *Motacilla,* wagtail; Gr. *-oides,* resembling.

motitensis After the settlement of Motita, near Old Latakoo, 135 miles north of the Orange River, South Africa, although according to Smith (1848), *Illustr. zool. south Africa, aves,* the original specimens were collected about sixty miles to the south of the Orange River.

motmot see **Momotus.**

mouki No expl. (Mathews (1912), *Novit. Zool.,* **18**, 310) or in a subsequent subspecific use; probably an Australian aboriginal name.

Moupinia After the town of Muping, Chinese Tibet (Paohing, Hsikang, China).

mouroniensis After the town of Moroni, Grand Comoro Island, Comoro Islands.

moussieri After Jean Moussier (b. 1795) French military surgeon and naturalist.

mozambicus After Mozambique or Moçambique, formerly Portuguese East Africa.

muelleri After **1** Salomon Müller (1804–1864) German ornithologist, collector, and author (*Ueber den charakter der thierwelt auf den inseln des Indischen archipels,* 1846; *Reizen en onderzoekingen in den Indischen archipel,* 1857) (*Heleia).* **2** Baron Johann Wilhelm von Müller (1824–1866) German naturalist and collector in tropical Africa, North America, and Mexico (*Merops).* **3** Baron Sir Ferdinand von Müller (1825–1896) German/Australian botanist and explorer (*Rallus).* **4** Lorenz Müller (1868–1953) German herpetologist and collector in Brazil (*Cranioleuca).*

muelleriana see **muelleri 1**.

mufumbiri Erroneous toponym after the Mufumbiri volcanoes (Mfumbiro or Virunga volcanoes), Zaïre/Rwanda border.

mugimaki Japanese name *mugimaki* for the Mugimaki Flycatcher *Ficedula.*

mulleri /mullerii After **1** Salomon Müller (1804–1864)(*Ducula)* (v. *muelleri* **1**) and his wife **2** Susan Müller (fl. 1841) (*Tanygnathus).*

Mulleripicus After Salomon Müller (v. *muelleri* **1**); genus *Picus,* woodpecker.

mulsant After Martial Étienne Mulsant (1797–1880) French naturalist, collector, and author (*Lettres à Julie sur l'ornithologie,* 1868; *Histoire naturelle des oiseaux-mouches ou colibris,* 1874).

multicolor L. *multicolor,* multi-coloured, many coloured (*multus,* many; *color,* colour).

multistriata /multistriatus L. *multi-,* many (*multus,* many, much); Mod. L. *striatus,* striped, streaked (L. *stria,* a furrow).

munda /mundus L. *mundus,* elegant, neat.

munroi After George C. Munro (1866–1963) US ornithologist, collector, and author (*Birds of Hawaii,* 1960).

mupinensis see **Moupinia.**

muraria Med. L. *murarius,* of walls (L. *murus,* a wall); in winter the Wallcreeper *Tichodroma* leaves its montane breeding haunts and occurs on the walls of castles, cathedrals, and similar man-made habitats.

murina /murinus Mod. L. *murinus,* mouse-grey, mouse-coloured (L. *murinus,* of mice, murine; *mus, muris,* a mouse).

murivora L. *mus,* a mouse; *-vorus,* eating.

murphyi After Robert Cushman Murphy (1887–1973) US ornithologist and author (*Bird islands of Peru,* 1925; *Oceanic birds of South America,* 1936).

Muscicapa Mod. L. *muscicapa,* a flycatcher (L. *musca,* a fly; *capere,* to seize).

Muscicapella Dim. of genus *Muscicapa,* flycatcher.

muscicapinus Genus *Muscicapa,* flycatcher; L. *-inus,* resembling.

Muscicapula Dim. of genus *Muscicapa,* flycatcher.

Muscigralla Genus *Muscicapa*, flycatcher; L. *grallae*, stilts; ref. long legs of the terrestrial Short-tailed Field Tyrant *M. brevicauda*.

Muscipipra Genus *Muscicapa*, flycatcher; genus *Pipra*, manakin.

Muscisaxicola Genus *Muscicapa*, flycatcher; genus *Saxicola*, chat (formerly a generic name for the wheatears *Oenanthe* and so used here).

Muscivora L. *musca*, a fly; *-vorus*, eating.

musculus L. *musculus*, a little mouse; ref. small size and grey coloration of the Mouse-coloured Penduline Tit *Anthoscopus*.

musica /musicus L. *musicus*, musical, a musician (Gr. *mousikos*, a musician; *mousa*, music).

Musophaga Botanical genus *Musa*, plantain, banana (Arabic *mauz*, banana); Gr. *phagos*, a glutton, an eater (*phagein*, to eat).

musschenbroekii After Samuel Cornelius Jan Willem van Musschenbroek (1827–1883) Dutch colonial administrator in the East Indies.

mustelina L. *mustelinus*, like a weasel (*mustela*, a weasel); ref. rich chestnut and white coloration of the Red-and-white Spinetail *Certhiaxis* and the Wood Thrush *Hylocichla*.

mutata L. *mutatus*, a change, an alteration; ref. polymorphic colour phases of the Madagascar Paradise Flycatcher *Terpsiphone*.

muticus L. *muticus*, curtailed, docked (*mutilus*, mutilated); from apparent silence of the Green Peafowl *Pavo*

muttui Malayalam (Indian) native name *muttupilla* for the Brown-breasted Flycatcher *Muscicapa*.

mutus L. *mutus*, silent, dumb.

Myadestes Gr. *mua*, a fly; *edestes*, an eater (*edo*, to eat).

myadestinus Genus *Myadestes*, solitaire; L. *-inus*, resembling; the Hawaiian thrushes *Phaeornis* are now considered allied to the American solitaires.

Mycerobas Gr. *mukter*, the snout (i.e. the bill); *keroboas*, sounding with horn; ref. loud whistling calls of the massive-billed Spot-winged Grosbeak *M. melanozanthos*.

Mycteria Gr. *mukter*, the snout or trunk (*mukterizo*, to turn up the nose); L. *-ius*, resembling, connection; ref. long, heavy bills of the American Wood Stork *M. americana* and the Jabiru *Jabiru*.

Myiagra Gr. myth. Myiagros, the flycatcher, an Elean god invoked against flies (*muia*, a fly; *agreo*, to seize).

Myiarchus Gr. *muia*, a fly; *arkhos*, a lord or prince.

Myiobius Gr. *muia*, a fly; *bios*, living, livelihood (i.e. eating).

Myioborus Gr. *muia*, a fly; *boros*, devouring (*bora*, food).

Myioceyx Gr. *muia*, a fly (i.e. small); genus *Ceyx*, kingfisher.

Myiodynastes Gr. *muia*, a fly; *dunastes*, a lord or ruler.

Myiopagis Gr. *muia*, a fly; *pagis*, a snare or trap (*pegnumi*, to make firm).

Myioparus Gr. *muia*, a fly; genus *Parus*, tit.

Myiophobus Gr. *muia*, a fly; *phobos*, a terror (*phebomai*, to flee).

Myiophoneus Gr. *muia*, a fly; *phoneus*, a murderer (*phonos*, murder) (cf. syn. genus *Phoneus*, shrike).

Myiopsitta Gr. *mus*, *muos*, a mouse; Mod. L. *psitta*, a parrot (Gr. *psittake* or *psittakos*, a parrot); ref. mouse-grey face and breast of the Monk Parakeet *M. monachus*.

Myiornis Gr. *muia*, a fly; *ornis*, a bird; the Eared Pygmy Tyrant *M. auricularis* is one of the smallest tyrant flycatchers.

Myiotheretes Gr. *muia*, a fly; *therates*, a hunter (*therao*, to hunt) (cf. *muiotheras*, a fly-hunter).

Myiotriccus Gr. *muia*, a fly; Mod. L. *triccus*, tyrant flycatcher.

Myiozetetes Gr. *muia*, a fly; *zetetes*, a searcher, an inquisitor (*zeteo*, to seek).

Myopornis Gr. *muops*, a horsefly; *ornis*, a bird.

myoptilus Gr. *mus, muos,* a mouse (i.e. grey-coloured); *ptilon,* plumage, feathers.

Myornis Gr. *mus,* a mouse; *ornis,* a bird; ref. mouse-grey plumage and furtive, scuttling habits of the Ash-coloured Tapaculo *M. senilis.*

Myospiza Gr. *mus,* a mouse; *spiza,* a finch; ref mousey, grey-brown plumage of the Grassland Sparrow *M. humeralis.*

myotherinus Gr. *muiotheras,* a fly hunter; L. *-inus,* resembling.

myristicivora Botanical genus *Myristica,* nutmeg (Gr. *murizein,* to anoint); L. *vorus,* eating.

Myrmeciza Gr. *murmex, murmekos,* the ant; *izein,* to ambush (*izo,* to sit still).

Myrmecocichla Gr. *murmex,* the ant; *kikhle,* thrush.

myrmecophoneus Gr. *murmex,* the ant; *phoneus,* a killer.

Myrmia No expl. (Mulsant (1876), *Ann. Soc. Linn. Lyon,* **22** [Catalogue des Oiseaux-Mouches ou Colibris], 228); doubtless ref. the small size and active behaviour of the Short-tailed Woodstar *M. micrura* (Gr. *murmex,* the ant; *mua,* a fly; *muron,* anything sweet or charming) (cf. Gr. myth. Myrina, a queen of the Amazons; Myrmidones, an industrious and ancient people of Greece).

Myrmoborus Gr. *murmex,* the ant; *boros,* devouring (*bora,* food).

Myrmochanes Gr. *murmex,* the ant; *khaino,* to gape greedily.

Myrmoderus Gr. *murmex,* the ant; *dero,* to batter or cudgel.

Myrmophylax Gr. *murmex,* the ant; *phulax,* a watcher, a governor.

Myrmorchilus Gr. *murmex,* the ant; Mod. L. *orchilus,* wren (v. *Hylorchilus*); the Stripe-backed Antbird *M. strigilatus* is related to the antwrens.

Myrmornis Gr. *murmex,* the ant; *ornis,* a bird.

Myrmothera Gr. *murmex,* the ant; *-thera,* a hunter, hunting.

Myrmotherula Dim. of genus *Myrmothera,* antpitta.

Myrtis After Myrtis, a Greek poetess of the fourth century BC.

mysolensis After the island of Mysol or Misol (Misool), Irian Jaya (Indonesian New Guinea).

mysorensis After the island of Mysore or Misori (Biak), Geelvink Bay, Irian Jaya.

mystacalis Mod. L. *mystacalis,* moustached (Gr. *mustax, mustakos,* a moustache; *mastax,* the jaws; L. *-alis,* pertaining to).

mystacea/mystaceus Mod. L. *mystaceus,* moustached (Gr. *mustax,* a moustache).

mystacophanos Gr. *mustax,* a moustache; *phanos,* bright (*phaino,* to shine); the Red-throated Barbet *Megalaima* has bright yellow moustaches.

Mystacornis Gr. *mustax,* a moustache; *ornis,* a bird; the Masked Babbler *M. crossleyi* has white moustachial streaks.

Myza Gr. *muzo,* to suck; ref. nectar-feeding habits of the Sulawesi streaked honeyeaters.

Myzomela Gr. *muzo,* to suck; *meli,* honey.

Myzornis Gr. *muzo,* to suck; *ornis,* a bird; the Fire-tailed Myzornis *M. pyrrhoura* is an aberrant babbler that feeds on nectar and minute insects.

Megapodius freycinet Dusky Scrubfowl

N

nabouroup Namaqua (South African) native name *nabouroup* for the Pale-winged Starling *Onychognathus*. This is one of the many names coined originally in Levaillant (1799–1808). The naturalist François Levaillant was one of the French school that included George Comte de Buffon and rejected the Linnaean binominal system. Unlike the Count, however, Levaillant was an adventurer and explorer who described his birds at first hand and not from cabinet specimens, although he was not above fabricating new species to sell his work! In his books he gave his discoveries coined French names or names selected from the local native vernacular, and it was left to later authors to perpetuate them in binominal nomenclature.

nacunda Güaraní (Paraguayan) Indian name *ñacundá,* big-mouth, for the Nacunda Nighthawk *Podager*.

naevia L. *naevius,* spotted (*naevus,* a spot or mole).

naevioides Specific name *Hylophylax naevia;* Gr. *-oides,* resembling.

naevius see **naevia.**

naevosa L. *naevus,* a spot; *-osus,* abundance.

nagaensis After the Naga Hills, Kachin, Burma (Myanmar).

nahani After Commandant P. F. Nahan (1867–1930) Belgian explorer in tropical Africa.

naina Misspelling of *nana* (q.v.).

nais Gr. myth. Nais, a naiad, a water- or river-nymph.

namaqua /namaquus After Namaqualand, a coastal region of western Cape Province, South Africa, and Namibia divided by the Orange River.

Namibornis After the Namib Desert, Namibia (from the aboriginal Nama Hottentots); Gr. *ornis,* a bird; the Herero Chat *N. herero* is confined to the arid fringes of the Namib Desert.

nana L. *nanus,* a dwarf (Gr. *nanos,* a dwarf).

Nandayus Güaraní (Paraguayan) Indian name *ñendáy* for the Black-hooded Conure *N. nenday*.

Nannopsittaca Gr. *nanos,* a dwarf; *psittakos,* a parrot.

Nannopterum Gr. *nanos,* a dwarf; *pteron,* a feather, a wing; ref. vestigial flight-feathers of the Galapagos Flightless Cormorant *N. harrisi*.

nanum /nanus see **nana.**

napensis After the Río Napo, Loreto, Peru.

Napothera Gr. *napos,* a woody dell, a glen; *-thera,* a hunter; ref. forest habitat of the Indomalayan wren-babblers.

narcissina L. *narcissinus,* narcissine (i.e. bright yellow)(Gr. *narkissinos,* narcissine; *narkissos,* the narcissus; *narkao,* to become numb or narcotic).

narcondami After Narcondam Island, Andaman Islands, Bay of Bengal.

narethae After the town of Naretha, Western Australia.

narina Hottentot (South African) name *narina,* flower, given to the Narina Trogon *Apaloderma* by Levaillant (1806) because of its colourful plumage. Lesson

(1830), *Traité d'ornithologie*, subsequently claimed that Levaillant had named the trogon after a Hottentot servant girl.

narokensis After Mount Narok (Ol Doinyo Orok), Rift Valley, Kenya.

Nasica L. *nasica*, large-nosed (i.e. long-billed) (*nasus*, the nose).

nasutus L. *nasutus*, having a large nose (i.e. long-billed or large-billed) (*nasus*, the nose).

natalensis After the settlement of Port Natal (Durban), Natal, South Africa. The area was discovered by Vasco da Gama on Christmas Day, 1497, and named by him *Terra Natalis*, Land of the Holy Birth (L. *natalis*, birthday).

natalis L. *natalis*, birthday (*nasci*, to be born); after Christmas Island, Indian Ocean, named on Christmas Day, 1643, by Captain William Mynors of the East India Company.

nationi After W. Nation (1826–1907) British botanist and collector in Peru, 1849.

nativitatis L. *nativitas*, birth; after Christmas Island (Kiritimati Atoll), Line Islands, Pacific Ocean, discovered by Captain Cook on Christmas Eve, 1777.

nattereri /nattererii After Johann Natterer (1787–1843) Austrian zoologist and collector, resident in Brazil, 1817–1835.

natunensis After the Natuna Islands, west of Borneo, Indonesia.

naumanni After **1** Johann Andreas Naumann (1744–1826) German farmer, naturalist, and author (*Naturgeschichte der vögel Deutschlands*, 1804) (*Turdus*) and his son **2** Johann Friedrich Naumann (1780–1857) artist, ornithologist, and author (*Die eier der vögel Deutschlands*, 1818) (*Falco*).

ndussumuensis After the Ndussuma country, Kivu, Zaïre.

Neafrapus Gr. *neo-*, new, different (*neos*, new); L. *Afer*, African; genus *Apus*, swift.

nebouxii After Adolphe Simon Néboux (fl. 1840) French surgeon, naturalist, and explorer.

nebularia L. *nebula*, a mist, a cloud; *-arius*, possessing; ref. cloudy grey upperparts of the Greenshank *Tringa*.

nebulosa /nebulosus L. *nebulosus*, misty, foggy; ref. grey mottled or clouded plumage.

necopinata L. *necopinatus*, unexpected.

necopinus L. *necopinus*, unexpected.

Necropsar Gr. *nekros*, dead (*nekus*, a corpse) (i.e. extinct); *psar*, the starling; the Rodrigues Starling *N. rodericanus* has not been recorded since the early 1700s.

Necropsittacus Gr. *nekros*, dead; *psittakos*, a parrot; the extinct Rodrigues Parrot *N. rodericanus* was described from bones and contemporary accounts.

Necrosyrtes Gr. *nekros*, a corpse; *suro*, to drag; ref. the scavenging behaviour of the Hooded Vulture *N. monachus*.

Nectarinia L. *nectar, nectaris*, nectar, honey; *-ius*, characteristic of; the sunbirds feed on nectar and minute insects.

nectarinioides Genus *Nectarinia*, sunbird; Gr. *-oides*, resembling; the Black-bellied Sunbird *Nectarinia* was originally described in the genus *Cinnyris*.

neergardi After P. Neergard (fl. 1907) South African laboratory assistant in Moçambique.

neglecta /neglectus L. *neglectus*, neglected, ignored (i.e. recently described).

nehrkorni After Adolf Nehrkorn (1841–1916) German oologist and collector.

Neisna Gr. myth. Neis (a form of Nais), a naiad or river-nymph.

nelicourvi No expl. (Scopoli (1786), *Deliciae florae et faunae insubricae*, fasc. 2, p. 96); doubtless from the Tamil (Sri Lanka) name *nellukuruvi* for a finch or waxbill, the Madagascan Nelicourvi Weaver *Ploceus* mistakenly thought to have come from Indomalaya.

Nelicurvius Specific name *Ploceus nelicourvi*.

nelsoni After Edward William Nelson (1855–1934) US ornithologist and founding President of the American Ornithologists' Union.

nematura Gr. *nema, nematos,* a thread; *-ouros,* tailed; the tail feathers of the Sharp-tailed Streamcreeper *Lochmias* are soft with thin, filamentous tips.

nemoricola L. *nemus, nemoris,* a forest, an open wood; *-cola,* inhabitant.

Nemosia French name *némosie* given to the Hooded Tanager *N. pileata* by Vieillot (1816), *Analyse,* p. 32 (Gr. *nemos,* a grove; ref. habitat).

nenday see **Nandayus.**

Neneba After the village of Neneba, Mount Scratchley, Papua New Guinea.

nengeta Tupí (Brazilian) Indian name *guiraru nheengeta* for a thrush-like bird.

Neochelidon Gr. *neo-,* new, different (*neos,* new, strange; in orn. used to indicate similarity or relationship); *khelidon,* a swallow.

Neochen Gr. *neo-,* new, different; *khen,* a goose.

Neochloe Gr. *neo-,* new; *khloe,* green (ref. genus *Vireo).*

Neochmia Gr. *neokhmos,* novel, unusual.

Neocichla Gr. *neo-,* new, different; *kikhle,* a thrush.

Neocossyphus Gr. *neo-,* new, different; genus *Cossypha,* robin-chat.

Neocrex Gr. *neo-,* new, different; genus *Crex,* crake.

Neoctantes Gr. *neo-,* new, different; *ktantes,* a murderer (i.e. a shrike); ref. heavy shrike-like bill of the Black Bushbird *N. niger.*

Neodrepanis Gr. *neo-,* new, different; genus *Drepanis,* mamo; ref. sickle-shaped bill of the sunbird asities.

Neolalage Gr. *neo-,* new, different; genus *Lalage,* triller.

Neolesbia Gr. *neo-,* new, different; genus *Lesbia,* trainbearer hummingbird.

Neolestes Gr. *neo-,* new, different; *lestes,* a robber (i.e. a shrike); the Black-collared Bulbul *N. torquatus* was formerly believed allied to the shrikes.

Neomixis Gr. *neo-,* new, different; syn. genus *Mixornis,* tit-babbler (Gr. *mixo-,* mixed, mingling; *ornis,* a bird).

Neomorpha /Neomorphus Gr. *neo-,* new, different; *morphe,* a form, kind, or sort; bringing attention to the large size, terrestrial habits, and non-parasitic behaviour of the Rufous-vented Ground Cuckoo *N. geoffroyi,* and the very different bill-shapes of the male and female Huia *Heteralocha acutirostris.*

Neoparadisea Gr. *neo-,* new, different; genus *Paradisaea,* bird of paradise.

Neopelma Gr. *neo-,* new, different; *pelma,* the sole of the foot; ref. the basally fused toes of the tyrant manakins.

Neophema Gr. *neo-,* new, different; *pheme,* a report (*phemi,* to speak).

Neophron Gr. myth. Neophron, a trickster metamorphosed into a vulture.

Neopipo Gr. *neo-,* new, different; Mod. L. *pipo,* manakin.

Neopsittacus Gr. *neo-,* new, different; *psittakos,* a parrot.

Neorhynchus Gr. *neo-,* new, different; *rhunkhos,* the bill; ref. thick, curved yellow bill of the Parrot-billed Seedeater *Sporophila peruviana.*

Neositta Gr. *neo-,* new, different; genus *Sitta,* nuthatch.

Neospiza Gr. *neo-,* new, different; *spiza,* a finch.

Neothraupis Gr. *neo-,* new, different; genus *Thraupis,* tanager.

Neotis Gr. *neo-,* new, different; genus *Otis,* bustard.

neoxena /neoxenus Gr. *neo-,* new, different; *xenos,* stranger.

Neoxolmis Gr. *neo-,* new, different; genus *Xolmis,* monjita.

Nephelornis Gr. *nephele,* a cloud; *ornis,* a bird; ref. cloud-forest habitat of the Pardusco *N. oneilli.*

Nephoecetes Gr. *nephos,* a cloud; *oiketes,* a dweller (*oikeo,* to inhabit).

nereis L. *nereis,* a nereid, a sea-nymph.

Nesasio Gr. *nesos,* an island; genus *Asio,* owl; the Fearful Owl *N. solomonensis* is confined to the Solomon Islands.

Nesierax Gr. *nesos*, an island; *hierax*, a hawk or falcon; the New Zealand Falcon *Falco novaezeelandiae* is confined to that country.

Nesillas Gr. *nesos*, an island; *illas* (a form of *ilias*), a thrush.

nesiotis Gr. *nesiotes*, an islander; the extinct Tristan Moorhen *Porphyriornis* formerly inhabited Tristan da Cunha Island.

Nesoceleus Gr. *nesos*, an island; genus *Celeus*, woodpecker; Fernandina's Flicker *Colaptes fernandinae* is found only on Cuba.

Nesocharis Gr. *nesos*, an island; *kharis*, grace, loveliness; the Mountain Oliveback *N. shelleyi* was described from the island of Fernando Póo (Bioko).

Nesocichla Gr. *nesos*, an island; *kikhle*, a thrush.

Nesoclopeus Gr. *nesos*, an island; *klopeus*, a thief; no expl. (Peters (1932), *Auk*, **49**, 348). Perhaps a general allusion to the Pacific islands being havens for pirates (v. *Cleptornis*), or a ref. to the removal of the Bar-winged Rail *N. poecilopterus* from the genus *Eulabeornis*.

Nesoctites Gr. *nesos*, an island; *ktites*, an inhabitant (*ktizo*, to occupy); the Antillean Piculet *N. micromegas* is found only on the island of Hispaniola.

Nesoenas Gr. *nesos*, an island; *oinas*, a pigeon; the Pink Pigeon *Columba mayeri* is confined to Mauritius.

Nesofregetta Gr. *nesos*, an island; genus *Fregetta*, storm-petrel; the polymorphic White-throated Storm Petrel *N. fuliginosa* breeds on islands in the tropical western Pacific.

Nesomimus Gr. *nesos*, an island; genus *Mimus*, mockingbird; the mockingbirds of this superspecies are confined to the Galapagos Islands.

Nesopelia Gr. *nesos*, an island; *peleia*, a dove or pigeon; the Galapagos Dove *Zenaida galapagoensis* is confined to the Galapagos Islands.

Nesophylax Gr. *nesos*, an island; *phulax*, a watcher, a guard; the Sooty Crake *N. ater* is found only on Henderson Island in the Pacific.

Nesopsar Gr. *nesos*, an island; *psar*, the starling (v. *Gnorimopsar*); the Jamaican Blackbird *N. nigerrimus* is confined to Jamaica.

Nesospingus Gr. *nesos*, an island; *spingos*, a finch; the Puerto Rican Tanager *N. speculiferus* is confined to Puerto Rico.

Nesospiza Gr. *nesos*, an island; *spiza*, a finch; the Tristan Bunting *N. acunhae* and Wilkins' Bunting *N. wilkinsi* inhabit the Tristan da Cunha group.

Nesotriccus Gr. *nesos*, an island; Mod. L. *triccus*, tyrant flycatcher; the Cocos Flycatcher *N. ridgwayi* is found only on Isla del Coco in the Pacific Ocean south of Costa Rica.

Nestor Gr. myth. Nestor, wise old king of Pylos at the siege of Troy, formerly a great warrior and universally respected for his wisdom; ref. the grey, hoary head of the Kaka *N. meridonalis*.

Netta Gr. *netta*, a duck.

Nettapus Gr. *netta*, a duck; *ops*, *opos*, a voice (cf. *-ops*, appearance; *pous*, a foot); ref. the quacking and squeaky voices of the Palaeotropical pygmy geese.

Nettion Gr. *nettion*, a small duck (dim. of *netta*, a duck).

neumanni After Oskar R. Neumann (1867–1946) German ornithologist and collector in tropical Africa, 1892–1894.

neumayer After Franz Neumayer (d. 1840) Austrian botanist and collector in Dalmatia, Jugoslavia.

nevermanni After Hans Nevermann (fl. 1950) German ethnologist in New Guinea, 1932.

newelli After Brother Matthias Newell (1854–1939) missionary to Hawaii, 1886–1924.

newtoni After **1** Alfred Newton (1829–1907) British ornithologist and author (*Ootheca Wooleyana*, 1864; *A dictionary of birds*, 1893) (*Acrocephalus*, *Coracina*, *Falco*, *Fulica*). **2** Colonel Francesco Newton (1864–1909) Portuguese botanist and collector on São Tomé, 1888; Fernando Póo (Bioko), 1894; and Timor, 1896 (*Lanius, Nectarinia*).

Newtonia/newtoniana see **newtoni 1.**

newtonii see **newtoni.**

niansae After Victoria Nyanza (Lake Victoria), East Africa.

nicaraguensis After Nicaragua, named for Nicarao, an Indian chief whose people lived on the shores of Lake Cocibolca (Lake Nicaragua).

Nicator Gr. *nikator,* a conqueror (i.e. a shrike).

nicefori/niceforoi After Brother Nicéforo Maria (b. 1880) Colombian missionary and herpetologist.

nicobarica/nicobaricus/nicobariensis After the Nicobar islands, Bay of Bengal.

nicolli After Michael John Nicoll (fl. 1924) who helped Reginald Moreau (v. *moreaui)* during his long residence in Egypt.

nidipendulum L. *nidus, nidi,* a nest; *pendulus,* hanging (*pendere,* to hang down, be suspended); the globular nest of the Hangnest Tody Tyrant *Idioptilon* is suspended from a twig.

nieuwenhuisii After Anton Willem Nieuwenhuis (b. 1864) Dutch ethnologist, collector in Borneo, 1893–1900, and author (*In centraal Borneo*, 1900).

niger L. *niger,* black, shining black (cf. *ater,* matt black, dead black).

nigerrimus L. *nigerrimus,* very black (super. of *niger,* black).

nigra see **niger.**

nigrescens L. *nigrescens,* blackish (*nigrescere,* to become black).

nigricans L. *nigricans,* blackish, swarthy.

nigricapillus L. *niger,* black; *-capillus,* capped.

nigricauda L. *niger,* black; *cauda,* the tail.

nigriceps L. *niger,* black; *-ceps,* headed.

nigricincta L. *niger,* black; *cinctus,* girdled, banded (*cingere,* to encircle).

nigricollis L. *niger,* black; Mod. L. *-collis,* necked.

nigrifrons L. *niger,* black; *frons,* the forehead, the brow.

nigrigenis L. *niger,* black; *genis,* cheeks (*gena,* the cheek).

nigrilore/nigriloris L. *niger,* black; Mod. L. *loris,* lored, bridled (L. *lorum,* a strap of leather).

nigrimenta/nigrimentum L. *niger,* black; *mentum,* the chin.

nigripectus L. *niger,* black; *pectus,* the breast.

nigripennis L. *niger,* black; *pennis,* winged (*penna,* a feather).

nigripes L. *niger,* black; *pes,* the foot.

nigrirostris L. *niger,* black; *-rostris,* billed.

nigriscapularis L. *niger,* black; Mod. L. *scapularis,* scapular, of the shoulders (L. *scapulae,* shoulders).

Nigrita L. *nigritia,* black; this is a replacement name for the preoccupied genus *Aethiops* (L. *Aethiops,* an Ethiopian, a Negro) (cf. Mod. L. *nigritus,* black).

nigriventris L. *niger,* black; *venter, ventris,* the belly.

nigrivestis L. *niger,* black; *vestis,* a covering or garment, clothing.

nigrobrunnea L. *niger,* black; Mod. L. *brunneus,* brown.

nigrocapillus see **nigricapillus.**

nigrocapitata L. *niger,* black; *capitatus,* headed (*caput,* the head).

nigrocincta see **nigricincta.**

nigrocinereus L. *niger,* black; *cinereus,* ash-coloured, grey (*cinis, cineris,* ashes).

nigrocinnamomea L. *niger,* black; Mod. L. *cinnamomeus,* cinnamon-coloured (L. *cinnamum,* cinnamon).

nigrocristatus L. *niger,* black; *cristatus,* crested (*crista,* a crest).

nigrocyanea L. *niger,* black; *cyaneus,* dark blue (Gr. *kuaneos,* dark blue).

nigrofumosus L. *niger,* black; *fumosus,* smokey, smoked (*fumus,* smoke).

nigrogenis see **nigrigenis.**

nigrogularis L. *niger,* black; Mod. L. *gularis,* throated, of the throat (L. *gula,* the throat).

nigrolineata L. *niger,* black; *lineatus,* lined, marked with lines (*linea,* a line).

nigrolutea L. *niger,* black; *luteus,* saffron-yellow.

nigromaculata L. *niger,* black; *maculatus,* spotted (*maculare,* to spot, to mark).

nigromitratus L. *niger,* black; *mitratus,* crowned, wreathed (*mitra,* a head-dress).

nigropectus see **nigripectus.**

nigropunctatus L. *niger,* black; Mod. L. *punctatus,* spotted (L. *punctum,* a spot).

nigrorufa /nigrorufus L. *niger,* black; *rufus,* red, ruddy.

nigrorum Mod. L. *Nigrorum,* of Negros, Philippines (L. *niger,* black; Spanish *negro,* black).

nigrotectus L. *niger,* black; *tectus,* covered (*tegere,* to cover).

nigroventris see **nigriventris.**

nigroviridis L. *niger,* black; *viridis,* green.

Nilaus Anagram of genus *Lanius,* shrike, coined to indicate relationship.

nilghiriensis After the Nilghiri or Nilgiri Hills, Tamil Nadu, India.

niloticus L. *Niloticus,* of the River Nile (*Nilus*).

Niltava Nepalese name *niltau* for the Rufous-bellied Niltava *N. sundara.*

Ninox Syn. genus *Nisus,* sparrowhawk; syn. genus *Noctua,* owl (L. *noctua,* an owl; *nox, noctis,* the night); of the Brown Hawk Owl *N. scutulata,* Hodgson (1837), *Madras Journ. Lit. Sci.,* **5**, 23, wrote, 'The experienced Shikaree who brought me, recently, a fine male specimen of this bird, asked me, when he put it in my hand, whether it was a *Baaz* (hawk) or an *Ulu* (owl)?...It is an owl, undoubtedly, but the most accipitrine of owls...'

nipalensis After Nepal, widely written Nipal in former times (Nepalese *ni pal,* holy land).

nippon /Nipponia Japanese *Nippon* or *Nihon,* Japan (*nishi,* the sun; *pon,* the source; the land of the rising sun).

nisoria Med. L. *nisus,* the sparrowhawk; L. *-orius,* belonging to; ref. barred and grey plumage, and glaring yellow eye of the Barred Warbler *Sylvia.*

nisus Med. L. *nisus,* the sparrowhawk.

nitens L. *nitens,* shining, glittering (*nitere,* to shine).

nitida L. *nitidus,* brilliant, glittering (*nitere,* to shine).

nitidissima L. *nitidissimus,* most brilliant (super. of *nitidus,* brilliant).

nitidula L. *nitidulus,* somewhat brilliant (dim. of *nitidus,* brilliant).

nitidum /nitidus see **nitida.**

nivalis L. *nivalis,* snowy, of snow (*nix, nivis,* snow).

nivea L. *niveus,* of snow, snowy, snow-white (*nix,* snow).

niveicapilla L. *niveus,* snow-white; *-capillus,* capped.

niveigularis /niveogularis L. *niveus,* snow-white; Mod. L. *gularis,* throated, of the throat (L. *gula,* the throat).

niveoguttatus L. *niveus,* snow-white; *guttatus,* spotted, speckled (*gutta,* a spot, a drop).

niveoventer L. *niveus,* snow-white; *venter,* the belly.

nivosa /nivosus L. *nivosus,* snowy, abounding in snow (*nix,* snow).

njombe After the town of Njombe, Iringa, Tanzania.

noanamae After the town of Noanamá, Chocó, Colombia.

nobilis L. *nobilis,* admirable, celebrated.

noctis L. *nox, noctis,* the night.

noctitherus L. *nox,* the night; Gr. *-thera,* a hunter.

noctivagus L. *noctivagus,* wandering by night (*nox,* the night; *vagus,* roaming).

noctua L. *noctua,* an owl sacred to the goddess Minerva.

noevia L. *naevius,* spotted (v. *coelebs* ref. the ligature æ).

noguchii After T. Noguchi (fl. 1887) Japanese collector.

Nomonyx Gr. *nomos,* regular, orderly (*nemo,* to control, possess); *onux,* a nail, a claw; the nail on the upper mandible of the Masked Duck *N. dominica,* 'viewed from above, is extremely small, narrow and linear, the broader terminal half being bent very abruptly downward and backward' (Ridgway (1880), *Proc. U S Nat. Mus.*, **3**, 15).

Nonnula Dim. of Med. L. *nonna,* a nun; ref. the demure appearance and quiet habits of the nunlets and other puffbirds.

nordmanni After Alexander von Nordmann (1803–1866) Finnish/Russian naturalist and explorer.

norfolciensis After Norfolk Island in the south-western Pacific Ocean, discovered by Captain Cook in 1774, and named for Edward Howard, 9th Duke of Norfolk (1686–1777).

Northiella Dim. after Alfred John North (1855–1917) Australian jeweller, oologist, ornithologist, and author (*Descriptive catalogue of the nests and eggs of birds found breeding in Australia and Tasmania*, 1889).

notabilis L. *notabilis,* remarkable, striking.

notata/notatus L. *notatus,* marked, spotted (*nota,* a mark).

Notharchus Gr. *nothes,* sluggish, lazy; *arkhos,* chief, prince; ref. quiet, lethargic behaviour of these large puffbirds.

Nothocercus Gr. *nothos,* false, spurious; *kerkos,* the tail; the very short tail of the Tawny-breasted Tinamou *N. julius* is hidden by the tail coverts (v. *Crypturellus).*

Nothocrax Gr. *nothos,* false, adulterated (i.e. smaller); genus *Crax,* currasow.

Nothoprocta Gr. *nothos,* false, spurious; *proktos,* the tail; ref. very short tail of the Chilean Tinamou *N. perdicaria*.

Nothura Gr. *nothos,* false, spurious; *oura,* the tail; ref. very short tail of the White-bellied Nothura *N. boraquira* (v. *Crypturellus).*

Notiochelidon Gr. *notios,* southern (*notos,* the south); *khelidon,* a swallow; these swallows occur through tropical South America.

Notiomystis Gr. *notios,* southern; *mustes,* a mystery; the Stitchbird *N. cincta* is a rare honeyeater of uncertain position confined to Little Barrier Island, New Zealand.

Notiopsar Gr. *notios,* southern; *psar,* a starling (v. *Gnorimopsar);* the Austral Blackbird *Curaeus curaeus* is found only in Chile and southern Argentina.

Notiospiza Gr. *notios,* southern; *spiza,* a finch; the Bar-winged Weaver *Ploceus angolensis* occurs in southern tropical Africa.

Notornis Gr. *notos,* the south or south-west; *ornis,* a bird; the rare Takahe *N. mantelli* is confined to the south-western part of South Island, New Zealand.

notosticta Gr. *noton,* the back; *stiktos,* spotted; ref. black-streaked back of the Oaxaca Sparrow *Aimophila*.

nouhuysi After Captain J. W. Nouhuys (b. 1859) Dutch naval officer and explorer in New Guinea.

novacapitalis L. *novus,* new; *capitalis,* first, chief (i.e. capital city); after Brasília, DF, the new (1960) capital city of Brazil.

novaecaledoniae L. *novus,* new; *Caledonia,* Caledonia, the highlands of Scotland; after the island of New Caledonia (v. *caledonica).*

novaeguineae L. *novus,* new; Mod. L. *Guinea,* Guinea; after the island of New Guinea, first named *Nueva Guinea* in 1545 by the Spanish explorer Ortíz de Retes, because natives in the Mamberano region reminded him of those he had encountered earlier in African Guinea.

novaehollandiae L. *novus,* new; Mod. L. *Hollandia,* Holland, Netherlands; after New Holland, the name by which eastern Australia was known to Europeans during the seventeenth and early eighteenth centuries.

novaeseelandiae L. *novus,* new; Mod. L. *Zeelandia,* Zeeland, Netherlands; after New Zealand, which received its name from the explorer Abel Tasman who explored the west coast in 1642, naming it *Nova Zeelanda* after his Dutch home.

novaesi After Fernando de C. Novaes (b. 1927) Brazilian scientist and author.

novaezelandiae see **novaeseelandiae.**

noveboracensis L. *novus,* new; *Eboracum,* York, England; after the city of New York, USA.

novicius L. *novicius,* new, fresh.

Novipulsatrix L. *novus,* new; genus *Pulsatrix,* owl.

nuba L. *Nuba,* a Nubian; ancient Nubia, the land of Negroes, lay in the Nile valley to the south of Egypt.

nubica After Nubia (v. *nuba).*

nubicoides Specific name *Merops nubicus;* Gr. *-oides,* resembling.

nubicus see **nubica.**

nuchalis Mod. L. *nuchalis,* nuchal, of the nape (Med. L. *nucha,* the nape; Arabic *nukha,* spinal marrow).

Nucifraga Mod. L. *nucifraga,* the nutcracker (L. *nux, nucis,* a nut; *frangere,* to shatter); the diet of the Spotted Nutcracker *N. caryocatactes* includes nuts and pine seeds.

nudiceps L. *nudus,* bare, naked; *-ceps,* headed; ref. bare blue forehead and face of the Bare-crowned Antbird *Gymnocichla.*

nudicollis L. *nudus,* bare, naked; Mod. L. *-collis,* necked; ref. bare greenish throat of the Bare-throated Bellbird *Procnias.*

nudigenis L. *nudus,* bare, naked; *genis,* the cheeks (*gena,* the cheek); ref. bold, bare eye-ring of the Bare-eyed Robin *Turdus.*

nudigula L. *nudus,* bare, naked; *gula,* the throat.

nudipes L. *nudipes,* bare-footed (i.e. unfeathered legs).

nuditarsus L. *nudus,* bare, naked; Gr. *tarsos,* the sole of the foot; ref. unfeathered tarsus of the Schrader Swiftlet *Aerodramus.*

nugator L. *nugator,* a fool; the Grenadine Flycatcher *Myiarchus* is known locally as 'Tom Fool' on account of its vacant whistling calls.

Numenius Gr. *noumenios,* a bird mentioned by Hesychius, traditionally associated with the curlew by reason of its crescent-shaped bill (*neos,* new; *mene,* the moon).

Numida L. *Numida,* a Numidian, an African; the Helmeted Guineafowl *N. meleagris* is found in suitable habitats throughout tropical Africa.

nuna After Nouna-Koali, a graceful Indian virgin in Jean Ferdinand Denis' novel *Ismaël Ben Kaïzar ou la découverture du Nouveau Monde,* 1829.

nuttalli /nuttallii After Thomas Nuttall (1786–1859) English botanist and ornithologist resident in USA, collector, and author (*Manual of the ornithology of the United States and Canada*, 1832).

Nuttallornis After Thomas Nuttall (v. *nuttalli)*; Gr. *ornis,* a bird.

nuttingi After Charles Cleveland Nutting (1858–1927) US naturalist and collector.

Nyctagreus Gr. *nux, nuktos,* night; *agreus,* a hunter.

Nyctanassa Gr. *nux,* night; *nassa,* a kind of duck or waterbird.

Nyctea Gr. *nuktia,* nocturnal, of the night (*nux,* night); ref. to owls in general rather than the mainly diurnal Snowy Owl *N. scandiaca.*

nycthemera Gr. *nux,* night; *emera,* day; ref. the contrasting silver and black plumage of the male Silver Pheasant *Lophura.*

Nyctibius Gr. *nuktibios,* night-living, night-feeding (*nuktos,* night; *bios,* living) (cf. Brazilian Portuguese name *noitibo,* potoo).

Nycticorax Gr. *nuktikorax,* a bird of evil omen mentioned by Aristotle, Hesychius, and other authors, probably the Long-eared Owl *Asio,* but long associated with the night heron.

Nycticryphes Gr. *nux,* night; *kruphe,* in secret, secretly; ref. nocturnal habits and secretive behaviour of the American Painted Snipe *N. semicollaris.*

Nyctidromus Gr. *nux,* night; *dromos,* running (*trekho,* to run); ref. nocturnal habits of the Pauraque *N. albicollis.*

Nyctiphrynus Gr. *nux,* night; *phrune,* a toad; ref. nocturnal call-notes of the Ocellated Poorwill *N. ocellatus.*

Nyctiprogne Gr. *nux,* night; Mod. L. *progne,* a swallow; ref. small size, erratic flight and long pointed wings of the Band-tailed Nighthawk *N. leucopyga.*

Nyctyornis Gr. *nux,* night, darkness; *ornis,* a bird; ref. the solitary habits, gloomy forest home, and supposed nocturnal activities of the Red-bearded Bee-eater *N. amicta.*

nyika After the Nyika Plateau, Northern Province, Malawi.

nympha L. *nympha,* a nymph, a spirit (Gr. *numphe,* a maiden or nymph).

Nymphicus L. *nymphicus,* nymph-like (*nympha,* a nymph).

nyroca Russian name *nyrok* for a duck.

Nystalus Gr. *nustalos,* drowsy; ref. quiet, lethargic behaviour of the puffbirds.

Nycticorax nycticorax Black-crowned Night Heron

O

oahensis After the island of Oahu, Hawaiian Islands.

oatesi After Eugene Oates (1845–1911) British naturalist, civil servant in India, and author (*The fauna of British India*, 1889).

obbiensis After the town of Obbia or Obiat, Mudugh, Somalia.

oberholseri After Harry Church Oberholser (1870–1963) US ornithologist and author (*The bird life of Louisiana*, 1938).

oberi After Frederick Ober (1849–1913) US naturalist and collector in the West Indies.

oberlaenderi After P. Oberländer (d. 1911) Austrian naturalist and collector in tropical Africa.

obscura L. *obscurus*, dark, dusky; **1** ref. dull-coloured plumage. **2** after Dusky Sound, South Island, New Zealand (*Charadrius*).

obscurior L. *obscurior*, darker (comp. of *obscurus*, dark).

obscurus see **obscura.**

obsoleta /obsoletum /obsoletus L. *obsoletus*, shabby, ordinary, plain (*obsolescere*, to wear out).

obstinata L. *obstinatus*, persistent, obstinate (*obstinare*, to persist); Hartert (1900), *Nov. Zool.*, **7**, 238, complained that the type specimens of the Sullen White-eye *Zosterops* were all bad skins.

obtusa L. *obtusus*, dull, blunt (*obtundere*, to beat upon).

ocai After Rafael Montes de Oca (fl. 1878) Mexican painter and naturalist.

occidentalis L. *occidentalis*, western (*occidens*, the west, evening).

occipitalis Late L. *occipitalis*, occipital, of the back of the head (L. *occipitium*, the occiput).

occultus L. *occultus*, hidden, concealed (*occulere*, to cover).

oceanica /oceanicus Med. L. *oceanicus*, oceanic (L. *oceanus*, the ocean; Gr. *okeanos*, the ocean, the great sea).

Oceanites Gr. myth. Okeanites or Oceanites, the sea-nymphs, three thousand in number, the daughters of the sea god Okeanos.

Oceanodroma Gr. *okeanos*, the ocean; *dromos*, running (*trekho*, to run); the pelagic storm-petrels come ashore only for breeding.

ocellata /ocellatum /ocellatus L. *ocellatus*, ocellated, marked with small eyes or eyelets (*ocellus*, an eyelet).

Ochetorhynchus Gr. *okhetos*, a channel, a canal; *rhunkhos*, the bill; the bills of the earthcreepers in this genus are completely straight or, at most, very slightly curved.

ochotensis After the Sea of Okhotsk (German *Ochots'kische Meer)* (Okhota River), north-east Siberia, USSR.

ochracea Mod. L. *ochraceus*, ochraceous, ochreous (L. *ochra*, ochre).

ochraceiceps Mod. L. *ochraceus*, ochraceous; L. *-ceps*, headed.

ochraceifrons Mod. L. *ochraceus*, ochraceous; L. *frons*, the forehead, the brow.

ochraceiventris Mod. L. *ochraceus*, ochraceous; L. *venter, ventris*, the belly.

ochraceus see **ochracea.**

ochrocephala Gr. *okhros*, sallow, pale yellow; *-kephalos*, headed.

ochrogaster Gr. *okhros,* sallow, pale yellow; *gaster,* the belly.

ochrolaemus Gr. *okhros,* sallow, pale yellow; *laimos,* the throat.

ochroleucus Gr. *okhroleukos,* yellowish-white.

ochromalus /Ochromela /ochromelas Gr *okhromelas,* dark and sallow, suffering from jaundice (i.e. yellow, or yellow and black)

ochropectus Gr. *okhros,* sallow, pale yellow; L. *pectus,* the breast.

ochropus Gr. *okhros,* sallow, pale yellow; *pous,* the foot.

ochropyga Gr. *okhros,* sallow, pale yellow; *puge,* the rump.

ochruros Gr. *okhros,* sallow, pale yellow; *-ouros,* tailed.

Ochthoeca Gr. *okhthe,* a hill or mound; *oikos,* a dwelling; the chat tyrants are usually found in open areas.

Ochthornis Gr. *okhthe,* a hill or mound; *ornis,* a bird; the Drab Water Tyrant *O. littoralis* frequently perches on sandbars.

ocistus Gr. *okistos,* very swift (*okus,* speedy, quick).

ocreata /Ocreatus L. *ocreatus,* wearing greaves, booted (*ocrea,* greaves).

octosetaceus L. *octo-,* eight; *seta,* a thick hair or bristle; *-aceus,* resembling; the Brazilian Merganser *Mergus* was described as having a nuchal crest composed of eight plumes or bristles.

ocularis L. *ocularis,* ocular, of the eyes (*oculus,* the eye).

oculatus L. *oculatus,* having eyes, conspicuous.

oculea /oculeus L. *oculeus,* full of eyes; from appearance of spotted plumage.

Oculocincta L. *oculus,* the eye; *cinctus,* wreathed, encircled (*cingere,* to encircle); ref. thin white eye-ring of the Pygmy White-eye *O. squamifrons.*

Ocyalus Dim. from Gr. *okua,* a spear; ref. broad, thick bill of the Band-tailed Oropendola *O. latirostris.*

Ocyphaps Gr. *okua,* a spear; *phaps,* a pigeon; ref. distinctive thin crest of the Crested Pigeon *O. lophotes.*

odiosa L. *odiosus,* odious, troublesome.

odomae After Babette Odom (fl. 1980) US birdwatcher and sponsor of field programmes in Peru.

Odontophorus Gr. *odontophoros,* bearing teeth; ref. thick, notched or 'toothed' bills of the Neotropical wood quails.

Odontorchilus Gr. *odous, odontos,* a tooth; Mod. L. *orchilus,* wren (v. *Hylorchilus*).

Odontospiza Gr. *odous,* a tooth; *spiza,* a finch; the Grey-headed Silverbill *O. griseicapilla* has a short, thick, notched bill.

oedicnemus Gr. *oideo,* to swell; *kneme,* the shin or leg; ref. swollen tibiotarsal joints of the Stone Curlew or Northern Thick-knee *Burhinus.*

Oedistoma Gr. *oideo,* to swell; *stoma,* the mouth; the Pygmy Honeyeater *O. pygmaeum* has swollen tomia (mandibular cutting edges) near the angle of the mouth.

oemodium L. *Oemodium,* of the Himalayas (*Oemodi Montes)*; only the western ranges of the Himalayas were known to the Greeks and Romans.

Oena Gr. *oinas,* a pigeon, probably the Rock Dove *Columba.*

Oenanthe Gr. *oinanthe,* a bird mentioned by Aristotle, not otherwise identified, but associated with the wheatear by later authors.

oenanthoides Genus *Oenanthe,* wheatear; Gr. *-oides,* resembling.

oenas see **Oena.**

oenochlamys Gr. *oinos,* wine; *khlamus,* a cloak or mantle; ref. the lilac-blue coloured upperparts of the Sulphur-billed Nuthatch *Sitta.*

Oenoenas Gr. *oinos,* wine; *oinas,* a pigeon or dove; ref. purplish or dark wine-red plumage of the Plumbeous Pigeon *Columba plumbea.*

oenone Gr. myth. Oenone or Oinone, a Phrygian nymph beloved and deserted by Paris.

oenothorax Gr. *oinos,* wine; *thorax,* the breast.

Oestrelata Gr. *oistrelatos*, driven by the gadfly (*oistros*, the gadfly or horsefly); ref. the erratic, bounding flight of the gadfly petrels *Pterodroma*.

oglei After M. G. Ogle (1842–1892) British surveyor in India.

Ognorhynchus Gr. *onkos*, a mass or bulk; *rhunkhos*, the bill; ref. thick feathering covering the base of the the lower mandible of the heavy-billed Yellow-eared Conure *O. icterotis*.

Oidemia Gr. *oidema*, a swelling or tumour; ref. black knob at base of drake Black Scoter's *Melanitta nigra* bill.

okinawae After the island of Okinawa, Ryukyu Islands, Japan.

olax L. *olax*, odorous, having a smell (*olere*, to emit an odour); the Little Green Pigeon *Treron* was described as allied to the Grey-winged Green Pigeon *T. aromatica* (q.v.).

oleaginea L. *oleagineus*, olive-coloured.

oleaginus L. *oleaginus*, oleaginous, olive coloured (*olea*, the olive tree).

Oligura Gr. *oligos*, small; *oura*, the tail.

olivacea Mod. L. *olivaceus*, olivaceous, olive-green (L. *oliva*, the olive).

olivaceiceps Mod. L. *olivaceus*, olivaceous; L. *-ceps*, headed.

olivaceofuscus Mod. L. *olivaceus*, olivaceous; L. *fuscus*, dusky, dark.

olivaceogriseus Mod. L. *olivaceus*, olive-green; Med. L. *griseus*, grey.

olivaceum/olivaceus see **olivacea.**

olivascens Mod. L. *olivascens*, somewhat olive.

olivater L. *oliva*, olive; *ater*, black.

olivea Mod. L. *oliveus*, olive-coloured.

olivetorum L. *olivetum*, *olivetorum*, an olive grove.

oliviae After Lady Olive Archer (fl. 1918) wife of British ornithologist, diplomat, and author Sir Geoffrey Archer.

olivicyanea Mod. L. *oliveus*, olive-coloured; L. *cyaneus*, dark blue (Gr. *kuaneos*).

olivieri After G. Olivier (fl. 1927) French zoologist, collector in Madagascar, and author (*Monographie des pies-grièches du genre Lanius*, 1944).

olivii After E. A. C. Olive (fl. 1898) collector in Australia.

olivinus Mod. L. *olivinus*, olive, olivine (L. *oliva*, olive).

Olor L. *olor*, a swan.

olrogi After Claes Christian Olrog (1912–1985) Swedish ornithologist resident in Argentina, and author (*Las aves Argentinas*, 1959).

omeiensis After Mount Omei, Szechwan (Omei Shan, Sichuan), western China.

omissa L. *omissus*, negligent, remiss (i.e. previously overlooked) (*omittere*, to let go).

Oncostoma Gr. *onkos*, swollen, bulky; *stoma*, the mouth; the Neotropical bentbills are so-called for their thick, downcurved bills.

oneilli After John P. O'Neill (b. 1942) US ornithologist, field worker, and artist.

onocrotalus L. *onocrotalus*, a pelican (Gr. *onokrotalos*, a pelican).

onslowi After William Hillier, 4th Earl of Onslow (1853–1911) Governor of New Zealand, 1888–1892.

Onychognathus Gr. *onux*, *onukhos*, a claw or nail; *gnathos*, the jaw or mouth; the chestnut-winged starlings have a sharp, curved nail at the end of the bill.

Onychorhynchus Gr. *onux*, a claw or nail; *rhunkhos*, the bill; the royal flycatchers have long bills with a curved nail at the apex.

oorti After P. van Oort (1804–1834) Dutch naturalist and collector in the East Indies.

opaca/opacus L. *opacus*, dark, obscure.

Ophrysia Gr. *ophrus*, the eyebrow; the extinct Himalayan Mountain Quail *O. superciliaris* had a distinctive broad, white eyebrow.

ophthalmica/ophthalmicus L. *ophthalmicus*, of the eye (Gr. *ophthalmikos*).

opistherythra Gr. *opisthe*, the back, the rear; *eruthros*, red; the Long-tailed Fantail *Rhipidura* has a pale rufous rump.

Opisthocomus Gr. *opisthokomos*, wearing the hair long behind (*opisthe*, behind, the back; *kome*, the hair); ref. the long, untidy crest of the strange Hoatzin *O. hoazin*.

opisthomelas Gr. *opisthe*, the back, the rear; *melas*, black; ref. the black undertail coverts of the Black-vented Shearwater *Puffinus*.

Opisthoprora Gr. *opisthe*, behind; *prora*, the face or head; ref. coppery-red crown and cheeks of the Mountain Avocetbill *O. euryptera*.

Opopsitta Gr. *opos*, fig juice; Mod. L. *psitta*, a parrot; figs and other fruits form the diet of the Papuan fig parrots. According to Wolstenholme (1926) the name is an error for *Cyclopsitta* (Gr. *kuklops*, round faced).

Oporornis Gr. *opora*, autumn; *ornis*, a bird; these migrant Nearctic warblers are said to be more common in the autumn or fall than in the spring.

optata L. *optatus*, welcome, select (*optare*, to choose).

oratrix L. *oratrix*, a female speaker or suppliant (*orare*, to speak).

orbignyianus After Alcide Dessalines d'Orbigny (1802–1857) French naturalist and author (*Voyage dans l'Amérique méridionale*, 1835; *Galerie ornithologique ou collection d'oiseaux d'Europe*, 1836; *Ornithologie de l'île de Cuba*, 1839).

orbitalis Med. L. *orbitalis*, of the eye or eye-socket.

orbitatum Med. L. *orbitatus*, the eye-cavity (L. *orbis*, a circle).

orbygnesius see **orbignyianus.**

orcesi After Gustavo Orces V. (fl. 1988) Ecuadorean ornithologist.

Orchesticus Gr. *orkhestikos*, of dancing (*orkheomai*, to dance); the Brown Tanager *O. abeillei* was described as allied to the genus *Saltator* (L. *saltator*, a dancer).

ordii After George Ord (1781–1866) US naturalist, collector, and philologist.

oreas L. *oreas*, a mountain nymph (Gr. *oreias*, an oread, a mountain nymph; *oreios*, living on the mountains).

oreganus After the Oregon country or Columbia River region of North America, divided in 1853 as the Oregon and Washington Territories, both later admitted to statehood.

Oreicola Gr. *oros*, *oreos*, a mountain; L. *-cola*, inhabitant.

orenocensis After the Río Orinoco, Venezuela.

oreobates Gr. *oreobates*, mountain ranging.

Oreocharis Gr. *oros*, *oreos*, a mountain; *kharis*, grace, beauty.

Oreocincla Gr. *oros*, a mountain; Mod. L. *cinclus*, a thrush.

Oreoica Gr. *oreioikos*, mountain dwelling.

Oreomanes Gr. *oros*, a mountain; *-manes*, passionately fond of.

Oreomystis Syn. genus *Oreomyza*, spikebill (Gr. *oros*, a mountain; genus *Myza*, honeyeater); *mustes*, a mystery.

Oreonympha Gr. *oros*, a mountain; *numphe*, a nymph or maiden.

Oreopeleia Gr. *oros*, a mountain; *peleia*, a pigeon or dove.

Oreophasis Gr. *oros*, a mountain; Mod. L. *phasis*, pheasant.

oreophila/oreophilus Gr. *oros*, a mountain; *philos*, loving, fond of.

Oreopholus Gr. *oreopolus*, mountain haunting (*oros*, a mountain; *poleo*, to range about).

Oreophylax Gr. *oreophulax*, a mountain guard; the Itatiaia Spinetail *O. moreirae* is confined to the sierras of south-east Brazil.

Oreopsar Gr. *oros*, a mountain; *psar*, a starling.

Oreopsittacus Gr. *oros*, a mountain; *psittakos*, a parrot.

Oreopyra Gr. *oros*, a mountain; *pura*, a pyre, a burning or glowing mass.

Oreornis Gr. *oros*, a mountain; *ornis*, a bird.

Oreortyx Gr. *oros*, a mountain; *ortux*, the quail.

Oreoscoptes Gr. *oros*, a mountain;

skoptes, a mocker (*skopto,* to jeer); the Sage Thrasher *O. montanus* was formerly known as the Mountain Mockingbird.

Oreoscopus Gr. *oros,* a mountain; *skopos,* a watcher (*skeptomai,* to watch); the Fernwren *O. gutturalis* inhabits the dense mountain rainforests of northern Queensland, Australia.

Oreostruthus Gr. *oros,* a mountain; *strouthos,* a sparrow *Passer* or any small bird.

Oreothlypis Gr. *oros,* a mountain; Mod. L. *thlypis,* a warbler.

Oreothraupis Gr. *oros,* a mountain; genus *Thraupis,* tanager.

Oreotriccus Gr. *oros,* a mountain; Mod. L. *triccus,* tyrant flycatcher.

Oreotrochilus Gr. *oros,* a mountain; genus *Trochilus,* hummingbird.

oreskios Gr. *oreskios,* mountain bred.

orientalis L. *orientalis,* eastern (*oriens, orientis,* the east, morning, the rising sun).

origenes Gr. *orei-,* mountain- (*oros,* a mountain); *genes,* offspring, descent.

Origma Gr. *orugma,* a mine or tunnel; the Australian Origma or Rock Warbler *O. solitaria* builds its nest in caves and under overhanging rocks.

orina/orinus Gr. *oreinos,* mountain dwelling.

Oriolia French name *oriolie,* reflecting supposed affinity of the Black Vanga *O. bernieri* to the orioles *Oriolus.*

oriolina Genus *Oriolus,* oriole; L. *-inus,* resembling.

orioloides Genus *Oriolus,* oriole; Gr. *-oides,* resembling.

Oriolus Med. L. *oryolus,* the Golden Oriole (L. *aureolus,* golden) (cf. Old French *oriol,* said to be onomatopoeic).

oritis Gr. *oritis,* sideritis, steel, a precious stone.

Oriturus No expl. (Bonaparte (1851), *Consp. gen. avium,* Vol. **I**, p. 469); probably a misspelt dim. form from Gr. *oreiotes,* a mountaineer, but perhaps the first part is either from Gr. *oritis,* siderite, steel, or *orizo,* to mark out or border, and the second element from Gr. *oura,* the tail, both ref. the sooty tail, broadly edged with grey, of the Striped Sparrow *O. superciliosus.*

orix L. *oryza,* rice (Gr. *oruza,* rice); the Red Bishop *Euplectes* feeds on grain and seeds.

ornata/ornatus L. *ornatus,* adorned, ornate (*ornare,* to adorn, furnish).

Ornithion Gr. *ornithion,* a little bird (dim. of *ornis,* a bird).

Oroaetus Gr. *oros,* a mountain; *aetos,* an eagle.

Orochelidon Gr. *oros,* a mountain; *khelidon,* a swallow.

Orodytes Gr. *oros,* a mountain; *dutes,* a diver (i.e. an inhabitant) (*duo,* to plunge).

orostruthus Gr. *oros,* a mountain; *strouthos,* a sparrow *Passer* or any small bird.

orpheus After Orpheus, the celebrated Thracian minstrel.

orrhophaeus Gr. *orrhos,* the tail or rump; *phaios,* grey, dusky.

orru No expl. (Bonaparte (1851), *Consp. gen. avium,* Vol. **I**, p. 385); probably based on an onomatopoeic New Guinea native name.

Ortalis Gr. *ortalis,* a chicken.

Orthogonys Gr. *orthos,* straight; *gonus,* jaw (in orn. the gonys (Gr. *gonia,* an angle) is the ridge formed by the junction of the two halves of the lower mandible along its underside near the tip); ref. the straight lower mandible of the Olive-green Tanager *O. chloricterus.*

Orthonyx Gr. *orthos,* straight; *onux,* a nail or claw; from supposed straight claws of the Logrunner *O. temminckii.*

Orthorhamphus Gr. *orthos,* straight; *rhamphos,* the bill.

Orthorhyncus Gr. *orthos,* straight; *rhunkhos,* the bill.

Orthotomus Gr. *orthotomeo,* to cut straight; the Indomalayan tailorbirds are well known for their intricate nest building skills.

ortizi After Javier Ortíz de la Puente (fl. 1951) Peruvian ornithologist.

ortoni After James Orton (1830–1877) US zoologist, collector in tropical America, and author (*The Andes and the Amazons*, 1870).

Ortygocichla Gr. *ortux, ortugos,* the quail *Coturnix; kikhle,* a thrush.

Ortygonax Gr. *ortux,* the quail; *anax,* a king; in ancient times the Corncrake *Crex* was said to accompany the quail *Coturnix* on its migrations.

Ortygospiza Gr. *ortux,* the quail; *spiza,* a finch; ref. the small size, cryptic plumage, and ground-loving habits of the Afrotropical quail-finches.

Ortyxelos Dim. from Gr. *ortux,* the quail.

oryzivora /oryzivorus L. *oryza,* rice; *-vorus,* eating.

Oryzoborus Gr. *oruza,* rice; *boros,* devouring.

osburni After W. Osburn (d.1860) collector in Jamaica, 1858–1860.

oscillans L. *oscillans,* swinging, oscillant (*oscillare,* to swing); ref. the indeterminate position of the Russet-backed Jungle Flycatcher *Rhinomyias.*

oscitans L. *oscitans,* yawning (*oscitare,* to yawn); ref. the conspicuous gap, the 'yawn', between the mandibles of the Oriental Openbill Stork *Anastomus.*

osculans Misprint for L. *oscillans,* oscillant, swinging (i.e. between other species).

Osculatia After G. Osculati (1808–1884) Italian explorer and collector in Iran, 1841 and tropical America, 1846–1848.

osea Gr. *osia,* divine, religious; the Red-tufted Sunbird *Nectarinia* was described from the Holy Land or Palestine.

oseryi After Eugène Comte d'Osery (1818–1846) French traveller and collector in tropical America, 1843–1846.

osgoodi After Wilfred Hudson Osgood (1875–1947) US zoologist, collector, and author (*The mammals of Chile*, 1923).

Osmotreron Gr. *osme,* a scent or odour; genus *Treron,* pigeon (v. *aromatica).*

ossifragus L. *ossifragus,* bone-breaking (*os,* a bone; *frangere,* to break) (cf. *ossifragus,* the sea eagle *Haliaeetus* or osprey *Pandion).*

ostenta L. *ostentus,* showy, conspicuous (*ostendere,* to display).

Ostinops Gr. *ostinos,* bony, made of bone; *ops,* the face; ref. the large ivory-coloured bill and frontal casque of the Crested Oropendola *Psarocolius decumanus.*

ostralegus L. *ostrea,* an oyster; *legere,* to collect, pick.

ostrinus L. *ostrinus,* purple (*ostrum,* purple dye).

Othyphantes Gr. *ous, otos,* the ear; *huphantes,* a weaver (*huphaino,* to weave); the male Reichenow's Weaver *Ploceus reichenowi* has conspicuous black ear coverts surrounded by golden-yellow.

Otidiphaps Gr. *otis, otidos,* a bustard; *phaps,* a pigeon; ref. the large size, stately bearing and terrestrial habits of the New Guinea Pheasant Pigeon *O. nobilis.*

otiosus L. *otiosus,* calm, gentle.

Otis Gr. *otis,* a bustard with long ear feathers.

Otocoris Gr. *ous, otos,* the ear; syn. genus *Corys,* lark.

Otophanes Gr. *ous,* the ear; *phanes,* displaying (*phaino,* to show); the Eared Poorwill *O. mcleodii* has erectile ear tufts.

ottonis After Otto Garlepp (1864–1959) German collector in Peru, 1895–1912.

Otus Gr. *otos,* an eared owl.

oustaleti After Émile Oustalet (1844–1905) French zoologist and author (*Les oiseaux de la Chine*, 1877; *Les oiseaux du Cambodge, du Laos*, 1899).

ovampensis After the Ovampo (Okavango) River, Angola.

owenii After Richard Owen (1804–1892) British zoologist, anatomist, and opponent of Darwinism.

owstoni After Alan Owston (1853–1915) British oologist and collector in Japan, 1871–1915.

oxycerca Gr. *oxus,* sharp, pointed; *kerkos,* the tail.

Oxyechus Gr. *oxuekhos,* shrill-sounding, sharp-sounding.

Oxylabes Gr. *oxulabes,* seizing quickly.

Oxylophus Gr. *oxus,* sharp, pointed; *lophos,* a crest.

Oxypelia Gr. *oxus,* quick, speedy; *peleia,* a dove.

Oxypogon Gr. *oxus,* sharp, pointed; *pogon,* a beard.

Oxyruncus Gr. *oxus,* sharp, pointed; *rhunkhos,* the bill.

Oxyura Gr. *oxus,* sharp, pointed; *oura,* the tail.

Oceanodroma castro Madeiran Storm-Petrel

P

pabsti After Guido Pabst (fl. 1969) Brazilian botanist and founder of Herbarium Bradeanum, Rio de Janeiro.

Pachycare Gr. *pakhus,* thick, large; *kare,* the head (v. *Pachycephala).*

Pachycephala Gr. *pakhus,* thick, large; *kephale,* the head; the Indomalayan and Australasian whistlers are also known as thickheads from the appearance of their large, rounded heads.

pachycephaloides Genus *Pachycephala,* whistler; Gr. *-oides,* resembling.

Pachycephalopsis Genus *Pachycephala,* whistler; Gr. *opsis,* appearance.

Pachycoccyx Gr. *pakhus,* thick, large; *kokkux,* the cuckoo; ref. large size, dense rump feathers, and stout bill of the Thick-billed Cuckoo *P. audeberti.*

Pachyphantes Gr. *pakhus,* thick, large; *huphantes,* a weaver; ref. thick bill of the Compact Weaver *Ploceus superciliosus.*

Pachyptila Gr. *pakhus,* thick, dense; *ptilon,* feather, plumage; ref. soft, dense feathering of the prions or whale-birds.

Pachyramphus Gr. *pakhus,* thick; *rhamphos,* the bill.

pachyrhyncha /pachyrhynchus Gr. *pakhus,* thick; *rhunkhos,* the bill.

pacifica /pacificus L. *pacificus,* pacific, peaceful. After the Pacific Ocean, so named by Magellan in 1520 in contrast to the stormy waters of the Straits of Magellan. In *Cacicus, Otus, Pyriglena, Pyrrhura,* and *Rhynchocyclus* the name refers to the Pacific or western slopes of the Andes in Colombia or Ecuador.

Padda Chinese *padda,* rice; rice seed is a favourite food of the Java Sparrow *P. oryzivora.*

paena From a Chuana (Botswana) native name for the Sandy Scrub Robin *Erythropygia.*

pagodarum Portuguese *pagode,* a temple or pagodah; the Brahminy Mynah *Sturnus* is commonly found around temples and villages.

Pagodroma Gr. *pagos,* ice; *dromos,* running (*trekho,* to run); the Snow Petrel *P. nivea* is always found near icefloes and icebergs.

Pagophila Gr. *pagos,* ice; *philos,* loving, fond of; the Ivory Gull *P. eburnea* breeds in the wastes of the high Arctic.

palauensis After the Palau, formerly Pelew, Island group in the Caroline Islands, Pacific Ocean.

palawanense /palawanensis After the island of Palawan, most westerly of the Philippines.

pallasi /pallasii After Peter Simon Pallas (1741–1811) German naturalist in the Russian service, explorer in Siberia and Kirghizia, and author (*Reise durch verschiedene provinzen der Russischen reichs,* 1771; *Zoographia Rosso-Asiatica,* 1811).

pallatangae After the settlement of Pallatanga, Chimborazo, Ecuador.

pallens L. *pallens,* pale, pale yellow, pale green (*pallere,* to be pale).

pallescens L. *pallescens,* pallid, palish (*pallescere,* to grow pale).

palliata /palliatus L. *palliatus,* cloaked, mantled (*pallium,* a mantle).

pallida L. *pallidus,* pallid, pale.

pallidiceps L. *pallidus,* pale; *-ceps,* headed.

pallidicinctus L. *pallidus,* pale; *cinctus,* girdled, banded (*cingere,* to encircle).

pallidigaster L. *pallidus,* pale; *gaster,* the belly.

pallidimas L. *pallidus,* pale; *mas,* male; the male Bar-bellied Goshawk *Accipiter* is paler and much less barred than the female.

pallidinucha L. *pallidus,* pale; Med. L. *nucha,* the nape (Arabic *nukha,* spinal marrow).

pallidior L. *pallidior,* more pale, paler (comp. of *pallidus,* pale).

pallidipes L. *pallidus,* pale; *pes,* the foot.

pallidirostris L. *pallidus,* pale; *-rostris,* billed.

pallidiventris L. *pallidus,* pale; *venter, ventris,* the belly.

pallidus see **pallida.**

pallipes see **pallidipes.**

palliseri After Captain Edward Palliser (1826–1907) and his brother F. H. Palliser (b. 1826) collectors in Ceylon (Sri Lanka), 1845–1851.

palmarum L. *palmarum,* of palm trees.

palmeri After **1** Henry Palmer (fl. 1893) Australian collector in the Hawaiian Islands, 1890–1893 (*Phaeornis, Rhodacanthis*). **2** Mervyn George Palmer (d. 1955) English collector in Nicaragua and Colombia, 1904–1917, and author (*Through unknown Nicaragua,* 1945) (*Tangara*).

Palmeria After Henry Palmer (v. *palmeri* **1**).

palpebralis L. *palpebralis,* of the eyelids (*palpebra,* the eyelid).

palpebrata Mod. L. *palpebratus,* eyelidded, furnished with eyelids (L. *palpebra,* the eyelid).

palpebrosa Mod. L. *palpebrosus,* having large brows, prominent eyelids (L. *palpebra,* the eyelid).

paludicola L. *paludicola,* a marsh dweller (*palus, paludis,* a marsh; *colere,* to inhabit).

palumboides Specific name *Columba palumbus;* Gr. *-oides,* resembling.

palumbus L. *palumbes,* the Wood Pigeon *Columba.*

palustre /palustris L. *paluster,* marshy, boggy (*palus, paludis,* a swamp or marsh).

pamela Gr. *pammelas,* all black.

pammelaena /pammelaina Gr. *pammelaina,* all black.

pampa After the pampas (Quechua Indian *pampas,* the plains), the treeless plains of Argentina. An erroneous type locality (*Campylopterus*).

panamensis After Panama, a former province of Colombia which seceded finally in 1903, named after Panamá, meaning 'many fish', an old Guaymí fishing village.

panayensis After the island of Panay, between Mindoro and Negros, Philippines.

panderi After Heinrich Christian von Pander (1794–1865) Latvian/German geologist and palaeontologist.

Pandion Gr. myth. Pandion, king of Attica, whose tragic daughters, Philomela and Procne, were metamorphosed into a nightingale and a swallow respectively.

pandoo Marathi (Indian) native name *pandu* for the male Indian Blue Rock Thrush *Monticola.*

panini After the island of Panay, Philippines.

Panterpe Gr. *panterpes,* all delighting.

Panurus Gr. *panu,* exceedingly; *-ouros,* tailed; ref. the long tail of the reed-haunting Bearded Tit *P. biarmicus.*

panychlora Gr. *panu,* exceedingly; *khloros,* green.

Panyptila Gr. *panu,* exceedingly; *ptilon,* the wing; ref. the long, scimitar wings of the Neotropical swallow-tailed swifts.

papa Late L. *papa,* a bishop.

Papasula Gr. *pappas,* father; genus *Sula,* gannet; after William Abbott (v. *abbotti* **2**) (*abbot,* from Hebrew *abba,* father) and to indicate the primitive origins of Abbott's Booby *P. abbotti.*

Paphosia Gr. myth. Paphos, a son of Pygmalion by a statue which had been turned into a woman by the goddess Venus, and founder of a city in Cyprus whose inhabitants were considered effeminate and lascivious.

papillosa L. *papilla,* a pimple; *-osus,* abundance; the Red-naped Ibis *Pseudibis* has a patch of red warts on the crown.

papou /papua /papuana /papuensis After Papua or New Guinea (Malay *papuwa,* curly haired, ref. frizzy hair of the New Guinea aborigines). In the case of *Pygoscelis* the epithet is erroneous.

par L. *par,* equal; the sexes of Hartert's Whistler *Pachycephala* are indistinguishable except for a very small difference in size.

Parabuteo Gr. *para,* derived, near; genus *Buteo,* buzzard; the Bay-winged or Harris' Hawk *P. unicinctus* was formerly placed in the genus *Buteo.*

Paradigalla Genus *Paradisea,* bird of paradise; L. *gallus,* a farmyard cock; ref. the colourful facial wattles of the New Guinea paradigallas.

Paradisaea Late L. *paradisus,* paradise (Gr. *paradeisos,* paradise, a pleasure ground). Magellan's sailors were told by Moluccan natives that the birds of paradise came from heaven, being called 'birds of god' (v. *apoda).*

paradisea /paradiseus /paradisi Late L. *paradisus,* paradise.

paradoxa Gr. *paradoxos,* unexpected, strange, marvellous.

Paradoxornis Gr. *paradoxos,* strange; *ornis,* a bird.

paradoxus see **paradoxa.**

paraguaiae After Paraguay. According to Chambers's World Gazetteer and Geographical Dictionary (1959), p. 514, this name means 'parrot river'.

parambae After the Hacienda Paramba, Imbabura, Ecuador.

paramelanotos Gr. *para,* near, derived from; specific name *Calidris melanotos.*

Paramythia Gr. *paramuthia,* encouragement; the type specimen of the Crested Berrypecker *P. montium* was collected at Disappointment Rock, British New Guinea.

parasiticus L. *parasiticus,* parasitic (*parasitus,* a parasite).

Pardalotus Gr. *pardalotos,* spotted like a leopard (*pardalis* or *pardos,* a leopard).

Pardirallus Gr. *pardos,* a leopard; genus *Rallus,* rail; ref. spotted upperparts of the Spotted Rail *Rallus maculatus.*

parellina Mod. L. *parellinus,* violet-coloured, litmus violet (French *sans pareil,* matchless).

parens L. *parens,* evident.

pareola Dim. of genus *Parus,* tit.

Pareudiastes Gr. *pareudiastes,* of a kind of waterfowl that comes on land in fine weather; the extinct Black Wood Rail *P. pacificus* was found only on the island of Savaii in Samoa.

pariae After the Paria peninsula, Sucre, Venezuela.

parina Genus *Parus,* tit; L. *-inus,* resembling.

Parisoma Genus *Parus,* tit; Gr. *soma,* the body (i.e. appearance).

parisorum After the Paris brothers (fl. 1837) natural history dealers and collectors in Mexico.

parkeri After Theodore H. Parker III (fl. 1986) US ornithologist, field worker specializing in the Neotropics, and author (*An introduction to bird finding in Peru*, 1976; *An annotated check-list of Peruvian birds*, 1982).

parkesi After Kenneth C. Parkes (b.1922) US zoologist.

parkinsoni After Sydney Parkinson (1745–1771) English artist, collector, and explorer.

Parmoptila Gr. *parme,* a small, round shield; *ptilon,* a feather; ref. the scale-like head and throat feathers of the Brown Antpecker *P. woodhousei.*

parnaguae After the town of Parnaguá, Piauí, Brazil.

Paroaria Tupí (Brazilian) Indian name *tijeguacú paroara* for a small yellow, red, and grey bird.

parodii After José Parodi Vargas (fl. 1973) Peruvian landowner.

Parophasma Genus *Parisoma,* tit warbler; Gr. *phasma,* an apparition; the Abyssinian Catbird *P. galinieri* was originally described as a *Parisoma* because of its tit-like actions and habits.

Paroreomyza Gr. *para,* near; syn. genus *Oreomyza,* spikebill (v. *Oreomystis*) (cf. Gr. *paroreios,* near a mountain or mountains).

Parotia Gr. *parotis,* a lock or curl of hair by the ear; ref. the racquet-tipped head wires that spring from behind the eyes of the male parotias or six-wired birds of paradise.

Parula /parulus Dim. of genus *Parus,* tit.

Parus L. *parus,* a tit.

parva L. *parvus,* little, small.

parvirostris L. *parvus,* small; *-rostris,* billed.

parvula /parvulus L. *parvulus,* very small (dim. of *parvus,* small).

parvus see **parva.**

Paryphephorus Gr. *paruphe,* a border, hem or edge; *-phoros,* carrying.

parzudakii After Charles Parzudaki (fl.1847) French traveller and collector in Colombia, 1841–1845.

Passer L. *passer,* a sparrow.

Passerculus L. *passercula,* a little sparrow (dim. of *passer,* a sparrow).

Passerella Dim. of genus *Passer,* sparrow.

Passerherbulus L. *passer,* a sparrow; *herbula,* a little herb (dim. of *herba,* herb, grass); Le Conte's Sparrow *P. leconteii* is found in weedy fields, meadows, and rank grasses.

Passerina L. *passerinus,* passerine, sparrow-like.

passerinii After Carlos Passerini (1793–1857) Italian naturalist and collector.

passerinum /passerinus see **Passerina.**

pastinator L. *pastinator,* a trench digger; large flocks of the Long-billed Corella *Cacatua* feed on the ground, digging up roots, insects and their larvae.

Pastor L. *pastor,* a shepherd; the Rose-coloured Starling *Sturnus roseus* follows sheep to feed on their parasites and the insects disturbed by the grazing animals.

patachonicus /Patagona /patagonica /patagonicus After Patagonia, the arid southern plains of Argentina, southern Chile, and Tierra del Fuego. When Magellan wintered on its coasts in 1520 he traded with Indians, probably the giant Tehuelches who wore animal skins on their feet, whom he called *patagones,* the big footed, the name gradually attaching itself to the area.

patagonus Spanish name *el Patagón* given to the Burrowing Parrot *Cyanoliseus* by Azara (1805) because it migrated from Patagonia (v. *patachonicus).*

pauliani After Renaud Paulian (b. 1913) French zoologist in Madagascar, 1949–1960.

paulina After Pauline Knip (1781–1851) French bird artist and author (*Les pigeons,* 1811).

paulistus After the state of São Paulo, south-eastern Brazil.

pauper L. *pauper,* poor, scanty; the Medium Tree Finch *Camarhynchus* has plain plumage and a smaller bill than the Large Tree Finch *C. psittacula.*

Pauxi Spanish *paují,* the peacock *Pavo;* used for a variety of large gamebirds by early Spanish settlers in tropical America.

Pavo L. *pavo,* the peacock.

pavonina /pavoninus L. *pavoninus,* pavonine, coloured like the peacock's tail.

paykullii After Baron Gustav Paykull (1757–1826) Swedish naturalist.

paynteri After Raymond A. Paynter, Jr. (b. 1925) US ornithologist, author, and curator of the Museum of Comparative Zoology, Harvard, 1961.

paytensis After the coastal town of Payta (Paita), Piura, north-western Peru.

pealei /pealii After Titian Ramsay Peale (1800–1885) US naturalist, artist, and collector in the Pacific, 1838–1842.

pectardens L. *pectus,* the breast; *ardens,* glowing, burning (*ardere,* to burn).

pectorale /pectoralis L. *pectoralis,* pectoral, of the breast (*pectus,* the breast).

pecuarius L. *pecuarius,* a grazier; Kittlitz's Plover *Charadrius* inhabits grasslands and waterside habitats.

Pedilorhynchus Gr. *pedilon,* a boot, a covering for the foot; *rhunkhos,* the bill.

Pedioecetes Gr. *pedion,* the plains; *oiketes,* an inhabitant.

Pedionomus Gr. *pedionomos,* plain-dwelling, field-haunting.

pekinensis After the city of Pekin or Peking (Beijing), capital of China.

pelagica /pelagicus L. *pelagicus,* marine, pelagic (*pelagus,* the open sea; Gr. *pelagos,* the sea). In the case of the Chimney Swift *Chaetura* the name is thought to be a misprint for L. *Pelasgica,* Pelasgian (after the Pelasgi, one of the old nomadic tribes of Greece) and refers to the sp. migratory habits.

Pelagodroma Gr. *pelagodromos,* sea running (*pelagos,* the sea; *dromos,* running; *trekho,* to run).

Pelargopsis Gr. *pelargos,* the stork *Ciconia; opsis,* appearance; ref. the massive red bill of the Stork-billed Kingfisher *P. capensis.*

Pelecanoides Genus *Pelecanus,* pelican; Gr. *-oides,* resembling; ref. the distensible throat pouch and broad based, hook tipped bills of the diving petrels.

Pelecanus L. *pelicanus,* a pelican (Gr. *pelekan,* a pelican).

pelegrinoides Late L. *pelegrinus,* the Peregrine Falcon *Falco;* Gr. *-oides,* resembling.

pelewensis see **palauensis.**

peli After H. S. Pel (fl. 1854) Governor of Dutch Gold Coast, 1840–1850.

pelios Gr. *pelios,* bruised, livid; ref grey blotched breast and underparts of the African Thrush *Turdus.*

pella L. *pellos,* dark-coloured (Gr. *pellos,* dark-coloured, blackish).

Pellorneum Gr. *pellos,* dark-coloured, dusky; *orneon,* a bird.

peltata L. *peltatus,* armed with a light shield (*pelta,* a shield); ref. the shape of the red eye-wattles of the Black-throated Wattle-eye *Platysteira.*

Peltohyas Gr. *pelte,* a small, rimless shield; syn. genus *Hyas,* plover (Gr. myth. Hyas, son of Atlas and prince of Mauretania, killed whilst hunting); the Inland Dotterel *P. australis* has scales on both sides of the tarsus.

Peltops Gr. *pelte,* a small shield; *ops,* the face; the heavy billed Lowland Peltops or Clicking Shieldbill *P. blainvillii* was thought to be allied to the broadbills *Eurylaimus.*

pelzelni After August von Pelzeln (1825–1891) Austrian ornithologist, collector, and author (*Ornithologie Brasileiras,* 1871; *Beitrage zur ornithologie Süd Afrikas,* 1882).

pembae /pembaensis After the island of Pemba, Tanzania.

penduliger L. *pendulus,* pendant, hanging down; *-ger,* bearing; ref. the pendulous, feathered throat wattle of the male Long-wattled Umbrellabird *Cephalopterus.*

pendulinus L. *pendulus,* pendant, hanging down; *-inus,* possessing, resembling; ref. the distinctive hanging nest of the Masked Penduline Tit *Remiz.*

Penelope **1** Gr. myth. Penelope, a Greek princess, daughter of Icarius and wife of Ulysses, King of Ithaca; the reason for bestowing this name on the Neotropical guans is unknown. **2** Gr. *penelops,* a kind of duck (*Anas).*

Penelopides No expl. (Reichenbach (1849), *Av. syst. nat.,* pl.XLIX); Reichenbach was fond of employing classical names, and the genus may be misspelt from Gr. myth Pelopia, a daughter of Niobe, or after Pelopidas (d. 364 BC) Theban general and Governor of

Boeotia. However, perhaps from genus *Penelope,* guan, and Gr. *-oides,* resembling; ref. the bare facial skin and trumpeting call-notes of the otherwise unrelated hornbills and guans.

Penelopina Dim. of genus *Penelope,* guan.

Peneoenanthe L. *pene,* almost, nearly; genus *Oenanthe,* wheatear; the Mangrove Robin *P. pulverulenta* 'has the tail of a similar style to that of a wheatear' (Mathews (1920), *The birds of Australia,* Vol. **8**, p. 273).

Peneothello L. *pene,* almost, nearly; syn. genus *Othello,* antshrike (so called because the species described was black, like Shakespeare's character Othello the Moor); the White-winged Robin *P. sigillatus* is black with a white wing-patch.

penicillata/penicillatus Mod. L. *penicillatus,* tufted, with brush like tufts (L. *penicillus,* a painter's brush).

pennata/pennatus L. *pennatus,* feathered; ref. feathered upper tibia of Darwin's Rhea *Pterocnemia* and fully feathered tarsus of the Booted Eagle *Hieraeetus.*

Pennula L. *pennula,* a little wing; ref. the reduced wings of the extinct and flightless Hawaiian Rail *P. sandwichensis.*

pensylvanica After Pennsylvania, one of the original Thirteen Colonies which formed the embryo United States. Pennsylvania was named for the English Quaker, William Penn (1644–1718), who established the colony in 1681 and drew up its constitution.

Penthoceryx Gr. *penthos,* misfortune, grief; *kerux,* a herald; the cries of the Banded Bay Cuckoo *Cacomantis sonneratii* are supposed to foretell misfortune and bad weather.

Pentholaea Gr. *penthos,* mourning, grief; *laios,* a kind of thrush mentioned by Aristotle; ref. the black plumage of the White-fronted Black Chat *Myrmecocichla albifrons.*

pentlandii After Joseph Barclay Pentland (1798–1873) Irish traveller, diplomat in Bolivia, 1836–1839, and explorer.

peposaca Güaraní (Paraguayan) Indian name *peposaca* for the Rosybill *Netta.*

peracensis After the Sultanate of Perak, Malaya, later (1895) incorporated into the British Federated Malay States (Malaysia).

percivali After Arthur Blayney Percival (1875–1941) British game-ranger in East Africa, 1901–1928 and author (*A game rangers note book,* 1924; *A game ranger on safari,* 1928).

percnopterus Gr. *perknopteros,* a kind of vulture mentioned by Aristotle (*perknos,* dark, dusky; *-pteros,* winged).

Percnostola Gr. *perknos,* dark, dusky; *stolas,* a cloak; ref. the dark lead-grey plumage of the male Black-headed Antbird *P. rufifrons.*

percussus **1** L. *percussus,* slain, struck; ref. the bloody-red patches on head and breast of the Crimson-breasted Flowerpecker (*Prionochilus).* **2** L. *percussus,* a beating, a striking (*percutere,* to strike); the Cuban Green Woodpecker *Xiphidiopicus* searches for insects and grubs by tapping the bark of trees.

perdicaria L. *perdix, perdicis,* a partridge; *-arius,* pertaining to.

Perdicula Dim. of genus *Perdix,* partridge.

perdita L. *perditus,* lost, abandoned (*perdere,* to destroy); the Red Sea Cliff Swallow *Petrochelidon* is known only from a single migrant specimen.

Perdix L. *perdix,* a partridge (Gr. *perdix,* a partridge).

peregrina/peregrinus **1** L. *peregrinus,* a stranger, a wanderer (*peregre,* abroad); the Tennessee Warbler *Vermivora* breeds mainly in eastern Alaska and Canada and was described from a migrant specimen collected in Tennessee. **2** Med. L. *falco peregrinus,* the

Peregrine Falcon *Falco,* so called because young birds on their first long distance migration were better suited to falconry than those taken from the nest.

Pericrocotus Gr. *peri-,* very; *krokotos,* saffron-coloured.

perijana After the Sierra de Perijá, Zulia, Venezuela.

Periporphyrus Gr. *peri-,* very, exceedingly; *porphureos,* rosy (*porphura,* purple).

Perisoreus No expl. (Bonaparte (1831), *Giornale Arcadico di Scienze Lettere ed Arti,* **49**, 42). Bonaparte considered these jays allied to the nutcrackers *Nucifraga* and analogous to the tits *Parus,* and the name perhaps reflects their habit of storing or hoarding nuts, acorns, and other food (Gr. *perisoreuo,* to heap up, bury beneath). However, Coues (1882), p. 67, suggested the derivation L. *peri-,* very, exceedingly, and *sorix,* a bird of augury dedicated to Saturn, the god originally placated by human sacrifice (v. *infaustus*), and this seems to be reinforced by Bonaparte who nominated a totally dark grey (i.e. funereal) bird as the type of his new genus.

Perissocephalus Gr. *perissos,* extraordinary, marvellous; *kephale,* the head; ref. the grotesque bare face of the Calfbird or Capuchinbird *P. tricolor.*

Perissotriccus Gr. *perissos,* extraordinary, marvellous; Mod. L. *triccus,* tyrant flycatcher.

perkeo No expl. (Neumann (1907), *Journ. für Orn.,* **55**, 352), but the name is suggestive of diminution and apt for this smallest species of *Batis* flycatcher.

perlata /perlatum /perlatus Mod. L. *perlatus,* pearlated, marked with pearl-shaped or tear-shaped spots (Med. L. *perla,* a pearl).

permista L. *permixta,* mixed.

Pernis Gr. *pernes,* a kind of hawk.

pernix L. *pernix,* agile, nimble.

peronii After François Peron (1775–1810) French naturalist, collector, and voyager.

perousii After Captain Jean François de Galaup Comte de la Pérouse (1741–1788) French explorer in the Pacific, 1785–1788.

perquisitor L. *perquisitor,* a hunter, a seeker.

perreini After Jean Perrein (1750–1805) French naturalist and collector in tropical Africa and America.

perrotii After Jean Perrot (1790–1858) French taxidermist.

persa L. *Persa,* Persian; the Guinea Turaco *Tauraco* has a crowned or crested appearance like a Persian head-dress.

persicus L. *Persicus,* Persian.

personata /personatus L. *personatus,* masked (*persona,* a mask).

personus L. *personus,* resounding, ringing.

perspicax L. *perspicax,* sharp-sighted, acute.

perspicillata /perspicillatum /perspicillatus Mod. L. *perspicillatus,* spectacled (Mod. L. *perspicillum,* a lens or spectacle; L. *perspicere,* to see through).

perstriata L. *per-,* very; Mod. L. *striatus,* striped, streaked (L. *stria,* a furrow).

pertinax L. *pertinax,* tenacious, obstinate.

peruana /peruviana After Peru, until 1821 the Spanish Vice-Royalty of Peru. An erroneous toponym in the cases of *Tangara* and *Vini.*

perversa L. *perversus,* crooked, distorted (*pervertere,* to overturn).

petechia Italian *petecchia,* a small red spot on the skin; the Golden Warbler *Dendroica* has chestnut or rufous spots and streaks on yellow underparts.

petenica After the Petén region of northern Guatemala.

petersi After James Lee Peters (1889–1952) US ornithologist and author (*Check-list of birds of the world,* 1931).

petersoni After Roger Tory Peterson (b. 1908) US field ornithologist, artist, and author (*A field guide to the birds,* 1934; *A field guide to western birds,* 1941; *A field guide to the birds of Britain and Europe,* 1954; *A field guide to Mexican birds,* 1973).

petiti After Louis Petit (1856–1943) French naturalist and collector in Cabinda enclave, 1876–1884.

Petrochelidon Gr. *petros,* a rock; *khelidon,* a swallow; many of these swallows nest in caves.

Petroica Gr. *petros,* a rock; *oikos,* inhabiting (*oikeo,* to inhabit, dwell); ref. the rocky habitat of the Scarlet Robin *P. multicolor* on Norfolk Island.

Petronia L. *petronius,* of rocks or mountains (Gr. *petros,* a rock).

Petrophassa Gr. *petros,* a rock; *phassa,* a pigeon.

petrophila Gr. *petros,* a rock; *philos,* loving, fond of.

petrosus L. *petrosus,* rocky, full of rocks; the Stone Partridge *Ptilopachus* occurs in rocky habitats north of the equator in tropical Africa.

Peucedramus Gr. *peuke,* a fir tree; *dromos,* running (*trekho,* to run); the Olive Warbler *P. taeniatus* is found in montane coniferous forests of Middle America and extreme south-western USA.

Pezites Gr. *pezites,* a walker, a marcher.

Pezopetes Gr. *pezos,* walking; *petes,* flying, a bird; ref. the ground-dwelling habits of the Large-footed Finch *P. capitalis.*

Pezophaps Gr. *pezos,* walking; *phaps,* a pigeon; the extinct solitaires and dodos are considered related to the pigeons.

Pezoporus Gr. *pezoporos,* walking, pedestrian.

Phacellodomus Gr. *phakelos,* a bundle; *domos,* a house; ref. the large stick nests of the Neotropical thornbirds.

Phaenicophaeus Gr. *phoinikophaes,* of crimson appearance; ref. the strikingly coloured face of the Crimson-faced Malkoha *P. pyrrhocephalus.*

Phaenicophilus Gr. *phoinix, phoinikos,* the palm tree; *philos,* loving, fond of.

phaenicuroides Specific name *Phoenicurus phoenicurus;* Gr. *-oides,* resembling.

Phaenostictus Gr. *phaino,* to display (cf. *phoinix,* crimson); *stiktos,* spotted; the Ocellated Antbird *P. mcleannani* has large buff-edged black spots on the upperparts.

phaeocephalus Gr. *phaios,* dusky, brown; *-kephalos,* headed.

phaeocercus Gr. *phaios,* dusky, brown; *kerkos,* the tail.

Phaeochroa Gr. *phaios,* dusky, brown; *khroa,* skin colour.

phaeochromus Gr. *phaios,* dusky, brown; *khroma,* colour.

Phaeomyias Gr. *phaios,* dusky, grey-brown; Mod. L. *myias,* flycatcher.

phaeonotus Gr. *phaios,* dusky, brown; *-notos,* backed.

Phaeoprogne Gr. *phaios,* dusky, brown; genus *Progne,* martin.

phaeopus Gr. *phaios,* dusky, grey-brown; *pous,* the foot.

phaeopygia Gr. *phaios,* dusky, brown; Mod. L. *pygius,* rumped (Gr. *puge,* the rump).

phaeopygus Gr. *phaios,* dusky, brown; *-pugos,* rumped.

Phaeornis Gr. *phaios,* dusky, grey; *ornis,* a bird.

phaeosoma Gr. *phaios,* dusky, brown; *soma,* the body.

Phaeothlypis Gr. *phaios,* dusky, brown; Mod. L. *thlypis,* warbler.

Phaeotriccus Gr. *phaios,* brown, dusky; Mod. L. *triccus,* tyrant flycatcher.

Phaethon Gr. *phaethon,* the sun (*phaos,* light); in Gr. myth. Phaethon was the disobedient son of Phoebus or Sol, the sun, and the Oceanid Clymene. Newton and Gadow (1896), p. 990, wrote of the Red-billed Tropicbird *P. aethereus,* 'Linnaeus bestowed upon it the generic term ... in allusion to its attempt to follow the path of the sun'.

Phaethornis Gr. *phaethon,* the sun (cf. *phaios,* plain); *ornis,* a bird.

phaeton Gr. myth. Phaethon, son of Phoebus the sun, and who upset the chariot of the sun.

Phaetusa Gr. myth. Phaetusa, sister to Phaethon and one of the daughters of Phoebus or Sol and Clymene.

Phainopepla/phainopeplus Gr. *phaeinos,* brilliant, splendid; *peplos,* a cloak; ref. the black shining plumage of the male Phainopepla *P. nitens* and the glittering green upperparts of the Santa Marta Sabrewing *Campylopterus.*

Phainoptila Gr. *phaeinos,* brilliant, splendid; *ptilon,* a feather, plumage.

phaionota Gr. *phaios,* dusky, brown; *-notos,* backed.

Phalacrocorax L. *phalacrocorax,* the cormorant (Gr. *phalakros,* bald; *korax,* the raven).

Phalaenoptilus Gr. *phalaina,* a moth; *ptilon,* plumage; ref. the soft, brown, cryptic plumage of the Common Poorwill *P. nuttallii.*

Phalaropus Gr. *phalaris,* the coot *Fulica; pous,* the foot; ref. lobed, coot-like feet of the Grey Phalarope *P. fulicarius.*

Phalcoboenus Late Gr. *phalkon,* a falcon *Falco;* Gr. *baino,* to walk, to go; from the terrestrial habits of the caracaras.

phalerata L. *phaleratus,* adorned with bosses or medallions; ref. glittering patches on crown and throat of the White-tailed Star-frontlet *Coeligena.*

Phapitreron Genus *Phaps,* bronzewing pigeon; genus *Treron,* green pigeon.

Phaps Gr. *phaps,* the Wood Pigeon *Columba.*

pharetra L. *pharetra,* a quiver; ref. the triangular or arrow-shaped marks on the underparts of the Arrowhead Warbler *Dendroica.*

Pharomachrus Gr. *pharos,* a loose cloak or mantle; *makros,* long; ref. the spectacular flowing train of the Resplendent Quetzal *P. mocinno.*

phasianella/phasianellus Dim. of L. *phasianus,* a pheasant.

phasianinus L. *phasianinus,* of pheasants, pheasant-like.

Phasianus L. *phasianus,* a pheasant. On the banks of the River Phasis in ancient Colchis the Argonauts found numerous pheasants. In orn. *phasis* is often used in substitution for *phasianus.*

Phasidus Gr. *phasidos,* of a pheasant (*phasianos,* a pheasant).

phayrei After Lieutenant General Sir Arthur Purves Phayre (1812–1885) first Commissioner of British Burma, 1862–1867, Governor of Mauritius, 1874–1878, and author (*History of Burma,* 1883).

Phedina No expl. (Bonaparte (1857), *Riv. Contemp. Torino,* **9**, 210); perhaps from Gr. *phaios,* brown and Malagasy name *tsidisidina* for a martin (cf. Italian *fedina,* a side whisker; Gr. *phaidimos,* sleek, famous).

Phedinopsis Genus *Phedina,* martin; Gr. *opsis,* appearance.

Phegornis Gr. *phengos,* light, splendour; *ornis,* a bird.

phelpsi/Phelpsia After William H. Phelps (1875–1965) Venezuelan ornithologist, collector, and author (*Lista de las aves de Venezuela,* 1958).

Pheucticus Gr. *pheuktikos,* shy, inclined to avoid.

Phibalura Gr. *phibalos,* slender, thin; *oura,* the tail.

Phigys Syn. specific name *Psittacus phigy,* a phonetic rendering for the islands of Fiji, the home of the Collared Lory *P. solitarius.*

philadelphia/philadelphicus After the city of Philadelphia, Pennsylvania, USA.

philbyi After Harry St John Bridge Philby (1885–1960) British explorer, Arabist, and advisor to King Ibn Saud.

Philemon Gr. *philemon,* affectionate, kissing; Newton and Gadow (1896), p. 292 footnote, explained the name of the Black-necked Friarbird *P. moluccensis,* 'Commerson had said that the word Polochion, which expressed the cry and was the name of the bird, signified *baisons-nous* [let us kiss], and hence proposed to call it *Philemon …*'.

Philentoma Gr. *phil-,* loving, fond of; *entoma,* insects; ref. the flycatching habits of these Indomalayan monarch flycatchers.

Philepitta Syn. genus *Philedon,* honeyeater (Gr. *philedonos,* fond of pleasure; v. *Philemon)*; genus *Pitta*, pitta; the asities were considered to possess characters of both of these groups.

Philesturnus Syn. genus *Philedon*, friarbird, honeyeater; genus *Sturnus*, starling; from the supposed characters of the Saddleback *Creadion carunculatus*.

Philetairus Gr. *phil-*, loving, fond of; *etairos*, a companion (*etes*, a kinsman) (cf. *philaiteros*, more beloved [comp. of *philos*, loved]); the Sociable Weaver *P. socius* is a highly gregarious species, breeding in huge communal nests.

philippae After Philippa Williams (fl. 1955) wife of British ornithologist John Williams.

philippensis After the Philippine Islands, named for Felipe (Philip)II King of Spain (1527–1598: reigned 1556–1598).

philippii After Filippo de' Filippi (1814–1867) Italian zoologist.

philippinensis /philippinus see **philippensis.**

phillipsi After **1** E. Lort Phillips (1857–1944) English collector and big-game hunter in East Africa, 1885–1895 (*Oenanthe)*. **2** Allan R. Phillips (fl. 1987) US ornithologist and taxonomist specializing in the Neotropics (*Tangara)*.

Philodice Gr. myth. Philodice, wife of Leucippus and daughter of Inachus.

Philohela Gr. *philo-*, loving, fond of; *helos*, a marsh or meadow.

Philohydor Gr. *philo-*, loving, fond of; *hudor*, water (cf. *philhudros*, water-loving).

Philomachus Gr. *philomakhos*, pugnacious, fond of fighting; male Ruffs *P. pugnax* gather in leks to dance and fight with one another.

philomela /philomelas L. *philomela*, the Nightingale *Luscinia*. In Gr. myth. Philomela, a daughter of Pandion, was changed into a nightingale.

Philortyx Gr. *philos*, allied with; *ortux*, the quail.

Philydor Gr. *philhudros*, water-loving.

Phimosus Gr. *phimos*, a muzzle (*phimoso*, to silence); the Whispering Ibis *P. infuscatus* is silent away from the nest.

Phlegopsis Gr. *phlego*, to flame or burn; *opsis*, appearance; ref. distinctive red facial patches of the bare-eyes.

Phleocryptes Gr. *phleos*, a reed or rush; *kruptes*, a skulker or hider (*krupto*, to conceal).

Phloeoceastes Gr. *phloios*, tree bark; *keazo*, to split or cleave; a group of large Neotropical woodpeckers with powerful bills.

Phlogophilus Gr. *phlox*, *phlogos*, a flame; *philos*, allied with; ref. the glittering plumage of the piedtails.

Phlogothraupis Gr. *phlox*, a flame; genus *Thraupis*, tanager.

Phodilus Gr. *phos*, *photos*, light; *deilos*, timorous, timid; the Bay Owl *P. badius* is strictly nocturnal.

phoebe **1** Gr. myth. Phoebe, a name for the goddess Diana (*Metallura)*. **2** 'Phoebe Flycatcher', an alliterative name given to the Eastern Phoebe *Sayornis* by Latham (1787).

Phoebetria Gr. *phoibetria*, a prophetess or soothsayer (cf. *phobetron*, an object of terror); no expl. (Reichenbach (1853), *Av. syst. nat.*, p. 5), but Coues (1882), writes, 'These words are with great propriety and correct sentiment applied to albatrosses, the import of whose weird presaging will be felt by one who reads Coleridge's *Antient Mariner*, or himself goes down the deep in ships'.

phoenicea /phoeniceus Gr. *phoinikeos*, all shades of dark red, red.

Phoenicircus Gr. *phoinix*, *phoinikos*, crimson red; *kerkos*, the tail.

phoenicius see **phoenicea.**

phoenicobia Gr. *phoinix*, *phoinikos*, the palm tree; *bios*, manner of living.

phoenicomitra Gr. *phoinix*, crimson red; *mitra*, a head-band, a girdle.

Phoeniconaias Gr. *phoinix*, crimson red; *naias*, a water nymph; the pink and red Lesser Flamingo *P. minor* inhabits salt lakes of tropical Africa.

Phoenicoparrus Gr. *phoinix*, crimson red; syn. genus *Parra*, jacana (but used in ornithology for a variety of unrelated waterbirds); these flamingos inhabit Andean salt lakes.

phoenicoptera Gr. *phoinix, phoinikos,* crimson red; *-pteros,* winged.

Phoenicopterus L. *phoenicopterus,* the flamingo (Gr. *phoinix,* crimson red; *-pteros,* winged).

phoenicotis Gr. *phoinix,* crimson red; *-otis,* eared.

Phoeniculus Dim. from Gr. *phoinix, phoinikos,* crimson red; ref. the red bill of the Green Wood-hoopoe *P. purpureus.*

phoenicura/Phoenicurus L. *phoenicurus,* the redstart (Gr. *phoinikouros,* the redstart). Specifically the epithet refers to birds which, like the redstart, have red tails (Gr. *phoinix, phoinikos,* crimson red; *-ouros,* tailed).

Pholia Contraction of syn. genus *Pholidauges,* starling (Gr. *pholis, pholidos,* a scale or metal plate; *auges,* translucent; *augeo,* to glitter); ref. violet and blue spangled upperparts of Sharpe's Starling *Cinnyricinclus sharpii.*

Phonygammus Gr. *phone,* a cry or sound; *gamos,* wedlock, marriage (i.e. a bride of sound); the New Guinea Trumpetbird *P. keraudrenii* has distinctive loud calls, and although no etymology is given in the original diagnosis (Lesson and Garnot (1826), *Bull. Sci. Nat.*, **8**, 110) the tracheal modification is referred to.

Phormoplectes Gr. *phormos,* a basket; *plektos,* plaited, twined (*pleko,* to weave).

Phragmaticola Gr. *phragmites,* a kind of reed; L. *-cola,* inhabitant.

phryganophila Gr. *phruganon,* a dry stick; *philos,* loving, fond of; the Chotoy Spinetail *Schoeniophylax* makes a bulky nest of interwoven twigs and sticks.

phrygia L. *Phrygius,* Phrygian (Gr. *Phrugios)*; Phrygia was a country of Asia Minor whose inhabitants were skilled in embroidering with gold; the ref. is to the gold-edged wings and tail of the Regent Honeyeater *Xanthomyza.*

Phrygilus Gr. *phrugilos,* a bird mentioned by Aristophanes, not otherwise identified, but conjectured by later writers to be a sparrow *Passer.*

Phylidonyris French *phylédon,* honeyeater (substantive equivalent of syn. genus *Philedon,* honeyeater); syn. genus *Cinnyris,* sunbird.

Phyllanthus Gr. *phullon,* a leaf; genus *Anthus,* pipit; the Capuchin Babbler *P. atripennis* was considered to resemble a laughing-thrush *Garrulax* but with the tail of a pipit-like longclaw *Macronyx.*

Phyllastrephus Gr. *phullon,* a leaf; *strepho,* to toss or bend; these Afrotropical bulbuls scratch about in the forest leaf litter.

Phyllergates Gr. *phullon,* a leaf; *ergates,* a workman, a worker.

Phyllolais Genus *Phylloscopus,* warbler; genus *Hippolais,* warbler.

Phyllomyias Gr. *phullon,* a leaf; Mod. L. *myias,* flycatcher.

Phylloscartes Gr. *phullon,* a leaf; *skartes,* a skipper or dancer.

Phylloscopus Gr. *phullon,* a leaf; *skopos,* a watcher.

Phytotoma Gr. *phuton,* a plant; *temno,* to cut; the plantcutters have bills admirably suited to snipping and cutting buds and fruit.

piaggiae After Carlo Piaggi (1827–1882) Italian explorer and collector in tropical Africa, 1876–1882.

Pica L. *pica,* the magpie. Used specifically for birds which, like the magpie, are pied black and white.

picaoides L. *pica,* the magpie; Gr. *-oides,* resembling; ref. the very long black and white tail of the Long-tailed Sibia *Heterophasia.*

picata Mod. L. *picatus,* pied like a magpie (L. *pica,* the magpie) (cf. L. *picatus,* bedaubed with tar, tarred; *pix, picis,* pitch).

Picathartes Genus *Pica,* magpie; genus *Cathartes,* vulture; ref. the pied plumage and bald vulture-like heads of the Afrotropical picathartes or bald-crows.

picatus see **picata.**

picazuro Güaraní (Paraguayan) Indian name *picazúró*, sour pigeon (from the taste of its flesh after it has eaten certain fruits), for the Picazuro Pigeon *Columba*.

picina L. *picinus*, pitch black (*pix, picis*, pitch).

picirostris Genus *Picus*, woodpecker; L. *-rostris*, billed.

pickeringii After Charles Pickering (1805–1878) US ornithologist.

Picoides Genus *Picus*, woodpecker; Gr. *-oides*, resembling.

picta /pictum L. *pictus*, painted (*pingere*, to paint).

picturata L. *picturatus*, painted, variegated.

pictus see **picta.**

picui Güaraní (Paraguayan) Indian name *picúi* for a small dove.

Piculus Dim. of genus *Picus*, woodpecker.

Picumnus French *picumne*, piculet (coined by Temminck (1825)) (dim. of genus *Picus)*.

Picus L. *picus*, a woodpecker (Gr. *pikos*, a woodpecker).

Piezorhina Gr. *piezo*, to squeeze or crush; *rhis, rhinos*, the nose; ref. compressed nostrils in heavy, laterally slender or 'squeezed' bill of the Cinereous Finch *P. cinerea*.

pilaris **1** Mod. L. *pilaris*, of the head (L. *pilus*, the hair) (*Atalotriccus)*. **2** Mod. L. *pilaris*, a thrush; a confused coining mistaking Gr. *trikhas*, a thrush, with *thrix, trikhos*, hair (*Turdus)*.

pileata /pileatus L. *pileatus*, capped (*pileus*, a felt cap).

Pilherodias Gr. *pilos*, a cap; *erodios*, a heron.

Pilorhinus Gr. *pilos*, felt cloth; *rhis, rhinos*, the nose; ref. the small, soft feathers covering the nostrils of the White-billed Starling *P. albirostris*.

pinaiae After Mount Pinaia (Gunung Pinaia) on the island of Seram, in the Moluccas.

Pinarochroa Gr. *pinaros*, dirty, shabby; *khroos*, coloured (*khros*, colour).

Pinarocorys Gr. *pinaros*, dirty, shabby; syn. genus *Corys*, lark.

Pinaroloxias Gr. *pinaros*, dirty, shabby; genus *Loxia*, crossbill; ref. the black plumage and curved bill of the Cocos Finch *P. inornata*.

Pinarornis Gr. *pinaros*, dirty, shabby; *ornis*, a bird.

Pindalus No expl. (Gurney (1862), *Ibis*, 152); perhaps a misprint for Pindarus (fl. 400 BC), a lyric poet of Thebes, in allusion to the sweet song of the Yellow-throated Warbler *Seicercus ruficapillus*.

Pinguinus English *penguin*, the name given to the Great Auk *P. impennis* by fisherman in Newfoundland in the sixteenth century; the name was subsequently transferred by association to the unrelated penguins of the Antarctic.

Pinicola L. *pinus*, a pine tree; *-cola*, inhabitant.

pinnatus L. *pinnatus*, feathered, winged.

pinon After L. Pinon (fl. 1824) wife of French explorer Captain Louis Freycinet.

pintadeanus French *pintade*, a guineafowl (Portuguese *pintado*, spotted); L. *-anus*, relating to.

pinus L. *pinus*, a pine tree.

Pionites Genus *Pionus*, parrot; Gr. *-oides*, resembling.

Pionopsitta Genus *Pionus*, parrot; Mod. L. *psitta*, a parrot.

Pionus Gr. *pionos*, plump, sleek.

pipiens L. *pipiens*, chirping (*pipare*, to chirp).

Pipile L. *pipilare*, to pipe or chirp; the piping guans are so called from their piping songs.

Pipilo L. *pipilo*, I chirp or twitter (*pipilare*, to twitter).

pipixcan According to Hernandez (1615), *Historia avium Novae Hispaniae*, *pipizcan* was the name given by Mexican Indians to a sort of gull (cf. Spanish *apipizca*, a gull).

Pipra Gr. *pipra* (a variant of *pipo)*, a small bird mentioned by Aristotle and other authors, never properly identified, but thought by some later commentators to be a pied woodpecker *Dendrocopos*. Its association with the colourful Neotropical manakins is purely arbitrary.

Pipraeidea Genus *Pipra*, manakin; Gr. *eidos*, species, form; the Fawn-breasted Tanager *P. melanonota* was considered to have a manakin-like bill.

Pipreola Dim. of genus *Pipra*, manakin.

Piprites Genus *Pipra*, manakin; Gr. *-oides*, resembling.

Pipromorpha Genus *Pipra*, manakin; Gr. *morphe*, shape, appearance.

Piranga Tupí (Brazilian) Indian name *tijepiranga* for some small bird.

piscator L. *piscator*, a fishmonger; a name based on the French *pêcheur du Sénégal*, fisher of Senegal (d'Aubenton (1765), *Planches enluminéez d'histoire naturelle*, pl. 478). The plate, of an unidentifiable raptor, cannot, by any stretch of the imagination, be associated with the Western Grey Plantain-eater *Crinifer*.

pistrinaria L. *pistrinarius*, a miller (*pistrina*, a bakehouse); ref. 'dusted' appearance of plumage of the Island Imperial Pigeon *Ducula*.

pitangua /Pitangus Tupí (Brazilian) Indian name *pitanga guasu* for a large flycatcher.

Pithecophaga Gr. *pithekophagos*, ape-eating (*pithekos*, an ape; *-phagos*, eating; *phagein*, to devour).

Pithys No expl. (Vieillot (1818), *Nouv. dict. hist. nat.*, Vol. **24**, p. 112); perhaps misspelt after Pythis, the Greek sculptor, or after Pytho, the ancient name of Delphi, seat of the oracle of Apollo, or (more likely) a combination of genus *Pipra*, manakin, and Mod. L. *thlypis*, a warbler.

pitiayumi Güaraní (Paraguayan) Indian name *pitiayumi*, meaning 'little yellow chest', for the Tropical Parula *Parula*.

pitius Local Chilean names *pitiu* or *pitigue* for the Chilean Flicker *Colaptes*.

Pitohui No expl.(Lesson (1830), *Traité d'ornithologie*, p. 375); perhaps based on Maori name *pitoitoi* for the unrelated robin flycatchers *Petroica*.

Pitta Telegu (Malayan) name *pitta*, a pretty bauble or pet.

Pittasoma Genus *Pitta*, pitta; Gr. *soma*, body (i.e. appearance).

pittoides Genus *Pitta*, pitta; Gr. *-oides*, resembling.

Pitylus Gr. *pitulos*, a small bird mentioned by Hesychius.

pityophila Gr. *pitus*, *pituos*, a pine tree; *philos*, loving, fond of.

Pityriasis Gr. *pituriasis*, scurf, bran like eruptions on the head (*pituron*, bran); ref. the bare, bristle covered head of the Bornean Bristlehead *P. gymnocephala*.

piurae After the department of Piura, Peru.

placens L. *placens*, charming (*placere*, to please).

placentis L. *placentis*, charming, pleasing.

placida /placidus L. *placidus*, quiet, gentle; coined to emphasize the demure appearance and gentle nature of the Peaceful Dove *Geopelia*, and to contrast and reflect the relationship of the Long-billed Plover *Charadrius* to the Killdeer Plover *C. vociferus*.

plagiata L. *plaga*, a weal, a stripe; ref. heavily barred underparts of the Grey Hawk *Asturina*.

plancus L. *plancus*, a kind of eagle (Gr. *plangos*, a type of eagle mentioned by Aristotle).

Platalea L. *platalea*, the Spoonbill *P. leucorodia*.

platenae After **1** Carl Constantin Platen (1843–1899) German physician at Amoy, China, who collected in the East Indies, 1878–1894 (*Prioniturus*) and his wife **2** Margarete Platen (fl. 1894) (*Ficedula, Gallicolumba*).

plateni After Carl Constantin Platen (v. *platenae* **1**).

platensis After **1** the Río de la Plata, Argentina, so named after a silver knife found there by the Portuguese explorer Estebão Froes in 1513 (*Leptasthenura)*. **2** the Spanish Vice-royalty of Río de la Plata (roughly equivalent to modern Argentina, Uruguay and Paraguay) (*Cistothorus, Embernagra)*.

platurus Gr. *platus,* wide, broad; *-ouros,* tailed.

Platycercus Gr. *platukerkos,* broad-tailed, flat-tailed.

Platycichla Gr. *platus,* broad; *kikhle,* thrush; these Neotropical thrushes are distinguished by 'a short and very broadly cleft bill, much depressed at the base' (Baird (1864), *Rev. Amer. birds*, Vol. **1**, p. 32).

Platylophus Gr. *platus,* broad, flat; *lophos,* a crest.

Platypsaris Gr. *platus,* broad, wide; syn. genus *Psaris,* cotinga, becard (Gr. *psar,* the starling); the Neotropical becards have broad bills.

platypterus Gr. *platus,* broad, wide; *-pteros,* winged, feathered.

platyrhynchos /platyrhynchum / Platyrinchus Gr. *platurhunkhos,* broad-billed.

platyrostris Gr. *platus,* broad, wide; L. *-rostris,* billed.

Platysmurus Gr. *platusmos,* widening; *oura,* the tail; the Black Magpie *P. leucopterus* has a long, rounded tail which is partly spread in flight.

Platyspiza Gr. *platus,* broad, wide; *spiza,* a finch; the Vegetarian Finch *P. crassirostris* has a short, deep, broad bill.

Platysteira Gr. *platus,* broad, wide; *steira,* a ship's keel; ref. the broad bill of the Afrotropical wattle-eyes.

platyura see **platurus.**

Plautus L. *plautus,* flat-footed (cf. Gr. *plotos,* swimming; *ploo,* to float).

plebejus L. *plebeius,* common, vulgar (*plebs, plebis,* the common people).

Plectorhyncha Gr. *plektron,* a spear point; *rhunkhos,* the bill; ref. the fine, pointed bill of the Striped Honeyeater *P. lanceolata.*

Plectrophenax Gr. *plektron,* a cock's spur; *phenax,* imposter; this name replaced the syn. genus *Plectrophanes* (Gr. *phanes,* showing, exhibiting; *phaino,* to show); both refer to the long hind claw of the snow buntings.

Plectropterus Gr. *plektron,* a cock's spur; *-pteros,* winged; both sexes of the Spur-winged Goose *P. gambensis* possess long, sharp, carpal spurs.

Plegadis Gr. *plegas, plegados,* a sickle; ref. the sickle-shaped bill of the Glossy Ibis *P. falcinellus.*

plengei After Manuel A. Plenge (fl. 1973) Peruvian ornithologist.

pleschanka Russian name *pleshanka* for the Pied Wheatear *Oenanthe.*

pleskei After Theodor Pleske (1858–1932) Russian zoologist and author (*Ornithographia Rossica*, 1889).

pleurostictus Gr. *pleura,* the side; *stiktos,* spotted, dappled.

plicatus L. *plicatus,* folded (*plicare,* to fold); ref. the wrinkled casque on top of the bill of the Papuan Hornbill *Rhyticeros.*

Ploceella Dim. of genus *Ploceus,* weaver.

Plocepasser Genus *Ploceus,* weaver; genus *Passer,* sparrow.

Ploceus Gr. *plokeus,* a weaver or braider (*pleko,* to weave).

plumata L. *plumatus,* plumate, plumed (*pluma,* a small soft feather).

plumbea L. *plumbeus,* leaden (*plumbum,* lead) (Mod. L. *plumbeus,* lead-grey).

plumbeiceps L. *plumbeus,* leaden; *-ceps,* headed.

plumbeitarsus L. *plumbeus,* leaden; Gr. *tarsos,* the flat of the foot.

plumbeiventris L. *plumbeus,* leaden; *venter, ventris,* the belly.

plumbeus see **plumbea.**

plumifera L. *pluma,* a plume, a small soft feather; *-fera,* bearing.

plumosus L. *plumosus,* feathered, downy (*pluma,* a plume); of the Boulder Chat *Pinarornis* Sharpe (1876) in Layard's *The birds of South Africa,* p. 230, wrote, '...it differs in the long loose plumage of the rump, and in the long upper tail-coverts..'.

plumula/plumulus L. *plumula,* a little feather, a down feather.

pluricinctus L. *plus, pluris,* many (comp. of *multus,* much); *cinctus,* banded, girdled.

pluto Roman myth Pluto, lord of the lower world beyond the grave; ref. black plumage of the Pluto Flycatcher *Myiagra.*

Pluvialis L. *pluvialis,* relating to rain (*pluvia,* rain); various explanation have been put forward for associating the plovers with rain (plumage dappled as if with raindrops, migrations to avoid rain, wet habitats, etc.), but the true origin is now lost in time. The cries of the Chestnut-bellied Cuckoo *Hyetornis* are said to presage rain (v. *Hyetornis).*

Pluvianellus Dim. of genus *Pluvianus,* plover.

Pluvianus French name *le Pluvian du Sénégal,* given to the Egyptian Plover *P. aegyptius* by Buffon (1770) (*pluvier,* a plover).

Pluviorhynchus French *pluvier,* a plover; Gr. *rhunkhos,* the bill.

Pnoepyga Gr. *pnoe,* air; *puge,* the rump; the wren babblers have short, degenerate tail feathers covered by long, soft, puffy rump feathers.

Podager L. *podager,* a gouty man; the small feet of the Nacunda Nighthawk *P. nacunda* render walking difficult.

podargina Genus *Podargus,* frogmouth; L. *-inus,* resembling.

Podargus French name *podarge* given to the frogmouths by Cuvier in 1798 with ref. to their short, weak legs (Gr. *podargos,* a gouty man).

Podica L. *podicus,* of or belonging to a foot; ref. the lobed feet of the African Finfoot *P. senegalensis.*

Podiceps L. *podex, podicis,* the vent or anus; *pes,* the foot; the legs of grebes are set back to the rear of the body.

Podilymbus Genus *Podiceps,* grebe; syn. genus *Colymbus,* grebe (North American usage) or diver *Gavia* (European usage).

podobe French name *le Podobé du Sénégal* given to the Black Bush Robin *Cercotrichas* by Buffon (1770).

Podoces Gr. *podokes,* fleet, swift-footed.

poecilocercus Gr. *poikilos,* spotted, variegated; *kerkos,* the tail.

poecilochrous Gr. *poikilokhroos,* of various colours.

Poecilodryas Gr. *poikilos,* spotted, variegated; *druas,* a dryad or tree-nymph.

poecilolaemus Gr. *poikilos,* spotted; *laimos,* the throat.

poecilonota Gr. *poikilonotos,* with back of various hues.

poeciloptera/poecilopterus Gr. *poikilopteros,* with wings of changeful hue.

poecilorhyncha/poecilorhynchus Gr. *poikilos,* spotted, variegated; *rhunkhos,* the bill.

poecilorrhoa Gr. *poikilos,* spotted, variegated; *orrhos,* the tail or rump.

poecilosterna Gr. *poikilos,* spotted, variegated; *sternon,* the breast.

Poecilothraupis Gr. *poikilos,* spotted, variegated; genus *Thraupis,* tanager.

poecilotis Gr. *poikilos,* spotted, variegated; *-otis,* eared.

Poecilotriccus Gr. *poikilos,* spotted, variegated; Mod. L. *triccus,* tyrant flycatcher.

Poecilurus Gr. *poikilos,* spotted, variegated; *-ouros,* tailed.

poensis After the island of Fernando Póo, Spanish Guinea (Bioko, Equatorial Guinea); named after the Portuguese explorer Fernão do Póo, who discovered the island in 1470.

Poeoptera Gr. *phaios,* grey, dusky; *-pteros,* winged; the male Narrow-tailed Starling *P. lugubris* has an ash-grey wing patch.

Poephila Gr. *poe,* grass; *philos,* loving, fond of.

Pogoniulus Dim. of syn. genus *Pogonias,* barbet (Gr. *pogonias,* bearded).

Pogonocichla Gr. *pogon, pogonos,* a beard; *kikhle,* a thrush; ref. the white loral and breast patches of the Starred Bush Robin *P. stellata.*

Pogonotriccus Gr. *pogon,* a beard; Mod. L. *triccus,* tyrant flycatcher; ref. the prominent rictal bristles of the bristle tyrants.

Poicephalus Gr. *phaios,* grey, dusky; *-kephalos,* headed. Some later authors amend the name to *Poeocephalus,* as if from Gr. *poio-,* grassy (i.e. green) (*poia,* grass).

poiciloptilus Gr. *poikilos,* spotted; *ptilon,* a feather.

poicilotis see **poecilotis.**

poioicephala see **Poicephalus.**

Polemaetus Gr. *polemos,* battle, war; *aetos,* an eagle; ref. the large size and aggressive habits of the Martial Eagle *P. bellicosus.*

Polemistria Gr. *polemistria,* an Amazon, a female warrior (*polemistes,* a warrior).

Polihierax Gr. *polios,* grey; *hierax,* a hawk.

poliocephala /poliocephalum / Poliocephalus Gr. *polios,* grey; *-kephalos,* headed.

poliocerca Gr. *polios,* grey; *kerkos,* the tail.

poliogaster /poliogastra Gr. *polios,* grey; *gaster, gastros,* the belly.

poliogenys Gr. *polios,* grey; *genus,* the cheek.

Poliolais Gr. *polios,* grey; Mod. L. *lais,* warbler.

Poliolimnas Gr. *polios,* grey; Mod. L. *limnas,* a rail.

poliolopha /poliolophus Gr. *polios,* grey; *lophos,* a crest or tuft.

polionota /polionotus Gr. *polios,* grey; *-notos,* backed.

poliopareia Gr. *polios,* grey; *pareia,* the cheek.

poliophrys Gr. *polios,* grey; *ophrus,* the eyebrow.

poliopleura Gr. *polios,* grey; *pleura,* the side.

polioptera /poliopterus Gr. *polios,* grey; *-pteros,* winged.

Polioptila Gr. *polios,* grey, hoary; *ptilon,* a feather, plumage.

poliosoma Gr. *polios,* grey; *soma,* the body.

poliothorax Gr. *polios,* grey; *thorax,* the breast.

Polioxolmis Gr. *polios,* grey; genus *Xolmis,* monjita.

Polipicus Gr. *polos,* a maiden; genus *Picus,* woodpecker.

polleni /pollenii After François Pollen (1842–1866) French naturalist, collector in Madagascar, 1863–1866 and author (*Recherches sur la faune de Madagascar et de ses dépendances*, 1868).

pollens L. *pollens,* powerful, strong.

Polyboroides Genus *Polyborus,* caracara; Gr. *-oides,* resembling.

Polyborus Gr. *poluboros,* voracious, greedy.

polychopterus Gr. *polu-*, many, varied; *khros*, colour, complexion; *-pteros*, winged.

polychroa Gr. *polu-,* very much; *khroa,* colour; ref. the uniform colouration of the Brown Prinia *Prinia.*

Polyerata Gr. *polu-*, very much; *eratos,* lovely (cf. *poluerastos,* much-loved).

polyglotta /polyglottos Gr. *poluglottos,* many-tongued, harmonious.

polygramma Gr. *polugrammos,* white-streaked (originally a white-streaked jasper).

Polyonymus Gr. *poluonomos,* famous, renowned.

polyosoma Gr. *polu-*, many; *soma,* the body; ref. the five colour phases of the Red-backed Hawk *Buteo.*

Polyplancta Gr. *poluplanktos,* beguiling.

Polyplectron Gr. *polu-,* many; *plektron,* a cock's spur; male peacock pheasants have up to three sharp tarsal spurs.

Polysticta /Polystictus Gr. *polustiktos,* much spotted, dappled.

Polytelis Gr. *poluteles,* sumptuous, extravagant.

Polytmus Gr. *polutimos,* costly, valuable.

Pomarea After Pomare II King of Tahiti (1781–1821).

Pomareopsis Genus *Pomarea,* monarch flycatcher; Gr. *opsis,* appearance.

pomarina/pomarinus **1** After Pomerania, a former dukedom of Poland, bordering the Baltic Sea (*Aquila*). **2** Gr. *poma, pomatos,* a lid; *rhis, rhinos,* the nose; the Pomarine Skua *Stercorarius* developes a sheath or cover over the nostrils during the breeding season.

Pomatorhinus Gr. *poma,* a lid; *rhis,* the nose; Horsfield (1821), *Trans. Linn. Soc. London,* **13**, 164, wrote of the Chestnut-backed Scimitar Babbler *P. montanus,* 'The corneous covering of the nares and its sudden compression and equal breadth beyond these form the distinguishing character of this genus'.

Pomatostomus Gr. *poma,* a lid; *stoma,* the mouth; ref. relationship and similarity to genus *Pomatorhinus.*

pompadora After Jeanne Antoinette Poisson Marquise de Pompadour (1721–1764) mistress of Louis XV King of France and whose favourite colour was purple.

pondicerianus After the former French colony of Pondichéry (Puduchcheri, Tamil Nadu), south-west of Madras, India.

Pooecetes Gr. *poe,* grass; *oiketes,* a dweller or inhabitant (*oikeo,* to inhabit).

poortmani After W. Poortman (fl. 1843) Dutch ornithologist and collector in Colombia.

Poospiza Gr. *poe,* grass; *spiza,* a finch.

Poospizopsis Genus *Poospiza,* warbling finch; Gr. *opsis,* appearance.

Popelairia/popelairii After Baron Popelaire de Terloo (fl. 1845) Belgian naturalist and collector in Peru.

porphyraceus Gr. *porphureos,* purple, red; L. *-aceus,* resembling.

porphyrea Gr. *porphureos,* purple, red.

porphyreolophus Gr. *porphureos,* purple; *lophos,* a crest or tuft.

Porphyrio L. *porphyrio,* the Purple Swamphen *P. porphyrio* (Gr. *porphurion).*

Porphyriops Genus *Porphyrio,* swamphen; Gr. *ops,* face (i.e. appearance).

Porphyriornis Genus *Porphyrio,* swamphen; Gr. *ornis,* a bird.

porphyreocephala Gr. *porphureos,* purple, red; *-kephalos,* headed.

Porphyrolaema Gr. *porphura,* purple; *laimos,* the throat.

Porphyrospiza Gr. *porphura,* purple; *spiza,* a finch.

Porphyrula Dim. of genus *Porphyrio,* swamphen.

portoricensis After the island of Porto Rico (Puerto Rico), easternmost of the Greater Antilles, discovered by Columbus in 1493 and settled by the Spanish in 1510.

Porzana Local Venetian (Italian) names *sporzana* or *porzana* for the smaller crakes.

Porzanula Dim. of genus *Porzana,* crake.

praecox L. *praecox,* premature; of the Cocha Antshrike *Thamnophilus* Zimmer (1937), *Amer. Mus. Nov.,* **917**, 1, wrote, 'Possibly *praecox* will some day be found to be a representative of *nigriceps,* but with the male unknown it would be undesirable to do more at the present time than suggest the possibility of such relationship...'.

Praedo L. *praedo,* a robber.

praetermissa L. *praetermissus,* overlooked, neglected.

prasina L. *prasinus,* leek-green (Gr. *prasinos,* leek-green; *prason,* a leek).

prasinorrhous Gr. *prasinos,* leek-green; *orrhos,* the rump or tail.

prasinus see **prasina.**

pratensis L. *pratensis,* of a meadow, found in meadows (*pratum, prati,* a meadow).

pratincola L. *pratum, prati,* a meadow; *incola,* an inhabitant or resident. *Pratincola* was formerly a generic name for the stonechats *Saxicola.*

preciosa L. *pretiosus,* costly, precious.

Premnoplex Gr. *premnon,* a tree trunk; *plesso,* to strike.

Premnornis Gr. *premnon,* a tree trunk; *ornis,* a bird.

presbytes Gr. *presbutis,* an old woman; ref. grey and white streaked head of the Timor Leaf Warbler *Phylloscopus* (cf. *presbus,* a bird mentioned by Aristotle and Hesychius, identified as the wren *Troglodytes* by some recent authors).

pretiosa see **preciosa.**

pretrei After Jean Gabriel Prêtre (1800–1840) French museum artist.

preussi After Paul Preuss (1861–1926) German botanist and collector in West Africa, 1886–1888, 1892–1898 and New Guinea, 1903.

prevostii After Florent Prévost or Prévôt (d.1870) French museum artist and author (*Iconographie ornithologique,* 1845; *Histoire naturelle des oiseaux d'Europe,* 1864).

prigoginei After A. Prigogine (b. 1913) Belgian zoologist, collector, and author.

primigenius L. *primigenius,* original, primitive; the Mountain Sunbird *Aethopyga* was regarded as a primitive species linking *Aethopyga* with *Cinnyris.*

primolinus After the three sons of Charlotte Honorine Joséphine Pauline Contessa Primoli di Foglia (1832–1901), grandsons of French ornithologist Prince Charles Bonaparte.

princei After Edwin C. Prince (d. 1873) English naturalist and secretary to John Gould.

princeps L. *princeps,* chief, leader, most distinguished.

principalis L. *principalis,* princely, imperial (*princeps, principis,* chief).

pringlii After Colonel Sir John Wallace Pringle (1863–1938) Chief Inspector of Railways Uganda, 1891–1892.

Prinia Javanese name *prinya* for the Bar-winged Prinia *P. familiaris.*

Priocella Syn. genus *Prion,* prion (Gr. *prion,* a saw; ref. serrated edges of mandibles); genus *Procellaria,* petrel.

priocephalus Misspelt from Gr. *polios,* grey; *-kephalos,* headed.

Prioniturus Gr. *prion, prionos,* a saw; *oura,* the tail; the tails of the racquet-tailed parrots have a ragged appearance.

Prionochilus Gr. *prion,* a saw; *kheilos,* a lip; ref. the minutely serrated tomia (margins of the bill) of the flowerpeckers.

Prionodura Gr. *prionodes,* serrated; *oura,* the tail; of the Golden Bowerbird *P. newtoniana* De Vis (1883), *Proc. Linn. Soc. New South Wales,* **7**, 561–562, wrote, 'Tail moderate, of twelve feathers which are subspinose at the apex...An aberrant form of the family...exaggerating the exsertion of the tips of the tail feathers...'.

Prionops Gr. *prion,* a saw; *ops,* the eye; Vieillot (1816), *Analyse,* p. 41, described the White Helmet-shrike *P. plumata* as having *paupières dentelées* (notched eyelids), but he was referring to the jagged or serrate-edged eye-wattles of this sp.

Priotelus Gr. *prion,* a saw; *telos,* accomplished, complete; from the ragged appearance of the Cuban Trogon's *P. temnurus* tail.

pririt French name *le Pririt* given to the Pririt Batis *Batis* by Levaillant (1805) in imitation of its call notes.

pristoptera Gr. *pristos,* sawn (*prister,* a saw); *-pteros,* winged; ref. the barbed edges of the outer primaries whence the Afrotropical saw-winged swallows get their name.

Pristorhamphus Gr. *prister, pristeros,* a saw; *rhamphos,* the bill.

pritchardii After William Thomas Pritchard (1829–1909) British Consul to Fiji, 1857–1362.

Probosciger L. *proboscis,* a snout or trunk (Gr. *proboskis,* the elephant's trunk); *-ger,* bearing; ref. the huge, curved bill of the Palm Cockatoo *P. aterrimus.*

Procellaria L. *procella,* a storm or gale; *-arius,* pertaining to; the shearwaters seem at home in the most violent tempests at sea.

Procelsterna Genus *Procellaria,* shearwater; genus *Sterna,* tern.

procerus L. *procerus,* long, large; ref. the extremely long sickle-shaped bill of the Greater Akialoa *Hemignathus.*

Procnias Gr. myth. Prokne or Procne, who was metamorphosed into a swallow; genus formerly applied to the Swallow Tanager *Tersina*, the female of which resembles a miniature female bellbird.

Proctopus Gr. *proktos*, the rear, the anus; *pous*, the foot (v. *Podiceps*).

procurvoides Syn. specific name *Xiphorhynchus procurvus* (now *Campylorhamphus trochilirostris*) (L. *procurvus*, curved forwards); Gr. *-oides*, resembling.

Prodotiscus Dim. of Gr. *prodotes*, a traitor, a betrayer; ref. the honeyguide habit of leading man and mammals to bees nests, although this behaviour is not recorded for the honeybirds.

productus L. *productus*, long, lengthened; ref. the elongated upper mandible of the extinct Norfolk Island Kaka *Nestor*.

Progne L. *progne*, a swallow (Gr. myth. Prokne or Procne, daughter of Pandion who was transformed into a swallow).

promeropirhynchus Genus *Promerops*, sugarbird; Gr. *rhunkhos*, the bill; ref. curved bill of the Strong-billed Woodcreeper *Xiphocolaptes*.

Promerops Gr. *pro-*, related to, close to (in orn. denotes priority, proximity, or intensity); genus *Merops*, bee-eater; ref. the curved bill of the sugarbirds; formerly an assortment of curve-billed, but otherwise unrelated, birds were included here.

propinqua/propinquus L. *propinquus*, related, similar.

proprium L. *proprius*, peculiar, special.

Propyrrhula Gr. *pro-*, related to, close to; genus *Pyrrhula*, bullfinch.

proregulus Gr. *pro-*, related to, close to; specific name *Regulus regulus*.

Prosobonia No expl. (Bonaparte (1850), *Comptes Rendus de l'Académie des Sciences, Paris*, **3**, 562). The Gr. *pro-* indicates relationship, and perhaps the second part of the name is an oblique ref. to the syn. genus *Pisobia*, sandpiper (Gr. *pisos*, a meadow; *bios*, livelihood, life).

Prosopeia Gr. *prosopeion*, a mask.

Prosthemadera Gr. *prosthema*, an appendage, an addition; *dere*, the neck; ref. the curled feathers and filamentous plumes on the neck of the Tui *P. novaeseelandiae*.

prosthemelas Gr. *prosthe*, before, in front of; *melas*, black; ref. the black hood and mantle of the Black-cowled Oriole *Icterus*.

Protonotaria French name *Le Protonotaire*, given to the Prothonotary Warbler *P. citrea* by Buffon (1770) (Late L. *protonotarius*, a Byzantine court notary, later a protonotary or Vatican registrar who wore golden yellow robes).

Proturacus Gr. *pro-*, before, earlier; syn. genus *Turacus*, turaco (v. *Tauraco*).

provocator L. *provocator*, a gladiator, a challenger; the Kandavu Honeyeater *Foulehaio* was discovered by naturalists on the HMS *Challenger* expedition to the Pacific, 1873–1876.

Prunella Dim. of Med. L. *prunus*, brown; ref. the rufous brown upperparts of the familiar Dunnock *P. modularis*.

prunellei After Clément François Victor Gabriel Prunelle (1777–1853) French scientist.

pryeri After H. J. S. Pryer (1850–1888) lepidopterist, collector in Japan, and author (*Rhopalocera Nihonica*, 1886).

przewalskii After General Nikolai Mikhailovitch Przhevalsky (1839–1888) Russian explorer and naturalist in central Asia, and author (*Mongoliya i strana Tangutov*, 1875).

Psalidoprocne Gr. *psalis, psalidos*, a pair of shears or scissors; L. *progne*, a swallow; the saw-winged swallows have barbed edges to the outer primaries.

Psaltria L. *psaltria*, a female harp singer (Gr. *psaltria*).

Psaltriparus Genus *Psaltria*, pygmy tit; genus *Parus*, tit.

Psamathia Gr. myth. Psamathe, one of the nereids or sea-nymphs; ref. the remote island habitat of the Palau Warbler *P. annae*.

psammocromius Gr. *psammos*, sand; *khroma*, complexion, colour; ref. yellow shoulder patch of the Mountain Widow *Coliuspasser*.

Psarisomus Syn. genus *Psaris*, cotinga, becard (Gr. *psar*, the starling); Gr. *soma*, the body; of the Long-tailed Broadbill *P. dalhousiae* Swainson (1837), *Classif. birds*, Vol. **2**, p. 261, wrote, 'Never having seen a specimen, I can only judge of this remarkable type by the figure in Dr Royle's work on Hymalayan Natural History. Its striking resemblance to a *Psaris*, and the smallness of its bill, makes me conclude it is the most aberrant, or tenuirostral type'.

Psarocolius Gr. *psar*, the starling; *koloios*, the Jackdaw *Corvus*.

Pselliophorus Gr. *pseliophoros*, wearing bracelets (*pselion*, an armlet or bracelet; *-phoros*, bearing); ref. conspicuous yellow thigh tufts of the Yellow-thighed Finch *P. tibialis*.

Psephotus Gr. *psephos*, a pebble, a mosaic stone; ref. variegated mosaic patterns of these beautiful Australian parrots.

Pseudacanthis Gr. *pseud-*, false, pseudo (in orn. indicating relationship or close similarity); genus *Acanthis*, linnet.

Pseudaedon Gr. *pseud-*, false, pseudo; syn. genus *Aedon*, nightingale (Gr. *aedon*, the nightingale).

Pseudalaemon Gr. *pseud-*, false, pseudo; genus *Alaemon*, hoopoe lark.

Pseudastrapia Gr. *pseud-*, false, pseudo; genus *Astrapia*, long-tailed bird of paradise.

Pseudattila Gr. *pseud-*, false, pseudo; genus *Attila*, attila.

Pseudeos Gr. *pseud-*, false, pseudo; genus *Eos*, lory.

Pseudhirundo Gr. *pseud-*, false, pseudo; genus *Hirundo*, swallow.

Pseudibis Gr. *pseud-*, false, pseudo; *ibis*, an ibis.

Pseudoalcippe Gr. *pseudo-*, false, pseudo; genus *Alcippe*, hill babbler.

Pseudobias Gr. *pseudo-*, false, pseudo; genus *Bias*, flycatcher.

Pseudobulweria Gr. *pseudo-*, false, pseudo; genus *Bulweria*, petrel.

Pseudocalyptomena Gr. *pseudo-*, false, pseudo; genus *Calyptomena*, green broadbill.

Pseudochelidon Gr. *pseudo-*, false, pseudo; *khelidon*, a swallow.

Pseudocolaptes Gr. *pseudo-*, false, pseudo; genus *Colaptes*, flicker, woodpecker.

Pseudocolopteryx Gr. *pseudo-*, false, pseudo; genus *Colopteryx*, pygmy tyrant.

Pseudocossyphus Gr. *pseudo-*, false, pseudo; genus *Cossypha*, robin chat.

Pseudodacnis Gr. *pseudo-*, false, pseudo; genus *Dacnis*, dacnis.

Pseudogyps Gr. *pseudo-*, false, pseudo; genus *Gyps*, griffon vulture.

Pseudoleistes Gr. *pseudo-*, false, pseudo; genus *Leistes*, blackbird.

Pseudonestor Gr. *pseudo-*, false, pseudo; genus *Nestor*, kaka, parrot; ref. the heavy, parrot-like bill of the Maui Parrotbill *P. xanthophrys*.

Pseudonigrita Gr. *pseudo-*, false, pseudo; genus *Nigrita*, waxbill.

Pseudopodoces Gr. *pseudo-*, false, pseudo; genus *Podoces*, ground jay.

Pseudoptynx Gr. *pseudo-*, false, pseudo; syn. genus *Ptynx*, eagle owl (Gr. *ptunx*, a bird mentioned by Aristotle, probably the eagle owl *Bubo*).

Pseudoscops Gr. *pseudo-*, false, pseudo; syn. genus *Scops*, scops owl (Gr. *skops*, the little eared owl).

Pseudoseisura Gr. *pseudo-*, false, pseudo; genus *Seisura*, flycatcher (Gr. *seisoura*, a bird mentioned by Hesychius, subsequently identified as a wagtail *Motacilla*); reflecting a sham relationship of the Neotropical cacholotes with the genus *Motacilla*.

Pseudospingus Gr. *pseudo-*, false, pseudo; genus *Hemispingus*, tanager, hemispingus.

Pseudototanus Gr. *pseudo-*, false, pseudo; syn. genus *Totanus*, redshank (Italian *totano*, the redshank *Tringa*).

Pseudotriccus Gr. *pseudo-,* false, pseudo; Mod. L. *triccus,* tyrant flycatcher.

pseudozosterops Gr. *pseudo-,* false, pseudo; genus *Zosterops,* white-eye.

psilolaemus Gr. *ptilon,* a feather (cf. *psilos,* simple, plain); *laimos,* the throat; of the Moorland Francolin *Francolinus* Gray (1867), *List birds Brit. Mus., Gallinae,* p. 50, wrote, 'Allied to *F. gutturalis,* Rüpp., but it is easily distinguished by the throat being white, with numerous black spots, one of which is situated at the apex of each feather'.

Psilopogon Gr. *psilos,* bare, naked; *pogon,* a beard; ref. the threadbare whiskers or beard of the Fire-tufted Barbet *P. pyrolophus.*

Psilopsiagon Gr. *psilos,* bare, naked; *siagon,* the jaw bone; Ridgway (1912), *Proc. Biol. Soc. Washington,* **25**, 100, differentiated the Mountain Parakeet *Bolborhynchus aurifrons* '...in having the feathers of the cheeks short (normal), not covering sides of mandible..'. No reason is given in the original etymology for the euphonic 'p'; perhaps Gr. *ops,* appearance, was intended.

Psilorhamphus Gr. *psilos,* simple; *rhamphos,* the bill; the Spotted Bamboo-wren *P. guttatus* was originally given the syn. name *Leptorhynchus* (Gr. *leptos,* slender, delicate; *rhunkhos,* the bill).

Psilorhinus Gr. *psilos,* bare, smooth; *rhis, rhinos,* the nose; ref. the unfeathered nostrils of the Brown Jay *P. morio.*

psittacea Mod. L. *psitta,* parrot; L. *-aceus,* resembling.

Psittacella Dim. of genus *Psittacus,* parrot.

Psittacula Dim. of genus *Psittacus,* parrot.

Psittaculirostris Genus *Psittacula,* parakeet; L. *-rostris,* billed; ref. the large, notched bills of the New Guinea fig parrots.

Psittacus L. *psittacus,* a parrot (Gr. *psittakos,* also *psittake, sittas,* and *sittakos,* a parrot).

Psitteuteles Mod. L. *psitta,* parrot; Gr. *euteles,* paltry, small.

Psittinus Mod. L. *psitta,* a parrot; L. *-inus,* resembling.

Psittrichas Mod. L. *psitta,* a parrot; Gr. *thrix, trikhos,* the hair; ref. the rigid, finely barbed feathers of Pesquet's Parrot *P. fulgidus.*

Psittirostra Mod. L. *psitta,* a parrot; L. *rostrum,* the bill; ref. the curved, parrot-like bill of the Hawaiian Ou *P. psittacea.*

Psomocolax Gr. *psomos,* a morsel or scrap; *kolax,* a parasite, a fawner.

Psophia Gr. *psophos,* any inarticulate sound; the trumpeters make curious, deep, ventriloquial booming calls of uncertain source, and formerly they were considered to be breaking wind.

Psophocichla Gr. *psophos,* a noise; *kikhle,* a thrush.

Psophodes Gr. *psophodes,* noisy.

ptaritepui After the mountain Cerro Ptari-tepui, Bolívar, Venezuela.

pteneres Gr. *ptenos,* winged; *eresso,* to row (*eretes,* a rower); ref. the way in which the Flightless Steamer Duck *Tachyeres* ploughs through the water.

Pteridophora Gr. *pteris, pteridos,* a feathery leaved fern; *-phoros,* carrying; ref. the extraordinary head-wires of the male King of Saxony's Bird of Paradise *P. alberti.*

Pternistis Gr. *pternistes,* one who trips with the heel; ref. the double spurs of the cock spurfowl.

Pterocles Gr. *pteron,* a wing; *-kles* (in composition), noted for, endowed with; ref. the very long primaries and swift flight of the sandgrouse.

Pterocnemia Gr. *pteron,* a feather; *kneme,* the leg; Darwin's Rhea *P. pennata* has the upper tibia feathered, instead of bare as in *Rhea.*

Pterodroma Gr. *pteron,* a wing; *dromos,* running (*trekho,* to run); ref. the swift but erratic flight of the gadfly petrels over the sea.

Pteroglossus Gr. *pteron,* a feather; *glossos,* the tongue; ref. the long, feather-like tongue characteristic of the toucans and aracaris.

Pteronetta Gr. *pteron,* a wing; *netta,* a duck; Hartlaub's Duck *P. hartlaubii* has a distinctive china-blue wing patch.

Pterophanes Gr. *pteron,* a wing; *phanes,* shining (*phaino,* to shine forth); the Great Sapphirewing *P. cyanopterus* has shining blue wings.

Pteropodocys Gr. *pteron,* a bird; *podokes,* swift-footed; the Ground Cuckoo-shrike *P. maxima* is mainly terrestrial.

Pteroptochos Gr. *pteron,* a wing; *ptokhos,* mean, beggarly; ref. the very short wings of the secretive, terrestrial Moustached Turca *P. megapodius.*

Pteruthius Gr. *pteron,* a wing; *eruthriao,* to be red; ref. the chestnut wing-patch of the White-browed Shrike Babbler *P. flaviscapis.*

Ptilinopus Gr. *ptilon,* a feather; *pous,* the foot; the fruit doves have fully feathered tarsi.

Ptilocichla Gr. *ptilon,* a feather; *kikhle,* a thrush; ref. the fluffy, degenerate rump feathers of the wren babblers.

ptilocnemis Gr. *ptilon,* a feather; *kneme,* the leg; the Rock Sandpiper *Calidris* has the upper part of the leg feathered almost to the knee.

ptilogenys Gr. *ptilon,* a feather; *genus,* the cheek; the Ceylon Grackle *Gracula* has a fully feathered face, unlike its congener the yellow-wattled Hill Grackle *G. religiosa.*

Ptilogonys Gr. *ptilon,* a feather; *gonu,* the knee; Baird (1866), *Rev. Amer. birds*, Vol. **I**, p. 411, wrote of the Grey Silky-flycatcher *P. cinereus*, 'There has been much diversity among writers in the spelling of the name of this genus, Mr Swainson, its author, having himself written it very differently. His first rendering of the name, however – *Ptilogonys* – is more nearly correct than the subsequent ones, as compounded of *ptilon* and *gonu,* or feathered knee, in allusion to the tuft of cottony feathers on the tibiae.'

Ptilolaemus Gr. *ptilon,* a feather; *laimos,* the throat; unlike many related genera the throat of the Brown Hornbill *P. tickelli* is fully feathered.

Ptilonorhynchus Gr. *ptilon,* a feather; *rhunkhos,* the bill; feathers cover the base of the bill, including the nostrils, of the Satin Bowerbird *P. violaceus.*

Ptilopachus Gr. *ptilon,* a feather; *pakhus,* thick, dense; the Stone Partridge *P. petrosus* has 'Tail broad, rounded…the feathers very soft. Feathers of the back and rump with the shafts thickened,and apparently spinous…' (Swainson (1837), *Classif. birds*, Vol. **2**, p. 344).

Ptiloprora Gr. *ptilon,* a feather; *prora,* the face; the honeyeaters in this genus have feathered faces, unlike their relations in *Philemon.*

ptilorhynchus Gr. *ptilon,* a feather; *rhunkhos,* the bill; the Crested Honey Buzzard *Pernis* has small, dense feathers on the forehead and lores, a protective adaptation to its diet.

Ptiloris Gr. *ptilon,* a feather; *rhis,* the nose; the riflebirds have the nostrils covered by dense feathering.

Ptilorrhoa Gr. *ptilon,* a feather; *orrhos,* the rump; ref. the lax, lengthened rump feathers of the jewel babblers.

Ptiloscelys Gr. *ptilon,* a feather; *skelos,* the leg; the Andean Lapwing *Belonopterus resplendens* has the upper part of the leg fully feathered.

Ptilostomus Gr. *ptilon,* a feather; *stoma,* the mouth; short, dense feathers cover the nostrils of the Piapiac *P. afer.*

ptilosus Gr. *ptilon,* a feather; L. *-osus,* abundance (cf. Gr. *ptilosis,* feathered); ref. the full, lax rump and back feathers of the Fluffy-backed Tit Babbler *Macronous.*

Ptilotis Gr. *ptilon,* a feather; *-otis,* eared.

Ptiloxena Gr. *ptilon,* a feather; *xenos,* strange; Chapman (1892), *Bull. Amer. Mus. Nat. Hist.*, **4**, 307, wrote of the Cuban Blackbird *Dives atroviolacea,* 'The most striking difference...is in the structure of the contour feathers of the fore parts of the body, especially those of the breast...the barbicels are fasciculate, and except at its basal third, adhere to the barb. For the terminal half of the feather...the barbs have no connection with one another, and this gives...a finely streaked hairy appearance.'

Ptochoptera Gr. *ptokhos,* mean, poor; *pteron,* wings.

Ptychoramphus Gr. *ptux, ptukhos,* a wrinkle; *rhamphos,* the bill; ref. small transverse ridges on the bill of Cassin's Auklet *P. aleuticus.*

Ptyonoprogne No expl. (Reichenbach (1850), *Av. syst. nat.*, pl.87, fig. 6), but probably from Gr. *ptuon,* a fan, and genus *Progne,* martin, with ref. to tail shape of these small martins.

Ptyrticus Gr. *pturtikos,* timorous.

pubescens L. *pubescens,* pubescent, downy (*pubescere,* to become pubescent, to put on the down of puberty); ref. the stiff bristles covering the nostrils of the Downy Woodpecker *Dendrocopos,* a feature more greatly developed in the larger Hairy Woodpecker *D. villosus.*

pucherani /pucheranii After Jacques Pucheran (1817–1894) French zoologist and author (*Voyage au Pôle Sud et dans l'Océanie sur les corvettes l'Astrolabe et la Zélée*, 1842).

Pucrasia Nepali name *pokras* for the Koklas Pheasant *P. macrolopha.*

pudibunda L. *pudibundus,* modest, bashful (*pudere,* to be ashamed); ref. the reddish throat patch of the Cañon Canastero *Asthenes.*

puella /puellus L. *puella,* a girl; *puellus,* a boy.

Puffinus English *puffin,* originally applied to the cured carcass of the nestling shearwater, a delicacy until the end of the eighteenth century. By confusion the name was gradually applied to the puffin *Fratercula,* but retained in ornithology for the shear-waters.

pugnax L. *pugnax,* combatative, fond of fighting (*pugnare,* to fight).

pulacayensis After the settlement of Pulacayo, Potosí, Bolivia.

pulchella /pulchellum /pulchellus L. *pulchellus,* very pretty (dim. of *pulcher,* beautiful).

pulcher L. *pulcher,* beautiful, lovely.

pulcherrima /pulcherrimus L. *pulcherrimus,* very beautiful (super. of *pulcher,* beautiful).

pulchra see **pulcher.**

pulchricollis L. *pulcher,* beautiful; Mod. L. *-collis,* necked.

pulitzeri After Ralph Pulitzer (1879–1939) US publisher and sponsor.

pullaria L. *pullarius,* of young animals or birds; ref. small size and ruddy, healthy cheeks of the Red-faced Lovebird *Agapornis.*

pullicauda L. *pullus,* dusky, blackish; *cauda,* the tail; of the Orange-billed Lorikeet *Neopsittacus* Hartert (1896), *Nov. Zool.*, **3**, 17, wrote, 'tail...much darker green above, and without an indication of orange tips, below dark brownish green instead of ochre yellow or orange'.

pulpa L. *pulpa,* flesh; the Sagon Bush Lark *Mirafra* is known only by a single, fat specimen.

Pulsatrix L. *pulsator,* a beater (*pulsavi,* to strike); ref. the rhythmic hooting calls of these Neotropical owls.

pulverulenta /pulverulentus L. *pulverulentus,* dusty, covered with a fine powder or bloom (*pulvis, pulveris,* dust).

pumilio L. *pumilio,* a dwarf.

pumilus L. *pumilus,* a dwarf.

puna Spanish *puna* (Quechua Indian *puna),* the high bleak Andean plateau above about 10000 feet (3050 m).

Punanetta After the Andean puna (v. *puna)*; Gr. *netta,* a duck.

punctata /punctatum /punctatus Mod. L. *punctatus,* spotted (L. *puntum, puncti,* a spot).

puncticeps L. *punctum, puncti,* a spot; *-ceps,* headed.

punctifrons L. *punctum,* a spot; *frons,* the forehead or brow.

punctigula L. *punctum,* a spot; *gula,* the throat.

punctulata /punctulatus Mod. L. *punctulatus,* spotted (L. *punctulum,* a small point).

punctuligera L. *punctulum, punctuli,* a small spot; *-gera,* bearing.

punensis **1** After the department of Puno, south-eastern Peru (*Asthenes, Grallaria, Phrygilus).* **2** after the Andean puna (v. *puna)* (*Geositta).*

punicea /puniceus L. *puniceus,* purple, red.

purnelli After Herbert A. Purnell (fl. 1931) Australian naturalist and collector.

purpurascens L. *purpurascens,* purplish (*purpurascere,* to grow purple).

purpurata /purpuratus L. *purpuratus,* clad in purple (*purpura,* purple).

purpurea L. *purpureus,* purple-coloured (Gr. *porphureos,* purple, red).

purpureicauda L. *purpureus,* purple-coloured; *cauda,* the tail.

Purpureicephalus L. *purpureus,* purple-coloured, red-coloured; Gr. *-kephalos,* headed.

purpureiceps L. *purpureus,* purple-coloured; *-ceps,* headed.

purpureiventris L. *purpureus,* purple-coloured; *venter, ventris,* the belly.

purpureus see **purpurea.**

purpuropterus L. *purpureus,* purple-coloured; Gr. *-pteros,* winged.

purusianus After the Rio Purús, Amazonas, Brazil.

pusilla /pusillus L. *pusillus,* very small, tiny.

pusio L. *pusio,* a little boy.

pustulatus L. *pustulatus,* blistered, pimply (*pustula,* a pimple or pustule); ref. the spotted and streaked mantle of the Flame-headed Oriole *Icterus.*

puveli After Pierre Puvel (fl. 1901) French settler in Portuguese Guinea (Guinea Bissau).

Pycnonotus Gr. *puknos,* thick, compact; *noton,* the back; the bulbuls have thickly feathered backs.

Pycnoptilus Gr. *puknos,* thick, excessive; *ptilon,* a feather; the Pilotbird *P. floccosus* has dense, silky plumage.

Pycnopygius Gr. *puknos,* dense, thick; Mod. L. *-pygius,* rumped (Gr. *puge,* the rump); these New Guinea honeyeaters have thickly feathered rumps.

pycrofti After A. T. Pycroft (1875–1971) New Zealand naturalist and collector.

pygargus Gr. *pugargos,* a bird of prey mentioned by Aristotle, Pliny, and Hesychius, not further identified, but since associated with a variety of eagles and the Hen Harrier *Circus.*

Pygarrhichas Gr. *puge,* the rump; *arrhikos,* a basket; ref. the stiff shafts and projecting tail spines of the White-throated Treerunner *P. albogularis.*

Pygiptila Gr. *puge,* the rump; *ptilon,* a feather; ref. the thickly feathered rump and lower back of the Spot-winged Antshrike *P. stellaris.*

pygmaea /pygmaeum /pygmaeus / pygmeus L. *pygmaeus,* pygmean, pygmy (Gr. *pugmaios,* dwarfish).

Pygochelidon Gr. *puge,* the rump; *khelidon,* a swallow.

Pygoscelis Gr. *puge,* the rump; *skelos,* the leg; the legs of penguins are set far back on the body, admirable for swimming, but forcing them to adopt an upright posture on land.

pylzowi After Lieutenant Mikhail Pylzov (fl. 1870) Russian explorer in central Asia.

pyra Gr. *pura,* a funeral pyre, a burning mass.

Pyrenestes Gr. *puren,* fruit stones; *-estes,* an eater (*edo,* to devour); the seedcrackers have heavy bills adapted for eating seeds, but they favour soft seeds rather than fruit stones.

pyrgita Gr. *purgites,* a sparrow *Passer.*

pyrhopterum Gr. *purrhos,* flame-coloured, red (*pur,* fire); *-pteros,* winged.

Pyriglena Gr. *puriglenos,* fiery-eyed.

Pyrilia Dim. of Gr. *pura*, a funeral pyre, a burning mass (Med. L. *pyralia*, torches).

Pyrocephalus Gr. *purrhos*, flame-coloured, red; *-kephalos*, headed.

Pyroderus Gr. *purrhos*, flame-coloured, red; *deras*, the neck.

pyrohypogaster Gr. *puro-*, fire, red; *hupo*, beneath; *gaster*, the belly.

pyrolophus Gr. *puro-*, fire, red; *lophos*, a crest or tuft.

Pyrope Gr. *puropes*, fiery-eyed (cf. Med. L. *pyropus*, a ruby).

pyropygia Gr. *puro-*, fire, red; Mod. L. *pygius*, rumped (Gr. *puge*, the rump).

Pyrrherodia Gr. *purrhos*, purple, red; *erodios*, a heron.

pyrrhocephalus see **Pyrocephalus.**

Pyrrhocoma Gr. *purrhokomes*, red-haired.

Pyrrhocorax L. *pyrrhocorax*, the chough (Gr. *purrhos*, flame-coloured; *korax*, the raven).

pyrrhodes Gr. *purrhodes*, fiery.

pyrrhogaster Gr. *purrhos*, flame-coloured, red; *gaster*, the belly.

pyrrholaemum /Pyrrholaemus Gr. *purrhos*, flame-coloured, red; *laimos*, the throat.

pyrrholeuca Gr. *purrhos*, flame-coloured, red; *leukos*, white.

Pyrrhomyias Gr. *purrhos*, flame-coloured, red; Mod. L. *myias*, flycatcher.

pyrrhonota /pyrrhonotus Gr. *purrhos*, flame-coloured, red; *-notos*, backed.

pyrrhophanus Gr. *purrhos*, flame-coloured, red; *phanos*, shining, bright (*phaino*, to shine forth).

pyrrhophia Gr. *purrhos*, flame-coloured, red; *phaios*, brown, dusky.

pyrrhophrys Gr. *purrhos*, flame-coloured, red; *-ophrus*, eyebrowed.

Pyrrhoplectes Genus *Pyrrhula*, bullfinch; genus *Euplectes*, bishop weaver.

pyrrhops Gr. *purrhos*, flame-coloured, red; *ops*, eye.

pyrrhoptera /pyrrhopterum / pyrrhopterus Gr. *purrhos*, flame-coloured, red; *-pteros*, winged (cf. *purrhopterux*, red-winged).

pyrrhopygia see **pyropygia.**

pyrrhopygus Gr. *purrhos*, flame-coloured, red; *-pugos*, rumped.

pyrrhothorax Gr. *purrhos*, flame-coloured, red; *thorax*, the breast.

pyrrhotis Gr. *purrhos*, flame-coloured, red; *-otis*, eared (cf. *purrhotes*, red-haired).

pyrrhoura Gr. *purrhos*, flame-coloured, red; *-ouros*, tailed.

Pyrrhula Gr. *purrhoulas*, a worm-eating bird mentioned by Aristotle.

Pyrrhuloxia Genus *Pyrrhula*, bullfinch; genus *Loxia*, crossbill.

Pyrrhuphonia Genus *Pyrrhula*, bullfinch; genus *Euphonia*, euphonia, tanager.

Pyrrhura /Pyrrhurus Gr. *purrhos*, flame-coloured, red; *oura*, the tail.

pyrrogenys Gr. *purrhos*, flame-coloured, red; *genus*, the cheek (cf. *purrhogeneios*, red-bearded).

Pyrroglaux Gr. *purrhos*, flame-coloured, red; syn. genus *Glaux*, owl (Gr. *glaux*, the Little Owl *Athene)*.

Pytilia Dim. of genus *Pitylus*, grosbeak.

pytyopsittacus Gr. *pitus*, *pituos*, a pine tree; *psittakos*, a parrot; ref. the preferred habitat of the Parrot Crossbill *Loxia*, so named for its heavy bill.

Loxia pytyopsittacus Parrot Crossbill

Q

quadragintus L. *quadraginta,* forty; from the supposed number of white spots in the wings of the Forty-spotted Pardalote *Pardalotus.*

quadribrachys L. *quadri-,* four (*quattuor,* four); Gr. *brakhus,* little; ref. the small white patches on the head of the Shining-blue Kingfisher *Alcedo.*

quadricinctus L. *quadri-,* four; *cinctus,* girdled, banded (*cingere,* to encircle).

quadricolor Med. L. *quadricolor,* four-coloured.

quadrivirgata L. *quadri-,* four; *virgatus,* striped (*virga,* a streak).

quartinia L. *quartus,* the fourth; the Grey-headed Waxbill *Estrilda* was the fourth species of waxbill described by Bonaparte (1851) in *Consp. gen. avium* (cf. Botanic *quartine,* a fourth ovular integument).

Quelea Med. L. *qualea,* the quail; the relevance of this name to the Afrotropical queleas or weavers is unclear.

querquedula L. *querquedula,* a type of duck.

Querula L. *querulus,* querulous, complaining, noisy.

quinquestriata L. *quinque,* five; Mod. L. *striatus,* striped, streaked (L. *stria,* a furrow).

quinticolor L. *quintus,* fifth; *color,* colour.

quiscalina Genus *Quiscalus,* grackle; L. *-inus,* resembling.

Quiscalus Specific name *Gracula quiscala* (now *Quiscalus quiscula).*

quiscula L. *quis,* what; *qualis,* of what kind; Linnaeus's original description (*Systema Naturae* (1758),(Ed. 10), book **1**, p. 109) of the Purple Grackle *Quiscalus* was based on Catesby's (1731) *Monedula purpurea,* the Purple Daw, and Brown's (1725) *Merops niger,* the Black Bee-eater, and he was doubtless unsure of its relationships. Newton in Coues (1882), remarked,'I cannot find this word or any thing like it in any older author... [Linnaeus] did not invent names. From his printing the word...with a capital initial letter, it is obvious that he regarded it as a substantive, and I should think he must have found it in some book of travels as the local name of a bird. The word seems to me Spanish or quasi-Spanish – say Creole – and the regular Castilian *quisquilla,* which dictionaries explain to be a trifling dispute, suggests a meaning, especially when one reads of the noisy and fussy bickerings of your Boat-tails.' Later, however (Newton and Gadow (1896)), he believed it to be 'from the Low Latin *Quiscula* or *Quisquilla,* which like *Quaquila* are supposed to be renderings of *Quagila* or some such word, and to mean Quail'.

quitensis After the city of Quito, capital of Ecuador, and named after the local Quitu Indians.

quixensis After the Río Quixos (Río Quijos), Napo-Pastaza, Ecuador.

quoyi After Jean René Constant Quoy (1790–1869) French naturalist, collector, explorer in the Pacific, 1826–1829, and author.

R

rabieri After Lieutenant Colonel Paul Rabier (fl. 1904) French Army in Indo China, 1897–1904.

rabori After Dioscoro Rabor (b. 1911) Filipino zoologist, field ornithologist, and collector.

racheliae After Rachel Cassin (fl. 1857) daughter of US ornithologist John Cassin.

radiata /radiatum /radiatus Mod. L. *radiatus,* barred (L. *radiatus,* provided with spokes; *radius,* a spoke or rod).

radiceus Misspelt from Mod. L. *radiatus,* barred.

radiolatus Mod. L. *radiolatus,* barred.

radiolosus Mod. L. *radiolosus,* barred (L. *radiolus,* a small rod).

radjah Moluccan native name *radja* for the Radjah Shelduck *Tadorna* on the island of Buru.

rafflesii After Sir Thomas Stamford Raffles (1781–1826) English colonial administrator, Lieutenant-Governor of Java, 1811–1815; Lieutenant-Governor of Benkoelen, Sumatra, 1818–1824; founder of Singapore, 1819; and first President of the Zoological Society of London.

raggiana After Marchese Francesco Raggi (fl. 1873) Italian naturalist and collector in New Guinea.

raimondii After Antonio Raimondi (1825–1890) Italian naturalist, explorer, and collector in Peru, 1850–1870.

Rallicula Dim. of genus *Rallus,* rail.

Rallina Genus *Rallus,* rail; L. *-inus,* resembling.

ralloides Genus *Rallus,* rail; Gr. *-oides,* resembling; the Andean Solitaire *Myadestes* was so named because its coloration, grey beneath and brown above, was considered analogous to that of the rails.

Rallus Med. L. *rallus,* a rail (German *ralle;* Swedish *rall).*

rama Hindu myth. Rama, the incarnation of Vishnu.

ramonianus After Brother Ramón Busquet (fl. 1840) missionary to Peru.

ramphastinus Genus *Ramphastos,* toucan; L. *-inus,* resembling.

Ramphastos Gr. *rhamphos,* the bill; *-astes,* possessor.

Ramphocaenus Gr. *rhamphos,* the bill; *kainos,* strange, new; ref. the extraordinary long bill of the Long-billed Gnatwren *R. melanurus.*

Ramphocelus Gr. *rhamphos,* the bill; *kelis,* a spot or stain; ref. the conspicuous silvery base on the lower mandible of the Brazilian Tanager *R. bresilius.* Genus amended by some authors to *Ramphocoelus,* as if from Gr. *rhamphos* and *coelus,* heaven.

Ramphocinclus Gr. *rhamphos,* the bill; Mod. L. *cinclus,* thrush.

Ramphocoris Gr. *rhamphos,* the bill; syn. genus *Corys,* lark; ref. the remarkable heavy bill of the Thick-billed Lark *R. clotbey.*

Ramphodon Gr. *rhamphos,* the bill; *odous, odon,* a tooth; ref. the minute serrations or teeth on the mandible of the Saw-billed Hermit *R. naevius.*

Ramphomicron Gr. *rhamphos,* the bill; *mikron,* a little (*mikros,* small).

Ramphotrigon Gr. *rhamphos,* the bill; *trigonon,* a triangle; ref. the broad, flat bill of the Rufous-tailed Flatbill *R. ruficauda*.

ramsayi After Colonel Robert George Wardlaw Ramsay (1852–1921) British ornithologist in India and Burma, 1872–1882, and President of British Ornithologists' Union, 1913–1918.

Ramsayornis After E. P. Ramsay (1842–1916) Australian naturalist; Gr *ornis,* a bird.

Randia After Austin L. Rand (b. 1905) US ornithologist, collector, and author (*The distribution and habits of Madagascar birds*, 1936; *Handbook of New Guinea birds*, 1967).

ranfurlyi After Uchter John Mark Knox, 5th Earl of Ranfurly (1856–1933) Governor of New Zealand, 1897–1904.

ranivorus L. *rana,* a frog; *-vorus,* eating (*vorare,* to devour).

rapax L. *rapax,* rapacious, grasping.

Raphidura see **Rhaphidura.**

Raphus Gr. *raphos,* a name given to the bustard *Otis* by Galenius (probably a misreading for *outis* or *otis,* the bustard) (cf. Gr. *rhaphoi,* birds mentioned by Hesychius).

rara **1** Chilean onomatopoeic name *rara* for the Chilean Plantcutter *Phytotoma*. **2** L. *rarus,* rare (*Lagonosticta, Lampribis).*

rarotongensis After the island of Rarotonga, Cook Islands.

raveni After Henry Cushier Raven (1889–1944) US explorer and collector in the East Indies, 1914–1918.

ravidus L. *ravidus,* greyish, dark-coloured (*ravus,* grey).

raytal Hindi name *retal* for the Sand Lark *Calandrella*.

razae After Raza (Razo) Island, Desertas group, Cape Verde islands.

rectirostris L. *rectus,* straight (*regere,* to guide); *-rostris,* billed.

rectunguis L. *rectus,* straight; *unguis,* a claw; ref. the long, spiky hind-claw of the Short-toed Coucal *Centropus*.

Recurvirostra L. *recurvus,* bent, curved back; *rostrum,* the bill.

recurvirostris L. *recurvus,* bent, curved back; *-rostris,* billed.

redivivum L. *redivivum,* renewed, renovated; of the California Thrasher *Toxostoma* Gambel (1845), *Proc. Acad. Nat. Sci.*, **2**, 264, wrote, 'This curious bird was first seen by the naturalists in La Pérouse's expedition, as he tells us, who considered it a *Promerops*...No further account of this bird (as far as I can learn) has been given since the time of La Pérouse until now.'

reevei After J. P. Reeve (fl. 1870) US collector in Ecuador.

reevesii After John Reeves (1774–1856) English civil servant in China, 1812–1831, naturalist, and collector.

regalis **1** L. *regalis,* royal, regal (*rex, regis,* a king) (*Heliangelus).* **2** after the village of Real del Monte, Hidalgo, Mexico (Spanish *real,* royal) (*Buteo).*

regia L. *regius,* royal, magnificent (*rex, regis,* a king).

regina L. *regina,* a queen.

regius see **regia.**

Reguloides Genus *Regulus,* kinglet; Gr. *-oides,* resembling.

regulorum L. *regulorum,* of kings or princelings (i.e. like *Regulus).*

Regulus L. *regulus,* a prince, a kinglet (Med. L. *regulus,* the wren *Troglodytes).* Specifically the epithet has been given to small, warbler-like spp. with a red crown (*Machaeropterus)* or an erectile rump patch (*Prodotiscus),* as well as the true 'kinglet', the Goldcrest *Regulus*.

rehsei After Ernst Rehse (fl. 1880) German ornithologist and collector in the Pacific.

reichardi After Paul Reichard (1845–1938) German engineer, geographer, and collector in East Africa, 1880–1884.

reichenbachii After Heinrich Gottlieb Ludwig Reichenbach (1793–1879) German zoologist, nomenclaturist, and author (*Praktische naturgeschichte der vögel*, 1845; *Avium systema naturale, das natürliche system der vögel*, 1849).

reichenovii /reichenowi After Anton Reichenow (1847–1941) German ornithologist, collector in West Africa, 1872–1873, and author (*Die vögel Afrikas*, 1900).

Reinard After Claudia Reinard (fl. 1930) wife of German ornithologist Ernst Hartert.

reinhardti After Johannes Theodor Reinhardt (1816–1882) Danish zoologist.

reinwardt /reinwardtii After Caspar Georg Carl Reinwardt (1773–1854) Dutch ornithologist and collector in Java, 1817–1822.

Reinwardtipicus After Caspar Reinwardt (v. *reinwardt)*; genus *Picus,* woodpecker.

Reinwardtoena After Caspar Reinwardt (v. *reinwardt)*; genus *Oena,* dove.

reinwardtsi see **reinwardt.**

reiseri After Othmar Reiser (1861–1936) Austrian oologist and collector in Brazil, 1903.

relicta /relictus L. *relictus,* isolated, relict (*relinquere,* to leave behind).

religiosa L. *religiosus,* religious, sacred; a misnomer, Linnaeus having confused the Hill Grackle or Mynah *Gracula* with the Common Mynah *Acridotheres tristis* which is held sacred by the Hindus.

remifer L. *remus, remi,* an oar; *-fer,* bearing; ref. the shape of the Lesser Racquet-tailed Drongo's *Dicrurus* tail racquets.

Remiz Polish name *remiz* for the Masked Penduline Tit *R. pendulinus.*

renatae After Renate van den Elzen (fl. 1981) Dutch ornithologist and collector.

renauldi After Père J. N. Renauld (1839–1898) French missionary to Annam (Vietnam).

rendovae After the island of Rendova, New Georgia group, Solomon Islands.

rennelliana After Rennell Island, Solomon Islands, named for James Rennell (1742–1830) British geographer, Surveyor-General of Bengal, 1764–1777.

repressa L. *repressus,* restrained, confined (*reprimere,* to hold back); this is a replacement name for the White-cheeked Tern *Sterna* formerly known as *S. albigena.*

resplendens L. *resplendens,* glittering, shining (*resplendere,* to glitter back).

respublica L. *respublica,* the republic; in naming Wilson's Bird of Paradise *Diphyllodes* Prince Bonaparte expressed his disenchantment with the French Republic, ensuring that since there could not be a paradisean republic there should at least be a republican bird of paradise.

restricta L. *restrictus,* restricted, close; the Tana Cisticola *Cisticola* is confined to a small area of eastern Kenya.

reticulata L. *reticulatus,* reticulated, marked so as to resemble a net (*reticulum,* a net).

retrocinctum L. *retro,* on the back; *cinctus,* girdled, banded; the Red-bearded Flowerpecker *Dicaeum* has a scarlet band across the upper back.

retzii After Anders Adolf Retzius (1796–1860) Swedish anatomist.

revoilii After Georges Révoil (1852–1894) French naturalist, collector in Somaliland, 1878–1880 and author (*La vallée du Darror; voyage aux pays Çomalis*, 1882).

reyii After J. G. C. E. Rey (1838–1909) German oologist.

reynaudii After A. M. Reynaud (1804–1872) French naval surgeon, explorer, and collector in Burma and Madagascar.

rex L. *rex,* king, ruler, prince.

Rhabdornis Gr. *rhabdotos,* striped; *ornis,* a bird.

Rhadina Gr. *rhadinos,* slim, delicate.

Rhagologus Gr. *rhagologos,* berry-gathering (*rhax, rhagos,* grape; *lego,* to gather).

rhami After H. C. de Rham (fl. 1851) naturalist and collector in the USA,1836.

Rhamphocharis Gr. *rhamphos,* the bill (i.e. long billed); *kharis,* grace, beauty.

Rhamphococcyx Gr. *rhamphos,* the bill; *kokkux,* the cuckoo; the colourful, arched bill of the Fiery-billed Malkoha *R. calyorhynchus* was considered to be like that of a toucan *Ramphastos.*

Rhamphomantis Gr. *rhamphos,* the bill; *mantis,* a prophet; the irritating cries of the Long-billed Cuckoo *R. megarhynchus* are supposed to foretell disaster.

Rhaphidura Gr. *rhaphis, rhaphidos,* a needle; *oura,* the tail.

Rhea Gr. myth. Rhea, daughter of Uranus and mother of Zeus.

Rhegmatorhina Gr. *rhegma, rhegmatos,* a fracture, a fissure; *rhis, rhinos,* the nose; of the Bare-eyed Antbird *R. gymnops* Ridgway (1888), *Proc. US Nat. Mus.*, **10**, 525, wrote, 'Similar to *Phlogopsis* but differing very materially in the form of the nostril, which consists of a narrow longitudinal slit-like opening, overhung by a broad, rather thickened, membrane or operculum.'

Rheinardia After Captain Rheinard (fl. 1883) French Army in Annam, Indo China (Vietnam).

rhinoceros L. *rhinoceros,* a rhinoceros.

Rhinocorax Gr. *rhis, rhinos,* the nose; *korax,* the raven; the Fan-tailed Raven's *Corvus rhipidurus* nostrils are concealed by a fan-shaped tuft of bristles.

Rhinocrypta Gr. *rhis,* the nose; *kruptos,* hidden, secret; the nostrils of the Crested Gallito *R. lanceolata* are hidden by the feathers of the forehead.

Rhinomyias Gr. *rhis,* the nose; Mod. L. *myias,* flycatcher (cf. Gr. *muo,* to shut, close); the nostrils of the jungle flycatchers are covered by rictal bristles.

Rhinoplax Gr. *rhis,* the nose; *plax,* a slab, a flat-topped stone; from the appearance of the casque of the Helmeted Hornbill *R. vigil.*

Rhinopomastus Gr. *rhis,* the nose; *poma, pomatos,* a lid or cover; ref. the covered nares of the Afrotropical scimitarbills.

Rhinoptilus Gr. *rhis,* the nose; *ptilon,* a feather; these nocturnal coursers have partially covered nostrils.

Rhinoptynx Gr. *rhis,* the nose; syn. genus *Ptynx,* eagle owl (Gr. *ptunx,* a sort of owl mentioned by Aristotle); ref. the heavy, elongated bill of the Striped Owl *R. clamator.*

Rhinortha Gr. *rhis,* the nose; *orthos,* straight; the Little Malkoha *R. chlorophaea* has a small, straight bill.

Rhipidornis Gr. *rhips, rhipidos,* a fan; *ornis,* a bird.

Rhipidura/rhipidurus Gr. *rhipis,* a fan; *oura,* the tail; ref. the expressive, long, fanned tails of the fantails, and the short, fan-shaped tail of the Fan-tailed Raven *Corvus.*

Rhizothera Gr. *rhiza,* a root; *thera,* a seeker or hunter; the Long-billed Partridge *R. longirostris* feeds on seeds, roots, and insects.

Rhodacanthis Gr. *rhodo-,* rosy (*rhodon,* the rose); genus *Acanthis,* linnet.

Rhodinocichla Gr. *rhodinos,* pink; *kikhle,* a thrush.

rhodocephala Gr. *rhodo-,* rosy; *-kephalos,* headed.

rhodochlamys Gr. *rhodo-,* rosy; *khlamus,* a cloak or mantle.

rhodochrous Gr. *rhodokhrous,* rose-coloured.

rhodocorytha Gr. *rhodo-,* rosy; *korus, koruthos,* a helmet, the head.

rhodogaster Gr. *rhodo-,* rosy; *gaster,* the belly.

rhodolaema Gr. *rhodo-,* rosy; *laimos,* the throat.

Rhodonessa Gr. *rhodo-,* rosy (*rhodon,* the rose); *nessa,* a duck.

rhodopareia Gr. *rhodo-,* rosy; *pareia,* the cheek.

Rhodopechys Gr. *rhodopekhus,* rosy-armed (*rhodon,* the rose; *pekhus,* the arm); ref. the pink and black wings of the Crimson-winged Finch *R. sanguinea.*

rhodopeplus Gr. *rhodopeplos,* roseate-robed (*rhodon,* the rose; *peplos,* a robe).

Rhodophoneus Gr. *rhodo-,* rosy; syn. genus *Phoneus,* shrike (Gr. *phoneus,* a killer); ref. the 'blood stained' plumage of the Rosy-patched Bush-shrike *R. cruentus.*

Rhodopis After Rodopis (fl. 500 BC) a famous Greek hetaira or courtesan (cf. Gr. *rhodopis,* rosy-faced).

rhodopyga Gr. *rhodopugos,* rosy-rumped.

Rhodospingus Gr. *rhodon,* the rose; *spingos,* a finch.

Rhodospiza Gr. *rhodon,* the rose; *spiza,* a finch.

Rhodostethia Gr. *rhodon,* the rose; *stethos,* the breast.

Rhodothraupis Gr. *rhodon,* the rose; genus *Thraupis,* tanager.

Rhopocichla Gr. *rhops, rhopos,* a bush or shrub; *kikhle,* a thrush.

Rhopodytes Gr. *rhops,* a bush; *dutes,* a diver (*duo,* to sink, to plunge).

Rhopophilus Gr. *rhops,* a bush; *philos,* loving, fond of.

Rhopornis Gr. *rhops,* a bush; *ornis,* a bird.

Rhyacornis Gr. *rhuax, rhuakos,* a torrent; *ornis,* a bird.

Rhynchocyclus Anagram of syn. genus *Cyclorhynchus* (Gr. *kuklos,* a shield, a circle; *rhunkhos,* the bill); ref. the distinctive bill-shape of the Olivaceous Flatbill *R. olivaceus.*

Rhynchopsitta Gr. *rhunkhos,* the bill (i.e. thick-billed); Mod. L. *psitta,* a parrot.

Rhynchortyx Gr. *rhunkhos,* the bill (i.e. heavy billed); *ortux,* the quail.

Rhynchospiza Gr. *rhunkhos,* the bill (i.e. heavy billed); *spiza,* a finch.

Rhynchostruthus Gr. *rhunkhos,* the bill (i.e. heavy billed); *strouthos,* a sparrow *Passer* or any small bird.

Rhynchothraupis Gr. *rhunkhos,* the bill (i.e. cone-billed); genus *Thraupis,* tanager.

rhynchotis Gr. *rhunkhos,* the bill; *-oides,* resembling; ref. the spatulate bill of the Australian Shoveler *Anas.*

Rhynchotus Gr. *rhunkhos,* the bill; *-otes,* featuring (*eidos,* form, appearance); ref. the curved, elongated bill of the Red-winged Tinamou *R. rufescens.*

Rhynochetos Gr. *rhis, rhinos,* the nose or (plural form *rhines)* nostrils; *khaite,* long hair (cf. *khetos,* want, need); the tubular nostrils of the Kagu *R. jubatus* are contained in a deep furrow and bordered by long, stiff, slightly curved hairs.

Rhyticeros Gr. *rhutis,* a wrinkle; *keros,* a horn; ref. the wave-like markings on the bill and casque of the Wreathed Hornbill *R. undulatus.*

Rhytipterna Gr. *rhutis,* a wrinkle; *pterna,* the heel; the tarsi of the Greyish Mourner *R. simplex* are heavily scaled, especially behind the ankle.

richardi After M. Richard (fl. 1815) French collector about whom nothing else is known.

richardsii After Rear Admiral George Edward Richards (1852–1927) English surveyor and collector in the East Indies.

richardsoni After W. B. Richardson (fl. 1919) US collector in Mexico, 1887–1890, Guatemala, 1897, 1907–1917, and Colombia, 1900.

Richmondena After Charles Wallace Richmond (1868–1932) US ornithologist, bibliographer, and nomenclaturist.

richmondi After **1** the Richmond River, New South Wales, Australia (*Gerygone).* **2** Charles Richmond (v. *Richmondena) (Arremonops).*

ricketti After C. B. Rickett (1851–1943) British collector in China.

ricordii After Alexandre Ricord (fl. 1838) French naval officer, surgeon, and collector in tropical America, 1826–1834.

ridgwayi /Ridgwayia After Robert Ridgway (1850–1929) US ornithologist, Smithsonian curator of birds, 1880–1929, and author (*The birds of North and Middle America*, 1901).

ridibundus L. *ridibundus,* laughing (*ridere,* to laugh).

ridleyana After Henry Nicholas Ridley (1855–1956) British botanist and collector on Ilha Fernando de Noronha, 1887.

riedeli After Johan Gerard Friedrich Riedel (1832–1911) Dutch colonial administrator in the East Indies, 1853–1883, naturalist, and collector.

riefferii After Rieffer (fl. 1840) an otherwise unknown collector in tropical America.

rikeri After C. B. Riker (1863–1947) US collector in Brazil, 1884–1887.

Rimator Late L. *rimator*, an investigator (L. *rimari*, to examine).

riocourii After the Comte de Boys de Riocour (fl. 1823) French naturalist and collector.

Riparia L. *riparius*, bank nesting (*ripa*, a bank).

ripponi Lieutenant Colonel G. C. Rippon (1861–1927) British Army in India and Burma, 1880–1914.

risora Brazilian (Portuguese) name *risora*, the mocker, for the Strange-tailed Tyrant *Yetapa* (*riso*, laughter; *risota*, mocking laugh; L. *risus*, laughter).

risoria Med. L. *risorie*, mocking (L. *risor*, *risoris*, a laugher; *ridere*, to laugh).

Rissa Icelandic *rita*, the kittiwake (Old Norse *ryta*).

rivoli/rivolii After François Victor Masséna Prince d'Essling and Duc de Rivoli (1795–1863) French ornithologist and collector.

rivularis L. *rivulus*, a brook, a rivulet.

rixosus L. *rixosus*, quarrelsome, pugnacious (*rixari*, to quarrel; *rixa*, a fight).

roberti After **1** Major Robert Alexander (1873–1928) brother of British explorer Captain Boyd Alexander (*Sheppardia*). **2** Alphonse Robert (fl. 1901) French collector in Brazil (*Conopophaga*).

robertsi After Austin Roberts (1883–1948) South African ornithologist, collector, and author.

robinsoni After Herbert Christopher Robinson (1874–1929) British ornithologist, collector in Malaysia, 1908–1926, and author (*The birds of Singapore*, 1927; *The birds of the Malay peninsula*, 1927).

roboratus L. *roboratus*, strong, vigorous.

roborowskii After Captain Vladimir Roborovski (1856–1910) Russian explorer in central Asia and author (*Ekspeditsii v storonu ot pugey Tibetskoy ekspeditsii*, 1896; *Otchet nachalnika ekspeditskii*, 1900).

robusta L. *robustus*, strong, powerful (*robur*, hardwood).

robustipes L. *robustus*, strong; *pes*, the foot.

robustirostris L. *robustus*, strong; *-rostris*, billed.

robustus see **robusta.**

rochii After S. Roch (1829–1906) army surgeon on Mauritius, 1862.

rochussenii After Jan Jacob Rochussen (1797–1871) Governor-General of the Dutch East Indies, 1845–1851.

rockefelleri After J. Sterling Rockefeller (1904–1944) US collector in the Congo, 1928–1929, and sponsor.

rodericanus After the island of Rodrigues, Mascarene Islands, named for the Portuguese navigator Diogo Rodriguez who discovered it in 1528 (Mod. L. *Rodericus*, Roderick and European equivalents; L. *-anus*, belonging to).

rodinogaster Gr. *rhodinos*, pink; *gaster*, the belly.

rodolphei After Rodolphe Meyer de Schauensee (1901–1986) US ornithologist, collector, and author (*The birds of Colombia*, 1964; *The species of birds of South America with their distribution*, 1966; *A guide to the birds of South America*, 1971; *The birds of China*, 1984).

rogersi After Revd. H. M. C. Rogers (1879–1926) missionary to Tristan da Cunha, 1922.

Rogibyx After John Porter Rogers (1873–1941) Australian collector; Mod. L. *ibyx*, a lapwing or plover.

rolland/Rollandia After Master Gunner Rolland (fl. 1820) of the French corvette *L'Uranie* which circumnavigated the globe, 1817–1820.

rolleti After Bruno Rollet (1810–1855) French big-game hunter in the Sudan and Ethiopia, 1841–1851.

Rollulus Malay name *rouloul* for the Crested Wood Partridge *R. rouloul* (although the name appears neither in Winstedt (1949) nor Wilkinson (1959)).

roquettei After Edgard Roquette-Pinto (b. 1884) Brazilian anthropologist and ornithologist.

roraimae /Roraimia After the isolated mountain Cerro Roraima, on the borders of Venezuela and Guyana.

roratus L. *roratus,* sprinkled, bedewed (*rorare,* to bedew); ref. the sleek, glossy plumage of the Eclectus Parrot *Eclectus.*

rosacea L. *rosaceus,* rosy, rosaceous (Mod. L. *rosaceus,* rosy pink).

rosea L. *roseus,* rosy, rose-coloured (*rosa,* the rose).

roseata /roseatus Med. L. *roseatus,* roseate, rose-coloured (Late L. *rosatus,* roseate; L. *roseus,* rosy).

roseicapillus L. *roseus,* rosy; *-capillus,* capped.

roseicollis L. *roseus,* rosy; Mod. L. *-collis,* necked.

roseifrons L. *roseus,* rosy; *frons,* the forehead, the brow.

roseigaster L. *roseus,* rosy; *gaster,* the belly.

rosenbergi /rosenbergii After **1** Baron Carl Benjamin Hermann von Rosenberg (1817–1888) German naturalist, surveyor, and collector in the East Indies (*Gymnocrex, Myzomela, Scolopax, Tyto*). **2** W. F. H. Rosenberg (fl. 1948) British natural history dealer and collector in tropical America, 1898 (*Amazilia*).

roseogrisea L. *roseus,* rosy; Med. L. *griseus,* grey.

roseogularis L. *roseus,* rosy; Mod. L. *gularis,* throated (L. *gula,* the throat).

roseus see **rosea.**

rositae After Rosita Sumichrast (fl. 1874) wife of Mexican naturalist F. Sumichrast.

rossae After Lady Anne Ross (d. 1857) wife of British explorer Rear Admiral Sir James Ross.

rossii After Bernard Rogan Ross (1817–1874) Irish trader and chief factor of Hudson's Bay Company.

rostrata L. *rostratus,* beaked, large billed (*rostrum,* the bill).

Rostratula Dim. from L. *rostratus,* large billed; the Greater Painted Snipe *R. benghalensis* has a long, curved bill.

rostratum /rostratus see **rostrata.**

Rostrhamus L. *rostrum,* the bill; *hamus,* a hook; ref. the specialized, hooked bill of the Snail Kite *R. sociabilis.*

rotensis After the island of Rota, in the Marianas.

rothschildi After Lionel Walter, 2nd Baron Rothschild of Tring (1868–1937) English ornithologist, entomologist, collector, publisher (*Novitates Zoologicae*), and author.

rougetii /Rougetius After J. Rouget (d. 1840) French explorer and collector in Ethiopia, 1839–1840.

rouloul see **Rollulus.**

rourei After Jean de Roure (fl. 1870) French traveller in Brazil.

rovuma After the Rovuma or Ruvuma River, forming the border between Tanzania and Moçambique.

Rowettia After John Quiller Rowett (1876–1924) English businessman and sponsor of Shackleton-Rowett Expedition to the Antarctic, 1921–1922.

rowleyi After George Dawson Rowley (1822–1878) English ornithologist and author (*Ornithological miscellany*, 1875).

rubecula Med. L. *rubecula,* the Robin *Erithacus* (dim. from L. *ruber,* red).

rubeculoides Specific name *Erithacus rubecula;* Gr. *-oides,* resembling; epithet given to species with ruddy or red breasts like the familiar Robin.

ruber L. *ruber,* red, ruddy.

rubescens L. *rubescens,* reddish, blushed (*rubescere,* to become red).

rubetra Mod. L. *rubetra,* a small bird, not further identified but associated with the Gr. *batis.*

rubica Med. L. *rubicare,* to redden.

rubicilla L. *ruber,* red; Mod. L. *cilla,* the tail (v. *Motacilla*) (cf. Mod. L. *rubicilla,* the Bullfinch *Pyrrhula*).

rubicilloides Specific name *Carpodacus rubicilla;* Gr. *-oides,* resembling.

rubicunda L. *rubicundus,* red, ruddy.

rubida L. *rubidus,* red, reddish.

rubidiceps L. *rubidus,* red; *-ceps,* headed.

rubidiventris L. *rubidus,* red; *venter, ventris,* the belly.

rubiensis After Rubi Island, Geelvink Bay, Irian Jaya (Indonesian New Guinea).

rubifrons L. *rubeus,* red; *frons,* the forehead, the brow.

rubigastra L. *rubeus,* red; *gaster,* the belly.

rubiginosa /rubiginosus L. *rubiginosus,* rusty, ferruginous.

rubinoides Syn. specific name *Trochilus rubineus* (now *Clytolaema rubricauda)*; Gr. *-oides,* resembling.

rubinus Med. L. *rubinus,* ruby (Old French *rubi* or *rubin,* a ruby).

rubra see **ruber.**

rubratra L. *ruber,* red; *ater,* black.

rubricapilla /rubricapillus L. *ruber,* red; *-capillus,* capped.

rubricata /rubricatus L. *rubricatus,* red, red-ochred (*rubrica,* red ochre).

rubricauda L. *ruber,* red; *cauda,* the tail.

rubriceps L. *ruber,* red; *-ceps,* headed.

rubricera L. *ruber,* red; Gr. *keras,* a horn; the Red-knobbed Imperial Pigeon *Ducula* has a swollen, cherry-red cere.

rubricollis L. *ruber,* red; Mod. L. *-collis,* necked.

rubricomus L. *ruber,* red; Gr. *-komos,* haired (*kome,* the hair).

rubrifacies L. *ruber,* red; *facies,* the face.

rubrifrons L. *ruber,* red; *frons,* the forehead or brow.

rubrigularis L. *ruber,* red; Mod. L. *gularis,* throated, of the throat (L. *gula,* the throat).

rubripes L. *ruber,* red; *pes,* the foot.

rubrirostris L. *ruber,* red; *-rostris,* billed.

rubritorques /rubritorquis L. *ruber,* red; *torques,* a collar or necklace.

rubrobrunnea L. *ruber,* red; Mod. L. *brunneus,* brown.

rubrocanus L. *ruber,* red; *canus,* grey, hoary.

rubrocapilla see **rubricapilla.**

rubrocristata L. *ruber,* red; *cristatus,* crested (*crista,* a crest).

rubrogenys L. *ruber,* red; *genis,* the cheeks.

rubronotata L. *ruber,* red; *notatus,* marked (*nota,* a mark).

ruckeri After S. Rücker (1815–1890) British naturalist and collector in Ecuador, 1846.

ruddi After C. D. Rudd (1844–1916) collector in East Africa, 1860–1916.

rudis L. *rudis,* a fencing foil, a sword; ref. the long, pointed bill of the Pied Kingfisher *Ceryle.*

rudolfi /rudolphi After Archduke Rudolf (1857–1889) Crown Prince of Austria–Hungary.

ruecki After M. Rück (fl. 1880) French traveller and collector in Malaysia.

rueppellii After Wilhelm Peter Eduard Simon Rüppell (1794–1884) German zoologist, explorer, collector, and author (*Reisen in Nubien, Kordofan und dem Peträischen Arabien*, 1829; *Systematische übersicht der vögel nord-ost-Afrikas*, 1845).

rufa L. *rufus,* red, ruddy. In orn. this epithet covers a wide spectrum of colours from yellow, orange, and brown to crimson, scarlet, and purple.

rufalbus L. *rufus,* red; *albus,* white.

rufaxilla L. *rufus,* red; *axilla,* armpit.

rufescens L. *rufescens,* reddish (*rufescere,* to become reddish).

rufibarba L. *rufus,* red; *barba,* a beard.

ruficapilla /ruficapillus L. *rufus,* red; *-capillus,* capped.

ruficauda L. *rufus,* red; *cauda,* the tail.

ruficaudatus L. *rufus,* red; *caudatus,* tailed.

ruficeps L. *rufus,* red; *-ceps,* headed.

ruficervix L. *rufus,* red; *cervix,* the nape or neck.

ruficollaris L. *rufus,* red; *collaris,* of the neck, collared.

ruficollis L. *rufus,* red; Mod. L. *-collis,* necked.

ruficoronatus L. *rufus,* red; *coronatus,* crowned (*coronare,* to crown).

ruficrissa L. *rufus,* red; Mod. L. *crissum,* the vent or crissum.

ruficrista L. *rufus,* red; *crista,* a crest.

rufidorsa /rufidorsum L. *rufus,* red; *dorsus,* the back.

rufifrons L. *rufus,* red; *frons,* the forehead or brow.

rufigaster /rufigastra L. *rufus,* red; *gaster,* the belly.

rufigena /rufigenis L. *rufus,* red; *gena,* the cheek; *genis,* the cheeks.

rufigula L. *rufus,* red; *gula,* the throat.

rufigulare /rufigularis L. *rufus,* red; Mod. L. *gularis,* throated, of the throat.

rufilata L. *rufus,* red; *latus,* the flank.

rufimarginatus L. *rufus,* red; *marginatus,* marginned, bordered (*marginare,* to border); ref. bright chestnut primaries and secondaries of the Rufous-winged Antwren *Herpsilochmus.*

rufina L. *rufus,* red; *-inus,* resembling, possessing (cf. Med. L. *rufinus,* a representation of gold).

rufinucha L. *rufus,* red; Med. L. *nucha,* the nape (Arabic *nukha,* spinal marrow).

rufinus see **rufina.**

rufipectoralis L. *rufus,* red; *pectoralis,* pectoral, of the breast.

rufipectus L. *rufus,* red; *pectus,* the breast.

rufipenne /rufipennis L. *rufus,* red; *pennis,* winged (*penna,* a feather).

rufipes L. *rufus,* red; *pes,* the foot.

rufipileatus L. *rufus,* red; *pileatus,* capped.

rufitorques L. *rufus,* red; *torques,* a collar or necklace.

rufiventer /rufiventris L. *rufus,* red; *venter, ventris,* the belly.

rufivertex L. *rufus,* red; *vertex,* the crown of the head.

rufivirgatus L. *rufus,* red; *virgatus,* striped.

rufoaxillaris see **rufaxilla.**

rufobrunneus L. *rufus,* red; Mod. L. *brunneus,* brown.

rufociliatus L. *rufus,* red; *cilium,* the eyelid; *-atus,* possessing.

rufocinctus L. *rufus,* red; *cinctus,* banded, girdled (*cingere,* to encircle).

rufocinerea /rufocinereus L. *rufus,* red; *cinereus,* ash-coloured (*cinis, cineris,* ashes).

rufocinnamomea L. *rufus,* red; Mod. L. *cinnamomeus,* cinnamon-coloured (L. *cinnamum,* cinnamon).

rufocollaris see **ruficollaris.**

rufocrissalis L. *rufus,* red; Mod. L. *crissalis,* crissal, vented (Mod. L. *crissum,* the vent).

rufofuscus L. *rufus,* red; *fuscus,* dusky, brown.

rufogularis see **rufigulare.**

rufolarvatus L. *rufus,* red; *larvatus,* masked (*larva,* a mask).

rufolateralis L. *rufus,* red; *lateralis,* lateral, of the side.

rufolavatus L. *rufus,* red; *lavatus,* washed, bedewed (*lavare,* to bathe).

rufomarginatus see **rufimarginatus.**

rufomerus L. *rufus,* red; Gr. *meros,* the thigh; the juvenile Green-cheeked Bronze Cuckoo *Chrysococcyx* has brown barred flanks.

rufonuchalis L. *rufus,* red; Mod. L. *nuchalis,* nuchal, of the nape (Med. L. *nucha,* the nape).

rufopalliatus L. *rufus,* red; *palliatus,* mantled (*pallium,* a mantle).

rufopectus see **rufipectus.**

rufopicta /rufopictus L. *rufus,* red; *pictus,* painted (*pingere,* to paint).

rufopileatum L. *rufus,* red; *pileatus,* capped.

rufoscapulatus L. *rufus,* red; *scapulae,* the shoulders; *-atus,* possessing.

rufosuperciliaris L. *rufus,* red; Mod. L. *superciliaris,* the eyebrow.

rufosuperciliatus L. *rufus,* red; *supercilium,* the eyebrow; *-atus,* possessing.

rufula /rufulus L. *rufulus,* reddish (dim. of *rufus,* red).

rufum /rufus see **rufa.**

rugiensis /ruki /Rukia After the island of Ruk or Roug (Truk), Caroline Islands.

rumicivorus L. *rumex, rumicis,* sorrel; *-vorus,* eating (*vorare,* to devour).

rupestris Mod. L. *rupestris,* of rocks, montane (L. *rupes,* a rock).

Rupicola L. *rupes,* a rock; *-cola,* dweller, inhabitant.

rupicoloides Specific name *Falco rupicola* (now a subsp. of *F. tinnunculus)*; Gr. *-oides,* resembling.

Rupornis L. *rupes,* a rock; Gr. *ornis,* a bird.

rupurumii After the Rupurumi (Rupununi) River, British Guiana (Guyana).

rushiae After Mrs P. Rush (fl. 1853) wife of US psychologist James Rush.

ruspolii After Eugenio Prince Ruspoli (1866–1893) Italian explorer in Ethiopia, 1891–1893, where he was killed by an elephant.

russatum /russatus L. *russatus,* clothed in red (*russus,* red).

rustica L. *rusticus,* rural, rustic.

rusticola /rusticolus **1** L. *rusticula,* a gamebird mentioned by Pliny and the poet Valerius Martialis (*Scolopax).* **2** L. *rusticola,* a rustic, a countryman (*Falco).*

ruticilla Mod. L. *ruticilla,* a redstart *Phoenicurus* (L. *rutilus,* red; Mod. L. *cilla,* the tail (v. *Motacilla)).*

rutila L. *rutilus,* red, golden, auburn.

rutilans L. *rutilans,* of a reddish glowing colour, glowing.

rutilus see **rutila.**

ruwenzori /ruwenzorii After the Ruwenzori Mountains or Mountains of the Moon, forming the border between Kivu Province, Zaïre, and Toro Province, Uganda.

Ruwenzorornis After the Ruwenzori Mountains (v. *ruwenzori)*; Gr. *ornis,* a bird.

ruweti After J. C. Ruwet (fl. 1965) Belgian ornithologist.

ruysi After Theodor Ruys (fl. 1905) Dutch traveller in New Guinea.

Rynchops Gr. *rhunkhos,* the bill; *ops,* the face; ref. the extraordinary long bills of the pantropical skimmers.

rytirhynchos Gr. *rhutis,* a wrinkle; *rhunkhos,* the bill.

Corvus rhipidurus Fan-tailed Raven

S

sabini After **1** Joseph Sabine (1770–1837) English zoologist (*Rhaphidura*). **2** General Sir Edward Sabine (1788–1883) English scientist and explorer (*Dryoscopus, Xema*).

sabota Chuana (Botswana) native name *sabota* for the Sabota Lark *Mirafra*.

sabrina L. *Sabrina,* the River Severn, England; a general ref. to the riverine habitat of the Blue-crowned Paradise Kingfisher *Tanysiptera*.

sacerdotis L. *sacerdos, sacerdotis,* a priest; the Purple-backed Kingfisher *Ceyx* was described from a specimen in the collection of the Revd. George Brown (b. 1835), Australian missionary.

sacerdotum L. *sacerdotum,* of priests (*sacerdos,* a priest); the Flores Monarch *Monarcha* was named after the Revd. Fr. J. A. J. Verheijen (fl. 1971) and the Revd. Fr. E. Schmutz (fl. 1971), Dutch missionaries, naturalists, and collectors on the island of Flores, Indonesia.

sacra L. *sacer,* sacred, holy; the Pacific Reef Egret *Egretta* is regarded as sacred by the natives of Raiatea in Tahiti.

Sagittarius L. *sagittarius,* an archer; Arnout Vosmaer (1769), *Description d'un oiseau de proie nommé le Sagittaire,* pp. 4–6, recorded that the Secretary Bird *S. serpentarius* was known as the *Sagittaire,* archer, by settlers at the Cape of Good Hope, from its striding gait being thought to resemble that of a crossbow-man advancing to shoot. He also noted that the local peasants had corrupted the name to *Sécrétaire,* secretary.

sagittata /sagittatus L. *sagittatus,* shot with arrows (*sagitta,* an arrow).

sagittirostris L. *sagitta,* an arrow; *-rostris,* billed.

sagrae After Ramón de la Sagra (1801–1871) Cuban botanist, naturalist, and author (*Historia fisica, politica y natural de la isla de Cuba*, 1837).

saipanensis After the island of Saipan, Marianas Islands.

sakalava After the Sakalava, a native clan of Madagascar whose kingdom, established in the sixteenth century, was subdued in 1824 and incorporated into a united Madagascar under Imerina kings.

Sakesphorus Gr. *sakesphoros,* shield-bearing; the Black-crested Antshrike *S. canadensis* has a conspicuous black breast patch.

salamonis After the Solomon Islands (v. *solomonensis).*

salangana Filipino native name *salangana* for a cave swiftlet.

salinarum Mod. L. *salinarum,* of the salinas (Spanish *salina,* salt lagoon; L. *sal, salis,* salt; *salinae, salinarum,* salt-works); after the Salinas Grandes and Salina Ambargasta, Córdoba, Argentina.

sallaei /sallei After Auguste Sallé (1820–1896) French entomologist and collector in tropical America, 1846–1856.

salmoni After T. K. Salmon (1841–1878) Colombian state engineer, 1870–1877, naturalist, oologist, and author.

salomonis After Salomon Müller (1804–1864) German zoologist and collector.

salomonseni After Finn Salomonsen (1909–1983) Danish ornithologist and author (*Birds of Greenland*, 1950; *Notes on flowerpeckers*, 1960).

Salpinctes Gr. *salpinktes,* the wren *Troglodytes.*

Salpornis Genus *Salpinctes,* wren; *ornis,* a bird.

Saltator L. *saltator,* a dancer; ref. the way in which the saltators hop heavily with both feet together.

Saltatricula Dim. of genus *Saltator,* saltator.

saltuarius L. *saltuarius,* a forester or ranger (*saltus,* a glade).

salvadorii /Salvadorina After Conte Adelardo Tommaso Salvadori Paleotti (1835–1923) Italian ornithologist, collector, and author (*Catalogo sistematico degli uccelli di Borneo*, 1874; *Ornitologia della Papuasia e delle Molucche*, 1880).

salvini After Osbert Salvin (1835–1898) English ornithologist, collector, and author (*Exotic ornithology*, 1866; *Nomenclator avium Neotropicalium*, 1873; *Biologia centrali Americana*, 1879).

samamisicus After the Samamish Alps, Elburz Mountains, Gilan, northwestern Iran.

samarensis After the island of Samar, Philippines.

samoensis After the Samoan Islands.

sanblasiana After the town of San Blas, Nayarit, Mexico.

sancta L. *sanctus,* holy, sacred; throughout the Pacific kingfishers were regarded as gods by the natives and believed to have power over the waves and the sea.

sanctaecatarinae After the state of Santa Catarina, south-eastern Brazil (Late L. *sanctus,* a saint; *sancire,* to make sacred; Mod. L. *Catarina,* Catherine and European equivalents).

sanctaecrucis After the Santa Cruz Islands, north of the New Hebrides (Late L. *sanctus,* a saint; L. *crux, crucis,* a cross; Spanish *cruz,* a cross).

sanctaehelenae After the island of St Helena in the South Atlantic, discovered by Portuguese explorers in 1502 (Late L. *sanctus,* a saint; L. *Helena,* Helena or Helen).

sanctaemariae After the Río Madre de Dios, Beni, Bolivia (Late L. *sanctus,* a saint; L. *Maria,* Mary; Spanish *madre de Dios,* mother of God).

sanctaemartae After the Sierra Nevada de Santa Marta, Magdalena, Colombia (Late L. *sanctus,* a saint; Mod. L. *Marta,* Martha and European equivalents).

sanctihieronymi After the town of San Geronimo, Vera Paz, Guatemala (Late L. *sanctus,* a saint; L. *Hieronymus,* Jerome and European equivalents).

sanctithomae After the island of São Thomé (São Tomé) in the Gulf of Guinea, West Africa, discovered by Portuguese explorers in 1471 and named for St Thomas, the patron saint of Portugal.

sandvicensis After **1** the town of Sandwich, Kent, England (*Sterna).* **2** the Sandwich Islands (v. *sandwichensis* 1) (*Branta).*

sandwichensis After **1** the Sandwich Islands, a former name for the Hawaiian Islands, named for John Montagu, 4th Earl of Sandwich (1718–1792), First Lord of the Admiralty, 1771–1782 (*Chasiempis, Pennula, Pterodroma).* **2** Sandwich Bay, Labrador Coast, eastern Canada (*Passerculus).*

sanfordi After Leland Cutler Sanford (1868–1950) US zoologist.

sanguinea /sanguineus L. *sanguineus,* bloody, of blood (*sanguis, sanguinis,* blood).

sanguiniceps L. *sanguis,* blood; *-ceps,* headed.

sanguinolenta /sanguinolentum / sanguinolentus L. *sanguinolentus,* bloody, blood-stained.

sannio L. *sannio,* a buffoon or jester; ref. the raucous calls and clownish white face of the White-faced Laughingthrush *Garrulax.*

santabarbarae After the Sierra Santa Bárbara, Jujuy, Argentina.

santaecrucis After the department of Santa Cruz, Bolivia (Late L. *sanctus,* a saint; L. *crux, crucis,* a cross; Spanish *cruz,* a cross).

santovestris After the island of Santo (Espiritu Santo), New Hebrides; L. *vester,* of.

Sapayoa After the Río Sapayo (Río Zapallo Grande), Esmeraldas, Ecuador.

Sapheopipo Gr. *sapheos,* clear, distinct; *pipo,* a bird identified by some later authors as a type of woodpecker (v. *Pipra).*

saphirina see **sapphirina.**

sapphira L. *sapphirus,* a sapphire.

sapphirina L. *sapphirinus,* sapphirine, of sapphire.

Sappho After Sappho or Psappho (fl. 600 BC) lyric poetess of Lesbos who gathered around her a group of friends, all of her own sex, in a cult organisation devoted to Aphrodite.

saracura Tupí (Brazilian) Indian name *saracura* for a wood rail or some other marsh bird.

sarasinorum After Paul Sarasin (1856–1929) and his cousin Fritz Sarasin (1859–1942) Swiss zoologists, explorers, collectors, and authors (*Reisen in Celebes,* 1905; *Die vögel Neu-Caledoniens und der Loyalty Inseln,* 1913).

Sarciophorus Gr. *sarx, sarkos,* flesh; *-phoros,* carrying; the Black-headed Plover *Hoplopterus tectus* has red wattles on the forehead.

Sarcogyps Gr. *sarx,* flesh; genus *Gyps,* vulture; the Indian King Vulture *S. calvus* has extensive bare, red skin on the head and neck.

Sarcophanops Gr. *sarx,* flesh; *phanos,* bright; *ops,* the eye; the Wattled Broadbill *Eurylaimus steerei* has a large blue eye-wattle.

Sarcops Gr. *sarx,* flesh; *ops,* the face; the Coleto *S. calvus* has a large naked area of skin on the face.

Sarcoramphus Gr. *sarx,* flesh; *rhamphos,* the bill; the Neotropical King Vulture *S. papa* has fleshy wattles at the base of the bill.

sarda L. *Sarda,* a Sardinian woman (*Sardinia;* Italian *Sardegna).*

Sarkidiornis Dim. from Gr. *sarx,* flesh; *ornis,* a bird; the Greater Comb Duck *S. melanotos* has a fleshy wattle at the base of the bill.

Saroglossa Gr. *saron,* a broom; *glossa,* the tongue; the Spot-winged Starling *S. spiloptera* has a brush-tipped tongue.

Sarothrura Gr. *saron,* a broom; *oura,* the tail; ref. the fluffy, decomposed tail-feathers of the Afrotropical flufftails.

sasakensis Malay name *Tanah Sasak* for the island of Lombok, Indonesia.

Sasia Nepali name *sasya* for the Russet Piculet *S. ochracea.*

sasin Nootka (Canadian) Indian name *sasin* for Allen's Hummingbird *Selasphorus.*

Sathrocercus Gr. *sathros,* unsound, decayed; *kerkos,* the tail; the tail-feathers of the Afrotropical bush warblers have a decomposed appearance.

Satrapa L. *satrapa,* a satrap, a Persian viceroy or governor (Gr. *satrapes,* a satrap).

saturata L. *saturatus,* richly coloured (*saturare,* to satisfy).

saturatior L. *saturatior,* more richly coloured (comp. of *saturatus,* richly coloured).

saturatus see **saturata.**

saturninus Med. L. *saturninus,* gloomy, saturnine.

satyra L. *satyrus,* a satyr, a horned sylvan deity (Gr. *saturos,* a satyr).

Saucerottia /saucerrottei After Antoine Constant Saucerotte (1805–1884) French naturalist, collector, and physician.

saularis Hindi name *saulary* for the Magpie Robin *Copsychus.* Earlier authors suggest that Linnaeus, thinking that Latham's 'Dial-bird' (Hindi *dhaiyal,* magpie robin) had something to do with a sun-dial, meant to name it *solaris* (L. *solaris,* solar, of the sun), but by a slip of the pen wrote *saularis.*

saundersi /saundersii After Howard Saunders (1835–1907) British ornithoogist and author (*Manual of British birds,* 1889).

Sauromarptis Gr. *sauros,* a lizard; *marptis,* a seizer or catcher.

saurophaga Gr. *sauros,* a lizard; *phagos,* a glutton, an eater (*phagein,* to eat).

Saurothera Gr. *sauros,* a lizard; *thera,* a hunter.

sauzieri After Théodore Sauzier (fl. 1887) French palaeontologist specializing in Mauritian fossils.

savana French *savane,* the savanna; the Fork-tailed Flycatcher *Tyrannus* was considered to prefer flooded savannas.

savannarum English *savanna,* the savanna or savannah (Spanish *zavana,* a type of open treeless plain); the Grasshopper Sparrow *Ammodramus* prefers open habitats.

savensis After the Save River, southern Moçambique.

savesi After Théodore Savés (1855–1918) French naturalist and traveller in New Caledonia.

savilei After Lieutenant Colonel Robert Vesey Savile (1873–1947) British diplomat, Sudan civil service, 1901; Governor of Darfur province, 1917–1923.

sawtelli After Gordon H. Sawtell (fl. 1973) British administrator in the Cook Islands.

saxatalis L. *saxum,* a rock; *talis,* great, fine (cf. *-alis,* pertaining to); the White-throated Swift *Aeronautes* was described from Inscription Rock, New Mexico Territory, USA, in 1853.

saxatilis L. *saxatilis,* rock frequenting (*saxum, saxi,* a rock or large stone).

Saxicola L. *saxum,* a stone or rock; *-cola,* inhabitant.

saxicolina Genus *Saxicola,* chat (formerly a generic name for the wheatears *Oenanthe*); L. *-inus,* resembling.

Saxicoloides Genus *Saxicola,* chat; Gr. *-oides,* resembling.

saya After Thomas Say (1787–1834) US zoologist, collector, and author (*American entomology*, 1824; *American conchology*, 1830).

sayaca Tupí (Brazilian) Indian name *sayacu* for a type of finch.

Sayornis After Thomas Say (v. *saya)*; Gr. *ornis,* a bird.

saypani After the island of Saypan (Saipan), Marianas Islands.

scalaris L. *scalaris,* a ladder; ref. barred upperparts of the Ladder-backed Woodpecker *Dendrocopos.*

scandens L. *scandens,* climbing (*scandere,* to climb).

scandiaca Mod. L. *Scandia,* Scandinavia.

scansor Mod. L. *scansor,* a climber (L. *scandere,* to climb).

Scaphidura Gr. *skaphis, skaphidos,* a small boat; *oura,* the tail; ref. the flat, rounded tail of the Great Cowbird *S. oryzivora.*

scapularis Mod. L. *scapularis,* scapular, of the shoulders (L. *scapulae,* shoulders).

Scardafella Italian *sgardafella,* scaly appearance.

Sceloglaux Gr. *skelos,* the leg; *glaux,* an owl; the rare Laughing Owl *S. albifacies* has long feathered legs and toes.

Scelorchilus Gr. *skelos,* the leg; Mod. L. *orchilus,* wren (v. *Hylorchilus).*

Scenopoeetes Gr. *skene,* a house; *poietes,* a maker; the Tooth-billed Bowerbird *S. dentirostris* clears a space on the forest floor which it decorates with leaves, and attracts females by a remarkable song.

Scepomycter Gr. *skepas,* a covering; *mukter,* the nose; the nostril of the Chestnut-hooded Warbler *S. winifredae* is partially covered by a horny hood.

schach Onomatopoeic name *scack* coined for the Long-tailed Shrike *Lanius* by Pehr Osbeck (1757) (*Dagbok öfver en Ostindische resa aren 1750–1752)*, Swedish chaplain, author, and Linnaean disciple who collected plants and birds in China.

schalowi After Hermann Schalow (1852–1925) German ornithologist and author (*Beitrage zur vögelfauna der Mark Brandenburg*, 1919).

scheepmakeri After C. Scheepmaker (fl. 1875) Dutch civil servant and collector.

Schetba Malagasy native name *schet-bé* for the Rufous Vanga *S. rufa.*

Schiffornis After M. Schiff (1823–1896) German ornithologist; Gr *ornis,* a bird.

schistacea Late L. *schistaceus,* slate grey, slate-coloured (Late L. *schistus,* slate; L. *lapis schistos,* a kind of fissile stone).

schistaceigula Late L. *schistaceus,* slate-coloured; L. *gula,* the throat.

schistaceus see **schistacea.**

Schistes Gr. *skhizo,* to cleave; ref. the short, sharply pointed bill of the Wedge-billed Hummingbird *S. geoffroyi.*

schisticeps Late L. *schistus,* slate; L. *-ceps,* headed.

schisticolor Late L. *schistus,* slate; L. *color,* colour, hue.

schistisagus Late L. *schistus,* slate; L. *sagus,* a cloak (Gr. *sagos,* a mantle or blanket).

Schistochlamys Late L. *schistus,* slate; Gr. *khlamus,* a cloak.

schistogynus Late L. *schistus,* slate; Gr. *gune,* a woman; unlike other members of its genus the female Bluish-slate Antshrike *Thamnomanes* shares the male's dull grey upperparts.

Schizoeaca Gr. *skhiza,* twigs, firewood; *oikia,* a dweller (*oikeo,* to inhabit); the Eye-ringed Thistletail *S. palpebralis* inhabits damp thickets and shrub.

schlegeli /Schlegelia /schlegelii After Hermann Schlegel (1804–1884) German ornithologist and author (*Fauna Japonica - aves*, 1844; *Kritische übersicht der europäischen vögel*, 1844; these two works, the first co-authored with Siebold and Temminck, were the earliest to use trinominals in the modern subspecific sense).

schleiermacheri After H. Schleiermacher (fl. 1877) German zoologist.

schneideri After Gustav Schneider (b. 1867) Swiss zoologist, collector in Sumatra, 1897.

schoeniclus Gr. *skhoiniklos,* a small waterside bird mentioned by Aristotle, variously identified by later authors.

Schoenicola Gr. *skhoinos,* a reed or rush; L. *-cola,* inhabitant.

Schoeniophylax Gr. *skhoinia,* a bunch of rushes; *phulax,* a watcher or sentinel.

schoenobaenus Gr. *skhoinos,* a rush or reed; *baino,* to go, to walk (i.e. to inhabit).

schomburgkii After Sir Robert Hermann Schomburgk (1804–1865) German-born British explorer and author.

Schoutedenapus After Henri Schouteden (b. 1881) Belgian ornithologist, collector, and Afrotropical specialist; genus *Apus,* swift.

schoutedeni see **Schoutedenapus.**

schrankii After Franz von Paula von Schrank (1747–1835) German botanist and author.

schreibersii After Karl Franz Anton von Schreibers (1775–1852) Austrian zoologist and collector in Brazil, 1817.

schuetti After Otto Schütt (1843–1888) German railway engineer in Angola, 1877–1879.

schulzi After W. Schulz (d. 1933) German zoologist in Argentina, 1866–1933.

schwarzi After Ludwig Schwarz (1822–1894) German astronomer in the Russian service with Transbaykal Expedition, 1855–1858.

scintilla L. *scintilla,* a spark, a glimmer.

scirpaceus L. *scirpus,* a reed or bulrush; *-aceus,* resembling (i.e. dwelling).

Scissirostrum L. *scissura,* a cleft; *rostrum,* the bill; the nostrils of the Finch-billed Mynah *S. dubium* are contained in a long, furrow-shaped cleft.

scita L. *scitus,* approved.

sclateri After **1** Philip Lutley Sclater (1829–1913) English ornithologist, collector, and author (*A Monograph of the birds forming the tanagrine genus Calliste*,1857; *Exotic ornithology*,1866; *The curassows*, 1875; *A monograph of the jacamars and puffbirds*,1879; *Birds of the Challenger expedition*, 1881; *Argentine ornithology*, 1888) (*Cacicus, Eudyptes, Forpus, Hylophilus, Icterus, Lopho-*

phorus, Meliarchus, Myrmotherula, Myzomela, Nonnula, Parus, Picumnus, Pseudocolopteryx, Thryothorus, Xanthomyias). **2** William Lutley Sclater (1863–1944) English ornithologist; Director Cape Town Museum, 1896–1906; and author (*Systema avium Aethiopicarum*, 1924) (*Spizocorys).*

Sclateria /sclateriana see **sclateri 1**.

Sclerurus Gr. *skleros,* stiff; *oura,* the tail; the tails of the Neotropical leaf-tossers have stiff shafts.

scolopacea /scolopaceus Genus *Scolopax,* woodcock; L. *-aceus,* resembling.

Scolopax L. *scolopax,* the woodcock or snipe (Gr. *askalopas* or *skolopax,* the woodcock).

scopifrons L. *scopa,* a broom; *frons,* the forehead or brow; ref. the stiff chestnut feathers on the forehead of the Brush-fronted Helmet Shrike *Prionops.*

scops Gr. *skops,* the little eared owl.

Scoptelus Gr. *skoptelos,* a mocker (*skopto,* to mock); ref. the harsh chattering cries of the Afrotropical woodhoopoes (cf. Afrikaans name *kakelaar,* chatterer, for the Red-billed Woodhoopoe *Phoeniculus purpureus).*

scopulinus L. *scopulus,* a cliff, crag, or rock; *-inus,* belonging to; the Red-billed Gull *Larus* often nests on rocky cliffs and stacks.

Scopus L. *scopa,* a broom made of twigs; ref. the untidy crest of the Hammerkop *S. umbretta.*

scoresbii After William Scoresby (1789–1857) English Arctic explorer and author (*The Arctic regions*, 1820).

scotica /scoticus L. *Scoticus,* Scottish, a Scot.

Scotocerca Gr. *skotos,* dark, dusky; *kerkos,* the tail.

Scotopelia Gr. *skotos,* gloom, darkness; specific name *Strix peli,* fishing owl.

scotops Gr. *skotos,* dark, dusky; *ops,* the face or eye.

Scotornis Gr. *skotos,* gloom, darkness; *ornis,* a bird.

scouleri After John Scouler (1804–1871) Scottish zoologist.

scrippsi After Robert Paine Scripps (1895–1938) US newspaper publisher.

scripta L. *scriptus,* written, drawn (*scribere,* to write); the Squatter Pigeon *Petrophassa* has a distinctive black and white face pattern and scaly upperparts and breast, and the Letter-winged Kite *Elanus* has an M-shaped line on the underwing.

scriptoricauda L. *scriptorius,* writing; *cauda,* the tail; ref. the obscure barring on the tail of the Speckle-throated Woodpecker *Campethera.*

scriptus see **scripta.**

scrutator L. *scrutator,* a searcher or examiner (*scrutari,* to examine).

scutulata L. *scutulatus,* diamond-shaped, checker-shaped (*scutula,* a square dish).

scutatus L. *scutatus,* armed with a shield (*scutum,* a shield).

Scytalopus Gr. *skutalon,* a stick or cudgel; *pous,* the foot; ref. the strong, booted tarsi of the terrestrial tapaculos.

Scythrops Gr. *skuthros,* angry, sullen; *ops,* the face; ref. the huge, curved bill and grim visage of the Channel-billed Cuckoo *S. novaehollandiae.*

sechellarum /sechellensis After the Seychelles Islands (Republic of Seychelles), Indian Ocean.

seductus L. *seductus,* remote, distant (*seducere,* to separate).

seebohmi After Henry Seebohm (1832–1895) British businessman, ornithologist, oologist, traveller, and author (*A history of British birds*, 1883; *The geographical distribution of the family Charadriidae*, 1887; *The birds of the Japanese Empire*, 1890; *A monograph of the Turdidae*, 1898; *The birds of Siberia*, 1901).

sefilata L *sex,* six; *filum, fili,* a thread, fillet or filament; *-atus,* possessing (French name *le Sifilet* or *Manucode à six filets* given to the six-wired bird of paradise *Parotia* by Buffon (1770); *six,* six; *filet,* a thread).

segmentata L. *segmentatus,* ornamented with purple or gold borders or patches.

segregata L. *segregatus,* isolated, segregated (*segregare,* to separate); Dammerman's Flycatcher *Muscicapa* was described as an insular race of *Alseonax latirostris.*

Seicercus Gr. *seio,* to shake, move to and fro; *kerkos,* the tail.

seimundi After E. C. H. Seimund (fl. 1925) British taxidermist and collector in South Africa, 1899–1903, Fernando Póo, 1904 and Malaya,1916.

Seisura Gr. *seisoura,* the 'wag-tail', a waterside bird mentioned by Hesychius and other authors, variously identified by later writers (*seio,* to shake; *oura,* the tail).

Seiurus Gr. *seio,* to shake or wag to and fro; *oura,* the tail.

Selasphorus Gr. *selas,* a bright flame, flash of lightning; *-phoros,* carrying.

seledon German name *der Seledonkopf* given to the Green-headed Tanager *Tangara* by Müller (1776), *Des Ritters Carl von Linné vollständige naturgeschichte, Suppl.*, p. 158 (French *céladon,* pale green).

Selenidera Gr. *selene,* the moon; *deras,* the neck; the toucanets have crescent-shaped nuchal collars.

Seleucidis L. *seleucides,* migrant birds sent by the gods to destroy locusts (i.e. birds from paradise) (Gr. *seleukis,* a locust-eating bird, perhaps the Rose-coloured Starling *Sturnus roseus).*

seloputo Javanese name *selo-puto* for the Pagoda Owl *Strix.*

Semeiophorus Gr. *semeion,* a standard or flag; *-phoros,* carrying; ref. the greatly elongated secondary feathers of the male Pennant-winged Nightjar *S. vexillarius.*

semibadius L. *semi-,* half; *badius,* bay-coloured, reddish-brown.

semibrunneus L. *semi-,* half; Mod. L. *brunneus,* brown.

semicincta L. *semi-,* half, small; *cinctus,* girdled, banded (*cingere,* to encircle; *semicinctum,* a narrow girdle).

semicinerea /semicinereus L. *semi-,* half; *cinereus,* ash-coloured (*cinis, cineris,* ashes).

semicollaris L. *semi-,* half, small; *collaris,* pertaining to the neck, collared.

semifasciata L. *semi-,* small, half; Late L. *fasciatus,* banded, striped (L. *fascia,* a band or girdle).

semiflava /semiflavum /semiflavus L. *semi-,* half; *flavus,* golden-yellow.

semifuscus L. *semi-,* half; *fuscus,* dusky, dark.

semilarvata /semilarvatus L. *semi-,* half, small; *larvatus,* masked (*larva,* mask).

Semioptera Gr. *semeion,* a standard or flag; *ptera,* wings; ref. the flag-like wing plumes of the male Wallace's Standardwing *S. wallacei.*

semipalmata /semipalmatus L. *semi-,* small, half; *palmatus,* palmate, marked with the palm of the hand (i.e. webbed) (*palma,* the palm of the hand); ref. partial webbing between toes.

semipartitus L. *semi-,* half; *partitus,* divided, shared (*partire,* to divide); the Silverbird *Empidornis* is bright grey above and tawny below.

semiplumbea /semiplumbeus L. *semi-,* half; *plumbeus,* leaden, plumbeous.

semirubra L. *semi-,* half; *ruber,* red.

semirufa /semirufus L. *semi-,* half; *rufus,* red, ruddy.

semitorquata /semitorquatus L. *semi-,* half, small; *torquatus,* collared, necklaced.

semitorques L. *semi-,* small, half; *torques,* a collar or necklace.

Semnornis Gr. *semnos,* majestic, august, solemn; *ornis,* a bird.

semperi After **1** Carl Semper (1832–1893) German zoologist, anatomist (*Zosterops).* **2** Revd. J. E. Semper (fl. 1873) resident on St Lucia, West Indies (*Leucopeza).*

senator L. *senator,* a senator; the Woodchat Shrike's *Lanius* nape was considered to be purple, and the name reflects the purple robes of state which distinguished the Roman senators.

senegala **1** French name *Le Sénégali* given to the Red-billed Firefinch *Lagonosticta* by d'Aubenton (1765), *Planches enluminéez.* **2** after Senegal (*Poicephalus, Tchagra)* (v. *senegalensis).*

senegalensis /senegallus After Senegal (French *Sénégal*) in tropical West Africa. In early ornithology the name equated to Guinea, and was applied generally to birds from tropical West Africa, as well as those from Senegal itself. In the cases of *Dryoscopus* and *Pterocles* the toponym is erroneous.

senegaloides Specific name *Halcyon senegalus;* Gr. *-oides,* resembling.

senegalus see **senegala.**

senex L. *senex,* an old person (i.e. white- or grey-crowned or headed).

senilis L. *senilis,* senile, of an old man (*senex, senis,* an old person, aged) (i.e. white-crowned or capped).

seniloides Specific name *Pionus senilis;* Gr. *-oides,* resembling.

sephaena No expl. (Smith (1836), *Rep. exped. centr. Africa,* p. 55); probably based on an otherwise unrecorded Chuana native name.

sephaniodes /Sephanoides Misspelt from Gr. *stephane,* a diadem; *-oides,* resembling; ref. the iridescent crown feathers of the male firecrowns.

sepiaria Mod. L. *sepiarius,* living in hedges (L. *saepes* or *sepes,* a hedge; *-arius,* pertaining to); Horsfield's Babbler *Malacocincla* inhabits lowland forests of Malaysia.

sepium L. *sepium,* of hedges (L. *saepes* or *sepes,* a hedge).

septentrionalis L. *septemtrionalis,* northern.

septimus L. *septimus,* seventh; the Philippine Frogmouth *Batrachostomus* was the seventh species of frogmouth to be described.

sepulcralis L. *sepulcralis,* of the tomb, sepulchral; the melancholy cries of the Indonesian Cuckoo *Cacomantis,* frequently uttered by moonlight from graveyards, are regarded with superstition by the natives.

seranensis After the island of Seran or Ceram (Seram), in the Moluccas, Indonesia.

serena L. *serenus,* fair, bright.

sericea /sericeus Med. L. *sericeus,* silken (Gr. *serikos,* silken).

sericocaudatus Gr. *serikos,* silken; L. *caudatus,* tailed (*cauda,* the tail).

Sericornis Gr. *serikos,* silken; *ornis,* a bird; the scrubwrens have soft, silky plumage.

Sericossypha Gr. *serikos,* silken; *kossuphos,* the Blackbird *Turdus;* ref. the velvety black plumage of the White-capped Tanager *S. albocristata.*

Sericotes Gr. *serikos,* silken; *-oides,* resembling.

Sericulus Dim. of L. *sericum,* silk (Gr. *serikos,* silken); ref. the brilliant silky plumage of the male Regent Bowerbird *S. chrysocephalus.*

Serilophus Gr. *serikos,* silken; *lophos,* a crest; the Silver-breasted Broadbill *S. lunatus* has the 'head with a procumbent crest of silky feathers' (Swainson (1837), *Classif. birds,* Vol. **2**, p. 262).

serina Mod. L. *serinus,* serine, canary-yellow (cf. Mod. L. *cerinus,* waxy-yellow).

Serinus French *serin,* a canary or serin.

serpentarius L. *serpens, serpentis,* a snake; *-arius,* pertaining to; the Secretary Bird *Sagittarius* (q.v.) feeds mainly on snakes, lizards, small birds, and rodents (cf. Med. L. *serpentarius,* serpentine).

Serpophaga Gr. *serphos,* a gnat; *phagos,* an eater.

serrana /serranus Portuguese *serra,* a mountain range

serrator Med. L. *serrator,* a sawyer (L. *serra,* a saw); ref. saw-edged bills.

serriana After Victor Serrès (fl. 1844) French anatomist and author.

serripennis L. *serra,* a saw; *pennis,* winged; ref. the hooked primary feathers of the Northern Rough-winged Swallow *Stelgidopteryx.*

serrirostris L. *serra,* a saw; *-rostris,* billed; the White-vented Violetear *Colibri* has minute serrations on the outer upper mandible.

serva L. *servus,* servile, slave (i.e. small, inferior).

setaria L. *seta* or *saeta*, a bristle or stiff hair; *-arius*, possessing; ref. the very long, pointed central tail feathers of the Araucaria Tit Spinetail *Leptasthenura*.

sethsmithi After L. M. Seth-Smith (1879–1955) government surveyor and collector in Uganda.

setifrons L. *seta* or *saeta*, a bristle; *frons*, the forehead or brow; the Spiny-faced or Speckled Antshrike *Xenornis* has bristle-like feathers on the forehead.

Setochalcis Gr. *ses*, *setos*, a moth; *khalkis*, a rapacious but unidentified bird of prey.

Setopagis Gr. *ses*, a moth; *pagis*, a snare or trap.

Setophaga Gr. *ses*, a moth; *phagos*, an eater.

Setornis L. *seta* or *saeta*, a bristle; Gr. *ornis*, a bird; ref. the distinctive cheek bristles of the Hook-billed Bulbul *S. criniger*.

severa/severus L. *severus*, grave, stern, grim, cruel.

sewerzowii After Nicolai Severzov (1827–1885) Russian zoologist and explorer in central Asia.

sganzini After Captain V. Sganzin (fl. 1840) French entomologist and traveller in Madagascar.

Shama Hindi name *shama* for the White-rumped Shama *Copsychus malabaricus*.

sharpei After **1** Richard Bowdler Sharpe (1847–1909) British ornithologist and author (v. *Bowdleria*) (*Lalage, Macronyx, Sheppardia, Smithornis, Terenura*). **2** Sir Alfred Sharpe (1853–1935) Governor of Nyasaland (Malawi) and naturalist (*Turdoides*).

sharpii see **sharpei 1**.

shelleyi After **1** Sir E. Shelley (1827–1890) English traveller in tropical Africa (*Cryptospiza, Francolinus*). **2** Captain George Ernest Shelley (1840–1910) British ornithologist, collector and author (*A handbook to the birds of Egypt*, 1872; *A monograph of the Cinnyridae, or family of sun birds*,1876; *The birds of Africa*, 1896) (*Aethopyga, Bubo, Nectarinia, Nesocharis, Passer, Spreo*).

Sheppardia After P. A. Sheppard (1875–1958) British farmer, oologist, collector, and settler in Rhodesia and Moçambique, 1898–1916.

shorii After the Hon. Frederick John Shore (1799–1837) in East India Company civil service.

Sialia/sialis Gr. *sialis*, a bird mentioned by Athenaeus and Hesychius, not otherwise identified.

Sibia Nepalese name *sibya* for the Black-capped Sibia *Heterophasia capistrata*.

sibilans L. *sibilans*, whistling (*sibilare*, to whistle).

sibilator L. *sibilator*, a whistler, a whisperer.

sibilatrix L. *sibilatrix*, whistling, hissing.

sibirica/sibiricus After Siberia, the northern and largest part of Asiatic Russia (Cossack *Sibir*, a Tartar khanate, near modern Tobolsk', captured by the Cossacks in 1581).

Sicalis Gr. *sikalis* or *sukalis*, a small, black-headed bird mentioned by Epicharmus, Aristotle, and others, not further identified, but probably a sort of warbler *Sylvia*.

siccatus L. *siccatus*, of dry land (*siccum*, dry land); the Black-masked Babbler *Rhopocichla* is confined to the dry zone of Sri Lanka.

sicki After Helmut Sick (b. 1910) German ornithologist resident in Brazil, collector, and author (*Ornitologia brasileira,uma introdução*, 1984).

sidamoensis After the province of Sidamo, Ethiopia.

sieboldii After Philipp Franz Balthazar von Siebold (1796–1866) German naturalist, traveller, collector in Japan, 1823–1829 and author (*Fauna Japonica – aves*, 1844).

siemiradzkii After Józef Siemiradski (1858–1933) Polish geologist in Ecuador, 1883.

siemsseni After G. Siemssen (1857–1915) German Consul General in Fukien, China, 1903.

Sigelus Gr. *sigelos*, silent, taciturn.

sigillatus L. *sigillatus*, adorned with little images or figures (*sigilla*, an image); ref. the white wing patches of the White-winged Robin *Peneothello*.

Sigmodus Gr. *sigma*, the letter Σ (S); *odous*, a tooth; ref. the toothed bill of the Red-billed Helmet Shrike *Prionops caniceps*.

signata/signatus L. *signatus*, well marked, distinct (*signare*, to mark).

siju Local Cuban name *sijú* for the Cuban Pygmy Owl *Glaucidium*.

silens L. *silens*, silent, still (*silere*, to be silent).

silvestris L. *silvestris*, belonging to woods (*silva*, a wood or forest).

silvicola L. *silvicola*, inhabiting woods (*silva*, a wood; *colere*, to inhabit).

Silvicultrix L. *silvicultrix*, inhabiting woods.

similis L. *similis*, similar, resembling.

simillimus L. *simillimus*, very similar (super. of *similis*, similar, like).

simlaensis After the city of Simla, capital of Himachal Pradesh state, India, and, under the British Raj, the summer residence of the Viceroy.

simoni After Eugène Louis Simon (1848–1924) French ornithologist and author (*Histoire naturelle des Trochilidae*, 1921).

simonsi After P. O. Simons (1869–1901) US collector in Bolivia.

Simoxenops Gr. *simos*, snub-nosed; genus *Xenops*, xenops.

simplex L. *simplex*, simple, plain.

sinaloa/sinaloae After the state of Sinaloa, Mexico.

sindianus After the province of Sind, formerly in British India, now in Pakistan.

sinense/sinensis Mod. L. *Sinensis*, Chinese (Gr. *Sinai*, Chinese).

singalensis Erroneous type locality after Sri Lanka (Ceylon); the natives of Sri Lanka are called Singhalese (Sanskrit *Simala*, Sri Lanka).

singularis L. *singularis*, unique, remarkable, single.

sinica Med. L. *Sinicus*, Chinese (Gr. *Sinai*, Chinese).

sintillata L. *scintillatus*, sparkling (*scintillare*, to sparkle).

sinuata L. *sinuatus*, curved (*sinus*, a curve); ref. shape of bill (*Pyrrhuloxia*) or primaries (*Cercomela*).

sipahi Hindi *sipahi*, a soldier, a sepoy; ref. scarlet uniforms of the native Indian infantry and the plumage of the Scarlet Finch *Haematospiza*.

siparaja Malay name *sipa raja* for the Crimson Sunbird *Aethopyga*.

Siphia Nepalese name *siphya* for the Orange-gorgetted Flycatcher *Ficedula strophiata*.

Siphonorhis Gr. *siphon, siphonos*, a tube or straw; *rhis*, the nose; ref. the excessively elongated, tubular nostrils of the Jamaican Pauraque *S. americanus*.

Sipia After the Sipi area, Chocó, Colombia.

Sipodotus Euphonic anagram of genus *Todopsis*, fairy wren.

sipora After the island of Sipora (Sipura), Mentawi Islands, Sumatra.

Siptornis Gr. *sipte* (also *sippe*, variant forms of *sitte*), a bird allied to the woodpeckers mentioned by Aristotle, Callimachus, and Hesychius, latterly associated with the nuthatch *Sitta; ornis*, a bird.

Siptornopsis Genus *Siptornis*, prickletail; Gr. *opsis*, appearance.

siquijorensis After the island of Siquijor, Visayan Islands, Philippines.

sirintarae After Princess Sirindhorn Thepratanasuda (fl. 1968) third daughter of Bumibhol Adulyadej King of Thailand.

Sirystes Gr. *suristes*, a piper, a pipe player; ref. the noisy whistling calls of the Sirystes *S. sibilator*.

sissonii After Mr Sisson (fl. 1868) US Consul at Mazatlán, Sinaloa, Mexico.

Sitagra Gr. *sitos*, corn; *agra*, catching, hunting.

Sitta Gr. *sitte*, a bird like a woodpecker mentioned by Aristotle, Callimachus, and Hesychius, now identified with the nuthatch.

Sittasomus Genus *Sitta*, nuthatch; Gr. *soma*, the body (i.e. appearance).

sitticolor Genus *Sitta*, nuthatch; L. *color*, colour, hue.

sittoides Genus *Sitta*, nuthatch; Gr. *-oides*, resembling.

Siva No expl. (Hodgson (1837), *India Revue*, **2**, 88), but probably from Hindi *siva*, happy (whence Siva, the Hindu god of life and death).

sjoestedti After Bror Yngve Sjöstedt (1866–1948) Swedish entomologist and collector in tropical Africa.

skua Faeroese name *skuvur* (Old Norse *skūfr)* for the Great Skua *Catharacta*.

Skutchia After Alexander F. Skutch (b. 1904) US ornithologist resident in Costa Rica, collector, field naturalist, and author (*Life histories of Central American birds*, 1954; *Life of the woodpecker*, 1985).

sladeni After Major A. G. Sladen (1880–1960) British collector in the Congo, 1906.

sladeniae After Mrs W. Sladen (fl. 1904) wife of British ornithologist W. Percy Sladen.

sloetii After Baron Ludolf Anne Jan Wilt Sloet van de Beele (1806–1890) Dutch jurist, historian, Governor-General of the Dutch East Indies, 1860–1866.

Smaragdites L. *Smaragdites*, a mountain on which emeralds were found.

Smaragdolanius Gr. *smaragdos*, a bright green gem, the emerald; genus *Lanius*, shrike.

Smicrornis Gr. *smikros*, small; *ornis*, a bird.

Smilorhis Gr. *smile*, a knife; *rhis*, the nose.

smithi /smithii After **1** Sir James Smith (1759–1828) English botanist, first President of the Linnean Society of London (*Petrophassa*). **2** Lieutenant Colonel Charles Hamilton Smith (1776–1859) English naturalist, artist, and collector (*Hirundo*). **3** Sir Andrew Smith (1797–1872) Scottish herpetologist, ornithologist, first Director of the Cape Town Museum (*Anas*).

Smithornis After Sir Andrew Smith (v. *smithi* **3**); Gr. *ornis*, a bird.

smyrnensis After the city of Smyrna (Izmir), Turkey.

Snethlagea After Emilia Snethlage (1868–1929) German ornithologist, field worker in Amazonia, 1905–1929 and author (*Catálogo das aves amazonicas*, 1914).

sociabilis L. *sociabilis*, sociable.

socialis L. *socialis*, sociable, social.

socius L. *socius*, sharing, allied.

socorroensis After Socorro Island, Revillagigedo Islands, Mexico.

socotrana/socotranus After Socotra (Sokotra) Island, South Yemen, Indian Ocean (Arabic *Soqotra*, isle of bliss).

soederstromi After L. Söderstrom (1843–1927) Swedish Consul General to Ecuador, 1886.

soemmerringi After Samuel Thomas von Sömmerring (1755–1830) German anatomist and scientist.

sokokensis After the Sokoke Forest, Coastal Province, Kenya.

solandri After Daniel Solander (1733–1782) Swedish naturalist and collector, pupil of Linnaeus.

solangiae After Princesse Solange Murat (b. 1894) wife of French collector and naturalist Prince Paul Murat.

solaris L. *solaris*, solar, of the sun (*sol, solis*, the sun).

solitaria /solitaris /solitarius L. *solitarius*, solitary, alone.

soloensis After the city of Solo (Surakarta), Jawa Tengah, Java.

solomonensis /solomonis After the Solomon Islands, Melanesia, in the south-west Pacific (Spanish *Islas de Solomon*, named by the Spanish explorer Álvaro de Mendaña de Neira in 1568 in anticipation of their riches and after the opulent biblical monarch).

solstitialis L. *solstitialis*, solar, of summer.

somalica /somalicus After British Somaliland (Somalia), named after the nomadic Samaal inhabitants.

Somateria Gr. *soma, somatos,* the body; *erion,* wool; the eider down used to line the nests of the Common Eider *S. mollissima* was formerly collected commercially.

sonnerati /sonneratii After Pierre Sonnerat (1748–1814) French naturalist, collector, and author (*Voyage à la Nouvelle Guinée*, 1776; *Voyages aux Indes Orientales et à la Chine*, 1782).

sophiae No expl. (Severzov (1873), *Izvest. Imper., Moskva*, No.**8**, pt. 2, 66); doubtless after the Tzarina Maria Alexandrovna (1824–1880), formerly Princess Maximiliane Wilhelmine Auguste Sophie Marie daughter of Ludwig II Grand Duke of Hesse-Darmstadt, and wife of Alexander II (1818–1881: reigned 1855–1881) Tzar of all the Russias.

sordida L. *sordidus,* shabby, dirty, poor (i.e. dull-coloured, small).

sordidulus L. *sordidulus,* somewhat dirty (dim. of *sordidus,* shabby).

sordidus see **sordida.**

Sorella No expl. (Hartlaub (1880), *Journ. für Orn.*, **28**, 210), but probably from Italian *sorella,* a sister (L. *soror,* a sister).

sorghophilus Italian *sorgo,* cane-grass; Gr. *philos,* loving, fond of.

soror L. *soror,* a sister (i.e. closely related).

sororcula L. *sororcula,* a little sister.

souancei After Baron Charles de Souancé (1823–1896) French ornithologist, naval commissary, and author (*Iconographie des perroquets*, 1857).

soui Galibi (Cayenne) Indian name *soui* for the Little Tinamou *Crypturellus* imitative of its callnotes.

souimanga French name *soui-manga* for a sunbird, based on Malagasy names *sohimanga* or *soimangavola* for the Bronze-backed Sunbird *Nectarinia notata.*

souleyetii After F. L. A. Souleyet (1811–1852) French naval surgeon and naturalist in the Pacific, 1836–1837.

souliei After Père J. A. Soulie (1858–1905) French missionary to China.

soumagnei After M. Soumagne (fl. 1877) French Consul in Madagascar, 1863.

souzae After J. A. de Souza (1836–1889) Portuguese zoologist.

sowerbyi After J. W. Sowerby (fl. 1898) English trooper in British South African Mounted Police and collector.

spadicea/spadiceus Mod. L. *spadiceus,* bright brown (L. *spadix, spadicis,* chestnut-coloured; Gr. *spadikos,* bay-coloured; *spadix,* a palm frond).

spadix L. *spadix,* chestnut-coloured.

spaldingii After E. Spalding (d. 1894) Australian taxidermist.

sparganura Gr. *spargao,* to abound, to teem; *oura,* the tail; ref. the long, rose red and purple banded tail of the Red-tailed Comet *Sappho.*

sparsa L. *sparsus,* speckled, spotted (*spargere,* to scatter).

Spartonoica Gr. *spartos,* a kind of shrub; *oikos,* an inhabitant.

sparverioides Med. L. *sparverius,* the sparrowhawk *Accipiter;* Gr. *-oides,* resembling; ref. the hawk-like appearance of the Large Hawk Cuckoo *Cuculus.*

sparverius Med. L. *sparverius,* the sparrowhawk *Accipiter.*

Spatula L. *spatula,* a spoon or spatule; ref. the large, spoon-shaped bill of the Northern Shoveler *Anas clypeata.*

spatulata Mod. L. *spatulatus,* spatulatae (L. *spatula,* a spoon); the elongated outer tail shafts of the Racquet-tailed Roller *Coracias* end in spatulate tips.

speciosa /speciosum /speciosus L. *speciosus,* splendid, beautiful (*species,* beauty).

spectabilis L. *spectabilis,* remarkable, showy, worth seeing.

specularioides Specific name *Anas specularis;* Gr. *-oides,* resembling.

specularis L. *specularis,* like a mirror (*speculum,* a mirror); ref. the metallic green and purple speculum or wing-patch of the Bronze-winged Duck *Anas.*

speculifera /speculiferus L. *speculum, speculi,* a mirror; *-fer,* bearing; both the White-winged Diuca Finch *Diuca* and the Puerto Rican Tanager *Nesospingus* have distinctive white wing patches.

speculigera L. *speculum,* a mirror; *-ger,* bearing; ref. white wing patch of the Black-and-white Tanager *Conothraupis.*

Speculipastor L. *speculum,* a mirror; syn. genus *Pastor,* starling; the Magpie Starling *S. bicolor* has a white wing patch.

Speirops Gr. *speira,* a circle or coil; *ops,* the eye; ref. the narrow white eye-ring of the Black-capped Speirops *S. lugubris.*

spekei After Captain John Hanning Speke (1827–1864) British explorer, the first European to see Lake Victoria.

spekeoides Specific name *Ploceus spekei;* Gr. *-oides,* resembling.

Spelaeornis Gr. *spelaion,* a cave; *ornis,* a bird; the wren babblers were considered allied to the wrens *Troglodytes* (Gr. *troglodutes,* the cave dwellers).

speluncae L. *spelunca,* a cave (Gr. *spelunx, spelunkos,* a cave); the nest of the Mouse-coloured Tapaculo *Scytalopus* is well hidden in a crevice or burrow.

spencei After William Spence (1783–1860) English entomologist.

Speotyto Gr. *speos,* a cave; genus *Tyto,* owl; the Burrowing Owl *S. cunicularia* appropriates the burrows dug by prairie dogs *Cynomys.*

sperata L. *sperata,* a bride (*sperare,* to hope for).

Spermestes Gr. *sperma,* a seed; *-estes,* an eater (εδω, to devour).

Spermophaga Gr. *sperma,* a seed; *phagos,* an eater (*phagein,* to eat).

Sphecotheres Gr. *sphex, sphekos,* a wasp; *theras,* a hunter; the figbirds feed on fruit.

Spheniscus Dim. of Gr. *sphen,* a wedge; ref. streamlined shape of swimming penguins.

sphenocercus Gr. *sphen, sphenos,* a wedge; *kerkos,* the tail.

Sphenocichla Gr. *sphen,* a wedge; *kikhle,* thrush; ref. the wedge-shaped bill of the Wedge-billed Wren Babbler *S. humei.*

Sphenoeacus Gr. *sphen,* a wedge; *oiax, oiakos,* the helm; ref. the wedge-shaped tail feathers of the Cape Grassbird *S. afer.*

Sphenorhynchus Gr. *sphen,* a wedge; *rhunkhos,* the bill.

Sphenostoma Gr. *sphen,* a wedge; *stoma,* the mouth.

sphenura /Sphenurus Gr. *sphen,* a wedge; *oura,* the tail.

Sphyrapicus Gr. *sphura,* a hammer; genus *Picus,* woodpecker.

spillmanni After F. Spillmann (fl. 1925) Austrian zoologist resident in Ecuador.

spilocephala /spilocephalus Gr. *spilos,* a spot; *-kephalos,* headed.

spilodera Gr. *spilos,* a spot; *deras,* the neck.

spilogaster Gr. *spilos,* a spot; *gaster,* the belly.

spilonota /spilonotus Gr. *spilos,* a spot; *-notos,* backed.

spiloptera /spilopterus Gr. *spilos,* a spot; *-pteros,* winged.

Spiloptila Gr. *spilos,* a spot; *ptilon,* a feather; ref. the black and white tipped tail of the Cricket Warbler *S. clamans.*

Spilornis Gr. *spilos,* a spot; *ornis,* a bird.

spilorrhoa Gr. *spilos,* a spot; *orrhos,* the rump.

spilothorax Gr. *spilos,* a spot; *thorax,* the breast.

Spindalis Gr. *spindalos,* an Indian bird akin to *attagas.* No expl. (Jardine and Selby (1837), *Illus. orn.*, Vol. **4**, pl. 9), but, according to Newton and Gadow (1896), p. 655, it is an error for *spindasis* (syn. genus *Spinus,* siskin; *dasos,* a thicket).

spinescens Syn. genus *Spinus,* siskin; L. *-escens,* almost resembling.

spinicauda L. *spina,* a thorn; *cauda,* the tail.

spinicollis L. *spina,* a thorn; Mod. L. *-collis,* necked.

spinoides Specific name *Carduelis spinus;* Gr. *-oides,* resembling.

spinoletta Local Florentine (Italian) name *spipoletta* for the 'little titlark'.

spinosa / spinosus L. *spinosus*, prickly, thorny (*spina*, a thorn).

Spinus Gr. *spinos*, a bird mentioned by Aristophanes, Dionysius, Claudius Aelianus, Hesychius, and others, not further identified, but later considered identical to the *spingos* or Chaffinch *Fringilla*.

spirurus Gr. *speira*, a coil; *-ouros*, tailed; the tail feathers of the Wedge-billed Woodcreeper *Glyphorhynchus* are spirally twisted towards their ends.

spixi / spixii After Johann Baptist von Spix (1781–1826) German naturalist in Brazil, 1817–1820 and author (*Avium species novae itinere Brasiliam*, 1824).

Spiza Gr. *spiza*, a common type of finch, identifed by most later authors as the Chaffinch *Fringilla coelebs* (v. *Chlorospingus* and *Geranospiza)*.

Spizaetus Gr. *spizias*, a hawk; *aetos*, an eagle.

Spizastur Gr. *spizias*, a hawk; L. *astur*, a hawk.

Spizella Dim. of genus *Spiza*, dickcissel, bunting.

Spiziapteryx Gr. *spizias*, a hawk; *pterux*, a wing; ref. the short, rounded, hawk-like wings of the Spot-winged Falconet *S. circumcinctus*.

Spizitornis Gr. *spizites*, a sort of tit, probably the Great Tit *Parus major; ornis*, a bird.

Spizixos Gr. *spiza*, a finch; genus *Ixos*, bulbul; ref. the short, finch-like bills of the finch-billed bulbuls.

Spizocorys Gr. *spiza*, a finch; syn. genus *Corys*, lark; ref. the conical, finch-like bill of the Pink-billed Lark *S. conirostris*.

splendens L. *splendens*, glittering, brilliant (*splendere*, to glitter).

splendida L. *splendidus*, brilliant, splendid.

splendidissima L. *splendidissimus*, most brilliant, magnificent (super. of *splendidus*, splendid).

splendidus see **splendida.**

spodiogenys Gr. *spodios*, ash-coloured, grey; *genus*, the cheek.

spodionota Gr. *spodios*, ash-coloured; *-notos*, backed.

spodiops Gr. *spodios*, ash-coloured; *ops*, the face.

spodioptila Gr. *spodios*, ash-coloured; *ptilon*, the wing.

spodiopygia Gr. *spodios*, ash-coloured; Mod. L. *pygius*, rumped (Gr. *puge*, the rump).

Spodiornis Gr. *spodios*, ash-coloured, grey (*spodos*, ashes); *ornis*, a bird.

spodiurus Gr. *spodios*, ash-coloured; *-ouros*, tailed.

spodocephala / spodocephalus Gr. *spodos*, ashes; *-kephalos*, headed.

sponsa L. *sponsa*, a bride (*spondere*, to pledge oneself).

Sporaeginthus Gr. *sporos*, a seed; genus *Aegintha*, waxbill.

Sporophila Gr. *sporos*, a seed; *philos*, loving, fond of.

Sporopipes Gr *sporos*, a seed; *opipteuo*, to watch, look around after.

spragueii After Isaac Sprague (1811–1895) US botanical draftsman and collector.

Spreo French name *le Spréo* given to the Pied Starling *S. bicolor* by Levaillant (1799) (Cape Dutch/Afrikaans *spreuw*, a starling).

spurius L. *spurius*, illegitimate. The description of the Orchard Oriole *Icterus* was based on Catesby's (1731) 'Basterd Baltimore', so called from its resemblance to the Baltimore Oriole *I. galbula*, and the juvenile and adult Red-capped Parrot *Purpureicephalus* are so different as to appear unrelated.

squalidus L. *squalidus*, dirty, squalid.

squamata / squamatus L. *squamatus*, scaled (*squama*, a scale).

squameiceps / squamiceps L. *squameus*, scaly; *-ceps*, headed.

squamifrons L. *squameus*, scaly; *frons*, the forehead or brow.

squamiger / squamigera L. *squamiger*, scale-bearing, scaly.

squamigularis L. *squama*, a scale; Mod. L. *gularis*, throated, of the throat.

squamipila L. *squama,* a scale; *pileus,* a cap.

squammata see **squamata.**

squamosa /squamosus L. *squamosus,* scaly (*squama,* a scale).

squamulata /squamulatus Mod. L. *squamulatus,* squamulated (L. *squamula,* a little scale).

Squatarola Local Venetian (Italian) name *sgatarola* for a kind of plover.

Stachyris Gr. *stakhus,* an ear of wheat; *rhis,* the nose; ref. the operculum, scale or flap almost closing the nostrils of these Indomalayan babblers.

Stactolaema Gr. *staktos,* trickling, dropping; *laimos,* the throat; ref. the tear- or drop-like spots on the throat and chest of the Yellow-headed Barbet *S. anchietae.*

stagnatilis L. *stagnatilis,* belonging to pools or marshes (*stagnum, stagni,* a pool, marsh or pond).

Stagonopleura Gr. *stagon, stagonos,* a drop, a spot; *pleura,* the side, the rib.

stairi After the Revd. John Bettridge Stair (1815–1898) missionary to Samoa.

stalkeri After W. Stalker (1879–1910) Australian ornithologist and collector.

stanleyi After the Hon. Edward Smith Stanley (1775–1851) 13th Earl of Derby, zoologist, collector, and founder of Derby Museum, Knowsley Park.

starki After Arthur C. Stark (1846–1899) British naturalist and collector, killed at the siege of Ladysmith during the Anglo-Boer War.

Starnoenas Syn. genus *Starna,* partridge (Italian *starna,* a partridge); Gr. *oinas,* a pigeon.

Steatornis Gr. *stear, steatos,* fat, tallow; *ornis,* a bird; the fat young of the Guacharo or Oilbird *S. caripensis* were formerly slaughtered in their thousands to provide oil for cooking and lamps.

steerei /steerii After J. B. Steere (1842–1940) US ornithologist and collector in the Philippines, 1874–1875, 1887–1888.

Steganopus Gr. *steganopous,* web-footed (*steganos,* webbed; *pous,* the foot).

Steganura Gr. *steganos,* covered, webbed; *oura,* the tail.

steinbachi After José Steinbach (d. 1930) collector in Argentina.

steindachneri After Franz Steindachner (1834–1919) Austrian zoologist, Director of Vienna Museum, 1882.

steini After G. H. W. Stein (b. 1897) German naturalist and collector in the East Indies.

stejnegeri After Leonhard Stejneger (1851–1943) Norwegian ornithologist resident in USA and collector.

Stelgidillas Gr. *stelgis, stelgidos* (variant form of *stlengis),* a strigil or scraper; *illas,* a thrush; the Slender-billed Greenbul *S. gracilirostris* has 'the maxilla with serrations on the cutting edge just posterior to the subterminal notch' (Oberholser (1899), *Proc. US Nat. Mus.,* **22**, 30).

Stelgidocichla Gr. *stelgis,* a strigil or scraper; *kikhle,* a thrush; the Yellow-whiskered Greenbul *S. latirostris* has the '...cutting edge of maxilla with several subterminal notches...by reason of its serrate maxilla...it scarcely needs comparison with other genera' (Oberholser (1905), *Smithson. Misc. Coll., Quarterly Issue,* **48** [2], 169).

Stelgidopteryx Gr. *stelgis,* a scraper or strigil; *pterux,* wing; in the rough-winged swallows the barbs of the outer web of the outermost primary end in recurved hooks

stellaris L. *stellaris,* starry, set with stars (*stella,* a star).

stelleri After Georg Wilhelm Steller (1709–1746) German naturalist and explorer in the Russian service who took part in Vitus Bering's expedition to Alaska, 1740–1742.

Stellula Dim. of L. *stella,* a star.

Stenostira Gr. *stenos,* narrow; *steira,* a ship's beak; ref. the thin, narrow bill of the Fairy Flycatcher *S. scita.*

stentoreus L. *stentoreus,* loud-voiced.

stenura Gr. *stenos,* narrow; *-ouros,* tailed.

stephani After Étienne Stephan Jacquinot (fl. 1840) father of French explorer Vice Admiral C. H. Jacquinot.

stephaniae After Princess Stephanie of Belgium (1864–1945), daughter of Leopold II King of the Belgians, and wife of Crown Prince Rudolf of Austria–Hungary.

Stephanibyx Gr. *stephanos,* a crown; Mod. L. *ibyx,* lapwing or plover.

Stephanoaetus Gr. *stephanos,* a crown; *aetos,* an eagle.

Stephanophorus Gr. *stephane,* a diadem; *-phoros,* carrying.

Stephanoxis Gr. *stephanos,* a crown; *oxus,* sharp, pointed; ref. the thin, pointed crest of the Black-breasted Plovercrest *S. lalandi.*

stepheni After Alfred Ernest Stephen (1879–1961) Australian government officer, businessman, and collector.

Stercorarius L. *stercorarius,* belonging to dung (*stercus, stercoris,* dung, muck); ref. the skuas' pursuit of other seabirds until they disgorge their food; the disgorged food was once thought to be excrement.

Sterna Old English *stearn* or *starn,* a tern.

Sternoclyta Gr. *sternon,* the breast; *klutos,* glorious, splendid.

sterrhopteron Gr. *sterrhos,* stiff; *pteron,* a feather; the Orange-throated Tanager *Wetmorethraupis* has stiff, waxy throat feathers.

stewarti After Surgeon General L. C. Stewart (1819–1888) British Army in India, 1848.

stictigula Gr. *stiktos,* spotted, dappled (*stizo,* to mark, to make spotted); L. *gula,* the throat.

stictocephalus Gr. *stiktos,* spotted, dappled; *-kephalos,* headed.

stictolaema Gr. *stiktos,* spotted, dappled; *laimos,* the throat.

stictolopha Gr. *stiktos,* spotted, dappled; *lophos,* a crest or tuft.

Stictonetta Gr. *stiktos,* spotted, dappled; *netta,* a duck.

stictoptera/stictopterus Gr. *stiktos,* spotted, dappled; *-pteros,* winged.

stictothorax Gr. *stiktos,* spotted, dappled; *thorax,* the chest.

sticturus Gr. *stiktos,* spotted, dappled; *-ouros,* tailed.

stierlingi After N. Stierling (fl. 1901) German traveller and collector in Nyasaland (Malawi) and Tanganyika (Tanzania).

Stigmatura Gr. *stigma, stigmatos,* a spot or mark; *oura,* the tail.

stigmatus Gr. *stigma,* a spot or mark (*stizo,* to mark).

Stilbopsar Gr. *stilbon,* glittering, glistening; *psar,* the starling.

Stiltia Med. English *stilte,* stilts; ref. the long legs of the Australian Pratincole *S. isabella.*

Stiphrornis Gr. *stiphros,* stout, compact; *ornis,* a bird; ref. the stout, booted tarsi of the Forest Robin *S. erythrothorax.*

Stipiturus L. *stipes, stipitis,* a stem or branch; Gr. *oura,* the tail; ref. the coarse, stiff tail-feathers of the emu-wrens.

Stizorhina Gr. *stizo,* to mark or distinguish; *rhis, rhinos,* the nose; ref. the distinctive keel-shaped culmen of the Rufous Jungle Thrush *S. fraseri.*

stolidus L. *stolidus,* dull, stupid, foolish.

stolzmanni After J. Stolzmann or Sztolcman (1854–1928) Polish zoologist and collector in tropical America, 1875–1883.

Stomiopera Gr. *stomion,* a bridle-bit; *pera,* excessive, greater; ref. the conspicuous, white, wedge-shaped gape mark of the White-gaped Honeyeater *Lichenostomus unicolor.*

stormi After Captain H. Storm (fl. 1901) German mariner, collector in East Indies, 1889.

strausae After Mrs Oscar Straus (fl. 1929) wife of US explorer and collector Oscar Straus.

strenua L. *strenuus,* brisk, active, restless.

Strepera Mod. L. *streperus,* noisy (L. *strepitare,* to be noisy).

strepitans L. *strepitans,* clattering, noisy (*strepitare,* to be noisy).

Streptocitta Gr. *streptos,* a collar; *citta,* the magpie (v. *Cissa);* ref. the broad white collars and long tails of the Wallacean magpie-mynahs.

Streptopelia Gr. *streptos,* a collar; *peleia,* a dove.

streptophorus Gr. *streptophoros,* wearing a collar or necklace.

Streptoprocne Gr. *streptos,* a collar; Mod. L. *procne,* a swallow.

stresemanni /Stresemannia After Erwin Friedrich Theodor Stresemann (1889–1972) German ornithologist, collector, and author (*Aves,* 1927; *Die entwicklung der ornithologie von Aristoteles bis zur gegenwart,* 1951).

striata Mod. L. *striatus,* striped, streaked (L. *stria,* a furrow or line).

striaticeps Mod. L. *striatus,* striped; L. *-ceps,* headed.

striaticolle /striaticollis Mod. L. *striatus,* striped; *-collis,* necked.

striatigula Mod. L. *striatus,* striped; L. *gula,* the throat.

striatigularis Mod. L. *striatus,* striped; *gularis,* throated, of the throat.

striatipectus Mod. L. *striatus,* striped; L. *pectus,* the breast.

striativentris Mod. L. *striatus,* striped; L. *venter, ventris,* the belly.

striatus see **striata.**

stricklandi /stricklandii After Hugh Edwin Strickland (1811–1853) British ornithologist, zoologist, and compiler of the Strickland Code (*Series of propositions for rendering the nomenclature of zoology uniform and permanent,* 1842) (v. Introduction, p. xiii).

strigiceps L. *striga,* a furrow; *-ceps,* headed.

strigilatus Mod. L. *strigilatus,* striped, streaked, furrowed (L. *striga,* a furrow).

strigirostris L. *strix, strigis,* an owl; *-rostris,* billed; the Tooth-billed Pigeon *Didunculus* has 'the maxilla, strigine-like, hooked and sharp pointed...' (Jardine (1845), *Ann. & Mag. Nat. Hist.,* **16,** 175).

Strigops L. *strix,* an owl; Gr. *ops,* the face; the rare Kakapo *S. habroptilus* has an owl-like face.

strigoides L. *strix,* an owl; Gr. *-oides,* resembling.

strigula L. *stria,* a furrow or line; *gula,* the throat; ref. the black bars on the throat of the Bar-throated Siva *Minla.*

strigulosus Mod. L. *strigulosus,* furrowed (L. *striga,* a furrow).

striifacies L. *stria,* a furrow or line; *facies,* the face.

striolata /striolatus Mod. L. *striolatus,* striolated, striped (L. *stria,* a furrow or line).

Strix L. *strix,* a screech owl (Gr. *strix* or *strinx,* an owl).

stronachi After B. W. H. Stronach (fl. 1962) collector.

strophianus L. *strophium,* a band; *-anus,* possessing.

strophiata L. *strophium,* a band; *-atus,* possessing.

strophium L. *strophium,* a band, a headband, a breastband.

struthersii After John Struthers (fl. 1855) Scottish physician.

Struthidea Dim. of Gr. *strouthos,* a sparrow *Passer* or any small bird.

Struthio L. *struthio,* the Ostrich *S. camelus* (a contracted form of *struthiocamelus)* (Gr. *strouthos* or *strouthiokamelos,* the ostrich).

stuarti After A. M. Stuart (fl. 1897) collector in Bolivia.

stuhlmanni After Franz Stuhlmann (1863–1928) German naturalist and collector in East Africa, 1888–1900.

sturmii After Johann Heinrich Christian Friedrich Sturm (1805–1862) German bird artist and collector.

Sturnella Dim. of genus *Sturnus,* starling.

sturninus L. *sturninus,* starling-like, speckled like a starling.

Sturnus L. *sturnus,* a starling.

stygia /stygius L. *stygius,* stygian, dismal (i.e. dark, black).

suahelicus After Swahili (also Kiswahili), an Arabic influenced Bantu language of East Africa (Arabic *sawahil*, the coast; the name was originally given to the peoples of Zanzibar and the opposite coast).

Suaheliornis After Swahili (v. *suahelicus*); Gr. *ornis*, a bird; the Greenbul Warbler *S. kretschmeri* was described from Mount Kilimanjaro in East Africa.

subaffinis L. *sub-*, related to, somewhat; specific name *Phylloscopus affinis*.

subalaris L. *subalaris*, under the arms (*subalares*, feathers under the wing).

subaureus L. *sub-*, underneath; *aureus*, golden.

subbrunneus L. *sub-*, underneath; Mod. L. *brunneus*, brown.

subbuteo L. *sub-*, related to; genus *Buteo*, buzzard; a Mod. L. translation of Gr. *hupotriorkhes*, a kind of hawk mentioned by Aristotle, not further identified, but by tradition associated with the European Hobby *Falco* (Gr. *hupo-*, somewhat, near; *triorkhes*, a bird of prey (v. *Dryotriorchis*)).

subcaeruleum Mod. L. *subcaeruleus*, blue-grey, bluish.

subcinnamomea L. *sub-*, underneath; Mod. L. *cinnamomeus*, cinnamon-coloured (L. *cinnamum*, cinnamon).

subcorniculatus L. *sub-*, related to; specific name *Philemon corniculatus*.

subcristata L. *sub-*, somewhat; *cristatus*, crested.

subcylindricus L. *sub-*, related to; specific name *Bycanistes cylindricus*.

subflammulatus L. *sub-*, underneath; Mod. L. *flammulatus*, ruddy, with flame-like markings.

subflava L. *subflavus*, yellowish, blonde.

subfrenatus L. *sub-*, related to; specific name *Lichenostomus frenatus*.

subgularis L. *sub-*, underneath; Mod. L. *gularis*, throated, of the throat.

subhimachala L. *sub-*, beneath, at the foot of; Hindi *himachal*, snow, snowy (i.e. the Himalayas); the Red-headed Rose-finch *Propyrrhula* was described from Nepal.

subis L. *subis*, a kind of bird that breaks eagles eggs.

Sublegatus L. *sub-*, related to; genus *Legatus*, piratic flycatcher.

subminuta L. *sub-*, related to; specific name *Calidris minuta*.

subniger L. *subniger*, blackish, somewhat black.

subochraceus L. *sub-*, underneath; Mod. L. *ochraceus*, ochreous, ochraceous.

subpersonatus L. *sub-*, related to; *personatus*, masked; ref. relationship of the Thin-billed Weaver *Ploceus* to other masked weavers.

subplacens L. *sub-*, related to; specific name *Myiopagis placens* (now regarded as a subsp. of *M. viridicata*).

subpudica L. *sub-*, underneath; *pudicus*, modest; ref. plain grey-brown underparts of the Silvery-throated Spine-tail *Synallaxis*.

subrubra L. *subruber*, reddish (*subrubere*, to blush or redden).

subruficapilla L. *sub-*, related to; specific name *Cisticola ruficapilla* (now regarded as a subsp. of *C. fulvicapilla*).

subruficollis L. *subrufus*, reddish; Mod. L. *-collis*, necked.

subrufus L. *subrufus*, reddish.

substriata/substriatus **1** L. *sub-*, underneath; Mod. L. *striatus*, striped (*Prinia*). **2** L. *sub-*, related to; specific name *Pardalotus striatus* (*Pardalotus*).

subsulphureus L. *sub-*, underneath; *sulfureus*, sulphurous.

subtilis L. *subtilis*, simple, plain.

subulata/subulatus Mod. L. *subulatus*, awl-shaped (L. *subula*, a shoemaker's awl); ref. bill shapes.

subunicolor L. *sub-*, underneath; *unicolor*, uniform, unicoloured.

subvinacea L. *sub-*, underneath; *vinaceus*, belonging to wine, vinaceous (Mod. L. *vinaceus*, pale vinaceous).

subviridis L. *subviridis*, somewhat green, greenish.

sueurii After Charles Alexandre Lesueur (1778–1846) French naturalist and explorer.

Suiriri Güaraní (Paraguayan) Indian name *suirirí* for two common species of flycatcher.

sukatschewi After V. P. Sukachev (fl. 1889) Russian merchant, explorer, and collector in China, 1884–1887.

Sula Norwegian *sula*, the Northern Gannet *S. bassana*.

sulaensis After the Sula Islands (Kepulauan Sula), in the Moluccas, Indonesia.

sulcatus L. *sulcatus*, furrowed, wrinkled; ref. the marks on the bill of the Groove-billed Toucanet *Aulacorhynchus*.

sulcirostris L. *sulcus, sulci*, a furrow; *-rostris*, billed.

sulfuratus L. *sulfuratus*, sulphurated.

sulfureopectus L. *sulfureus*, sulphurous; *pectus*, the breast.

sulfuriventer L. *sulfureus*, sulphurous; *venter*, the belly.

sulphurata/sulphuratus L. *sulphuratus*, sulphurated.

sulphurea L. *sulphureus*, sulphurous, sulphureous.

sulphureipygius L. *sulphureus*, sulphurous; Mod. L. *pygius*, rumped.

sulphuriventer L. *sulphureus*, sulphurous; *venter*, the belly.

sulphurescens L. *sulphureus*, sulphurous; *-escens*, becoming (i.e. sulphurish).

sulphurifera L. *sulphureus*, sulphurous; *-fer*, bearing.

sultanea Arabic *sultan*, a king or sultan.

sumatrana/sumatranus After Sumatra (Malay *Sumatera*). In the case of *Tanygnathus* the toponym is erroneous.

sumichrasti After Francis E. Sumichrast (1828–1882) Mexican naturalist and field worker.

sundara Hindi *sundar*, beautiful.

sundevalli After Carl Sundevall (or Sundewall) (1801–1875) Swedish ornithologist and author (*Svenska foglarna*, 1856).

sunensis After the Río Suno, Pastaza, Ecuador.

sunia Nepalese name *sunya kusial* for the Oriental Scops Owl *Otus*.

superba/superbus L. *superbus*, magnificent, splendid.

superciliaris Mod. L. *superciliaris*, eyebrowed (L. *supercilium*, the eyebrow).

superciliosa/superciliosus L. *superciliosus*, supercilious, eyebrowed.

superflua L. *superfluus*, overflowing, abundant.

surda L. *surdus*, still, silent.

surinamensis/surinamus After Surinam, or Dutch Guiana, named for its aboriginal Surinen Indians.

Surnia No expl. (Duméril (1806), *Zoologie analytique*, p. 34); perhaps a contracted anagram of the syn. specific name *Strix funerea* was intended, although Langton (1920) writes, 'Schwaner [?] gives *Surnion* as a Greek word meaning an owl' (cf. syn. genus *Syrnium*, owl; Russian *sova*, an owl; Nepalese *sunya*, an owl).

Surniculus Dim. of French name *le Surnicou* given to the Drongo Cuckoo *S. lugubris* by Lesson (1830), *Traité d'Ornithologie*, Vol. **2**, p. 151, with ref. to its sombre plumage (genus *Surnia*, owl; genus *Cuculus*, cuckoo).

surrucura Güaraní (Paraguayan) Indian name *surucuá* for a trogon or similar sized bird.

suscitator L. *suscitator*, an awakener (*suscitare*, to arouse); in the breeding season the hen Barred Buttonquail *Turnix* drums at all hours of the day and night.

susurrans L. *susurrans*, whispering (*susurrare*, to murmur).

Suthora Nepalese name *suthora* for the Ashy-eared Parrotbill *Paradoxornis nipalensis*.

sutorius L. *sutorius*, of a shoemaker (*sutor*, a shoemaker); the nest of the Long-tailed Tailorbird *Orthotomus* is built inside the fold created by sewing two leaf edges together.

Suya Nepalese name *suya* for the Nepal Fulvetta *Alcippe nipalensis*.

svecica Mod. L. *Suecicus,* Swedish.

swainsoni /swainsonii After William Swainson (1789–1855) English naturalist, artist, collector, and author (*Zoological illustrations*, 1820; *Fauna boreali Americana*, 1831; *The natural history of the birds of western Africa*, 1837; *The natural arrangement and relations of the family of flycatchers*, 1838; *A selection of the birds of Brazil and Mexico*, 1841).

swalesi After Bradshaw Swales (1875–1928) US ornithologist and author (*Birds of Haiti and the Dominican Republic*, 1931).

swierstrai After C. J. Swierstra (fl. 1940) South African zoologist and Director of the Transvaal Museum, 1940.

swinderniana After Theodorus van Swinderen (1784–1851) Dutch naturalist.

swinhoii After Robert Swinhoe (1836–1877) British diplomat in China and naturalist.

swynnertoni /Swynnertonia After Charles Francis Massey Swynnerton (1877–1938) first game warden in Tanganyika Territory (Tanzania), 1919–1929 and Director of Tsetse Research, 1929–1938.

sybillae No expl. (Salvin & Godman (1892), *Ibis*, 327); perhaps after the Sibyllae or Sibyls, the inspired prophetesses of the Roman world.

Sycobrotus Gr. *sukon,* a fig; *bibrosko,* to eat.

sylvanus L. *silvanus,* of the woods (*silva,* a wood).

sylvatica /sylvaticus L. *silvaticus,* of woods or trees.

sylvestris L. *silvestris,* belonging to woods.

Sylvia Mod. L. *sylvia,* a warbler (L. *silva,* a wood).

sylvicola L. *silva,* a wood; *-cola,* inhabitant.

sylviella Dim. of genus *Sylvia,* warbler.

Sylvietta Dim. of genus *Sylvia,* warbler.

sylviolus Dim. of genus *Sylvia,* warbler.

Sylviorthorhynchus Genus *Sylvia,* warbler; Gr. *orthos,* straight; *rhunkhos,* the bill; ref. the small size, warbler-like appearance, and straight bill of the Brown Wiretail *S. desmursii.*

Sylviparus Genus *Sylvia,* warbler; genus *Parus,* tit.

Syma Gr. myth. Suma or Syma, a sea nymph.

symonsii After Roden E. Symons (b. 1884) South African game warden.

Symplectes Gr. *sun-* (*sum-* before *p),* together, connected with (i.e. related to); *plektes,* a weaver (cf. *sumpleko,* to entwine, to plait).

Synallaxis French name *synallaxe* given to the spinetails by Vieillot (1818), *Nouv. Dict. Nat. Hist.*, Vol. **24**, 117, with ref. to their distinctive characters warranting generic separation (Gr. *sunallaxis,* exchange).

Syndactyla Gr. *sun-,* together; *daktulos,* the finger (i.e. the toe); the middle toes of the foliage-gleaners are joined for part of their length.

Synoicus **1** Gr. *sunoikos,* living together; the Swamp Quail *S. ypsilophorus* occurs in groups of up to one hundred. **2** after the Sinai Peninsula, Egypt (*Carpodacus).*

Synthliboramphus Gr. *sunthlibo,* to compress; *rhamphos,* the bill; ref. the laterally compressed bills of the murrelets.

Sypheotides No expl. (Lesson (1839), *Rev. Zool.*, 47); probably from Gr. *siphon,* a reed or straw; *ous, otos,* the ear; *-ides,* resembling; ref. the distinctive, lengthened, spoon-tipped ear plumes of the breeding male Lesser Florican *S. indica.*

syriacus L. *Syriacus,* Syrian.

Syrigma Gr. *surigma,* a whistling (*surizo,* to whistle).

syrinx Gr. myth Syrinx or Surinx, a water nymph metamorphosed into a reed to escape the amorous attentions of Pan. The transformation was to no avail, since Pan plucked the reed to make his pipe.

Syrmaticus Gr. *surma, surmatos,* a robe with a long train, anything trailed or dragged behind; ref. the long tails of these beautiful pheasants.

syrmatophorus Gr. *surma,* a robe with a long train; *-phoros,* bearing; ref. the long central tail-feathers of the Tawny-bellied Hermit *Phaethornis.*

Syrrhaptes Gr. *surrhaptos,* sewn together; these Palaearctic sandgrouse have the front toes fused together and densely feathered.

szalayi After Freiherr Imre von Szalay (1846–1918) Hungarian naturalist, traveller in Asia, Director of the National Museum, Budapest, 1894–1916, and author.

szechenyii After Graf Béla Szécheny (1837–1918) Hungarian explorer in central Asia and author.

Sylvia melanocephala Sardinian Warbler

T

tabuensis After the island of Tonga Tabu (Tongatapu), largest of the Tongan islands.

tacarcunae After the Cerro Tacarcuna, Darién, Panama.

tacazze After the River Tacazzé (Takazzé or Tekezé), Welo, Ethiopia.

Taccocua French name *tacco* given to the lizard cuckoos *Saurothera* by Lesson (1830), *Traité d'Ornithologie*, Vol. **2**, 143; genus *Coua*, coua.

tachardus French name *le Tachard* given to the Mountain Buzzard *Buteo* by Levaillant (1799) because it was the only sp. shot not by himself, but by his servant Klaas (*tâcher*, to endeavour; *busard*, buzzard).

tachiro French name *le Tachiro* given to the African Goshawk *Accipiter* by Levaillant (1799)(*tache*, a spot or blotch; *rond*, round; ref. brown and rufous blotched underparts).

Tachornis Gr. *takhus*, fleet, fast; *ornis*, a bird.

Tachuris Güaraní (Paraguayan) Indian names *tachurí* or *tarichú* for various small flycatchers.

Tachybaptus Gr. *takhus*, fast; *baptos*, dipped (*bapto*, to sink under).

Tachycineta Gr. *takhukinetos*, moving quickly.

Tachyeres Gr. *takhueres*, fast rowing; the steamer ducks plough through the water.

Tachymarptis Gr. *takhus*, fast; *marptis*, a seizer.

Tachyphonus Gr. *takhuphonos*, fast speaking.

taciturnus L. *taciturnus*, silent, quiet.

tacsanowskius/taczanowskii After Wladyslaw Taczanowski (1819–1890) Polish ornithologist and author (*Ornithologie du Pérou*, 1884).

Tadorna French name *tadorne* for the Common Shelduck *T. tadorna*.

tadornoides Specific name *Tadorna tadorna;* Gr. *-oides*, resembling.

taeniata/taeniatus L. *taenia*, a headband or ribbon (Gr. *tainia*, a band); *-atus*, possessing.

taeniolaema Gr. *tainia*, a band; *laimos*, the throat.

taeniopterus Gr. *tainia*, a band; *-pteros*, winged.

Taeniopygia Gr. *tainia*, a band; Mod. L. *-pygius*, rumped.

Taeniotriccus Gr. *tainia*, a band; Mod. L. *triccus*, tyrant flycatcher; the Black-chested Tyrant *T. andrei* has a broad, black chestband.

taha Sotho (Lesotho) native name *thakha* for the bishop weavers *Euplectes*.

tahapisi Chuana (Botswana) native name *tahapitsi* for various finches and widowbirds.

tahitica/tahitiensis After Tahiti, largest of the Society Islands, French Polynesia.

taitensis Former French spelling *O-Taïti* for the island of Tahiti, Society Islands.

taivanus After the island of Taiwan, formerly Formosa.

takatsukasae After Prince Nobusuke Taka-Tsukasa (1889–1959) Japanese ornithologist and collector.

talaseae After the town of Talasea, New Britain Island, Bismarck Archipelago, Papua New Guinea.

talatala Chuana (Botswana) native name *talatala* for the White-bellied Sunbird *Nectarinia*.

Talegalla French name *talève,* a gallinule (Malagasy *talavana)*; L. *gallus,* a cock; ref. the mixed characteristics of the New Guinea brush-turkeys.

tallmanorum After D. A. and Mrs E. J. Tallman (fl. 1981) US ornithologists and collectors.

talpacoti No expl. (Temminck in Knip (1811), *Les pigeons*, p. 22, plate 12).

tamae After the Páramo de Tamá, Táchira, Venezuela.

tamarugense After the plantations of tamarugo *Prosopis tamarugo,* a native tree, in Tarapacá, Chile, where the Tamarugo Conebill *Conirostrum* was discovered.

tamatia Tupí (Brazilian) Indian name *tamatia* for a spotted, thrush-like bird.

Tanagra Tupí (Brazilian) Indian name *tangara* for a multicoloured, finch-like bird.

Tanagrella Dim. of syn. genus *Tanagra,* tanager.

tanagrinus Syn. genus *Tanagra,* tanager; L. *-inus,* resembling.

tanganjicae After Lake Tanganyika, Zaïre.

Tangara Tupí (Brazilian) Indian name *tangara* for a multicoloured, finch-like bird.

Tangavius French name *le Tangavio* given to the Shiny Cowbird *Molothrus bonariensis* by Buffon (1770) (contraction of *tangara violet,* violet tanager).

tanki No expl. (Blyth (1843), *Journ. Asiat. Soc. Bengal*, **12**, 180), except, 'Among Dr Buchanan Hamilton's drawings is that of a species named by him *Turnix tanki'* (cf. Lepcha (Indian) name *tanki* for the sparrowhawk *Accipiter).*

tannensis After Tanna Island (Tana), New Hebrides (Vanuatu).

tanneri After Lieutenant Commander Zera L. Tanner (fl. 1897) US navy, deep-sea explorer, and collector.

Tanygnathus Gr. *tanu-,* long (*tanuo,* to stretch); *gnathos,* the jaw; ref. the very large and heavy bill of the Great-billed Parrot *T. megalorynchos*.

Tanysiptera Gr. *tanusipteros,* long-winged.

tao Gr. *tao,* the peacock *Pavo* (cf. Brazilian name *jão* for the larger tinamous).

Taoniscus Dim. of specific name *Tinamus tao*.

Tapera **1** Tupí (Brazilian) Indian name *matim tapirera* for a cuckoo whose cries represented the voices of ghosts. **2** Tupí (Brazilian) Indian name *taperá,* for a kind of swallow (*Phaeoprogne).*

Taphrolesbia Gr. *tephra,* ashes; genus *Lesbia,* trainbearer; ref. the pale grey underparts of the long-tailed Grey-bellied Comet *T. griseiventris*.

Taphrospilus Gr. *tarphus,* close, dense; *spilos,* a spot; ref. the thickly spotted underparts of the Many-spotted Hummingbird *T. hypostictus*.

Taraba Anagram of genus *Batara,* antshrike, coined to indicate relationship (cf. Tupí Indian name *tarabe,* a green parrot).

taranta After the Taranta Pass, near Massawa (Mits'rwa), Eritrea, Ethiopia.

tarda L. *tarda,* an Hispanic name for the Great Bustard *Otis*.

tarnii After Surgeon J. Tarn (1794–1877) English collector.

Tarsiger Gr. *tarsos,* the flat of the foot (i.e. the visible 'leg' of a bird); L. *-ger,* bearing; ref. the long, slender, booted legs of the Golden Bush Robin *T. chrysaeus*.

tasmanicus After the island state of Tasmania, Australia (named for the Dutch navigator Abel Janszoon Tasman 1603–1659).

tataupa Güaraní (Paraguayan) Indian name *ynambú tatáupá,* house tinamou, for the Tataupa Tinamou *Crypturellus,* because it occurs near villages and country haciendas.

tatei After George Henry Hamilton Tate (1894–1953) British mammalogist, collector in Ecuador, 1921–1924; Venezuela, 1925–1928; and author.

Tauraco English name *touraco,* turaco, coined by Edwards (1743)(*Natural history of birds; Gleanings of natural history),* supposedly based on *turakoo,* a West African native name. Buffon (1783) coined *tourocco* from French *tourterelle,* turtle dove, and *hocco,* curassow.

Tchagra French name *tchagra,* given to the Southern Bush-shrike *T. tchagra* by Levaillant (1797), in imitation of its incessant calls 'tcha tcha tcha gra'.

Tchitrea Malagasy name *tchitrec* for the paradise flycatcher *Terpsiphone.*

tectes No expl. (Gmelin (1789), *Systema naturae,* (Ed. 13), Vol.**1** [2], p. 940); probably based on local Réunion name *tec-tec* for the Réunion Stonechat *Saxicola.*

tectus L. *tectus,* covered, concealed (*tegere,* to cover).

teerinki After Major C. G. J. Teerink (1897–1942) Dutch Army in the East Indies.

teesa Hindi name *tisa* for the White-eyed Buzzard *Butastur.*

tegimae After S. Tegima (fl. 1887) Japanese naturalist and Director of the Educational Museum, Tokyo.

Telacanthura Gr. *telos,* accomplished; *akantha,* a thorn; *oura,* the tail; ref. the projecting tail spines of the Afrotropical spinetailed swifts.

telasco After Telasco, an Inca warrior in Jean François Marmontel's novel *Les Incas ou la destruction de l'Empire du Pérou,*1777.

Teledromas Gr. *tele,* far (cf. *telos,* accomplished); *dromas,* running (*trekho,* to run).

Teleonema Gr. *teleo-,* perfect; *nema,* a thread; ref. the thread- or wire-like tail filaments of the Wire-tailed Manakin *T. filicauda.*

telescophthalmus Gr. *teleskopos,* conspicuous; *ophthalmos,* the eye; ref. the naked pale blue eye-ring of the Frilled Monarch *Arses.*

Telespyza Gr. *tele,* far; *spiza,* a finch; ref. the remote island range of the Laysan Finch *T. cantans,* a Hawaiian honey-creeper.

Telmatodytes Gr. *telma, telmatos,* a marsh or pool; *dutes,* a diver (i.e. an inhabitant) (*duo,* to plunge).

Telophorus No expl. (Swainson (1831), *Fauna Bor.-Amer.,* Vol. **2**, p. 481); doubtless misspelt from either Gr. *telos,* perfect, accomplished; *phone,* the voice or cry, or syn. genus *Phoneus,* shrike; ref. the colourful appearance and unusual duetting songs of the bush-shrikes.

Temenuchus Gr. *temenoukhos,* a priest (*temenos,* a temple precinct); the Brahminy Mynah *Sturnus pagodarum* is frequently found around temples.

temia French *le Temia* given to the Racquet-tailed Treepie *Crypsirina* by Levaillant (1799) (Gr. *temei,* cut; *temno,* to cut; ref. tail shape).

temminckii After Coenraad Jacob Temminck (1778–1858) Dutch ornithologist, collector, and author (*Nouveau recueil de planches coloriées d'oiseaux,* 1820).

Temnotrogon No expl. (Bonaparte (1854), *Ateneo Italiano,* **2** [8], 129); probably from genus *Temnurus,* ratchet-tailed treepie, and genus *Trogon,* trogon.

Temnurus Gr. *temno,* to cut; *oura,* the tail; ref. long, notched tail of the Ratchet-tailed Treepie *T. temnurus.*

temperatus L. *temperatus,* moderate, quiet, temperate (*temperare,* to control oneself).

temporalis L. *temporalis,* of the temples of the head (*tempus,* the temples).

tenebricosa L. *tenebricosus,* gloomy, dark (*tenebrae,* darkness).

tenebrosa /tenebrosus L. *tenebrosus,* dark, gloomy.

tenella L. *tenellus,* tender, delicate (dim. of *tener,* delicate).

tenellipes L. *tenellus,* delicate; *pes,* the foot.

tenellus see **tenella.**

tener L. *tener,* delicate.

teneriffae After the island of Teneriffe (Tenerife), Canary Islands.

tenuirostris L. *tenuis,* slender, thin; *-rostris,* billed.

tephrocotis Gr. *tephra,* ashes; *-otis,* eared.

Tephrodornis Gr. *tephra,* ashes; *ornis,* a bird.

tephrolaema Gr. *tephra,* ashes; *laimos,* the throat.

tephronota /tephronotum /tephronotus Gr. *tephra,* ashes; *-notos,* backed.

Tephrophilus Gr. *tephra,* ashes; *philos,* loving, fond of; the Masked Mountain Tanager *Buthraupis wetmorei* was discovered on the slopes of the volcano Sangay in Ecuador.

Tephrozosterops Syn. genus *Tephras,* white-eye (Gr. *tephra,* ashes); genus *Zosterops,* white-eye.

teraoki After N. Teraoki (1885–1955) Japanese collector.

Terathopius Gr. *teras, teratos,* a meteor, a marvel; *ops, opos,* appearance.

terek After the Terek River, Dagestan, USSR.

Terenotriccus Gr. *teren, terenos,* soft, delicate; Mod. L. *triccus,* tyrant flycatcher.

Terenura Gr. *teren,* soft, delicate; *oura,* the tail.

Teretistris Gr. *teretisma,* whistling, twittering (*teretizo,* to twitter).

ternatana After the island of Ternate, near Halmahera, in the Moluccas, Indonesia.

Terpsiphone Gr. *terpsi-,* delighting in (*terpo,* to delight); *phone,* the voice; ref. the pleasant warbling songs of some of the paradise flycatchers, although this replacement name must have been given in expectation rather than fulfilment, most species having harsh calls.

terraereginae L. *terra,* land; *regina,* a queen; after the state of Queensland, Australia (named for Queen Victoria).

terrestris L. *terrestris,* terrestrial, of the ground (*terra,* earth, ground).

Terricolumba L. *terra,* ground, earth; *columba,* a pigeon.

terrisi After Terris Moore (b. 1908) US zoologist and collector.

Tersina French name *la Tersine* given to the Swallow Tanager *T. viridis* by Buffon (1770) (dim. of L. *tersus,* burnished, neat).

Tesia Nepalese name *tisi* for the Grey-bellied Ground Warbler *T. cyaniventer.*

tessmanni After Günter Tessmann (b. 1889) German botanist and collector in Guinea and the Cameroons, 1904–1914 and Peru, 1923.

testacea L. *testaceus,* brick-coloured (*testa,* a brick or tile).

tethys L. *tethys,* the sea, a marine goddess.

Tetrao L. *tetrao,* a gamebird, probably the Black Grouse *Lyrurus* (Gr. *tetraon,* a gamebird, perhaps the Pheasant *Phasianus).*

Tetraogallus Genus *Tetrao,* capercaillie; genus *Gallus,* fowl; ref. the large size of the Palaearctic snowcocks.

Tetraophasis Genus *Tetrao,* capercaillie; Mod. L. *phasis,* pheasant; ref. the grouse-like appearance of the monal-partridges.

Tetrastes Gr. *tetras,* a gamebird mentioned by Aristophanes; *-astes,* resembling.

Tetrax Gr. *tetrax,* an edible gamebird mentioned by Aristophanes, Athenaeus, Alexander Myndios, and Epicharmus.

tetrix Gr. *tetrix,* a ground nesting bird mentioned by Aristotle, identified as a sort of grouse by mediaeval writers.

texensis Erroneous toponym after the Republic of Texas (admitted to statehood 1845); the Vermilion-crowned Flycatcher *Myiozetetes* ranges from eastern Mexico to northern Costa Rica.

textilis L. *textilis,* woven, plaited; ref. the loosely woven, domed nest of the Thick-billed Grasswren *Amytornis.*

textrix L. *textrix,* a female weaver.

teydea After Mount Teyde or Teide, Tenerife, Canary Islands.

teysmanni After J. E. Teysmann (1808–1882) Dutch botanist, collector, and curator Buitenzorg Gardens, Java.

Thalasseus Gr. *thalassios,* maritime, of the sea (*thalassa,* the sea).

thalassina /thalassinus L. *thalassinus,* sea green (Gr. *thalassinos,* sea green; *thalassa,* the sea).

Thalassoica Gr. *thalassa,* the sea; *oikos,* inhabiting (*oikeo,* to inhabit).

Thalassornis Gr. *thalassa,* the sea (can also mean any body of salt or brackish water); *ornis,* a bird; the White-backed Duck *T. leuconotus* inhabits quiet pools, backwaters, marshes, and swamps.

Thalurania Gr. *thalia,* abundance, wealth; *ouranios,* heavenly.

Thamnistes Gr. *thamnos,* a bush or shrub; *ktistes,* an occupier (*ktizo,* to occupy).

Thamnolaea Gr. *thamnos,* a bush; *laios,* a thrush.

Thamnomanes Gr. *thamnomanes,* passionately fond of bushes.

Thamnophilus Gr. *thamnos,* a bush; *philos,* loving, fond of.

Thamnornis Gr. *thamnos,* a bush; *ornis,* a bird.

Thaumastura Gr. *thaumastos,* wonderful, marvellous; *oura,* the tail.

Thaumatibis Gr. *thaumatos,* wonderful, marvellous; *ibis,* an ibis.

thayeri After Colonel John Eliot Thayer (1862–1933) US ornithologist and collector.

theklae After Thekla Brehm (1832–1858) daughter of German ornithologist Christian Ludwig Brehm.

thenca Local Chilean names *tenca* or *trenca* for the Chilean Mockingbird *Mimus.*

theomacha Med. L. *theomacha,* a witch.

theresae After Theresa Clay (fl. 1954) British parasitologist and ornithologist, niece and companion of ornithologist Colonel Richard Meinertzhagen.

theresiae After **1** Empress Theresa (1822–1889), wife of Pedro II Emperor of Brazil (*Polytmus*). **2** Princess Therese of Bavaria (1850–1925), daughter of Luitpold Prince Regent of Bavaria (*Metallura*).

Theristicus Gr. *theristikos,* of reaping (*theristron,* a sickle); ref. sickle-shaped bill of the Buff-necked Ibis *T. caudatus.*

Thescelocichla Gr. *theskelos,* marvellous; *kikhle,* a thrush; Oberholser (1905), *Smithsonian Misc. Coll., Quarterly Issue,* **48**, 155, wrote of the White-tailed Greenbul *T. leucopleura,* 'This genus is so different from *Bleda* that it seems rather remarkably to have escaped separation until now.'

thibetanus After Thibet, a former spelling of Tibet (v. *tibetanus*).

thilius Local Chilean name *trile* for the Yellow-winged Blackbird *Agelaius.*

Thinocorus Gr. *this, thinos,* sand, a beach; syn. genus *Corys,* lark; ref. the semi-desert habitat and lark-like appearance of the Neotropical seedsnipe.

Thinornis Gr. *this,* sand, a beach; *ornis,* a bird.

Thlypopsis Mod. L. *thlypis,* warbler; Gr. *opsis,* appearance.

tholloni After M. Thollon (d. 1896) collector in the Congo.

Thomasophantes After the island of São Thomé (São Tomé) (Portuguese for St Thomas); Gr. *huphantes,* a weaver; the Buff-faced Weaver *Ploceus sanctaethomae* is endemic to São Tomé.

thomensis After the island of São Tomé, Equatorial Guinea, discovered in 1471 by Portuguese explorers (v. *sanctithomae*). In the case of *Estrilda* the toponym is erroneous.

thompsoni After Henry Nilus Thompson (d. 1938) British forest officer in Burma and Director of Forestry, Nigeria.

thoracica /thoracicus Med. L. *thoracicus,* pectoral, of the thorax (Gr. *thorakikos,* pectoral; *thorax,* the breast).

Thraupis Gr. *thraupis,* a kind of finch mentioned by Aristotle.

Threnetes Gr. *threnetes,* a mourner; ref. the sombre plumage of the barbthroats.

threnothorax Gr. *threnos,* wailing, funereal; *thorax,* the breast.

Threskiornis Gr. *threskos,* religious; *ornis,* a bird; the Sacred Ibis *T. aethiopicus* was held in veneration by the ancient Egyptians.

Thripadectes Gr. *thrips, thripos,* a woodworm; *dektes,* a biter (*dakno,* to bite).

Thripias Gr. *thrips,* a woodworm (coined on the analogy of *spizias,* a hawk or finch catcher, from *spiza,* a finch).

Thripophaga Gr. *thripophagos,* eating woodworms.

Thryomanes Gr. *thruon,* a reed or rush; *-manes,* passionately fond of.

Thryophilus Gr. *thruon,* a reed or rush; *philos,* loving, fond of.

Thryorchilus Gr. *thruon,* a reed or rush; Mod. L. *orchilus,* wren (v. *Hylorchilus).*

Thryothorus Gr. *thruon,* a reed or rush; *thouros,* leaping (*throsko,* to leap).

thula Araucano (Chilean) Indian name *thula* for the Black-necked Swan *Cygnus melanocoryphus,* but given to the Snowy Egret *Egretta* in error.

thunbergi After Carl Peter Thunberg (1743–1828) Swedish botanist, collector, disciple of Linnaeus, and author (*Flora Japonica*, 1784; *Prodromus plantarum Capensium*, 1794).

thura After Thura Nilsson (fl. 1850) daughter of Swedish zoologist Sven Nilsson.

Thyellodroma Gr. *thuella,* a storm or tempest; *dromos,* running (*trekho,* to run).

thyroideus Gr. *thureos,* a large oblong shield; *-oides,* resembling; ref. the broad, black breast patch of Williamson's Sapsucker *Sphyrapicus.*

tianduana After Tiandou (Tayandu) Island, west of Kai Islands, Indonesia.

Tiaris Gr. *tiara,* a tiara, the Persian royal head-dress; ref. the brightly patterned head of the Yellow-faced Grass-quit *T. olivacea.*

tibetanus After Tibet, formerly an independent state, now a province (Xizang Zizhiqu) of China.

tibialis L. *tibialis,* pertaining to the shin (*tibia,* the shin).

tibicen L. *tubicen,* a trumpeter.

Tichodroma Gr. *teikhos,* a wall; *dromos,* running (*trekho,* to run) (v. *muraria)*

tickelli /Tickellia /tickelliae After Colonel Samuel Richard Tickell (1811–1875) British Army in India and Burma, field ornithologist, author, and artist. The termination of the binomen *Cyornis tickelliae* might suggest that the sp. was named for Mrs Tickell, but it was the colonel who originally noticed the specific characters of this flycatcher.

tigrina /tigrinus L. *tigrinus,* tigrine, barred or striped like a tiger (*tigris,* a tiger).

Tigriornis Gr. *tigris,* a tiger; *ornis,* a bird; ref. the heavily barred plumage of the White-crested Tiger Heron *T. leucolophus.*

Tigrisoma Gr. *tigris,* a tiger; *soma,* the body; ref. the heavily barred and striped plumage of the Rufescent Tiger Heron *T. lineatum.*

Tijuca No expl. (Férussac (1829), *Bull. Sci. Nat.*, **19**, 324), or in Lesson (1830), *Centurie zoologique*, pl. 6, where the Black-and-gold Cotinga *T. atra* is referred to as *le Tijuca noir*. Marcgrave (1648) gave *tijeguacú* as a Tupí Indian name for a type of manakin or cotinga, and Buffon (1770) writes that *tijé* is a Tupí Indian generic term and uses *le Tijé* for the Blue-backed Manakin *Chiroxiphia*. In south-east Brazil, where this cotinga is found, *tijuca* is a component of various placenames.

Tilmatura Gr. *tilma, tilmatos,* plucked, shredded; *oura,* the tail; ref. the forked tail of the Sparkling-tailed Hummingbird *T. dupontii.*

Timalia No expl. (Horsfield (1821), *Trans. Linn. Soc. London*, **13**, 150), and although probably based on an otherwise unrecorded Malay name it is not listed amongst the Javanese names recorded therein. Newton and Gadow (1896) accepted the amended name *Timelia* and suggested the derivation Gr. *timao,* to honour, and *helios,* the sun.

Timeliopsis Syn. genus *Timelia*, babbler; Gr. *opsis*, appearance; the Tawny Straightbill *T. fulvigula* was formerly considered to be a babbler.

timorensis/timoriensis After the island of Timor, Indonesia.

Tinamotis Genus *Tinamus*, tinamou; genus *Otis*, bustard.

Tinamus Galibi (Cayenne) Indian name *tinamu* for the tinamous, Gallicized as *tinamou* by Barrère (1745).

tinniens L. *tinniens*, tinkling (*tinnire*, to tinkle).

tinnunculus L. *tinnunculus* or *tinnungulus*, the kestrel, a kind of hawk (*tinnulus*, shrill sounding).

tintillon Local Canarian (Spanish) name *chinchillón* for the Atlantic Chaffinch *Fringilla* (cf. Portuguese *tentilhão*).

tiphia No expl. (Linnaeus (1758), *Systema Naturae*, (Ed.10), book **1**, p. 186); perhaps after Tiphys, the pilot of the Argonauts.

tirica Tupí (Brazilian) Indian names *tirica* or *tiriva* for a small parakeet.

tithys No expl. (Taczanowski (1877), *Proc. Zool. Soc. London*, 323); perhaps from Gr. *titis*, a small chirping bird.

Tityra L. *tityrus*, an aristocratic hooligan; ref. the aggressive, bullying behaviour of the Neotropical tityras.

Tmetothylacus Gr. *tmetos*, cut, furrowed; *thulakos*, pantaloons; ref. the unfeathered tibia of the Golden Pipit *T. tenellus*.

tobaci After the island of Tobago in the West Indies.

Tockus French name *le Tock* given to the hornbills by Buffon (1770), and said to be based on an onomatopoeic Senegalese native name.

toco Güaraní (Paraguayan) Indian name *tucá* for a toucan.

tocuyensis After the town of El Tocuyo, Lara, Venezuela.

Todiramphus Genus *Todus*, tody; Gr. *rhamphos*, the bill.

Todirostrum Genus *Todus*, tody; L. *rostrum*, the bill; ref. the long, narrow, flat bills of the tody-flycatchers.

Todopsis Genus *Todus*, tody; Gr. *opsis*, appearance; ref. the long, flat, broad bill of the Emperor Wren *T. cyanocephala*.

Todus L. *todus*, a small bird mentioned by Plautus and Festus, not further identified.

Tolmarchus Gr. *tolma*, bold, daring; *arkhos*, chief, ruler.

tolmiei After William Fraser Tolmie (1812–1886) Scottish naturalist and collector resident in Canada, 1832–1886.

Tolmomyias Gr. *tolma*, bold, daring; Mod. L. *myias*, flycatcher.

tombacea French *tombac*, a copper alloy of a yellowish brown colour (Portuguese *tambaca;* Malay *tambaga*, copper); L. *-aceus*, resembling.

tomentosa Mod. L. *tomentosus*, covered with dense, short hairs.

tonsa L. *tonsus*, shaved, clipped (*tondere*, to shave); the female White-spotted Wattle-eye *Platysteira* has a black crown patch resembling a monk's tonsure.

Topaza L. *topazus*, topaz, green jasper (Gr. *topazos*, topaz).

torda Swedish name *tordmule* for the Razorbill *Alca* (cf. German *tordalk*).

Torgos Gr. *torgos*, a vulture.

tormenti After Point Torment, near Derby, Western Australia.

torosus L. *torosus*, brawny, muscular (*torus*, muscle).

torotoro Onomatopoeic New Guinea aboriginal name *torotoro* for the Yellow-billed Kingfisher *Halcyon*.

torquata/torquatus L. *torquatus*, collared, necklaced (*torquis* or *torques*, a collar).

torqueola Med. L. *torqueola*, a twisted necklace.

torquilla Dim. from L. *torquere*, to twist; ref. the serpentine head movements of the Northern Wryneck *Jynx* when caught.

Torreornis After Carlos de la Torre (1858–1950) Cuban naturalist and professor at Havana University; Gr. *ornis*, a bird.

torridus L. *torridus,* scorched, parched, torrid.

torringtoni After George Byng, 7th Viscount Torrington (1812–1884) Governor of Ceylon (Sri Lanka).

totanus Italian name *totano* for a redshank *Tringa.*

totta After the Hottentots (v. *hottentotta)* because the Cape Siskin *Serinus* was found in their villages.

Touit Tupí (Brazilian) Indian name *tuiete* for the small parrotlets.

toulou Malagasy name *toloho* for the Black Coucal *Centropus.*

toulsoni After A. Toulson (fl. 1878) collector in Angola, 1870.

toussenelii After Alphonse Toussenel (1803–1885) French author (*L'Esprit des bêtes.Le monde des oiseaux, ornithologie passionnelle,* 1853).

townsendi After **1** John Kirk Townsend (1809–1851) US ornithologist, collector, and author (*Dendroica, Myadestes).* **2** Charles Wendell Townsend (1859–1934) US field ornithologist (*Thalurania).*

toxopei After Lambertus Johannes Toxopeus (b. 1894) Dutch zoologist, lecturer in Indonesia, and collector.

Toxorhamphus Gr. *toxon,* a bow; *rhamphos,* the bill.

Toxostoma Gr. *toxon,* a bow; *stoma,* the mouth; ref. the curved bills of the Nearctic thrashers.

tracheliotus Gr. *trakhelia,* gristle, meat scraps; *ous, otos,* the ear; ref. the head and neck wattles of the Lappet-faced Vulture *Torgos.*

Trachylaemus Gr. *trakhus,* rough, harsh; *laimos,* the throat; ref. the bristled throat feathers of the Yellow-billed Barbet *Trachyphonus purpuratus.*

Trachyphonus Gr. *trakhuphonos,* harsh-voiced.

tractrac French name *le Tractrac* given to the Tractrac Chat *Cercomela* by Levaillant (1805), in imitation of its callnote.

Tragopan L. *tragopan,* a fabulous, horned purple-headed bird mentioned by Pliny and Pomponius Mela.

traillii After Thomas Stewart Traill (1781–1862) Scottish physician, philosopher, and zoologist.

tranquebarica After the town of Tranquebar, Coromandel Coast (Tamil Nadu), India.

transfasciatus Mod. L. *transfasciatus,* cross-barred (L. *trans,* across; Mod. L. *fasciatus,* barred).

traversi/Traversia After Henry H. Travers (1844–1928) New Zealand ornithologist.

traviesi After Édouard Traviès (b. 1800) French naturalist and artist.

traylori After Major Melvin Alvah Traylor (b. 1915) US ornithologist, collector, and author (*Birds of Angola,* 1960).

Tregellasia After Thomas H. Tregellas (1864–1938) Australian field ornithologist.

Treron Gr. *treron,* a pigeon or dove.

triangularis L. *triangularis,* triangular; the Olive-backed Woodcreeper *Xiphorhynchus* has triangular, buff-coloured spots on the underparts.

Tribonyx Gr. *tribe,* wearing away; *onux,* a claw; ref. short, blunt hind claw of the Tasmanian Native Hen *T. mortierii.*

tricarunculata L. *tri-,* three; Mod. L. *carunculatus,* wattled (L. *caruncula,* a small piece of flesh).

trichas Gr. *trikhas,* a kind of thrush.

Trichastoma Gr. *thrix, trikhos,* the hair; *stoma,* the mouth; the White-chested Babbler *T. rostratum* has 'very long slender setae at the gape, affording a ready distinguishing character' (Blyth (1842), *Journ. Asiat. Soc. Bengal,* **11**, 795).

Trichocichla Gr. *thrix,* the hair; *kikhle,* a thrush; ref. the full-feathered rump of the Long-legged Warbler *T. rufa.*

Trichodere Gr. *thrix,* the hair; *dere,* the neck or throat; the White-streaked Honeyeater *T. cockerelli* is 'readily distinguished...by the hairlike appearance of the sides of the feathers on the throat and foreneck' (North (1912), *Ibis,* 120).

Trichoglossus Gr. *thrix,* the hair; *glossa,* the tongue; ref. the brush-tipped tongues of the lorikeets.

Tricholaema Gr. *thrix,* the hair; *laimos,* the throat; the throat feathers of the Hairy-breasted Barbet *T. hirsuta* end in lengthened little bristles.

Tricholimnas Gr. *thrix,* the hair; Mod.L. *limnas,* a rail; ref. the soft, loose-textured plumage of the flightless Forest Rail *T. lafresnayanus.*

Trichopicus Gr. *thrix,* the hair; genus *Picus,* woodpecker; no expl. (Bonaparte (1854), *Ateneo Italiano,* **2**[8], 129), but doubtless ref. the long stiff bristles covering the nostrils of the White-fronted Woodpecker *T. cactorum* (v. *pubescens).*

trichopsis Gr. *thrix,* the hair; *opsis,* the face.

Trichothraupis Gr. *thrix,* the hair; genus *Thraupis,* tanager; ref. the hair-like feathers or hackles on the nape of the Black-goggled Tanager *T. melanops.*

trichroa Gr. *tri-,* three; *khroa,* colour, complexion.

Triclaria Gr. myth. Triclaria, an epithet of the goddess Diana.

tricollaris L. *tri-,* three; *collaris,* collared.

tricolor Late L. *tricolor,* tricoloured, of three colours.

Tricophorus Gr. *thrix,* the hair; *-phoros,* bearing; ref. the long, fine bristles springing from the nape and backs of these Indomalayan bulbuls.

tridactyla/tridactylus Gr. *tridaktulos,* having three digits (i.e. three toed).

trifasciatus Mod. L. *trifasciatus,* three-striped, three-banded.

Trigonoceps Gr. *trigonos,* triangular; L. *-ceps,* headed; ref. the angular appearance of the head of the White-headed Vulture *T. occipitalis.*

trigonostigma Gr. *trigonos,* triangular; *stigme,* a mark; the Orange-bellied Flowerpecker *Dicaeum* has a triangular orange mark on the back and rump.

Tringa Gr. *trungas,* a thrush-sized waterbird mentioned by Aristotle, not further identified, but taken by later authors to be a sandpiper, a wagtail *Motacilla,* or a dipper *Cinclus.*

trinitatis After the island of Trinidad (Spanish, the Trinity; Late L. *trinitas,* the Trinity).

trinotatus L. *tri-,* three; *notatus,* marked, noted; the Spot-tailed Sparrowhawk *Accipiter* has three white bars on the tail.

tristigma Gr. *tri-,* three; *stigme,* a spot or mark; ref. white marks on the first three primaries of the Freckled Nightjar *Caprimulgus.*

tristigmata L. *tri-,* three; *stigmata,* spots; the Sulawesi Ground Dove *Gallicolumba* has a golden forehead, grey crown and purple nape.

tristis L. *tristis,* sad, gloomy (Mod. L. *tristis,* dull-coloured).

tristissima L. *tristissimus,* dismal, very gloomy (super. of *tristis,* sad).

tristrami/tristramii After the Revd. Henry Baker Tristram (1822–1906) Canon of Durham, traveller, naturalist, antiquarian, and early supporter of Darwin.

tristriatus L. *tri-,* three; Mod. L. *striatus,* striped.

triurus Gr. *tri-,* three; *-ouros,* tailed; in flight the White-banded Mockingbird *Mimus* appears to have three tails, the illusion given by its black and white tail feathers.

trivialis L. *trivialis,* common, ordinary.

trivirgata/trivirgatus L. *tri-,* three; *virgatus,* striped.

trocaz Local Madeiran (Portuguese) name *pombo trocaz* for the Trocaz Pigeon *Columba.*

Trochalopteron Gr. *trokhalos,* round; *pteron,* wing, feather.

trochileum Dim. of L. *trochilus,* a small bird identified with the wren *Troglodytes* or the goldcrest *Regulus.*

trochilirostris L. *trochilus,* a semicircle or hoop; *-rostris,* billed.

trochiloides Specific name *Phylloscopus trochilus;* Gr. *-oides,* resembling.

Trochilus Gr. *trokhilos,* a small bird mentioned by Aristotle, identified by later writers with the wren *Troglodytes.*

Trochocercus Gr. *trokhos,* round; *kerkos,* the tail.

Troglodytes Gr. *troglodutes,* the wren (*troglodutes,* a cave-dweller; *trogle,* a hole; *duo,* to plunge into).

troglodytoides Genus *Troglodytes,* wren; Gr. *-oides,* resembling.

Trogon Gr. *trogon,* gnawing, fruit eating (*trogo,* to eat fruits).

Trogonurus Genus *Trogon,* trogon; Gr. *oura,* the tail (i.e. long-tailed).

Tropicoperdix L. *tropicus,* tropical; *perdix,* a partridge.

tropica L. *tropicus,* tropical.

Tropicranus Gr. *tropis,* a ships keel; *kranos,* a helmet; ref. the heavy, sharp-ridged bill of the Long-crested Hornbill *T. albocristatus.*

tropicus see **tropica.**

trudeaui After James de Bertz Trudeau (1817–1887) US physician, explorer, collector, and painter.

Trugon Gr. *trugon,* the Turtle Dove *Streptopelia.*

Tryngites Genus *Tringa,* sandpiper; Gr. *-oides,* resembling.

tschegrava Russian name *chegrava* for the Caspian Tern *Hydroprogne caspia.*

tschudii After Baron Johann Jacob von Tschudi (1818–1889) Swiss explorer, collector in Peru, 1838–1842 and author (*Untersuchungen über die fauna Peruana. Ornithologie,*1844).

tschutschensis German phonetic rendering after the Chukotskiy Peninsula, Siberia, USSR.

tuberculifer L. *tuberculum,* a small bump or swelling; *-fer,* carrying; the Dusky-capped Flycatcher *Myiarchus* has two small, horny swellings at the bend of each wing.

tuberosa L. *tuberosus,* full of lumps or swellings; ref. the casqued bill of the Razor-billed Curassow *Crax.*

tucanus Güaraní (Paraguayan) Indian name *tucán* for a toucan.

tucinkae No expl. (Seilern (1913), *Verhandlungen der ornithologischen gesellschaft in Bayern,* **11**, 276); perhaps the name of a Peruvian Indian tribe or based on an onomatopoeic tribal name for the Scarlet-hooded Barbet *Eubucco.*

tucumana After the province of Tucumán, Argentina.

tuerosi After Justo Tueros Aldana (fl. 1983) Peruvian miner, amateur collector, and naturalist.

tukki Sumatran (Malay) name *tukik* or *tukek* for a woodpecker.

tullbergi After Tycho F. Tullberg (1842–1920) Swedish zoologist and collector.

Tumbezia After the town and port of Tumbes (formerly Puerto Pizarro), northwestern Peru.

tumultuosus L. *tumultuosus,* restless, uproarious (*tumultus,* confusion).

Turacoena Syn. genus *Turacus,* turaco (v. *Tauraco);* Gr. *oinas,* a dove; ref. the iridescent green upperparts of the black cuckoo-doves.

turatii After Conte Ercole Turati (1829–1881) Italian naturalist and collector.

turcosa Mod. L. *turcosus,* turquoise blue (French *turquoise;* Old French *turkeis,* Turkish).

turdinus Mod. L. *turdinus,* thrush-like (L. *turdus,* a thrush) (cf. Med. L. *turdinus,* the quail *Coturnix).*

Turdoides Genus *Turdus,* thrush; Gr. *-oides,* resembling.

Turdus L. *turdus,* a thrush.

Turnagra Genus *Turdus,* thrush; syn. genus *Tanagra,* euphonia, tanager; the New Zealand Piopio *T. capensis* was originally described as a species of tanager from Cape Town, South Africa. The Swedish botanist and author, Anders Sparrman, was based at Cape Town, and accompanied the Forsters on Cook's *Resolution* voyage around the world, 1772–1775, but, with the passage of time, became confused over the correct provenance of many of the specimens obtained on the voyage (*Resa till Goda Hopps-Udden, Sodra Pol-kretsen och Omkring Jordklotet,* 1783).

turneri After H. V. Allan Turner (1876–1953) resident in Kenya, 1909–1953.

Turnix Contraction of genus *Coturnix,* quail; reflects the smaller size and lack of hind-toe of the Palaeotropical button-quails.

Turtur L. *turtur,* the Turtle Dove *Streptopelia.*

tuta L. *tutus,* revered, regarded (*tueri,* to regard); the White-fronted Kingfisher *Halcyon* is regarded as sacred by the natives of Bora-bora, Society Islands.

Tychaedon Gr. *tukhe,* good fortune; *aedon,* the nightingale *Luscinia.*

Tylas Gr. *tulas,* a kind of thrush mentioned by Alexander Myndios.

Tylibyx Gr. *tulas,* a kind of thrush; Mod. L. *ibyx,* a lapwing or plover; ref. thrush-like markings on the breast of the Spot-breasted Plover *Hoplopterus melanocephalus.*

tympanistria Gr. *tumpanistria,* a female drummer; ref. the rapid, regular drumming song of the Tambourine Dove *Turtur.*

tympanistrigus Gr. *tumpanion,* a pearl; L. *striga,* a furrow; ref. white mottled underparts of the Olive-crowned Bulbul *Pycnonotus.*

Tympanuchus Gr. *tumpanon,* a kettledrum; *ekho,* a reverberating sound; ref. the booming courtship calls of the Greater Prairie Chicken *T. cupido.*

typica /typicus L. *typicus,* typical (i.e. the type species of the genus; see *Int. Code Zool. Nomen.,* (Ed. 3), art.68[c]).

typus L. *typus,* a type (v. *typica).*

Tyranneutes Gr. *turanneuo,* to be a tyrant or despot (*turannos,* a tyrant, despot or king).

tyrannina Mod. L. *tyranninus,* like a tyrant flycatcher, tyrannine.

Tyranniscus Dim. of genus *Tyrannus,* kingbird, tyrant flycatcher.

Tyrannopsis Genus *Tyrannus,* kingbird, tyrant flycatcher; Gr. *opsis,* appearance.

Tyrannulus Dim. of genus *Tyrannus,* kingbird, tyrant flycatcher.

Tyrannus L. *tyrannus,* a tyrant, despot, or king (Gr. *turannos,* a tyrant; also a name given to the goldcrest *Regulus* because of its golden crown); the name was given to the kingbirds because of their aggressive behaviour and bright crowns (cf. Mod. L. *tyrannus,* a shrike *Lanius).*

tyrianthina L. *tyrianthinus,* of a colour between purple and violet, Tyrian red or purple.

tyro Gr. myth. Tyro, a beautiful nymph who fell in love with the river Enipeus.

tytleri After Colonel R. C. Tytler (1818–1872) British Army in India, naturalist, and collector.

Tyto Gr. *tuto,* an owl.

tzacatl After Tzacatl, a warrior chief of Aztec mythology.

Trogon collaris Collared Trogon

U

ucayalae After the Río Ucayali, Loreto, Peru.

ugiensis After the island of Ugi (Uki), near San Cristóbal Island, Solomon islands.

ulietanus /ulietensis After the island of Ulieta (Raiatea), Society islands.

ultima L. *ultimus,* farthest, most distant (super. of *ultra,* beyond).

ultramarina Med. L. *ultramarinus,* ultramarine colour, shades of blue (L. *ultra,* beyond; *marinus,* of the sea; ref. the lapis lazuli brought from beyond the sea).

uluguru After the Uluguru Mountains, Morogoro, Tanzania.

ulula L. *ulula,* a screech owl mentioned by Pliny, Terentius Varro, and Vergilius Maro.

umbellus L. *umbella,* a parasol; the Ruffed Grouse *Bonasa* erects neck ruffs and crest in display.

umbra L. *umbra,* a shadow or phantom.

umbratilis L. *umbratilis,* retiring, remaining in the shade (*umbra,* the shade).

umbretta French name *ombrette* given to the Hammerkop *Scopus* because of its uniform umber or earthy-brown colour (Italian *terra d'ombra,* shadow earth).

umbrovirens L. *umbra,* a shadow, darkness; *virens,* green.

unappendiculatus L. *unus,* one; *appendix, appendicis,* an appendage; the Single-wattled Cassowary *Casuarius* has a single blue wattle on the foreneck.

unchall Javanese (Malay) name *burong unchall* for the Bar-tailed Cuckoo Dove *Macropygia.*

uncinatus L. *uncinatus,* hooked, curved (*uncus, unci,* a hook).

undata /undatus Med. L. *undatus,* wavy, with undulating wave-like markings (L. *unda,* a wave).

underwoodii After Underwood (fl. 1832) the otherwise unknown collector of the type specimen of the Booted Racquet-tail *Ocreatus.*

undulata /undulatus L. *undulatus,* wavy, furnished with wave-like markings (*unda,* a wave).

unduligera Mod. L. *undulus,* a small wave; L. *-ger,* carrying.

unicincta /unicinctus L. *uni-,* single; *cinctus,* banded (*cingere,* to encircle).

unicolor L. *unicolor,* of one colour, uniform.

unicornis L. *unicornis,* single-horned; ref. the long, cylindrical 'helmet' of the Horned Curassow *Pauxi.*

uniformis L. *uniformis,* simple, uniform.

unirufa /unirufus L. *uni-,* single; *rufus,* red (i.e. completely red or brown).

Upucerthia Genus *Upupa,* hoopoe; genus *Certhia,* treecreeper.

Upupa L. *upupa,* the Hoopoe *U. epops.*

Uraeginthus Gr. *oura,* the tail (i.e. long-tailed); genus *Aegintha,* waxbill.

Uragus Gr. *oura,* the tail; L. *-ago,* possessing; ref. the very long tail of the Long-tailed Rosefinch *U. sibiricus* (cf. Gr. *ouragos,* leader of the rear guard).

uralensis After the Ural Mountains, USSR, traditionally forming the boundary between Europe and Asia.

uraniae After the French corvette *L'Uranie* which, with *La Physicienne,* circumnavigated the globe, 1817–1820.

Uratelornis Gr. *oura,* the tail (i.e. long-tailed); genus *Atelornis,* ground roller.

urbica L. *urbicus,* urban, frequenting the town or city (*urbs, urbis,* a city or town); the Western House Martin *Delichon* builds its nest under the eaves of houses and other man-made structures.

Uria Gr. *ouria,* a kind of waterbird mentioned by Athenaeus, not further identified.

urile German name *urile* given to the Red-faced Cormorant *Phalacrocorax* by Steller (1751) (L. *urinare,* to dive).

urinatrix L. *urinator,* a diver (*urinare,* to dive).

Urochroa Gr. *oura,* the tail; *khroa,* skin colour, complexion; ref. white and bronze-coloured tail of the White-tailed Hillstar *U. bougueri.*

urochrysia Gr. *oura,* the tail; *khrusos,* gold.

Urocissa Gr. *oura,* the tail (i.e. long-tailed); genus *Cissa,* green magpie.

Urocolius Gr. *oura,* the tail (i.e. long-tailed); genus *Colius,* mousebird.

Urocynchramus Gr. *oura,* the tail (i.e. long-tailed); syn. genus *Cynchramus,* bunting (Gr. *kinkhramas,* a migratory bird mentioned by Aristotle and Hesychius, not further identified, but associated with a bunting by later commentators).

Urogalba Gr. *oura,* the tail (i.e. long-tailed); contraction of genus *Galbula,* jacamar.

urogallus Mod. L. *urogallus,* the mountain cock or capercaillie (partial phonetic rendering of German *auerhuhn* or *orhun,* the mountain cock) (cf. Gr. *oura,* the tail; L. *gallus,* a cock).

Uroglaux Gr. *oura,* the tail (i.e. long-tailed); *glaux,* an owl.

Urolais Gr. *oura,* the tail (i.e. long-tailed); *laios,* a thrush.

Urolestes Gr. *oura,* the tail (i.e. long-tailed); *lestes,* a robber (i.e. a shrike).

Uroleuca Gr. *oura,* the tail; *leukos,* white.

Uromyias Gr. *oura,* the tail (i.e. long-tailed); Mod. L. *myias,* flycatcher.

Uropelia Gr. *oura,* the tail (i.e. long-tailed); *peleia,* a dove.

urophasianus Gr. *oura,* the tail (i.e. long-tailed); *phasianos,* a pheasant.

Uropsalis Gr. *oura,* the tail; *psalis,* a pair of shears; ref. the extraordinary tail of the Lyre-tailed Nightjar *U. lyra.*

Uropsila Gr. *oura,* the tail; *psilos,* thin, slender; ref. the short, thin tail of the White-bellied Wren *U. leucogastra.*

uropygialis Med. L. *uropygium,* the rump (Gr. *ouropugion,* a birds rump).

Urosphena Gr. *oura,* the tail; *sphen,* a wedge; the bush warblers have short, stubby tails.

urosticta /Urosticte /urostictus Gr. *oura,* the tail; *stiktos,* spotted, dappled.

Urothraupis Gr. *oura,* the tail (i.e. long-tailed); genus *Thraupis,* tanager.

Urotriorchis Gr. *oura,* the tail (i.e. long-tailed); *triorkhes,* a bird of prey (v. *Dryotriorchis).*

ursulae After Ursula Davies (fl. 1903) niece of British explorer Captain Boyd Alexander.

urubambae /urubambensis After the Río Urubamba, Cuzco, Peru.

urubitinga Tupí (Brazilian) Indian name *urubutsin* or *urubu* for a vulture.

Urubitornis Syn. genus *Urubitinga,* black hawk; Gr. *ornis,* a bird.

urumutum Tupí (Brazilian) Indian name *urumutum* for a curassow, named for its hooting cries.

usambiro After the village of Usambiro, Mwanza, Tanzania.

usheri After H. B. Usher (b. 1893) British ornithologist.

ussheri After H. T. Ussher (1836–1880) British diplomat Gold Coast (Ghana), 1866–1872; Governor of Labuan, 1875–1879; Governor of Gold Coast,1879–1880.

usticollis L. *ustus,* burnt, scorched (*urere,* to burn); Mod. L. *-collis,* necked.

ustulatus L. *ustulatus,* burnt (*ustulare,* to burn; dim. of *urere,* to burn).

V

vaalensis After the River Vaal, Natal, South Africa.

vagabunda L. *vagabundus,* wandering, vagabond.

vagans L. *vagans,* wandering (*vagare,* to wander).

vaillantii After Commandant Jean Jacques Rousseau Levaillant (1790–1876) French Army in Algeria, naturalist, and author (*Introduction à l'histoire des mammifères et des oiseaux du nord de l'Afrique,* 1851).

validirostris L. *validus,* strong, powerful; *-rostris,* billed.

validus L. *validus,* strong, powerful (*valere,* to be strong).

valisneria After Antonio Vallisnieri de Vallisnera (1661–1730) Italian naturalist and botanist, and author (*Istoria del camaleonte Africano,* 1715; *Raccolta di varj trattati,* 1715). Newton and Gadow (1896) say the name comes from a food plant (*Valisneria,* wild celery) beloved of the Canvasback *Aythya.*

vana No expl. (Hartert (1930), *Nov. Zool.,* **36**, 42); perhaps the name of a Papuan village on the Anggi-gidji Lake, Irian Jaya.

vanderbilti After George Washington Vanderbilt (1914–1961) US philanthropist and sponsor.

Vanellus Med. L. *vanellus,* the Lapwing *V. vanellus* (dim. from L. *vannus,* a winnowing fan; ref. the floppy wing action).

Vanga Malagasy name *vanga* for the Hook-billed Vanga *V. curvirostris.*

vanicorensis / vanikorensis After the Vanikoro Islands, Santa Cruz group, Solomon Islands.

varia L. *varius,* variegated, diverse.

variabilis L. *variabilis,* changeable, variable.

variegata L. *variegatus,* variegated (*varare,* to variegate).

variegaticeps L. *variegatus,* variegated; *-ceps,* headed.

variegatus see **variegata.**

variolosus Med. L. *variolus,* a pustule, a spot; L. *-osus,* abundance.

varius see **varia.**

varzeae Brazilian Portuguese *várzea,* the seasonally flooded areas adjacent to the Amazon River.

vasa Malagasy name *vaza* for the Greater Vasa Parrot *Coracopsis.*

vassali After J. J. Vassal (fl. 1924) French collector in China, 1910.

vassorii After M. Vassor (fl. 1840) French collector in Colombia.

vaughani/vaughanii After **1** Commander R. E. Vaughan (1874–1937) English naval officer in the Pacific, 1893 (*Conopoderas).* **2** John Henry Vaughan (1892–1965) English jurist, ornithologist, and author (*Birds of Zanzibar,* 1930) (*Zosterops).*

vauxi After William Sansom Vaux (1811–1882) US archaeologist and mineralogist.

vegae After the exploration vessel *Vega* used by Nils Adolf Erik Baron Nordenskjöld (1832–1901) Swedish Arctic explorer.

velata/velatum/velatus L. *velatus,* veiled, covered (*velare,* to cover).

Veles L. *veles,* a skirmisher.

velia No expl. (Linnaeus (1758), *Systema Naturae,* (Ed. 10), book **1**, p. 188); perhaps a mistake for Gr. *elea,* a small bird mentioned by Aristotle.

vellalavella After the island of Vella Lavella, New Georgia group, Solomon Islands.

velox L. *velox,* rapid, swift.

velutina French *velouté,* velvety (*velours,* velvet; Late L. *velvettum,* velvet); L. *-inus,* resembling.

venerata L. *veneratus,* venerated (*venerari,* to worship); the Olive-backed Kingfisher *Halcyon* is regarded as sacred by the natives of Tahiti.

venezuelana/venezuelanus/ venzuelensis After Venezuela ('Little Venice'), a name first given to the settlement of Coro in 1499 by Amerigo Vespucci in allusion to the Indian villages built on stilts or poles upon the waters of Lake Maracaibo. The name gradually attached itself to the region.

Veniliornis Syn. genus *Venilia,* woodpecker (L. *venilia,* a wave, a band); Gr. *ornis,* a bird; ref. the white barred primaries of the Smoky-brown Woodpecker *V. fumigatus.*

ventralis L. *ventralis,* ventral, of the belly (*venter, ventris,* the belly).

venusta L. *venustus,* beautiful, lovely, charming.

venustulus L. *venustulus,* charming, lovely (dim. of *venustus,* charming).

venustus see **venusta.**

veraepacis After the department of Vera Paz, Guatemala (L. *verus,* true; *pax, pacis,* peace).

veraguensis After the province of Veragua, New Granada (Veraguas, Panama).

veredus L. *veredus,* a swift horse (i.e. fast running).

vermiculatus L. *vermiculatus,* vermiculated, wrought with markings resembling the sinuous tracks of worms (*vermis,* a worm).

Vermivora L. *vermis,* a worm; *-vorus,* eating (*vorare,* to devour).

vermivorus Mod. L. *vermivorus,* worm eating.

vernalis L. *vernalis,* of spring, vernal (i.e. green) (*ver,* spring).

vernans Mod. L. *vernans,* green (L. *ver,* spring).

veroxii see **verreauxi 1.**

verreauxi After **1** Édouard Verreaux (1810–1868) and his brother Jules Pierre Verreaux (1808–1873) French naturalists, collectors, and natural history dealers (*Aquila*). **2** Jules Pierre Verreaux (v. 1) (*Coua, Leptoptila, Lophornis, Paradoxornis, Rhipidura, Tauraco*).

Verreauxia see **verreauxi 1.**

verreauxii see **verreauxi.**

verrucosus L. *verrucosus,* warty, tubercular (*verruca,* a wart).

versicolor L. *versicolor,* of various colours, particoloured.

versicolurus L. *versicolor,* versicoloured; Gr. *-ouros,* tailed; the tail of the Canary-winged Parakeet *Brotogeris* is green above and blue below (cf. L. *versicolorus,* of various colours, particoloured).

versteri After Florentius Abraham Verster van Wulverhorst (1826–1923) Dutch zoologist, administrator Rijksmuseum van Natuurlijke Historie, Leiden, 1860–1920.

verticalis Late L. *verticalis,* crowned (L. *vertex, verticis,* the crown of the head).

vesper L. *Vesper,* the planet Venus, the brightest star in the evening sky.

vespertinus L. *vespertinus,* of evening or twilight (*vesper, vesperis,* evening).

Vestiaria L. *vestiarius,* clothes, of clothes; ref. the bright red plumage of the Iiwi *V. coccinea.*

vestitus L. *vestitus,* clothing, clothes; ref. glittering plumage of the Glowing Puffleg *Eriocnemis.*

vetula L. *vetula,* an old woman; given to spp. with shrill, cackling cries.

vexillarius L. *vexillarius,* a standard-bearer.

viaticus L. *viaticus,* belonging to a road or journey (i.e. cursorial) (*via,* a road).

vicina L. *vicina,* a neighbour; the Tagula Meliphaga *Meliphaga* was originally described as an island subsp. of the Mimic Meliphaga *M. analoga*.

vicinior L. *vicinior*, very near (comp. of *vicina*, near) (i.e. related or similar to).

victor L. *victor*, victor, conqueror.

victoria After Queen Victoria of Great Britain (1819–1901: reigned 1837–1901) and Empress of India.

victoriae After **1** Victoire Mulsant (fl. 1846) wife of French naturalist Martial Mulsant (*Lesbia*). **2** Queen Victoria (v. *victoria*) (*Ptiloris*). **3** Princess (later Empress) Victoria of Prussia (1840–1901) (*Lamprolia*).

victorini After Johan Fredrik Victorin (1831–1855) Swedish collector in South Africa and author (*Resa i Kaplandet aren 1853–1855, jagt och naturbilder*, 1863).

Vidua L. *vidua*, a widow; ref. the black caps and upperparts of the male whydahs.

viduata L. *viduata*, widowed.

viduus L. *viduus*, widowed, bereaved.

vieilloti After Louis Pierre Vieillot (1748–1831) French naturalist, collector, and author (*Oiseaux dorés ou à reflets métalliques*, 1800; *Histoire naturelle des plus beaux oiseaux chanteurs de la zone torride*, 1805; *La galerie des oiseaux*, 1820).

vieillotioides Specific name *Pyrrhomyias vieillotii* (now a subsp. of *P. cinnamomea*); Gr. *-oides*, resembling.

vigil L. *vigil*, watchful, wakeful.

vigorsii After Nicholas Aylward Vigors (1785–1840) British naturalist and editor of the Zoological Journal, 1827–1834.

viguieri After Camille Viguier (fl. 1888) French zoologist and collector in Panama.

vilasboasi After the brothers Orlando Villas Boas (b. 1916), Claudio Villas Boas (b. 1918) and Leonardo Villas Boas (1920–1961) Brazilian pioneers, anthropologists, and authors (*Xingu; the Indians, their myths*, 1975).

vilcabambae After the Cordillera de Vilcabamba, Cuzco, Peru.

vilissimus L. *vilissimus*, worthless (i.e. very small) (super. of *vilis*, cheap).

villanovae After the town of Vila Nova (Senhor do Bonfim), Bahia, Brazil.

villaviscensio After Manuel Villavicencio (fl. 1865) collector in Ecuador, 1850–1865.

villosa/villosus L. *villosus*, hairy, shaggy (*villus*, shaggy hair).

vinacea L. *vinaceus*, belonging to wine (i.e. wine-coloured) (*vinum, vini*, wine).

vinaceigula L. *vinaceus*, vinaceous; *gula*, the throat.

vinaceus see **vinacea.**

Vinago Med. L. *vinago*, a kind of pigeon which appeared at the time of the wine-harvest (L. *vinum, vini*, wine; *-ago*, resembling); ref. the purple or mauve shoulder patches of the African green pigeons *Treron*.

vincens L. *vincens*, surpassing (*vincere*, to surpass).

Vini Tahitian native name *vini* for the Ultramarine Lory *V. ultramarina*.

vinipectus L. *vinum, vini*, wine; *pectus*, the chest.

vintsioides French name *Le Vintsi* given to the Madagascar Pygmy Kingfisher *Ispidina madagascariensis* by Buffon (1770) (Malagasy name *vintsy* for the Madagascar Malachite Kingfisher *Alcedo*); Gr. *-oides*, resembling.

viola L. *viola*, violet colour (*viola*, the violet).

violacea/violaceus L. *violaceus*, violaceous, violet-coloured.

violiceps L. *viola*, violet colour; *-ceps*, headed.

violifer L. *viola*, violet colour; *-fer*, bearing.

vipio L. *vipio*, a Balearic name for the lesser edible crane mentioned by Pliny.

virata L. *viratus*, manly, spirited.

virens L. *virens*, green (*virere*, to be green).

virenticeps L. *virens, virentis*, green; *-ceps*, headed.

Vireo L. *vireo*, a small green bird, perhaps the Greenfinch *Carduelis* (*virescere*, to become green).

Vireolanius Genus *Vireo*, vireo; genus *Lanius*, shrike.

virescens L. *virescens,* greenish (*virescere,* to become green).
virgata L. *virgatus,* striped, streaked (*virga,* a streak).
virgaticeps L. *virgatus,* striped; *-ceps,* headed.
virgatus see **virgata.**
virginiae After Mary Virginia Anderson (fl. 1858) wife of US army naturalist and collector Surgeon William Anderson.
virginianus After **1** the Virginia colonies, North America, named for Elizabeth I of England, the Virgin Queen (1533–1603: reigned 1558–1603); the term Virginia was originally applied to all those parts of North America not held by the French or Spanish (*Colinus*). **2** the states of Virginia, USA (*Bubo*).
virgo L. *virgo,* a maiden or virgin.
viridanus L. *viridans,* green.
viridescens L. *viridescens,* greenish (*viridescere,* to become green).
Viridibucco L. *viridis,* green; Mod. L. *bucco,* a barbet.
viridicata L. *viridicatus,* green, made green.
viridicauda L. *viridis,* green; *cauda,* the tail.
viridiceps L. *viridis,* green; *-ceps,* headed.
viridicollis L. *viridis,* green; Mod. L. *-collis,* necked.
viridicyana L. *viridis,* green; *cyaneus,* dark blue (Gr. *kuaneos,* dark blue).
viridifacies L. *viridis,* green; *facies,* the face.
viridiflavus L. *viridis,* green; *flavus,* yellow (Mod. L. *viridiflavus,* greenish-yellow).
viridifrons L. *viridis,* green; *frons,* the forehead or brow.
viridifuscus L. *viridis,* green; *fuscus,* dusky, brown.
viridigaster L. *viridis,* green; *gaster,* the belly.
viridigenalis L. *viridis,* green; *gena,* the cheek; *-alis,* pertaining to.
viridigula L. *viridis,* green; *gula,* the throat.
viridipallens L. *viridis,* green; *pallens,* pale, wan.
viridirostris L. *viridis,* green; *-rostris,* billed.
viridis L. *viridis,* green. Erroneous epithet in the cases of *Frederickena* and *Terpsiphone.*
viridissima/viridissimum L. *viridissimus,* very green (super. of *viridis,* green).
Viridonia L. *viridis,* green.
viscivorus L. *viscum, visci,* mistletoe; *-vorus,* eating.
vitellina/vitellinus Med. L. *vitellinus,* vitelline, a deep yellow colour tinged red (L. *vitellus,* egg-yolk).
Vitia/vitiensis Fijian native name *Viti* for their home islands.
vitiosus L. *vitiosus,* defective, faulty (*vitium,* a fault); of the Double-banded Pygmy Tyrant *Lophotriccus* Bangs and Penard (1921), *Bull. Mus. Comp. Zool.*, **64**, 373, wrote, '...the bird to which the name *speciferum* has been erroneously applied by authors...We would have preferred a freshly collected specimen for the type, rather than the old Lafresnaye specimen, but we have no other'.
vitriolina L. *vitreolus,* of glass, shining, brilliant; *-inus,* resembling.
vittata/vittatus L. *vittatus,* banded, striped (*vitta,* a ribbon or band).
vitticeps L. *vitta,* a head band; *-ceps,* headed.
Vivia Nepalese name *wiwi* for the Speckled Piculet *Picumnus innominatus.*
vivida L. *vividus,* vivid, lively.
vocifer French name *Le Vocifer* given to the African Fish Eagle *Haliaeetus* by Levaillant (1799) (L. *vox, vocis,* a voice; *-fer,* carrying; *vociferari,* to shout); the loud and distinctive cries of this splendid raptor evoke the spirit of Africa.
vociferans L. *vociferans,* shouting.
vociferoides Specific name *Haliaeetus vocifer;* Gr. *-oides,* resembling.
vociferus L. *vociferus,* clamorous, vociferous (*vociferari,* to shout).
Volatinia Dim. of L. *volatus,* flying (i.e. a bird).

Vosea After C. R. Vose (1890–1957) US businessman, explorer, and sponsor.

vosseleri After Julius Vosseler (1861–1933) German scientist in German East Africa (Tanzania), 1903–1909.

vulcani L. *vulcani,* of the volcano or Vulcan the fire god; after the Volcán de Irazú, Costa Rica.

vulcania L. *vulcanius,* volcanic.

vulgaris L. *vulgaris,* common, ordinary.

vulnerata/vulneratum L. *vulneratus,* wounded (i.e. blood- or red-stained) (*vulnerare,* to wound).

vulpina L. *vulpinus,* vulpine, fox-like (i.e. tawny-coloured) (*vulpes, vulpis,* a fox).

Vultur L. *vultur,* a vulture.

vulturina/vulturinum L. *vulturinus,* vulturine, like a vulture (*vultur, vulturis,* a vulture); ref. bare, naked or thinly feathered heads.

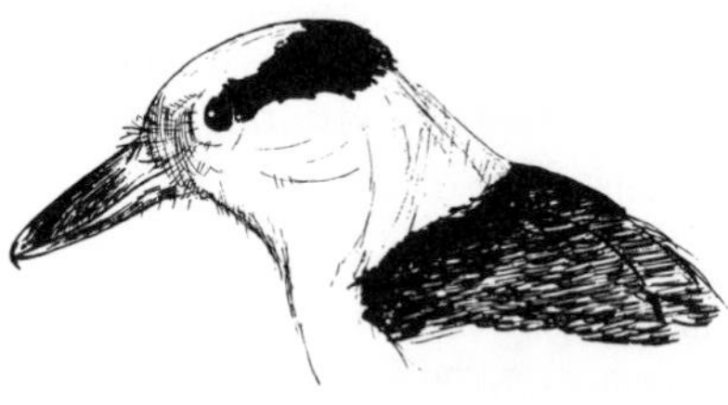

Vanga curvirostris Hook-billed Vanga

W

waalia Abyssinian (Ethiopian) name *waalia* for Bruce's Green Pigeon *Treron*.

waddelli After Lieutenant Colonel Laurence Austine Waddell (1854–1938) British explorer, collector in Tibet, 1903–1904 and author (*The birds of Sikhim*, 1893; *Lhasa and its mysteries*, 1905).

wagleri After Johann Georg Wagler (1800–1832) German herpetologist, systematist, and author (*Monographia psittacorum*, 1832).

wahlbergi After Johan August Wahlberg (1810–1856) Swedish naturalist and collector in South Africa, 1838–1856 (killed by an elephant).

wahnesi After Carl Wahnes (1835–1910) German naturalist and collector in New Guinea.

waigiuensis After the island of Waigiu (Waigeo), Irian Jaya, Indonesian New Guinea.

wakensis After Wake Island in the north Pacific, named for British mariner William Wake (fl. 1796).

wakoloensis After Lake Wakolo, Buru Island, Moluccas.

waldeni/waldenii After Arthur Hay Viscount Walden later 9th Marquis of Tweeddale (1824–1878) British ornithologist, collector, and author.

Waldronia After Waldron de Witt Miller (1879–1929) US zoologist.

wallacei/wallacii After Alfred Russel Wallace (1823–1913) British zoologist, father of zoogeography, collector in Brazil, 1848 and the East Indies, 1854, and author (*The Malay archipelago*, 1869; *Contributions to the theory of natural selection*, 1870; *Island life*, 1880).

walleri After Gerald Waller (fl. 1880) English naturalist and collector in East Africa.

wallichii After Nathaniel Wallich (1786–1854) Danish botanist; Superintendent of East India Company gardens, Calcutta, 1815–1846; and collector in Nepal, 1820–1822.

wallicus After the state of New South Wales, Australia (Mod. L. *Wallia*, Wales). Historically the name New South Wales was first applied to the whole eastern half of the continent of Australia (i.e. excluding Western Australia).

wardi/wardii After **1** S. N. Ward (1813–1897) British colonial administrator in India, 1832–1863 (*Zoothera*). **2** Swinburne Ward (1830–1897) British civil commissioner Seychelles, 1862–1868 (*Psittacula*). **3** C. Ward (fl. 1886) English naturalist and collector (*Pseudobias*). **4** Captain Francis Kingdon-Ward (1885–1958) English botanist, explorer, collector, and author (*The land of the blue poppy*, 1913; *Plant hunting on the edge of the world*, 1930) (*Harpactes*).

Warsanglia After the Warsangli area of north-eastern British Somaliland (Somalia).

warszewiczi After J. Warszewicz (1812–1866) Polish botanist and collector in Peru, 1860.

watersi After Thomas Waters or Watters (1840–1904) British consular service.

waterstradti After Johannes Waterstradt (b. 1869) Danish entomologist, botanist, and collector.

watertonii After Charles Waterton (1782–1865) English naturalist and collector.

watkinsi After **1** Henry George Watkins (fl. 1912) English collector in Peru, 1910–1912 *(Grallaria)* and his brother **2** C. Watkins (fl. 1912) *(Incaspiza)*.

watsonii After Gavin Watson (1796–1858) US ornithologist and author.

webbianus After Philip Barker Webb (1793–1854) botanist, topographer, and collector.

websteri After Captain Herbert Cayley Webster (fl. 1897) English explorer in New Guinea and author (*Through New Guinea*, 1898).

weddellii After Hugh Algernon Weddell (1819–1877) botanist; explorer in tropical America, 1843–1847; and author (*Voyage dans le nord de la Bolivie et dans les parties voisines du Pérou*, 1853; *Chloris Andina.Essai d'une flore de la région alpine des cordillères de l'Amérique du Sud*, 1855).

weiskei After Emil Weiske (1867–1950) Austrian collector in Australia and New Guinea, 1895–1900 and Argentina, 1910.

welchmani After the Revd. H. J. Welchman (1850–1908) missionary to the Solomon Islands.

wellsi After J. Wells (fl. 1902) resident on Grenada, West Indies, 1886–1902.

westermanni After Gerard Fredrick Westermann (1807–1890) Dutch zoologist and author (*De toerakos*, 1860).

westlandica After Westland Province, South Island, New Zealand.

wetmorei After Alexander Wetmore (1886–1978) US ornithologist, palaeornithologist, collector, and author (*A systematic classification for the birds of the world*, 1930; *Birds of Haiti and the Dominican Republic*, 1931; *A revised classification for the birds of the world*, 1951; *A classification for the birds of the world*, 1960; *The birds of the Republic of Panama*, 1965).

Wetmorethraupis After Alexander Wetmore (v. *wetmorei)*; genus *Thraupis,* tanager.

weynsi After Lieutenant Colonel A. F. G. Weyns (1854–1944) Belgian explorer and collector in tropical Africa, 1888–1903.

whartoni After Rear Admiral Sir William James Lloyd Wharton (1843–1905) British hydrographer Indian Ocean, 1872–1880.

whiteheadi After John Whitehead (1860–1899) British explorer Borneo, 1885–1888; the Philippines, 1893–1896; Hainan, 1899; collector and author (*Exploration of Mount Kina Balu, North Borneo*, 1893).

whitei After Captain Samuel Albert White (1870–1954)English naturalist, collector, settled in Australia, 1906–1916.

whitelyi After Henry Whiteley, Jr. (1844–1892) English collector in Japan and Peru.

whitemanensis After the Whiteman Range, New Britain, Bismarck Archipelago, Papua New Guinea.

whitii After Ernest William White (1858–1884) English collector in Argentina and author (*Cameos from the silver-land*, 1881).

whitneyi After **1** Josiah Dwight Whitney (1819–1896) US mining geologist; California state geologist; Founder of the Harvard School of Mining, 1868 *(Micrathene)*. **2** Harry Payne Whitney (1873–1930) US philanthropist who sponsored a series of expeditions to the Pacific, 1921–1922 *(Cichlornis, Pomarea)*.

whytii After Alexander Whyte (1834–1905) government naturalist in Nyasaland (Malawi), 1891–1897 and author.

wilhelminae After Wilhelmina Meyer (fl. 1873) wife of German zoologist Adolf Meyer.

wilkinsi After Captain Sir George Hubert Wilkins (1888–1958) polar explorer; Arctic, 1913–1917; Antarctic, 1920–1922.

willcocksi After General Sir James Willcocks (1847–1926) commander of Ashanti expedition 1900 and author (*From Cabul to Kumassi*, 1904).

williami After William Wilson (fl. 1846) French amateur ornithologist.

williamsi After John G. Williams (fl. 1963) British ornithologist; curator of Coryndon Museum, Nairobi; and author (*A field guide to the birds of east Africa*, 1980).

williamsoni After Kenneth Williamson (b. 1914) English ornithologist, field researcher, and author (*Identification for ringers*, 1960).

wilsoni After **1** Thomas B. Wilson (1807–1865) US ornithologist (*Chondrohierax, Coeligena*). **2** Scott Burchard Wilson (d. 1923) English ornithologist, collector, and author (*Aves Hawaiienses. The birds of the Sandwich Islands,* 1890) (*Heterorhynchus*). **3** Captain M. Wilson (1869–1900) British Army in Ashanti, 1900 (*Hypochera*).

Wilsonia After Alexander Wilson (1766–1813) Scottish/US ornithologist and author (*American ornithology*, 1808).

wilsonii see **wilsoni.**

winchelli After Newton Horace Winchell (1839–1914) US geologist, archaeologist, and author (*The aborigines of Minnesota*, 1911).

winifredae After Winifred Moreau (b. 1891) wife of English ornithologist Reginald Moreau.

wolfi After Theodor Wolf (b. 1841) Ecuadorean state geologist.

wollweberi After Wollweber (fl. 1840) an unknown traveller and collector in Mexico.

woodfordi After **1** Colonel E. J. Woodford (1761–1835) British Army in Europe, 1794–1817 and collector (*Ciccaba*). **2** Charles Morris Woodford (1852–1927) Resident Commissioner in Solomons Protectorate, 1896–1914 (*Nesoclopeus, Corvus*).

Woodfordia see **woodfordi 2**.

woodfordii see **woodfordi.**

woodhousei After Samuel W. Woodhouse (1821–1904) US surgeon and naturalist in tropical Africa.

woodi After Major General Leonard Wood (1860–1927) US Governor General of the Philippines, 1906–1908.

woodwardi After **1** Bernard Henry Woodward (1846–1912) curator of Perth Museum, Western Australia (*Amytornis*). **2** Revd. R. B. Woodward (fl. 1885) British naturalist in South Africa, 1881–1899 and author (*Natal birds*, 1899) (*Stactolaema*).

woosnami After R. B. Woosnam (1880–1915) game-ranger in East Africa.

worcesteri After Dean Conant Worcester (1866–1924) US ornithologist, zoologist, collector, and author.

wortheni After Charles K. Worthen (1850–1909) US collector.

wrightii After Charles Wright (1811–1885) US botanist, teacher, and collector.

wumizusume Japanese name *umisuzume,* meaning 'sea sparrow' (*umi,* the sea; *suzume,* a sparrow), for the Crested Murrelet *Synthliboramphus.*

wyatti After Claude Wyatt (1842–1900) English ornithologist, collector in Colombia, 1870 and author (*A monograph of the Hirundinidae*, 1885).

wyvilliana After Sir Charles Wyville Thompson (1830–1882) English naturalist who led the *Challenger* expedition, 1873–1876.

Synthliboramphus wumisuzume Crested Murrelet

X

xanthetraea Gr. *xanthos,* yellow; *etron,* the belly.

xanthocephala /xanthocephalus Gr. *xanthos,* yellow; *-kephalos,* headed.

xanthochlorus Gr. *xanthos,* yellow; *khloros,* green.

xanthochroa Gr. *xanthokhroos,* yellow-skinned (cf. *xanthokhros,* brown).

xanthogaster /xanthogastra Gr. *xanthos,* yellow; *gaster,* the belly.

xanthogenys Gr. *xanthos,* golden; *genus,* the cheek.

xanthogonys Gr. *xanthos,* yellow; Mod. L. *gonys,* the gonys (v. *Orthogonys).*

xanthogramma Gr. *xanthos,* yellow; *gramme,* a line, a stroke.

xantholaema /xantholaemus Gr. *xanthos,* yellow; *laimos,* the throat.

xantholophus Gr. *xantholophos,* yellow-crested.

xantholora Gr. *xanthos,* yellow; Mod. L. *lorum,* the lores.

xanthomus Gr. *xanthos,* yellow; *omos,* the shoulder.

Xanthomyias Gr. *xanthos,* yellow; Mod. L. *myias,* flycatcher.

Xanthomyza Gr. *xanthos,* golden, yellow; Mod. L. *myza,* honeyeater; ref. the golden-edged wing and tail feathers of the Regent Honeyeater *X. phrygia.*

xanthonotus Gr. *xanthos,* yellow; *-notos,* backed.

xanthonura Gr. *xanthos,* yellow; *oura,* the tail.

Xanthophilus Gr. *xanthos,* yellow; *philos,* dear, pleasing.

xanthophrys Gr. *xanthos,* yellow; *ophrus,* the eyebrow.

xanthophthalmus Gr. *xanthos,* yellow; *ophthalmos,* the eye.

Xanthoploceus Gr. *xanthos,* yellow; genus *Ploceus,* weaver.

xanthoprymna Gr. *xanthos,* yellow; *prumnos,* rear, hindmost; the Red-tailed Wheatear *Oenanthe* has a reddish-chestnut rump.

xanthops Gr. *xanthos,* yellow; *ops,* the face or appearance.

Xanthopsar Gr. *xanthos,* yellow; *psar,* a starling (v. *Gnorimopsar*).

xanthopterus Gr. *xanthos,* yellow; *-pteros,* winged.

xanthopterygius Gr. *xanthos,* yellow; *pteruges,* the wings.

xanthopygaeus /xanthopygius Gr. *xanthos,* yellow; Mod. L. *-pygius,* rumped.

xanthopygos /xanthopygus Gr. *xanthos,* yellow; *-pugos,* rumped.

xanthorhynchus Gr. *xanthos,* yellow; *rhunkhos,* the bill.

xanthornus Gr. *xanthos,* yellow; *ornis,* a bird.

xanthorrhous Gr. *xanthos,* yellow; *orrhos,* the rump or tail.

xanthoschistos Gr. *xanthos,* yellow; Mod. L. *schistos,* slate colour.

xanthosterna Gr. *xanthos,* yellow; *sternon,* the breast.

Xanthotis Gr. *xanthos,* yellow; *-otis,* eared; ref. the small yellow ear tuft of the Tawny-breasted Honeyeater *X. flaviventer.*

Xanthoura Gr. *xanthos,* yellow; *oura,* the tail.

xantusii After Lajos Jánosz Xántus de Vesey, also known as John Xantus (fl.

1892) Hungarian exile, adventurer, collector, and diplomat.

xavieri After Xavier Dybowski (fl. 1892) French explorer and collector in the Congo.

Xema No expl. (Leach (1819) in Ross' *A Voyage of Discovery*, App. 2, p. lvii), where Sabine's Gull *X. sabini* is first referred to as Sabine's Xeme; perhaps from Gr. *xene*, a female guest or stranger.

Xenerpestes Gr. *xenos*, strange, foreign; *herpestes*, a creeper (strictly a snake or insect).

Xenicus Gr. *xenikos*, strange, different.

Xenocopsychus Gr. *xenos*, strange, different; genus Copsychus, magpie robin.

Xenodacnis Gr. *xenos*, strange, different; genus *Dacnis*, dacnis.

Xenoglaux Gr. *xenos*, strange, different; *glaux*, an owl.

Xenoligea Gr. *xenos*, strange, different; syn. genus *Ligea*, warbler (Gr. myth. Ligea, one of the Nereids; *ligeia*, clear toned, an epithet for the nightingale).

Xenopipo Gr. *xenos*, strange, different; Mod. L. *pipo*, manakin.

Xenopirostris Genus *Xenops*, xenops; Gr. *-rostris*, billed; ref. the high, laterally compressed bill, with lower mandible recurved, of the Razor-billed Vanga *X. xenopirostris*.

Xenops Gr. *xenos*, strange; *ops*, the face; the xenops have short, upturned bills.

Xenopsaris Gr. *xenos*, strange, different; syn. genus *Psaris*, becard, cotinga (Gr. *psar*, the starling).

xenopterus Gr. *xenos*, strange, different; *-pteros*, winged; the Horqueta Crake *Laterallus* has white barred wing coverts.

Xenorhynchus Gr. *xenos*, strange, different; *rhunkhos*, the bill; ref. the huge, black bill of the Black-necked Stork *X. asiaticus*.

Xenornis Gr. *xenos*, strange, different; *ornis*, a bird.

Xenospingus Gr. *xenos*, strange, different; *spingos*, a finch.

Xenospiza Gr. *xenos*, strange, different; *spiza*, a finch.

xenothorax Gr. *xenos*, strange, different; *thorax*, the breast; ref. conspicuous black and white throat patch of the White-browed Tit Spinetail *Leptasthenura*.

Xenotriccus Gr. *xenos*, strange, different; Mod. L. *triccus*, tyrant flycatcher.

Xenus Gr. *xenos*, a stranger or foreigner; the migratory Terek Sandpiper *X. cinereus* breeds in northern Siberia and winters mainly south of the equator in Africa, Asia, and Australia.

Xiphidiopicus Gr. *xiphidion*, a dagger; genus *Picus*, woodpecker; the Cuban Green Woodpecker *X. percussus* has a short, dagger-like bill.

Xiphidiopterus Gr. *xiphidion*, a dagger; *-pteros*, winged; ref. the large carpal spurs of the White-headed Lapwing *Hoplopterus albiceps*.

Xiphirhynchus Gr. *xiphos*, a sword (in classical times this word referred to the long straight sword, and *makhaira* was the curved scimitar); *rhunkhos*, the bill; ref. the very long decurved bill of the Slender-billed Scimitar Babbler *X. superciliaris*.

Xiphocolaptes Gr. *xiphos*, a sword; genus *Colaptes*, woodpecker; ref. the heavy, somewhat curved bill of the White-throated Woodcreeper *X. albicollis* (the woodcreepers were originally considered to be allied to the woodpeckers).

Xipholena Gr. *xiphos*, a sword; *olene*, the lower arm; ref. the stiff, elongated upper wing-coverts of the Pompadour Cotinga *X. purpurea*.

Xiphorhynchus Gr. *xiphos*, a sword; *rhunkhos*, the bill; these woodcreepers have heavy, usually straight, bills.

Xolmis No expl. (Boie (1828), Isis von Oken, col. 318), apart from a ref. to an otherwise unknown Gr. *xolmis* in an earlier issue (Isis von Oken (1826), col. 973) where further ref. is made to Azara's *peposaza* (sic); perhaps based on an unrecorded Güaraní (Paraguayan) Indian name, although not found under *pepoaza* in Azara (1805).

Y

Yanacea/yanacensis After the village of Yánac, Ancash, Peru.

yarrelli/yarrellii After William Yarrell (1784–1856) English ornithologist, bookseller, and author (*History of British birds*, 1843).

yaruqui After the town of Yaruqui, Pichincha, Ecuador.

yatii After Colonel Sir Charles Edward Yate (1849–1940) English political officer Afghan Boundary Commission; Consul General Seistan, 1894–1898; and author (*Khorasan and Sistan*, 1900).

yelkouan Turkish name *yelkovan* for a shearwater.

yeltoniensis After Lake Yelton (Ozero Elton), Volgograd, USSR.

yemenensis After the Yemen in south-western Arabia, ancient Arabia Felix.

yersini After Alexandre Yersin (1863–1943) Swiss bacteriologist and Director of the Pasteur Institute in French Indo-China, 1919.

yessoensis After Yesso (Hokkaido), the most northerly main island of Japan.

Yetapa Güaraní (Paraguayan) Indian name *güirá yetapu*, 'cutter-bird', for the Strange-tailed Tyrant *Y. risora*, with ref. to the manner in which it opens and closes the tail in flight like shears.

yncas After the Yncas, a former spelling for Incas (Quechua Indian *inca*, lord, king), the native rulers of pre-Conquest Peru (v. *inca*).

youngi After John Young (1838–1901) English civil servant, ornithologist, and collector in the Pacific, 1886–1887.

ypecaha Güaraní (Paraguayan) Indian name *ypacahá* for the Giant Wood Rail *Aramides*.

ypiranganus After the town of Ypiranga, São Paulo (Ipiranga, Paraná), Brazil.

ypsilophorus Gr. *upsilon*, the letter *Y* (lower case *u*); *-phoros*, carrying; ref. the appearance of the markings on the underparts of the Brown Quail *Synoicus*.

yucatanensis/yucatanica/yucatanicus After the Yucatán Peninsula, lying between the Gulf of Mexico and the Caribbean Sea, mainly in Mexico but partly also in Belize and northern Guatemala.

Yuhina Nepalese name *yuhin* for the Stripe-throated Yuhina *Y. gularis*.

yunnanensis After the province of Yunnan in south-western China.

yuracares After the Yuracarés, an Indian tribe of Bolivia.

Carduelis yarrellii Yellow-faced Siskin

Z

zabele Local Brazilian name *zabelé* for the Tawny-browed Tinamou *Crypturellus*.

zambesiae After the River Zambesi or Zambezi, rising near the junction of Zaïre, Angola and Zambia, and flowing south-eastwards to the Moçambique Channel.

Zanclostomus Gr. *zanklon,* a sickle; *stoma,* the mouth; ref. the curved, red bill of the Red-billed Malkoha *Z. javanicus*.

Zanda No expl. (Mathews (1913), *Austral Avian Record*, **1**, 196), or where otherwise used subspecifically (e.g. *Northiella)*; elsewhere Mathews also coined *zarda, zamba,* and *bamba;* according to Conisbee (1957) *zanda* is an aboriginal name.

zantholeuca Gr. *xanthos,* yellow; *leukos,* white.

zanthopygia see **xanthopygaeus.**

zappeyi After W. R. Zappey (1878–1914) collector for Harvard University.

Zaratornis After the town of Zárate, Lima, Peru (for Agustín de Zárate (fl. 1577) Spanish historian and Comptroller of Finances for Castile and Peru); Gr. *ornis,* a bird.

Zarhynchus Gr. *za-,* very much; *rhunkhos,* the bill; ref. the swollen frontal shield and large bill of the Chestnut-headed Oropendola *Z. wagleri*.

Zavattariornis After E. Zavattari (b. 1883) Italian zoologist; Gr. *ornis,* a bird.

zealandicus After New Zealand (v. *novaeseelandiae)*; an erroneous toponym in the case of *Cyanoramphus*.

Zebrilus Dim. from French *zêbre,* a zebra (Italian *zebra,* from a Congolese native name); ref. the finely striped plumage of the Zig-zag Heron *Z. undulatus*.

zelandica After New Zealand (v. *novaeseelandiae)*; an erroneous toponym in the case of *Aplonis*.

zeledoni /Zeledonia After José C. Zeledón (1846–1923) Costa Rican naturalist and collector.

zelichi After Mateo Ricardo Zelich (fl. 1969) Argentinian naturalist.

zena Gr. *zene,* a type of finch, probably the Goldfinch *Carduelis*.

Zenaida After Zénaïde Laetitia Julie Princesse Bonaparte (1804–1854) wife of French ornithologist Prince Charles Bonaparte.

Zenaidura Genus *Zenaida,* dove; Gr. *oura,* the tail (i.e. long-tailed).

zenkeri After Georg Zenker (1855–1922) German botanist and collector in tropical Africa.

zeylanica /zeylanicus /zeylonensis / zeylonica /zeylonus Mod. L. *Zeylona,* Ceylon (Sri Lanka). In the cases of *Pycnonotus* and *Telophorus* the toponyms are erroneous.

zimmeri /Zimmerius After John Todd Zimmer (1889–1957) US ornithologist and author (*Birds of the Marshall Field Peruvian expedition*, 1930; *Studies of Peruvian birds*, 1931).

zimmermani After R. Zimmerman (d. 1903) German botanist and collector.

Zodalia No expl. (Mulsant (1877), *Histoire naturelle des oiseaux-mouches*, Vol. **3**, p. 281), where the Purple-tailed Comet *Z. glyceria* is otherwise known as *Zodalie;* perhaps a dim. from Gr. *zodion,* a small painted figure.

zoeae After Zoë Lesson (fl. 1810) wife of French naturalist and author René Lesson.

Zonaeginthus Gr. *zone*, a girdle or band; genus *Aegintha*, waxbill.

zonaris L. *zona*, a girdle or band; *-aris*, pertaining to.

zonarius L. *zonarius*, of a girdle (*zona*, a girdle; Gr. *zone*, a girdle).

zonatus Mod. L. *zonatus*, banded, ringed (L. *zona*, a girdle).

Zonerodius Gr. *zone*, a girdle or band; *erodios*, a heron; ref. the heavily barred plumage of the Forest Bittern or New Guinea Tiger Heron *Z. heliosylus*.

Zonibyx Gr. *zone*, a girdle or band; Mod. L. *ibyx*, a lapwing or plover; the Rufous-chested Dotterel *Charadrius modestus* has a rich rufous breast bordered beneath by a black band.

Zonifer L. *zona*, a girdle or band; *-fer*, carrying; the Banded Lapwing *Hoplopterus tricolor* has a broad black breastband.

zoniventris L. *zona*, a girdle or band; *venter, ventris*, the belly.

Zonotrichia Gr. *zone*, a girdle or band; *thrix, trikhos*, the hair; ref. the black and white striped head pattern of the White-throated Sparrow *Z. albicollis*.

zonurus Gr. *zone*, a girdle or band; *-ouros*, tailed.

Zoonavena Dim. of syn. genus *Zoonava*, swiftlet (according to Conisbee (1957) *zoonav* is an aboriginal name).

Zoothera Gr. *zoon*, an animal; *-thera*, a hunter; the ground thrushes feed mainly on worms and insects.

Zosterops Gr. *zoster*, a belt or girdle; *ops*, the eye; the white-eyes take their name from the ring of white feathers around each eye.

Zoothera dauma White's Thrush

Bibliography

Alexander, W. B. (1963). *Birds of the ocean.* Putnams, New York.

Ali, S. (1949). *Indian hill birds.* Oxford University Press, Bombay.

Ali, S. (1980). Indian ornithology: the current trends. *Bull. Brit. Orn. Cl.*, **100**[1], 80–3.

Ali, S., and Ripley, S. D. (1968–1974). *Handbook of the birds of India and Pakistan,* Vols. 1–10. Oxford University Press, Bombay.

Aliandro, H. (1972). *Dicionário de bolso dos linguas Portuguêsa e Inglêsa.* McGraw Hill, Rio de Janeiro.

Amadon, D. and Bull, J. (1988). Hawks and owls of the world: a distributional and taxonomic list. *Proc. Western Foundation of Vertebrate Zoology*, **3**(4), 294–357.

American Ornithologists' Union. (1889). *Checklist of North American birds.* (Abridged edn). American Ornithologists' Union, Washington DC.

American Ornithologists' Union. (1983). *Checklist of North American birds.* (6th edn). American Ornithologists' Union, Lawrence, Kansas.

Andrew, D. G. (1990). Shearwaters, sirens and halcyons. *British Birds,* **83**, 334–35.

Aristophanes. (c.400 BC). *Ornithes.*

Aristotle. (c.340 BC). *Historia animalium* (*Peri ta zoa historiai*).

Azara, F. M. de (1805). *Apuntiamentos para la historia natural de los páxaros de Paragüay y Río de la Plata.* Madrid.

Bannerman, D. A. (1963). *Birds of the Atlantic Islands.* Vol. 1: *A history of the birds of the Canary Islands and the Salvages.* Oliver and Boyd, Edinburgh.

Bannerman, D. A. (1980). Reminiscences of the Club. *Bull. Brit. Orn. Cl.*, **100**[1], 22–5.

Bannerman, D. A., and Bannerman, W. M. (1965). *Birds of the Atlantic Islands.* Vol. 2: *A history of the birds of Madeira, the Desertas and the Porto Santo Islands.* Oliver and Boyd, Edinburgh.

Bannerman, D. A., and Bannerman, W. M. (1966). *Birds of the Atlantic Islands.* Vol. 3: *A history of the birds of the Azores.* Oliver and Boyd, Edinburgh.

Bannerman, D. A., and Bannerman, W. M. (1968). *Birds of the Atlantic Islands.* Vol. 4: *A history of the birds of the Cape Verde Islands.* Oliver and Boyd, Edinburgh.

Barclay, G. (1978). *A history of the Pacific.* Sidgwick and Jackson, London.

Barclay-Smith, P. (1980). Recollections of personalities of the Club. *Bull. Brit. Orn. Cl.*, **100**[1], 15–22.

Barnard, M. (1976). *A history of Australia.* Angus and Robertson, Sydney.

Barraclough, G. (ed.) (1984). *The Times atlas of world history.* Times Books, London.

Barrère, P. (1745). *Ornithologiae specimen novum.* Perpignan.

Beehler, B. M., Pratt, T. K., and Zimmerman, D. A. (1986). *Birds of New Guinea.* Princeton University Press, Princeton, NJ.

Benson, C. W. (1980). Fifty years of ornithology in the Malagasy faunal region. *Bull. Brit. Orn. Cl.*, **100**[1], 76–80.

Benson, C. W., Brooke, R. K., Dowsett, R. J., and Irwin, M.P.S. (1971). *The birds of Zambia.* Collins, London.

Betteridge, H. T. (1978). *Cassell's German dictionary.* Cassell, London.

Bezzel, E. (1980). Ornithological advances in western Europe during the last 50 years. *Bull. Brit. Orn. Cl.*, **100**[1], 47–50.

Blake, E. R. (1969). *Birds of Mexico : a guide for field identification.* University of Chicago Press, Chicago.

Blake, E. R. (1977). *Manual of Neotropical birds,* Vol.1. University of Chicago Press, Chicago.

Blakers, M., Davies, S. J. J. F., and Reilly, P. N. (1984). *The atlas of Australian birds.* Melbourne University Press, Carlton, Victoria .

Blunt, W. (1984). *The compleat naturalist : a life of Linnaeus.* Collins, London.

Bochenski, Z. (1980). Some trends in ornithology in east European countries during the last fifty years. *Bull. Brit. Orn. Cl.*, **100**[1], 50–5.

Bock, W. J. (1990). A Special Review: Peters' 'Check–list of Birds of the World' and a History of Avian Checklists. *Auk* ,**107**[3], 629–48.

Bond, J. (1974). *Birds of the West Indies.* (4th edn). Collins, London.

Bourne, W. R. P. (1990). Ancient bird names. *British Birds,* **83**, 555–6.

Brisson, M. J. (1760). *Ornithologie.* Jean–Baptiste Bauche, Paris.

British Ornithologists' Union Records Committee. (1988). Suggested changes to the English names of some Western Palearctic birds. *Ibis*, **130** [4], [Supplement] / *British Birds*, **81**, 355–77.

Britton, P. L. (1980). Ornithological progress in eastern Africa during the past 50 years. *Bull. Brit. Orn. Cl.*, **100**[1], 68–73.

Brooke, R. K. (1980). Ornithology in southern Africa, 1930–1980. *Bull. Brit. Orn. Cl.*, **100**[1], 73–6.

Brown, L. and Amadon, D. (1968). *Eagles, hawks and falcons of the world,* Vols. 1–2. Country Life Books, Feltham, Middx.

Brown, L. H., Urban, E. K. and Newman, K. (1982). *The birds of Africa,* Vol. 1. Academic Press, London.

Brudenell–Bruce, P. G. C. (1975). *The birds of New Providence and the Bahama Islands.* Collins, London.

Buffon, G. L. Comte de. (1770–1783). *Histoire naturelle des oiseaux.* Paris.

Burke's Royal Families of the World, Vol. 1. (1977). Burke's Peerage Ltd., London.

Burton, J. A. (ed.) (1973). *Owls of the world.* Peter Lowe (Eurobook), London.

Campbell, B. and Lack, E. (ed.) (1985). *A dictionary of birds.* Poyser, Calton, Staffs.

Carpenter, C. (1978). *The Guinness book of kings, rulers and statesmen.* Guinness Superlatives, Enfield, Middx.

Catesby, M. (1731). *The natural history of Carolina, Florida and the Bahama Islands.* London.

Cave, F. O. and Macdonald, J. D. (1955). *Birds of the Sudan.* Oliver and Boyd, Edinburgh.

Cayley, N. W. (1966). *What bird is that ? (*4th edn). Angus and Robertson, Sydney.

Chambers biographical dictionary (1974). Chambers, Edinburgh.

Chambers's world gazetteer and geographical dictionary (1959). Chambers, Edinburgh.

Chambers Twentieth Century dictionary (1977) Chambers, Edinburgh.

Chambers world gazetteer: an A – Z of geographical information. 5th edn. (1988). Chambers, Edinburgh / Cambridge University Press, Cambridge.

Cleasby, R., Vigfusson, G., and Craigie, W. A. (1962). *An Icelandic–English dictionary.* Oxford University Press.

Clusius, C. (1605). *Exoticorum libri decem.* Leiden.

Coates, B. J. (1985 and 1990). *The birds of Papua New Guinea,* Vols. I–II. Dove Publications Pty., Alderley, Qld.

Cocker, P. M. and Inskipp, C. (1988). *A Himalayan ornithologist. The life and work of Brian Houghton Hodgson.* Oxford University Press, Oxford.

Collar, N. J. and Andrew, P. (1988) *Birds to watch. The ICBP world checklist of threatened birds.* ICBP, Cambridge.

Conisbee, L. R. (1953). *A list of the names proposed for genera and subgenera of recent mammals, 1904–1951.* British Museum (Natural History), London.

Conisbee, L. R. (1957). *Index generum avium, 1901–1950: names proposed for genera and subgenera of recent birds during the first half of the twentieth century.* MS.

Coombes, A. J. (1985). *The Collingridge dictionary of plant names.* Collingridge, Feltham, Middx.

Cooper, W. T. and Forshaw, J. M. (1977). *The birds of paradise and bower birds.* Collins, Sydney.

Corbet, G. B. and Hill, J. E. (1980). *A world list of mammalian species.* British Museum (Natural History), London.

Coues, E. (1882). *The Coues check list of North American birds.* (2nd edn). Estes and Lauriat, Boston.

Cramp, S. (ed.) (1977–1988). *The birds of the western Palearctic,* Vols. 1–5. Oxford University Press.

Cranbrook, Earl of (1980). The state of ornithology in eastern Asia. *Bull. Brit. Orn. Cl.,* **100**[1], 84–9.

David Evans' Catalogues (1974–1977). Fine bird books for the connoisseur and collector (Unnumbered). David Evans, Winchester, Hants.

Davis, L. I. (1972). *A field guide to the birds of Mexico and Central America.* University of Texas Press, Austin.

Delacour, J. (1947). *Birds of Malaysia,* Macmillan, New York.

Delacour, J. (1966). *Guide des oiseaux de la Nouvelle–Calédonie.* Éditions Delachaux et Niestlé, Neuchatel.

Delacour, J. (1973). *The waterfowl of the world,* Vols. 1–4. Hamlyn, London.

Delacour, J. (1977) *The pheasants of the world.* (2nd edn). Spur Publications, Hindhead, Surrey.

Delacour, J. and Amadon, D. (1973). *Curassows and related birds.* American Museum of Natural History, New York.

Delacour, J. and Mayr, E. (1946). *Birds of the Philippines.* Macmillan, New York.

Devlin, J. C. and Naismith, G. (1978). *The world of Roger Tory Peterson.* David and Charles, Newton Abbott, Devon.

Diccionario de la lengua Española. (1970). Real Academia Española, Madrid.

Dictionary of National Biography. (1885–1971). Oxford University Press.

Dionysius. ? BC. *De Avibus* (*Ornithika*).

Dowsett, R. J. (1988). The genera of Afrotropical birds, and their etymology. Part 1. *Tauraco,* **1**: 139–47.

Dresser, H. E. (1881). *A list of European birds including all species found in the Western Palaearctic region.* H. E. Dresser, London.

duPont, J. E. (1971). *Philippine birds.* Delaware Museum of Natural History, Greenville, Del.

duPont, J. E. (1975). *South Pacific birds.* Delaware Museum of Natural History, Greenville, Del.

Edwards, E. P. (1972). *A field guide to the birds of Mexico.* Ernest P. Edwards, Sweet Briar, Va.

Eisenmann, E. (1955). The Species of Middle American Birds. *Transactions of the Linnaean Society of New York* (Vol. VII).

Etchécopar, R. D. and Hüe, F. (1967). *The birds of North Africa,* (trans. P. Hollom). Oliver and Boyd, Edinburgh.

Etchécopar, R. D. and Hüe, F. (1978). *Les oiseaux de Chine, de Mongolie et de Corée : non passereaux.* Les Éditions du Pacifique, Papeete, Tahiti.

Etchécopar, R. D. and Hüe, F. (1983). *Les oiseaux de Chine, de Mongolie et de Corée : passereaux.* Société Nouvelle des Éditions Boubée, Paris.

Falla, R. A., Sibson, R. B., and Turbott, E. G. (1970). *A field guide to the birds of New Zealand.* Collins, London.

ffrench, R. (1973). *A guide to the birds of Trinidad and Tobago.* Livingston, Wynnewood, Pa.

Fisher, J., Simon, N., and Vincent, J. (1969). *The red book : wildlife in danger.* Collins, London.

Fjeldså, J. and Krabbe, N. (1990). *Birds of the high Andes.* Zoological Museum, University of Copenhagen/Apollo Books, Svendborg.

Fleming, C. A. (1983). *George Edward Lodge : unpublished bird paintings.* Michael Joseph, London.

Fry, C. H. (1984). *The bee-eaters.* Poyser, Calton, Staffs.

Fry, C. H. Keith, S., and Urban, E. K. (eds.) (1988). *The birds of Africa,* Vol. 3. Academic Press, London.

Fuller, E. (1987). *Extinct birds.* Viking/Rainbird, London.

Gaza, T. (1476). *Aristotelis de natura animalium.* Venice.

Gesner, C. (1555). *Historia animalium.* Zurich.

Gibb, J. A. (1980). New Zealand ornithology during the past 50 years. *Bull. Brit. Orn. Cl.,* **100**[1]: 93–6.

Gilliard, E. T. (1958). *Living birds of the world.* Hamish Hamilton, London.

Gilliard, E. T. (1969). *Birds of paradise and bower birds.* Weidenfeld and Nicolson, London.

Godman, C. E. (1980). Recollections. *Bull. Brit. Orn. Cl.,* **100**[1]: 29–34.

Goodwin, D. (1967). *Pigeons and doves of the world.* British Museum (Natural History), London.

Goodwin, D. (1976). *Crows of the world.* British Museum (Natural History), London.

Goodwin, D. (1982). *Estrildid finches of the world.* British Museum (Natural History), London.

Gotch, A. F. (1979). *Mammals — their Latin names explained.* Blandford Press, Poole.

Gotch, A. F. (1981). *Birds — their Latin names explained.* Blandford Press, Poole.

Gould, S. J. (1990). Bully for Brontosaurus. *Natural History,* 2/90: 16–24.

Grandsaignes d'Hauterive, R. (1947). *Dictionnaire d'Ancien Français (Moyen Age et Renaissance).* Librairie Larousse, Paris.

Greenoak, F. (1981). *All the birds of the air.* Penguin, Middx.

Greenway, J. C. (1967). *Extinct and vanishing birds of the world.* Dover Publications, New York.

Gruson, E. S. (1972). *Words for birds — a lexicon of North American birds with biographical notes.* Quadrangle Books, New York.

Gruson, E. S. (1976). *A checklist of the birds of the world.* Collins, London.

Hall, B. P. and Moreau, R. E. (1970). *An atlas of speciation in African passerine birds.* British Museum (Natural History), London.

Hancock, J. and Elliott, H. (1978). *The herons of the world.* London Editions.

Harris, A., Tucker, L., and Vinicombe, K. (1989). *The Macmillan field guide to bird identification.* Macmillan, London.

Harris, M. (1974). *A field guide to the birds of Galapagos.* Collins, London.

Harrison, C. (1982). *An atlas of the birds of the western Palaearctic.* Collins, London.

Harrison, P. (1983). *Seabirds — an identification guide.* Croom Helm, Beckenham, Kent.

Harvey, P. and Heseltine, J. E. (1959). *The Oxford companion to French literature.* Oxford University Press, Oxford.

Haverschmidt, F. (1968). *Birds of Surinam.* Oliver and Boyd, Edinburgh.

Hayman, P., Marchant, J., and Prater, T. (1986). *Shorebirds — an identification guide to the waders of the world.* Croom Helm, Beckenham, Kent.

Hemming, J. (1978). *Red gold : the conquest of the Brazilian Indians.* Macmillan, London.

Henry, G. M. (1971). *A guide to the birds of Ceylon.* Oxford University Press, London.

Hernandez, F. (1615). *Historia avium Novae Hispaniae.* Mexico.

Herring, H. (1961). *A history of Latin America.* Alfred A. Knopf, New York.

Heseltine, N. (1971). *Madagascar.* Pall Mall Press, London.

Hesychius. (c. 400 AD). *Lexicon.*

Hilty, S. L. and Brown,W. L. (1986). *A guide to the birds of Colombia.* Princeton University Press.

Hollom, P. A. D. (1968). *The popular handbook of British birds.* Witherby, London.

Hollom, P. A. D., Porter, R. F., Christensen, S., and Willis, I. (1988). *Birds of the Middle East and North Africa.* Poyser, Calton, Staffs.

Hopkins, J. G. E. (ed.) (1964). *Concise dictionary of American biography.* Chas. Scribner's Sons, New York.

Horne, E. C. (1974). *Javanese—English dictionary.* Yale University Press, New Haven, Conn.

Howard, R. and Moore, A. (1980). *A complete checklist of the birds of the world.* Oxford University Press.

Hüe, F. and Etchécopar, R. D. (1970). *Les oiseaux du Proche et du Moyen Orient.* Éditions Boubée, Paris.

Hyamson, A. M. (1951). *A dictionary of universal biography of all ages and of all peoples.* Routledge and Kegan Paul, London.

*International Code of Zoological Nomenclature, (*3rd edn). (1985). International Trust for Zoological Nomenclature, London.

Isakov, Y. A. (1980). Some of the results of ornithological investigations in the Soviet Union for the past fifty years. *Bull. Brit. Orn. Cl.*, **100**[1]: 55–61.

Isler, M. L. and Isler, P. R. (1987). *The tanagers : natural history, distribution and identification.* Oxford University Press.

Jeffrey, C. (1977). *Biological nomenclature.* Edward Arnold, London.

Jobling, J. A. (1974–1989). *A list of recent bird species, with suggested English names and brief distribution outlines.* MS.

Jobling, J. A. (1985). *Aalge to Zosterops: an etymology of described bird names.* MS.

Johnsgard, P. A. (1988). *The quails, partridges, and francolins of the world.* Oxford University Press.

Johnson, A. W. (1965–1972). *The birds of Chile and adjacent regions of Argentina, Bolivia and Peru.* Vols. 1–2 and supplement. Platt Establecimientos Gráficos, Buenos Aires.

Johnson, T. H. and Stattersfield, A. J. (1990). A global review of island endemic birds. *Ibis*, **132** [2]: 167–80 (Proceedings of the 1989 BOU Annual Conference on 'Applied Ornithology').

Jonsson, L. (1978). *Birds of sea and coast.* Penguin Books, Harmondsworth, Middx.

Jonsson, L. (1982). *Birds of the Mediterranean and Alps.* Croom Helm, London.

Jorgensen, H. J. (1958). *Nomina avium Europaearum.* Ejnar Munksgaard, Copenhagen.

Judd, H. P., Pukui, M. K., and Stokes, J. F. G. (1955). *Hawaiian English vocabulary.* Tongg Publishing, Honolulu.

King, B. F. and Dickinson, E. C. (1975). *A field guide to the birds of South-East Asia.* Collins, London.

King, W. B. (compiler) (1981). *Endangered birds of the world : the ICBP bird red data book.* Smithsonian Institution, Washington.

Koepcke, M. (1970). *The birds of the department of Lima, Peru.* Livingston, Wynnewood, Pa.

Lack, D. (1968). *Ecological adaptations for breeding in birds.* Chapman and Hall, London.

Lack, D. (1971). *Ecological isolation in birds.* Blackwell Scientific Publications, Oxford.

Lack, D. (1974). *Evolution illustrated by waterfowl.* Blackwell Scientific Publications, Oxford.

Lack, D. (1976). *Island biology illustrated by the land birds of Jamaica.* Blackwell Scientific Publications, Oxford.

Lambourne, M. (1987). *John Gould — bird man.* Osberton Publications, Milton Keynes.

Land, H. C. (1970). *Birds of Guatemala.* Livingston, Wynnewood, Pa.

Landsborough Thomson, A. (ed.) (1964). *A new dictionary of birds.* Nelson, London.

Langrand, O. (1990). *Guide to the birds of Madagascar.* Yale University Press.

Langton, H. (1920). *Dictionary of ornithological names and terms.* MS.

Latham, R. E. (1965). *Revised Medieval Latin word-list.* Oxford University Press, Oxford.

Lemprière's Classical Dictionary (1984). (Facsimile of 1865 ed.). Bracken, London.

Levaillant, F. (1799–1808). *Histoire naturelle des oiseaux d'Afrique.* Vols. 1–6. Paris.

Lewis, C. T. and Short, C. (1907). *A Latin dictionary.* Clarendon Press, Oxford.

Lewis, C. T. and Short, C. (1962). *A Latin dictionary.* Oxford University Press.

Liddell, H. G. and Scott, R. (1961). *A Greek–English lexicon.* Oxford University Press.

Liddell, H. G., Scott, R., and Jones, H. S. (1968). *Greek–English lexicon. A Supplement.* Clarendon Press, Oxford.

Linnaeus, C. (1758). *Systema naturae per regna tria naturae,* (ed. X), *Tom.* ***1****, Regnum animale.* Salvius, Stockholm (1956 facsimile, British Museum (Natural History), London).

Lister, M. (1962). *A glossary for bird watchers.* Phoenix House, London.

Lockwood, W. B. (1984). *The Oxford book of British bird names.* Oxford University Press.

Macdonald, J. (1973). *Birds of Australia.* Witherby, London.

Mackworth-Praed, C. W. and Grant, C. H. B. (1952–1955). *Birds of eastern and north eastern Africa.* Vols. 1–2. Longmans Green, London.

Mackworth-Praed, C. W. and Grant, C. H. B. (1962–1963). *Birds of the southern third of Africa.* Vols. 1–2. Longmans Green, London.

Mackworth-Praed, C. W. and Grant, C. H. B. (1970–1973). *Birds of west central and western Africa.* Vols. 1–2. Longmans Group, London.

McLachlan, G. R. and Liversidge, R. (1970). *Roberts birds of South Africa.* Central News Agency, Cape Town.

Macleod, R. D. (1952). *Key to the names of British plants.* Pitman, London.

Macleod, R. D. (1954). *Key to the names of British birds.* Pitman, London.

Macleod, R. D. (1956). *Key to the names of British fishes, mammals, amphibians and reptiles.* Pitman, London.

Macleod, R. D. (1959). *Key to the names of British butterflies and moths.* Pitman, London.

Madge, S. and Burn, H. (1988). *Wildfowl — an identification guide to the ducks, geese and swans of the world.* Christopher Helm Publishers, London.

Mansion, J. E. (1972). *Harraps new standard French and English dictionary. Part 1.* Harrap, London.

Marcgrave, G. (1648). *Historiae rerum naturalium Brasiliae.* Liber V.

Mathews, G. M. (1927 and 1930). *Systema avium Australasianarum.* Parts I–II. British Ornithologists' Union / Taylor and Francis, London.

Mayr, E. (1945). *Birds of the southwest Pacific.* Macmillan, New York.

Mayr, E. and Amadon, D. (1951). *A classification of recent birds.* American Museum Novitates no. 1496, New York.

Mearns, B. and Mearns, R. (1988). *Biographies for birdwatchers.* Academic Press, London.

Meinertzhagen, R. (1954). *Birds of Arabia.* Oliver and Boyd, Edinburgh.

Meyer de Schauensee, R. (1966). *The species of birds of South America and their distribution.* Livingston, Narberth, Pa.

Meyer de Schauensee, R. (1971). *A guide to the birds of South America.* Oliver and Boyd, Edinburgh.

Meyer de Schauensee, R. (1984). *The birds of China.* Oxford University Press.

Meyer de Schauensee, R. and Phelps, W. H., Jr. (1978). *A guide to the birds of Venezuela.* Princeton University Press.

Milon, P., Petter, J.-J., and Randrianasolo, G. (1973). *Faune de Madagascar, XXXV, oiseaux.* Office Rech. Scient. Tech. Outre-Mer, Tananarive, Madagascar / Centre Nat. Rech. Scientifique, Paris.

Möhring, H. G. (1752). *Avium genera.* Aurich.

Molina, G. I. (1782). *Saggio sulla storia naturale del Chili.* Bologna.

Moreau, R. E. (1972), *The Palaearctic-African bird migration systems.* Academic Press, London.

Morel, G. J. (1980). Fifty years of ornithology in west Africa. *Bull. Brit. Orn. Cl.*, **100**[1]: 66-8.

Morony, J. J., Jr., Bock, W. J., and Farrand, J., Jr. (1975). *Reference list of the birds of the world.* American Museum of Natural History, New York.

Mountfort, G. (1988). *Rare birds of the world.* Collins, London.

National Geographic Society. (1983). *Field guide to the birds of North America.* National Geographic Society, Washington.

Neave, S. A. (ed.) (1939-1940). *Nomenclator zoologicus, a list of the names of genera and subgenera in zoology, 1758-1935.* Vols. 1-4. The Zoological Society of London, London.

Newman, K. (1984). *Newman's birds of southern Africa.* Macmillan South Africa, Johannesburg.

Newton, A. and Gadow, H. (1896). *A dictionary of birds.* Black, London.

Nicholson, E. M. (1980). Co-operative ornithology and conservation in western Europe. *Bull. Brit. Orn. Cl.*, **100** [1]: 44-7.

North-Coombes, A. (1971). *The island of Rodrigues.* North-Coombes/ Mauritius Advertising Bureau, Port Louis.

Ouellet, H. and Godfrey, W. E. (1980). Ornithology in Canada in the 20th century : a capsule overview. *Bull. Brit. Orn. Cl.*, **100**[1]: 115-8.

Oxford English Dictionary, Vols. 1-13 plus supplements. (1961-1982). Oxford University Press.

Palmer, T. S. (1928). Notes on persons whose names appear in the nomenclature of California birds. *Condor*, **30** (5): 261-307.

Paynter, R. A., Traylor, M. A., and Winter, B. (1975). *Ornithological gazetteer of Bolivia.* Harvard University Press, Cambridge, Mass.

Peck, R. McC. (1983). *A celebration of birds : the life and art of Louis Agassiz Fuertes.* Collins, London.

Penny, M. (1974). *The birds of Seychelles and the outlying islands.* Collins, London.

Peters, J. L. (1934-1986). *Check-list of birds of the world.* Vols. 1-15. Harvard University Press / Museum of Comparative Zoology, Cambridge, Mass.

Peterson, R. T. (1947). *A field guide to the birds.* Houghton Mifflin, Boston.

Peterson, R. T. (1961). *A field guide to western birds.* Houghton Mifflin, Boston.

Peterson, R. T. and Chalif, E. L. (1973). *A field guide to Mexican birds.* Houghton Mifflin, Boston.

Peterson, R. T., Mountfort, G., and Hollom, P. A. D. (1954). *A field guide to the birds of Britain and Europe.* Collins, London.

Pizzey, G. (1983). *A field guide to the birds of Australia.* Collins, Sydney.

Pliny. (c.65AD). *Naturalis historia,* (trans. H. Rackham) (1983). *Pliny natural history III* (books VIII -XI) Harvard University Press, Cambridge, Mass.

Potter, S., and Sargent, L. (1973). *Pedigree : essays on the etymology of words from nature.* Collins, London.

Pough, R. H. (1957). *Audubon western bird guide.* Doubleday, New York.

Pratt, H. D., Bruner, P. L., and Barrett, D. G. (1987). *A field guide to the birds of Hawaii and the tropical Pacific.* Princeton University Press.

Prescott, W. (1949). *History of the conquest of Mexico.* Allen and Unwin, London.

Prestwich, A. A. (1963). *I name this parrot...* A. A. Prestwich, Edenbridge, Kent.

Rackham, H. *see* Pliny.

Rand, A. L. and Gilliard, E. T. (1967). *Handbook of New Guinea birds.* Weidenfeld and Nicolson, London.

Ray, J. (1678). *The ornithology of Francis Willughby of Middleton.* Royal Society, London (1972 facsimile, Paul Minet, Newport Pagnell, Bucks.).

Reed, A. W. (1968). *Concise Maori dictionary.* A. H. and A. W. Reed, Wellington.

Reid–Henry, D. and Harrison, C. (1988). *The history of the birds of Britain.* Collins, London.

Ricklefs, R. E., and Gill, F. B. (1980). Fifty years of American ornithology. *Bull. Brit. Orn. Cl.*, **100**[1]:118–22.

Ridgely, R. S. (1976). *A guide to the birds of Panama.* Princeton University Press.

Ridgely, R. S. and Tudor, G. (1989). *The birds of South America : the oscine passerines.* Oxford University Press.

Ripley, S. D. (1961). *A synopsis of the birds of India and Pakistan.* Bombay Natural History Society, Bombay.

Ripley, S. D. (1977). *Rails of the world : a monograph of the family Rallidae.* David Godine, Boston.

Rose, H. J. (1961). *A handbook of Latin literature.* Methuen, London.

Rose, H. J. (1964). *A handbook of Greek literature.* Methuen, London.

Rothschild, M. (1983). *Dear Lord Rothschild.* Balaban, Philadelphia.

Sansoni–Harrap. (1970). *Standard Italian and English dictionary.* Harrap, London.

Sauer, G. (1982). *John Gould, the bird man : a chronology and bibliography.* Lansdowne, Melbourne.

Sclater, W. L. (1924 and 1930). *Systema avium Aethiopicarum.* Parts I– II. British Ornithologists' Union, London.

Seltzer, L. E. (ed.) (1962). *The Columbia Lippincott gazetteer of the world.* Columbia University Press, New York.

Serventy, D. L. (1980). Developments in Australian ornithology. *Bull. Brit. Orn. Cl.*, **100**[1]: 89– 93.

Sharpe, R. B. (1899–1909). *A hand-list of the genera and species of birds.* Vols 1–5 and General Index. British Museum (Natural History), London.

Sharrock, J. T. R. (compiler) (1976). *The atlas of breeding birds in Britain and Ireland.* British Trust for Ornithology, Tring, Herts.

Short, L. L. (1982). *Woodpeckers of the world.* Delaware Museum of Natural History, Greenville, Del.

Sibley, C. G. and Monroe, Jr., B. L. (1990). *Distribution and taxonomy of birds of the world.* Yale University Press.

Sibley, C. G., Ahlquist, J. E., and Monroe, Jr., B. L. (1988). A classification of the living birds of the world based on DNA–DNA hybridization studies. *Auk*, **105** (3): 409–23.

Simpson, D. P. (1979). *Cassell's Latin–English, English–Latin dictionary.* Cassell, London.

Simpson, K. and Day, N. (1989). *Field guide to the birds of Australia.* Christopher Helm (Publishers), London.

Slater, P. (1971). *A field guide to Australian birds — non-passerines.* Oliver and Boyd, Edinburgh.

Slater, P. (1975). *A field guide to Australian birds — passerines.* Scottish Academic Press, Edinburgh.

Smythies, B. E. (1953). *The birds of Burma* (2nd edn). Oliver and Boyd, Edinburgh.

Smythies, B. E. (1981). *The birds of Borneo.* (3rd edn). Sabah Society and Malayan Nature Society, Kuala Lumpur.

Snow, D. W. (ed.) (1971). *The status of birds in Britain and Ireland.* Blackwell, Oxford.

Snow, D. W. (ed.) (1978). *An atlas of speciation in African non-passerine birds.* British Museum (Natural History), London.

Snow, D. (1980). Ornithological research in tropical America — the last 35 years. *Bull. Brit. Orn. Cl.*, **100**[1]: 123–31.

Snow, D. W. (1982). *The cotingas.* British Museum (Natural History), London/Oxford University Press.

Stearn, W. T. (1983). *Botanical Latin.* David and Charles, London.

Steller, G. W. (1751). *De bestiis marinis.* St Petersburg.

Stresemann, E. (1975). *Ornithology from Aristotle to the present.* Harvard University Press, Cambridge, Mass.

Stutley, M. and Stutley, J. (1977). *A dictionary of Hinduism.* Allied Publishers, Bombay.

Summers–Smith, J. D. (1988). *The sparrows.* Poyser, Calton, Staffs.

Sutton, G. M. (1951). *Mexican birds, first impressions.* University of Oklahoma Press.

Tate, P. (1986). *Birds, men and books.* Henry Sotheran Ltd., London.

The Times atlas of the world : comprehensive edition. (1978). Times Books, London.

Thompson, D'A. W. (1895). *A glossary of Greek birds.* Oxford University Press.

Treharne, R. F. and Fullard, H. (ed.) (1974). *Muir's historical atlas : ancient, medieval and modern.* Book Club Associates, London.

Turner, A. and Rose, C. (1989). *A handbook to the swallows and martins of the world.* Christopher Helm (Publishers), London.

Urban, E. K. Fry, C. H., and Keith, S. (ed.) (1986). *The birds of Africa,* Vol.2. Academic Press, London.

Van Marle, J. G. and Voous, K. H. (1988). *The birds of Sumatra.* British Ornithologists' Union (Checklist no. 10), Tring, Herts.

Van Tyne, J. and Berger, A. J. (1965). *Fundamentals of ornithology.* John Wiley, New York.

Vaurie, C. (1971). *Classification of the ovenbirds (Furnariidae).* Witherby, London.

Vaurie, C. (1972). *Tibet and its birds.* Witherby, London.

Vaurie, C. (1980). *Taxonomy and geographical distribution of the Furnariidae.* American Museum of Natural History, New York.

Vernon, J. D. R. (1980). 50 years of ornithology in north–west Africa 1930–1980. *Bull. Brit. Orn. Cl.*, **100**[1]: 61–6.

Vianna, F. de M. (ed.) (1979). *The international geographic encyclopaedia and atlas.* Macmillan, London.

Villas Boas, O., and Villas Boas, C. (1975). *Xingu : the Indians, their myths.* Souvenir Press, London.

Voous, K. H. (1960). *Atlas of European birds.* Nelson, London.

Voous, K. H. (1978). *List of recent Holarctic bird species.* Academic Press/British Ornithologists' Union, London.

Wadley, N. J. P. (1980). Random recollections of the BOC 1950 – 1960. *Bull. Brit. Orn. Cl.*, **100**[1]: 36–40.

Walters, M. (1980). *The complete birds of the world.* David and Charles, Newton Abbott, Devon.

Warham, J. (1980). Recent trends in sub–Antarctic ornithology. *Bull. Brit. Orn. Cl.*, **100**[1]: 96–102.

Wetmore, A. (1951). A revised classification for the birds of the world. *Smithsonian Miscellaneous Collections*, **117**(4) (pub. no. 4057), Washington.

Wetmore, A. (1960). A classification for the birds of the world. *Smithsonian Miscellaneous Collections*, **139**(11) (pub. no. 4417), Washington.

Wheldon and Wesley's Catalogues. (1978–1990). Books on Natural History (New (Fifth) Series, Nos.**142–191**). Wheldon and Wesley Ltd., Codicote, Herts.

White, C. M. N., and Bruce, M. D. (1986). *The birds of Wallacea.* British Ornithologists' Union (Checklist no. 7), London.

White, G. (1789). *The natural history of Selborne.* Benjamin White, London (Reissue (1984).*The illustrated natural history of Selborne.* Macmillan, London).

Whittell, H. M. (1954). *The literature of Australian birds. A history and a bibliography of Australian ornithology.* Paterson Brokensha Pty. Ltd., Perth.

Who Was Who. Vols. 1–5. (1941–1962). Black, London.

Wilcox, D. (1975). *Explorers.* Book Club Associates, London.

Wilkinson, R. J. (1959). *A Malay—English dictionary (Romanised),* Vols. 1–2. Macmillan, London.

Winstedt, R. (1949). *An English Malay dictionary.* Kelly and Walsh, Singapore / Kegan Paul Trench Trubner, London.

Wolstenholme, H. (1926). *The scientific names of Australian birds. Appendix to the official checklist of the birds of Australia.* (2nd edn). Royal Australasian Ornithologists Union, Melbourne.

Wright, T. (1884). *Anglo–Saxon and Old English vocabularies.* Trubner, London.

Wynne, O. E. (1956). *Key–list of the Palaearctic and Oriental passerine birds.* T. Buncle and Co., Arbroath, Angus.

Wynne, O. E. (1969). *Biographical key – names of birds of the world.* O. E. Wynne, Fordingbridge, Hants.

Yamashina, Y. (1961). *Birds in Japan: a field guide.* Tokyo News Service, Tokyo.

Young, E. C. (1980). The present status of Antarctic ornithology. *Bull. Brit. Orn. Cl.*, **100**[1]:102–15.

Zimmer, J. T. (1948). *Studies of Peruvian birds. No.53.* American Museum Novitates no.1380, New York.

Zimmer, J. T. (1953). *Studies of Peruvian Birds. No.65.* American Museum Novitates no.1649, New York.

Appendix

WHILST the Dictionary was in final proof Sibley and Monroe (1990), with its 'provocative new taxonomy', was received, too late to incorporate the nomenclatural notes and all the specific and generic names referred to therein. This appendix includes such names as were identified during a brief scan of that work, together with other names and clarifications. Entries not found in the main Dictionary are identified by (*).

Acredula (*) L. *acredula*, a bird mentioned by Tullius Cicero (*De Divinatione*), not otherwise identified, but subsequently associated with an owl *Strix*, a thrush *Turdus*, or, in orn., a long-tailed tit *Aegithalos*.

afraoides (*) Specific name *Eupodotis afra*; Gr. *-oides*, resembling.

albaria (*) L. *albarius*, whitened, plastered.

annae The Tawny-booted Racquet-tail *Ocreatus* was named after Anna Gräfin von Branicka (fl. 1894), wife of the Polish ornithologist Xavier Graf von Branicki (d. 1926).

athensis (*) After the Athi River, south-central Kenya.

atrinucha (*) L. *ater*, black; Med. L. *nucha*, the nape.

aucupum (*) L. *aucupari*, to watch for, to lie in wait.

augustaevictoriae (*) After Auguste Viktoria (1859–1921) first wife of Wilhelm II Emperor of Germany and King of Prussia (abdicated 1918), and daughter of Friedrich VIII Duke of Schleswig-Holstein (d. 1880).

aurantiiventris (*) Mod. L. *aurantius*, orange-coloured; L. *venter*, the belly.

balim (*) After the Balim River, Oranje Mountains, New Guinea (Baliem River, Irian Jaya).

biarmicus According to Newton and Gadow (1896) the Bearded Tit *Panurus* was named, erroneously, after Biarmia, an old name for the district of Perm, Russia.

bicknelli (*) After E. P. Bicknell (1859–1925), US ornithologist and a founder of the American Ornithologists' Union.

Blagrus (*) Probably from Gr. *blax, blakos*, inactive, sluggish.

boreus (*) L. *boreus*, northern.

bornea This is one of many names rejected by Sharpe (1899–1909), and other conservative authors 'as conveying a wrong notion as to the habitat of the species.' He was inconsistent in his opposition to such names, however, and elswhere allowed erroneous toponyms to stand.

brachylopha (∗) Gr. *brakhus*, short; *lophos*, a crest.
bracei (∗) After L. J. K. Brace (1852–1938), Bahamian botanist.

cacabata (∗) L. *cacabatus*, sooty, black (*cacabus*, a cooking-pot).
cashmirensis (∗) After the princely state of Kashmir, now Jammu and Kashmir, India.
chii (∗) Onomatopoeic Spanish name *chii* given to a pipit *Anthus* by Azara (1805). Dispute surrounds the proper identity of Azara's bird, and the name is now regarded as unidentifiable.
Coloburis (∗) Gr. *kolobos*, docked, shortened; *-ouros*, tailed.
confusa (∗) L. *confusus*, confused, perplexed (*confundere*, to confuse, disturb).
Corthylio (∗) Gr. *korthilos*, a small wren-like bird mentioned by Hesychius.

Daulias (∗) L. *Daulias avis*, the Daulian bird, the nightingale *Luscinia*. Ancient Daulis, in Phocis, was the scene of the gruesome feast where the dishonoured sisters Philomela (q.v.) and Procne tricked the Thracian king Tereus into eating the flesh of his own son, Itylus.
dichrocephalus This specific name was erroneously listed as *dicrocephala* in the *General Index* (1912) to Sharpe (1899–1909), a mistake perpetuated by subsequent authors, including Sclater (1930), Peters (1962), Vol. XV, and Howard and Moore (1980). The original spelling, and reference, is *Hyphantornis dichrocephalus* Salvadori, 1896, *Ann. Mus. Civ. Genova* **16** (2), 45.
drygalskii (∗) After Erich Dagobert von Drygalski (1865–1949) German polar explorer and author (*Die deutsche Südpolar-Expedition auf dem schiff* Gauss, 1902).

elisabeth The Southern Blue-eared Glossy Starling *Lamprocolius* was named after Elisabeth Deniger (fl. 1924), first wife of German ornithologist Erwin Stresemann.
Emarginata Sharpe (1899–1909) remarked that *Emarginata* was an adjective and could not be employed in a generic sense, and replaced it with *Poliocichla* (Gr. *polios*, grey; Mod. L. *cichla*, a thrush).
emiliae The Western Negrofinch *Nigrita* was named after Emily Sharpe (fl. 1870) wife of English ornithologist R.B. Sharpe (v. *Bowdleria*).

fascinans (∗) L. *fascinans*, enchanting, charming (*fascinare*, to bewitch).
forbesi The White-collared Kite *Leptodon* was named for W. A. Forbes (v. *forbesi* **2**).
franklinii Franklin's Grouse *Dendragapus* is named for Sir John Franklin (1786–1847), English Arctic explorer.

gingalensis (∗) After the Gin Ganga, a river of southern Ceylon (Sri Lanka) (v. *singalensis*).

guineensis (*) After Guinea (v. *guinea*).
gustavi The Blue-margined Parakeet *Brotogeris* was named for Gustav Garlepp (v. *garleppi*).

hardyi (*) After John William Hardy (b. 1930), US ornithologist.
Hemixos (*) Gr. *hemi*, half (i.e. related); genus *Ixos*, bulbul.
heterolaemus (*) Gr. *heteros*, different; *laimos*, the throat.
Heteroxolmis (*) Gr. *heteros*, different; genus *Xolmis*, monjita.
hoyi (*) After Gunnar Arthur Hoy (fl. 1961), Argentinian ornithologist.

ibis The Swedish explorer Fredrik Hasselqvist (d. 1752) was persuaded by his Egyptian dragoman that the Cattle Egret *Bubulcus* was, in fact, the sacred ibis of the ancients. His papers eventually found their way to Linnaeus, who perpetuated the deception in the egret's specific name.
improbus (*) L. *improbus*, restless, impudent, bold.
interjecta (*) L. *interiectus*, placed between, interjected (*intericere*, to throw between).
iridescens (*) L. *iris, iridis*, the rainbow; *-escens*, almost resembling.

juninensis The Junin Grebe *Podiceps* is named for Lake Junín, Peru.

kandti (*) After R. Kandt (1867–1918) German collector in the Congo (Zaïre), 1897–1900.
kisserensis (*) After the island of Kisser (Kisar), near Timor, in the Lesser Sundas, Indonesia.
koesteri (*) After Paul Koester (fl. 1920), collector in Angola.
kungwensis (*) After Mount Kungwe, Tanganyika (Mount Nkungwe, Kigoma, Tanzania).

larvaticola (*) Specific name *Lagonosticta larvata*; L. *-cola*, inhabitant; the parasitic Barka Widowfinch *Hypochera* lays its eggs in the nests of the Masked Firefinch *L. larvata*.
Lemuresthes (*) Mod. L. *Lemuria*, Madagascar, the island of lemurs *Lemur* (L. *lemures*, ghosts); genus *Amauresthes*, munia.
Lewinia (*) After J. W. Lewin (v. *lewinii*).
lilianae Lilian's Meadowlark *Sturnella* was named for Lilian Baldwin (fl. 1930) wife of US patron of natural history S. P. Baldwin.
limbata (*) L. *limbatus*, edged, bordered.
litae (*) After the settlement of Lita, Esmeraldas, Ecuador.
loennbergi Sibley and Monroe (1990) pointed out that ICZN (1985), Edn. 3, article 32(d)(i)(2) regarding the replacement of the umlaut (v. *buergersi*) refers only to words of Germanic origin. This epithet should be properly spelt *lonnbergi*.
luapula (*) After the Luapula River, a feeder of Lake Bangweolo, Northern Rhodesia (Zambia).

ludlowi (*) After Frank Ludlow (1885–1972) English educationalist, botanist, ornithologist, and explorer in Chinese Turkistan, Tibet, and Bhutan.

lydiae (*) After Lydia Portenko (fl. 1929) wife of Russian zoologist L. A. Portenko.

marwitzi (*) After Captain A. von der Marwitz (fl. 1895) German Army in German East Africa (Tanzania), 1895–1906, and the Cameroons, 1913.

melas (*) Gr. *melas*, black.

molybdophanes (*) Gr. *molubdos*, lead, plumbago; *phanes*, showing, exhibiting; ref. the naked blue-grey neck and thighs of the Somali Ostrich *Struthio*.

monguilloti (*) After M. Monguillot (fl. 1926) French civil servant.

moseleyi (*) After Henry Nottidge Moseley (1844–1891), British naturalist and author (*Notes by a naturalist on the* Challenger, 1879; *Narrative of the cruise of HMS* Challenger, 1885).

Motacilla Specifically the epithet refers to the tail-bobbing habit of the Louisiana Waterthrush *Seiurus*.

Mulleripicus The student who delves into Prince Bonaparte's passion for coining new names will find some fascinating material, but will be frustrated by the lack of explanation accompanying a large proportion of them. Many of the Prince's contemporaries, and subsequent ornithologists of a less intuitive perception, were horrified and dismayed by his disregard for grammar and the classics. He was the darling of French ornithologists, who regarded him as 'l'un des ornithologistes les plus célèbres de l'Europe', but Cabanis raged about barbarisms which were neither Greek nor Latin, Coues dismissed his later work as worthless and pernicious, and Sharpe refused to recognize or use any of the Prince's 'nonsense-names' such as *Graydidascalus, Kaupifalco, Mulleripicus*, and *Reinwardtoena*.

ngamiense (*) After Lake Ngami in northern Bechuanaland (Botswana).

Notocarbo (*) Gr. *notos*, the south; syn. genus *Carbo*, cormorant.

nyassae (*) After the Nyassa district of south-western German East Africa (Ruvuma, Tanzania).

Nyctagreus cf. Gr. *agreus*, a black songbird skilled in mimicry, mentioned by Claudius Aelianus.

Ochthodiaeta (*) Gr. *okhthos*, a hill or mound; *diaeido*, to sing.

Ocyceros (*) Gr. *okua*, a spear; *keros*, a horn; ref. the sharp-pointed bill-casque of the Indian Grey Hornbill *Tockus birostris*.

Philacte (*) Gr. *phil-*, loving, fond of; *akte*, the beach, the sea-shore.

pretoriae (*) After the city of Pretoria, Transvaal, South Africa (named after Boer leader Andries Wilhelmus Jacobus Pretorius (d. 1853)).

Pseudelaenia (∗) Gr. *pseud-*, false, pseudo; genus *Elaenia*, elaenia.
punctatissima (∗) Mod. L. *punctatissimus*, very spotted, heavily spotted (super. of *punctatus*).
Pyrgilauda (∗) Gr. *purgites*, a sparrow; L. *alauda*, a lark.

raricola (∗) Specific name *Lagonosticta rara*; L. *-cola*, inhabitant; the parasitic Jambandu Widowfinch *Hypochera* lays its eggs in the nests of the Black-bellied Firefinch L. *rara*.
Rhinoploceus (∗) Gr. *rhis*, the nose; genus *Ploceus*, weaver.

scheffleri (∗) After Herr Scheffler (fl. 1911) an otherwise unknown collector in German East Africa (Tanzania).
somervillei (∗) After W. Somerville (1771–1860) Scottish army surgeon, civil commissioner and explorer in South Africa, and head of Army Medical Department, Scotland, 1813–1816.
somptuosus (∗) L. *sumptuosus*, sumptuous, costly, extravagant.
souimanga Sibley and Monroe (1990) confirm that the original orthography of this name is *sovimanga*, although they acknowledge that the author Johann Gmelin normally used 'v' for 'u'. Gmelin's name was based on the French substantive *souimanga*, which dates back to the seventeenth century, and this spelling has been retained by Peters (1967), Vol. XII, and other authors.
sparveroides (∗) Specific name *Falco sparverius*; Gr. *-oides*, resembling.
subflavescens (∗) L. *sub-*, underneath; *flavescens*, golden-yellowish.
submonileger (∗) L. *sub-*, related to; specific name *Ficedula monileger*.

tangorum (∗) After the Tang dynasty of northern China, whose emperors ruled 618–916 AD.
taverneri (∗) After P. A. Taverner (1875–1947) Canadian ornithologist and author (*Birds of Canada*, 1934).
tenuepunctatus (∗) L. *tenuis*, little, fine; Mod. L. *punctatus*, spotted.
Tetraenura (∗) Gr. *tetraino*, to pierce or perforate; *oura*, the tail.
thagus (∗) Local Chilean name *tagua* for a coot *Fulica* (Aymara Indian *ajoya*). This name was erroneously given to another waterbird, the Peruvian Pelican *Pelecanus*, by Molina (1782).
thruppi (∗) After J. G. Thrupp (fl. 1900) British surgeon in Somaliland, 1884–1885.
togoensis (∗) After the mandated territory of British Togoland (now divided between Ghana and Togo).
Trichixos (∗) Gr. *thrix*, the hair; genus *Ixos*, bulbul.
Tricholestes (∗) Gr. *thrix*, the hair; *lestes*, a robber.

umbrina (∗) Mod. L. *umbrinus*, brown, umber (L. *umbra*, a shade).
umbrinodorsalis (∗) Mod. L. *umbrinus*, brown; Med. L. *dorsalis*, of the back, backed.

usta (*) L. *ustus*, burnt, scorched.

vindhiana (*) After the Vindhya Hills, Mahratta States (Madhya Pradesh), India.

Xanthomyias Mod. L. *myias*, a flycatcher (Gr. *muia*, a fly), was created on the analogy of Gr. *spizias*, a hawk, a finch-catcher, from *spiza*, a finch.